Citrus bergamia

Bergamot and Its Derivatives

Medicinal and Aromatic Plants — Industrial Profiles

Individual volumes in this series provide both industry and academia with in-depth coverage of one major genus of industrial importance.

Series Edited by Dr. Roland Hardman

Citrus bergamia
Bergamot and Its Derivatives

Edited by **Giovanni Dugo**
Ivana Bonaccorsi

Medicinal and Aromatic Plants — Industrial Profiles

CRC Press
Taylor & Francis Group
Boca Raton London New York

CRC Press is an imprint of the
Taylor & Francis Group, an **informa** business

Dedication

To my friends—Alberto, Andrea, Antonella, Carmelo, Diego,
Enza, Eugenio, Felice, Francesco, Giacomo, Giulio, Loris,
Peppino, Pino, Umberto—*who Fortune has brought to me.
With you I shared joy and pain, enthusiasm and
delusion, principles, feelings, emotions.
Without you my life wouldn't have been as colorful.*

Giovanni Dugo

To my daughters Lidia *and* Adele:
*Love Nature,
appreciate Her gifts: they will be precious to you.
Love yourself and love each other:
You are the most precious gifts!*

Ivana Bonaccorsi

Contents

Series Preface

There is increasing interest in industry, academia, and the health sciences in medicinal and aromatic plants. The transition from plant to the eventual product used by the public involves many processes. This series brings together information that is currently scattered throughout an ever-increasing number of journals. Each volume gives an in-depth look at one plant genus about which an area specialist has assembled information ranging from the production of the plant to market trends and quality control.

Many industries are involved, such as forestry, agriculture, chemical, food, flavor, beverage, pharmaceutical, cosmetic, and fragrance. The plant's raw materials are roots, rhizomes, bulbs, leaves, stems, barks, wood, flowers, fruits, and seeds. These yield gums, resins, essential (volatile) oils, fixed oils, waxes, juices, extracts, and spices for medicinal and aromatic purposes. All these commodities are traded worldwide. A dealer's market report for an item may say "drought in the country of origin has forced up prices."

Natural products do not mean safe products, and every industry has to take this into account because they are subject to regulation. For example, a number of plants that are approved for use in medicine must not be used in cosmetic products. The assessment of "safe to use" starts with the harvested plant material, which has to comply with an official monograph. This may require absence of, or prescribed limits of, radioactive material, heavy metals, aflatoxin, and pesticide residue, as well as the required level of active principle. This analytical control is costly and tends to exclude small batches of plant material. Large-scale, contracted, mechanized cultivation with designated seed or plantlets is now preferable.

Today, plants are selected not only for the yield of active principle, but also for the plant's ability to overcome disease, climatic stress, and the hazards caused by mankind. Such methods as in vitro fertilization, meristem cultures, and somatic embryogenesis are used. The transfer of sections of DNA is giving rise to controversy in the case of some end uses of the plant material.

Some suppliers of plant raw material are now able to certify that they are supplying organically farmed medicinal plants, herbs, and spices. The Economic Union directive CVO/EU No. 2092/91 details the specifications for the obligatory quality controls to be carried out at all stages of production and processing of organic products.

Fascinating plant folklore and ethnopharmacology sometimes lead to medicinal potential. For example, there are muscle relaxants based on the arrow poison curare from species of *Chondrodendron*, and the antimalarials derived from species of *Cinchona* and *Artemisia*. The methods of detection of pharmacological activity have become increasingly reliable and specific, frequently involving enzymes in bioassays and avoiding the use of laboratory animals. By using bioassay-linked fractionation of crude plant juices or extracts, compounds can be specifically targeted. In this way it is clear, for example, which compounds inhibit blood platelet aggregation, or have antitumor, antiviral, or other activity. With the assistance of robotic devices, all the members of a genus may be readily screened. However, the plant material must be fully authenticated by a specialist.

The medicinal traditions of ancient civilizations such as China and India have a large armamentarium of plants in their pharmacopoeias that are used throughout Southeast Asia. A similar situation exists in Africa and South America. Thus, a very high percentage of the world's population relies on medicinal and aromatic plants for their medicine. Western medicine is also responding. Already in Germany all medical practitioners have to pass an examination in phytotherapy before being allowed to practice. It is notable that medical, pharmacy, and health-related schools throughout Europe and the United States are increasingly offering training in phytotherapy. Independent herbal medicine companies had been very strong in Germany but by 1995, 11 of these (the majority) had been acquired by the international pharmaceutical companies because they had neglected this area of their business—the pharmaceutical companies were now recognizing the growing demand for herbal medicines in the Western world. An important role in meeting this demand has been played by the American Botanical Council (ABC).

A flourishing herbal medicines industry had already arisen in the United States by the 1980s. A nonprofit organization was needed to control the situation, and this was met by the ABC. This was formed by its founder and executive director Mark Blumenthal, who also became editor of its quarterly journal, *HerbalGram*. This reached issue 97 in the spring of 2013. A monthly electronic issue entitled *HerbalEGram* now exists.

HerbalGram stands back from the cut and thrust of research. It keeps an eagle eye on legislation affecting herbs and the public, plant conservation, and other important plant matters. It also responds to inaccuracies and misrepresentations in the media.

ABC makes available important plant monographs, such as the "Expanded Commission E online, German Government Monographs" of approved herbs (Agrimony to Yarrow) and unapproved herbs (Alpine lady's mantle herb to Zedoary rhizome).

The *ABC Clinical Guide to Herbs*, edited by Mark Blumenthal et al., is a continuing online reference, which contains extensive monographs on the efficacy of the 30 most popular herbs used in the United States; "Healthy Ingredients" covers plants and related materials used in dietary supplements and natural cosmetics.

Depending on the level of membership paid, Herb-Clip and HerbMedPro are online plant reviews covering efficacy, safety, activity formulas, plant photos, distribution maps, and cultivation. ABC's staff reviews books and also sells books at a discount to members.

Since 1995, because of take-overs, the number of international pharmaceutical companies has been reduced. In contrast, there has been a marked increase in scientists engaged in plant science around the world. The explosion in plant science resulted in the need for a rapid dispersal of research results. This led to the publication of, for example, the abstracts *Review of Aromatic and Medicinal Plants* (RAMP), a subset of Cabi. Debbie Cousins, BSc, editor of RAMP, kindly informed me that the total number of articles in RAMP has increased from 4,700 (1995) to 27,000 (2011).

Existing journals have increased the number of articles per issue and also the frequency of publication. In addition, many new journals are being published including those now online.

The National Health Service of the United Kingdom has complained that in 2012 no new antibiotic or new affordable anticancer medicine had been produced by the international pharmaceutical industry. It is my belief that plant science will yield the next breakthrough anticancer drug.

For Volume 51, I thank its editors, Giovanni Dugo and Ivana Bonaccorsi for their dedicated work, and the chapter contributors for their authoritative information. My thanks are also due to John Sulzycki, plant science executive editor for CRC Press, and Joselyn Banks-Kyle, the project coordinator, for their ready help.

Roland Hardman, BPharm, BSc (Chem), PhD, FRPharmS

Preface

The origins of bergamot still remain mysterious and fascinating. This plant has been successfully cultivated in Calabria since the eighteenth century in a small area where it found the ideal habitat. Attempts to cultivate bergamot in different locations worldwide were not as successful.

In Calabria it is common to define bergamot as "the prince of the Citrus genus" and to assert that bergamot essential oil "scented the world." It is true, in fact, that bergamot oil has a very characteristic composition, different from any other citrus peel oil. The volatile fraction is dominated by oxygenated compounds, mainly linalool and linalyl acetate, together accounting for more than 50% of the oil. The essential oils extracted from the peel of other citrus fruits usually contain no more than 5% of total oxygenated volatile compounds. The profile of the oxygen heterocyclic compounds is simpler than in other citrus oils, such as lime, lemon, or bitter orange, but is characterized by the presence of noticeable amounts of a psoralen, bergapten. This component is considered phototoxic, but also possesses prodigious effects against different diseases.

Although the production of bergamot and of its derivatives is less than that of every other citrus used for the primary transformation, its chemical composition and biological properties have been a subject of great scientific interest. Even if the use of bergamot oil has faced periods of crisis, it continues to be considered essential in the formulation of highly appreciated perfumes. Today there is also increased demand for the bergamot oil for food flavorings and gastronomy.

This book represents a tribute to bergamot. We are emotionally and scientifically involved with this citrus, as we have dedicated most of our research activities to this subject. This book covers almost all aspects of bergamot: historical and botanic origins, cultural practices, transformation technologies, chemical composition and use of its derivatives, possible contaminations, and biological activity.

We hope that the information given here, never collected in a single place, will be a useful and accurate reference for all who are interested in bergamot: botanists, industrial and corporate operators, chemists and biologists, and both students and professionals.

We would like to conclude with the words written for this book by Antoine Maisondieu, famous creator of perfumes at the Givaudan France Fragrances:

> I love most of the natural products of the perfumer "palette," but when it comes to bergamot, she has a very special place. I think she's my preferred one!
>
> I like her for her simplicity, her crystalline freshness (but not evanescent!), her precision, and her perfect balance between acidity and sweetness.
>
> She makes me think of beautiful women of the Italian movies from the 1950s and the 1960s.
>
> I find her sunny, joyful, cheerful, and aristocratic. Her smell is like the crystalline laugh of a beautiful young woman (the laugh of Claudia Cardinale in *Il Gattopardo*).

I could go on forever on the aesthetic and emotional reasons that make me use it, but on the technical side bergamot is also perfect. I use it for her freshness, this top note which is immediately recognizable and agreeable. (I think that if we were giving the smell to the people in the street 80 percent would love it.)

The most beautiful quality is the true one from Calabria. I use it to create a contrast. She can fluidify a perfume or aerate it. For example, if you remove it from "Shalimar" or from most of the old Guerlain perfumes, they will loose all their charm.

Perfumery is a contrast between light, volatile raw materials like bergamot and heavier ones like vanilla, and most of the time this contrast gives birth to beautiful creations.

These words express the gratitude and love for bergamot essential oil felt by someone who makes the most noble use of it.

Giovanni Dugo and Ivana Bonaccorsi

Editors

Professor Giovanni Dugo obtained his Ph.D. in chemistry in 1964 from the University of Messina. The same year he became a professor of organic chemistry at Messina. Professor Dugo was later a professor of food chemistry and of chemistry and technology of citrus derivatives. Professor Dugo recently retired from his position of Full Professor of Food Chemistry at the University of Messina.

Dugo's scientific activity is focused on the development of innovative analytical methods and the study of food matrices using such innovative methodologies as multidimensional liquid chromatography (comprehensive LC); multidimensional gas chromatography (MDGC and comprehensive GC); ultrafast-GC and ultrafast-GC/MS; on-line SPME-GC/MS; micro-HPLC and micro-HPLC/API/MS; multidimensional HPLC and micro-HPLC; superheated HPLC; and LC × GC. This includes method validation using pure standard compounds and complex food samples and exploitation of the developed methods for the study of food matrices: essential oils, fruit juices of citrus and noncitrus origin, food lipids, wines, coffee, cheese, and vegetable products. In particular, he studies the following classes of compounds contained in these food products: triglycerides, fatty acids, sterols, tocopherols, monoterpenes, sesquiterpenes, coumarines, psoralens, polymethoxyflavones, carotenoids, anthocyanins, and other flavonoid-structured compounds, as well as pesticides, paraffins, aromatic hydrocarbons, and pyrazines, correlating the results attained with the genuineness, quality, and nutritional characteristics of the studied food samples.

This scientific activity, focused mainly on the development of separation methods and on the analysis of complex matrices, is reported in 350 national and international papers; approximately the same number of communications at national and international symposia; several chapters in scientific books and encyclopedias; he is also the editor of books dedicated to the chemistry and technology of citrus products and one on food toxicology.

Dugo participates in, chaired, and coordinated numerous committees and organizations whose activities and research projects focused on food chemistry, advanced analytical techniques, aromatic plants, and citrus chemistry and technology. In 2005 he founded The Mediterranean Separation Science Foundation Research and Training Center in Messina, which he cochaired with Professor Mondello; from 2002 to 2004 he was Director of the Ph.D. school on Food Chemistry and Safety, University of Messina. Professor Dugo is coordinator of the Food Chemistry Group of the Italian Chemistry Society (SCI) and a member of the board for the *Journal of Essential Oil Research*. He is a referee of numerous scientific journals.

Professor Dugo also served in the following academic positions: from 1995 to 1998 he was Vice-Rector of the University of Messina and President of the Evaluation Board "Nucleo di Valutazione" of the University of Messina; from 2003 to 2007 he was Vice-Rector and Delegate for the Scientific Research Activity of the same university.

Professor Dugo has received the Medal of Food Science of Italian Society and the medal awarded by the Flavour Science Italian Society. In 2009, Professor Dugo was awarded of the Liberti Medal by the Italian Society of Chemistry (SCI) for his contribution to the diffusion of science.

Dr. Ivana Lidia Bonaccorsi has been an assistant professor of food chemistry at the University of Messina since 1998. She graduated, with honors, with a Ph.D. in chemistry in 1995 from the University of Messina. After graduation, from January to July 1996, Dr. Bonaccorsi worked as an intern at the Ingredient Control Laboratories of the Coca-Cola Company under the supervision of Dr. Terence Radford, applying the analytical and sensorial evaluation of ingredients used in soft drinks. During the same period, she collaborated with Dr. Benjamin Clark and Theresa Chamblee at the Research and Development Department of the Coca-Cola Company in Atlanta, Georgia, exploiting GC-FTIR to study germacrenes in sweet orange essential oil.

While at the Coca-Cola Company, Dr. Bonaccorsi collaborated with the Flavour Development Department, in particular with Charles Stephens, in the creation of new flavors for soft drinks. She was later awarded a scholarship by a citrus company for a project to destroy pesticide residue in citrus essential oils. In 1997 she was granted a one year postdoctoral fellowship by the University of Messina to work at Virginia Tech University under the supervision of Dr. Harold McNair. During this period, in addition to the research activity focused on development of new chromatographic methods for the analyses of citrus derivatives, Dr. Bonaccorsi collaborated as assistant professor for various ACS Short Courses. She later organized and taught similar courses at the University of Messina on basic GC, GC-MS, HPLC, and HPLC-MS.

Since 1998, Dr. Bonaccorsi has taught numerous official courses at the University of Messina, such as formulation of cosmetics, food chemistry, functional foods, biotechnology in food chemistry, and basic inorganic chemistry. Presently Dr. Bonaccorsi teaches courses of Functional Food and Food Chemistry at the School of Pharmacy of the SCIFAR department of the University of Messina. She is also teacher at the PhD school on Food Chemistry and Safety, University of Messina. She participated in congress organization activities and research projects on food chemistry, advanced analytical techniques, aromatic plants, and citrus chemistry and technology. Dr. Bonaccorsi has been a member of The Mediterranean Separation Science Foundation Research and Training Center in Messina since its founding in 2005. This organization is chaired by Professor Giovanni Dugo and Professor Mondello. Since 2012 she has been a member of the directory board of the Food Chemistry Group of the Italian Chemistry Society (SCI).

Dr. Bonaccorsi's research interests mainly focus on the determination of contaminants of citrus oils (pesticides and plasticizers); development of advanced chromatographic and spectroscopic techniques for the analysis of real samples (rapid RP-HPLC methods for the analyses of the nonvolatile fraction of citrus essential oils; determination of carotenoids in citrus essential oils and juices, olive oils, and other fruits; studies on the composition of the volatile fraction of numerous citrus essential oils, by GC, GC-MS, MDGC and by esGC); assessment of genuineness parameters by GC-C-IRMS and by enantioselective separations (esGC and MDGC) on natural complex matrices; and development of new methods for the analyses of different food matrices by innovative chromatographic techniques.

Dr. Bonaccorsi is the author of about 50 scientific papers published in national and international scientific journals. She is coauthor of numerous book chapters on the composition of citrus peel, leaves, and flower oils. She has been a lecturer at numerous national and international congresses and symposia. Dr. Bonaccorsi is a referee of numerous international journals.

Contributors

Giovanni Enrico Agosteo
Mediterranea University of Reggio
  Calabria
Reggio Calabria, Italy

Norbert Bijaoui
French Society of Perfumers
Paris, France

Giuseppe Bisignano
Department SCIFAR (Scienze del
  Farmaco e dei Prodotti per la
  Salute)
University of Messina
Messina, Italy

Ivana Bonaccorsi
Department SCIFAR (Scienze del
  Farmaco e dei Prodotti per la
  Salute)
University of Messina
Messina, Italy

Frederic Bourgaud
Laboratoire Agronomie & Enviroment
  UMR
Nancy Università INRA ENSAIA
Vandoeuvre, France

Santa Olga Cacciola
University of Catania
Catania, Italy

Pasquale Cavallaro
Consorzio del Bergamotto di Reggio
  Calabria
Reggio Calabria, Italy

Maria Assunta Chiacchio
Dipartimento di Scienze del Farmaco
University of Catania
Catania, Italy

Francesco Cimino
Department SCIFAR (Scienze del
  Farmaco e dei Prodotti per la Salute)
University of Messina
Messina, Italy

Antonella Cotroneo
Department SCIFAR (Scienze del
  Farmaco e dei Prodotti per la Salute)
University of Messina
Messina, Italy

Francesco Crispo
Consorzio del Bergamotto di Reggio
  Calabria
Reggio Calabria, Italy

Giuseppa Di Bella
Department SASTAS (Dipartimento
  di Scienze dell'Ambiente, della
  Sicurezza, del Territorio, degli
  Alimenti e della Salute) "G. Stagno
  d'Alcontres"
University of Messina
Messina, Italy

Angelo Di Giacomo
Consultant
Messina, Italy

Rosa Di Sanzo
Department of Agricultural Sciences
University of Reggio Calabria
Reggio Calabria, Italy

Giacomo Dugo
Department SASTAS (Dipartimento
  di Scienze dell'Ambiente, della
  Sicurezza, del Territorio, degli
  Alimenti e della Salute) "G. Stagno
  d'Alcontres"
University of Messina
Messina, Italy

Giovanni Dugo
Department SCIFAR (Scienze del
   Farmaco e dei Prodotti per la Salute)
University of Messina
Messina, Italy

Paola Dugo
Department SCIFAR (Scienze del
   Farmaco e dei Prodotti per la Salute)
University of Messina
Messina, Italy

Paul Forlot
Sofibio, Scientic Division
Monaco, Principality of Monaco

Giselda Gallucci
Department of Chemistry
University of Calabria
Arcavacata di Rende, Italy

Florin Gazea
Consultant
Reggio Calabria, Italy

Maria Paola Germanò
Department SCIFAR (Scienze del
   Farmaco e dei Prodotti per la Salute)
University of Messina
Messina, Italy

Rosario Lo Curto
Department SASTAS (Dipartimento
   di Scienze dell'Ambiente, della
   Sicurezza, del Territorio, degli
   Alimenti e della Salute) "G. Stagno
   d'Alcontres"
University of Messina
Messina, Italy

Gaetano Magnano di San Lio
University of Reggio Calabria
Reggio Calabria, Italy

Antoine Maisondieu
Givaudan France Fragrances
Paris, France

Naim Malaj
Department of Chemistry
University of Calabria
Arcavacata di Rende, Italy

Andreana Marino
Department SCIFAR (Scienze del
   Farmaco e dei Prodotti per la Salute)
University of Messina
Messina, Italy

Luigi Mondello
Department SCIFAR (Scienze del
   Farmaco e dei Prodotti per la Salute)
University of Messina
Messina, Italy

David A. Moyler
Consultant
Kent, United Kingdom

Antonia Nostro
Department SCIFAR (Scienze del
   Farmaco e dei Prodotti per la Salute)
University of Messina
Messina, Italy

Paul Pevet
Institute Fédéral des Neorosciences
   Cellulaires et Intégratives
Louis Pasteur University
Strasbourg, France

Francesco Pizzimenti
Department SCIFAR (Scienze del
   Farmaco e dei Prodotti per la Salute)
University of Messina
Messina, Italy

Terence Radford
Consultant
Atlanta, Georgia

Carla Ragonese
Department SCIFAR (Scienze del
   Farmaco e dei Prodotti per la Salute)
University of Messina
Messina, Italy

Antonio Rapisarda
Department SCIFAR (Scienze del
Farmaco e dei Prodotti per la
Salute)
University of Messina
Messina, Italy

Vilfredo Raymo
Simone Gatto
Messina, Italy

Maria Restuccia
Department SCIFAR (Scienze del
Farmaco e dei Prodotti per la
Salute)
University of Messina
Messina, Italy

Elvira Romano
Department of Chemistry
University of Calabria
Arcavacata di Rende, Italy

Roberto Romeo
Department SCIFAR (Scienze del
Farmaco e dei Prodotti per la
Salute)
University of Messina
Messina, Italy

Mariateresa Russo
Department of Agricultural Sciences
University of Reggio Calabria
Reggio Calabria, Italy

Marina Russo
Department SCIFAR (Scienze del
Farmaco e dei Prodotti per la Salute)
University of Messina
Messina, Italy

Antonella Saija
Department SCIFAR (Scienze del
Farmaco e dei Prodotti per la Salute)
University of Messina
Messina, Italy

Marcello Saitta
Department SASTAS (Dipartimento
di Scienze dell'Ambiente, della
Sicurezza, del Territorio, degli
Alimenti e della Salute) "G. Stagno
d'Alcontres"
University of Messina
Messina, Italy

Danilo Sciarrone
Department SCIFAR (Scienze del
Farmaco e dei Prodotti per la Salute)
University of Messina
Messina, Italy

Luisa Schipilliti
Department SCIFAR (Scienze del
Farmaco e dei Prodotti per la Salute)
University of Messina
Messina, Italy

Giovanni Sindona
Department of Chemistry
University of Calabria
Arcavacata di Rende, Italy

Germana Torre
Department SCIFAR (Scienze del
Farmaco e dei Prodotti per la Salute)
University of Messina
Messina, Italy

Peter Quinto Tranchida
Department SCIFAR (Scienze del
Farmaco e dei Prodotti per la Salute)
University of Messina
Messina, Italy

Alessandra Trozzi
Department SCIFAR (Scienze del
Farmaco e dei Prodotti per la Salute)
University of Messina
Messina, Italy

1 Origin, History, Diffusion

Angelo Di Giacomo and Giovanni Dugo

CONTENTS

1.1 DIFFUSION OF CITRUS IN ITALY

Almost all citrus species are native to the subtropical and tropical regions of Asia and Malaysia (Gallesio 1811; Tolkowsky 1938). However, it is not an easy task to trace back the path which led to their worldwide diffusion. It has long been confirmed that before it was introduced in Europe, the orange was already cultivated in China. This is proven by an ancient manuscript from 1178, the famous "Chü lu" written by Han Yen-Chih and translated in English by Hagerty (1923). This text describes 27 varieties of sweet oranges, bitter oranges, and mandarins, reporting on the cultivation practices, harvest procedures, and cures for diseases.

It is also possible to assume that citrus fruits, cultivated by the Chinese as well as other populations of the Far East, underwent significant modifications through natural evolution processes.

Citron was the first citrus known to European civilization. This information was reported by Theophrastus (300 B.C.), who named it "Median" or "Persian" apple due to the diffusion of this fruit in these areas. The introduction of Citron in Sardinia, in Naples, and even in northern Italy, cultivated in greenhouses, can be dated back to the third century. It is believed that Palestine built a bridge to transfer Citron in Italy, due to the moderate climate, which is particularly suitable for its cultivation (Di Giacomo 1970).

Other members of the Rutaceae family arrived in Europe during the Arab empire. Bitter orange and lemon were imported by the Arabs from India during the tenth century and successively introduced in different locations of the Mediterranean basin, in particular in Sicily and Sardinia.

The Crusades had an impact on the diffusion of citrus cultivation; species such as lemon and bitter orange are mentioned by historiographers only after the Crusades. The orange was introduced in Europe around 1400; the most interesting evidence in Italy is from Leandro Alberti (1550), who during a journey in Southern Italy in 1523, referred to significant plantations of oranges, lemons, and citrons in Sicily and in Calabria. Moreover, in the treatise written in 1510 by the Sicilian Antonino Venuto, four species of citrus are mentioned: bitter orange, citron, lemon, and grapefruit (Lanza 1949).

1.2 BERGAMOT IN ITALY

Bergamot started to be cultivated in Italy around the seventeenth century. The introduction of this plant occurred during two different phases, clearly independent of one another: the first, in Tuscany, was limited to some plants used solely for ornamental purposes in gardens, documented by historical evidence integrated by either botanical or artistic interests; the second, in Calabria, was in a small region close to the city of Reggio Calabria.

1.3 BERGAMOT IN TUSCANY

The pictorial work by the painter Bimbi (Bartolomeo del Bimbo, born in Settignano in 1648), appreciated for his peculiar descriptive fidelity, represents "a precious and reliable illustrative evidence of the interest at that time for natural science" (Strocchi 1982) and, we can add, for Citrus in particular.

The iconography (oil on canvas) we are referring to consists of four large canvases: three of them measure 174 × 233 cm and one 130 × 160 cm. The canvases have accurate legends and detailed charts. They represent numerous assorted subjects (oranges, chinottos, lumias, citrons, limes, bergamots, and melangolos) pendent from an espalier.

Among the 112 subjects identified, three are characterized by morphological characters which recall bergamot (Baldini et al. 1982):

- "Bergamotta scannellata": two spherical, medium-sized hesperides, with a slightly hollow apex, short neck and a ribbed, slightly corrugated epicarp.
- "Lumia similar to the bergamotta pear": two spherical, medium-sized hesperides with a smooth epicarp.
- "Bergamotta pear, orange race": three turbinate, medium-sized hesperides with smooth epicarps.

The canvas, once on display in the Castle of the Topaia, today is kept by the Superintendent for the Artistic and Historical Heritage of Florence and Pistoia. They integrate other bibliographical and iconographic sources of that era with the color information and pictorial details.

For example, the information on bergamot can be considered a significant support to the description of bergamot (Lumia bergamotta) made by Clarici (1726) approximately in the same period. If we consider that authors such as Del Riccio (1595) and Ferrari (1646) in the period immediately before never mentioned bergamot, the hypothesis of the introduction of bergamot in Tuscany during the time of Cosimo III dei Medici (1670–1725) at the end of the seventeenth century is validated (Targioni Tozzetti 1780). In addition, Micheli, who lived during the same period as

Clarici, referred to a strange fruit named "bergamotta apple," which he observed in the Tuscan garden owned by Marchesi Nicoli.

1.4 BERGAMOT IN REGGIO CALABRIA

Webber (1948) referred to two types of bergamot reported by Volckamer and five by Risso and Poiteau.

Volckamer (1708–1711), a physician and botanist from Nuremberg, indicates the Limon bergamotta as the "*gloria limonum et fructus inter omnes nobilissimus*" and ascribes the name to the shape of the fruit, similar to the pear bergamotta, from the Turkish term "*Beg-armudi*," which means "pear of the Prince." About this topic, it was recently noted (Kunkar 1997) that the exact Turkish term to indicate "pear of the Prince" should be "*bey armudu*."

Volckamer also notes that Italians used the peel of this fruit to obtain an extremely fine essence; he neglects the hypothesis that the term "bergamot" could derive from the city of Bergamo, since at that time none of the esperideae were cultivated in that region.

Risso and Poiteau (1818–1822), in Chapter 4 ("Description des Bergamottiers") of their treatise (*Histoire Naturelle des Orangers*), illustrated with rich botanical details the characteristics of the following fruits: Bergamottier ordinaire, *Citrus Bergamia Vulgaris,* Bergamotta ordinaria; Bergamottier a fruit toruleu, *Citrus Bergamia Torulosa,* Bergamotta striata; Bergamottier a petit fruit, *Citrus Bergamia Parva,* Bergamotta piccolo; Bergamotte Mellarose, *Citrus Bergamia Mellarosa,* Bergamotta Mellarosa; Bergamotte Mellarose a fleur double, *Citrus Bergamia Mellarosa Plena,* Bergamotta Mellarosa with a double flower. Boards were enclosed with the text, reproducing with accurate paintings the five types of bergamot. According to Engler (1931), the name "bergamot" came from the small city of Bergamia located in the northern part of Smyrna. In the fourth book of the treatise by Ferrari (1646), *Baptistae Ferrarii Senensis et Societate Iesu*, dedicated to citrus, 15 varieties of oranges are described; among these one, *Aurantium stellatum et roseum* (Ferrari 1646, chapter 6), was later identified (G. Spinelli 1968) as morphologically close to a variety of bergamot (*Citrus Bergamia Mellarosa*). However, from a careful reading of the original text and analyzing the picture (black and white) attached, it appears that Spinelli's hypothesis is not fully certain, since the description is also compatible with other citrus species. This is also supported by the fact that the presence of bergamot in Italy, at that time, was not recorded elsewhere.

Mottareale (1928) reported that Griso (1905) wrote in 1878 (and later published) that in Europe bergamot was transferred from the Canary Highlands by Christopher Columbus. It was later brought to Reggio Calabria from the city of Berga, which was located in the province of Barcelona (Spain), thus the name Bergamotto. A plant bought from a Spaniard for 18 escudos was grafted on a lemon at the beginning of 1500, in a garden of the Count Valentino of Santa Caterina (Reggio Calabria).

P. Spinelli (1951) derives the name bergamot from Bego, an Epirote prince. Other hypotheses, more or less reliable, have been given on the introduction of bergamot in Calabria. According to the historian Spanò Bolani (1857) it was Menzà, a canonical, who brought bergamot to Reggio Calabria, with the first cultivations registered in 1726. Bomboletti (1879), confirming the Spanish origin (Berga) of bergamot, refers to a local legend about Valentino selling a plant for 28 escudos to a gardener named Rovetto.

De Nava (1910) dates the introduction of the plant back to 1600, and to the middle of the 1700s the start of cultivation. Finally, G. Spinelli (1968) asserts: "It is highly probable that bergamot was introduced from Greece, where it was effectively proved the presence of these plants, and arrived in Reggio Calabria through Venice. However we cannot exclude that it originated in Calabria deriving from spontaneous mutation of different species."

Up to the present day, we cannot establish how bergamot arrived in Reggio Calabria, or even the origin of the name "bergamot." The only certainties are the fact that this species was one of the last Citrus imported in Italy, and that around 1750 bergamot was already cultivated close to Reggio Calabria, thanks to the interest and passion of some crafty farmer.

An important event linked to the history and development of bergamot is the first creation of an eau de toilet "Aqua admirabilis," later called "cologne." This name was derived from the city of Cologne, where it was produced, and from where it spread worldwide.

Cologne, patented in 1704, was formulated by an Italian peddler, Gian Paolo Feminis, who was born in Crana and emigrated to Cologne in 1680. When Feminis died, he left the recipe to Giovanni Antonio Farina, who then left it to Giovanni Maria Farina, from whom the Farina family descends, heirs of the secret of the production of cologne. Bergamot essential oil is a fundamental ingredient of cologne, as well as many other perfumery products.

With regards to the first steps of the bergamot cultural practices in Calabria and to the shy start of the commerce of the essential oil, the "memories" of Calabrò Anzalone (1804) practically represent a unique source of information. Calabrò reported,

> Bergamot fruit, from which is extracted the essential oil, belongs to a tree of the Genus Citrus, class Polyadelphia, order Icosandria, named by Linneus in his system *Citrus aurantium,* var. Bergamot, originated from the Barbados highlands, according to Tree Hughes.
>
> We do not know the origin of this tree, nor how it was introduced here for the first time: however some old inhabitants of this town refer that few trees of bergamot already existed since a long time, but these were located in the gardens of few owners and were cultivated for ornamental purposes, not for their use. They also reported that if the owners did make them to extract the essential oil from the fruit of this tree, most of the time this remained unsold, for the lack of demand. Only few quantities were sold to people from abroad who occasionally passed by. The price was four ducats per pound. The presence of the foreigners interested in finding bergamot essential oil, induced for the first time in 1750, Mr. D. Nicola Parisi to look at this plant with industrial interest. Thus in his large property located in Giunghi, he multiplied the number of bergamot plants. Other land owners followed Parisi, also in view of the good price paid for the essential oil, so they also multiplied the number of their trees. In few years the industry of bergamot spread in that area.

The bergamot plantations spread easily, thanks to the specific properties of the land in the area of Reggio Calabria, mainly "dry, not excessively pebbly, easy to irrigate more frequently than for other Citrus" (Giuffrè 1967). These conditions are found along the Tyrrhenian coast, between Reggio and Cannitello di Villa S. Giovanni, and on the Ionian coast, from Reggio to Palizzi.

At the beginning of the nineteenth century it was already possible to find, around the Reggio area, small gardens and modest plots of land with a limited number of bergamot plants. The cultivation developed during the second half of the century, with a larger extension and number of plants located on the Ionian coast. In fact, on the Tyrrhenian side the cultivation of bergamot was in competition with that of the oranges of Gallico and the lemons of Catona and Cannitello, well appreciated for their properties, while on the Ionian side bergamot was cultivated exclusively. In addition, the oil yield of the fruits cultivated on the Tyrrhenian coast was lower. At the end of the nineteenth century, the bergamot plantations were estimated to be close to 1500 hectares.

In the next 30 years (1900–1930), the expansion was extremely rapid, so that the surfaces cultivated with bergamot plants doubled, reaching 150 tons of essential oil production. In the same period, farmers tried to extend the cultivation area on the Ionian Coast (from Palizzi to Gioiosa Ionica), and on the Tyrrhenian Coast, toward Gioia Tauro, Rosarno and Pizzo. However, such attempts failed with the only exception the area of Brancaleone.

After 1930, a period of crisis occurred, and for 20 years the expansion of bergamot cultivation stopped. Contractions of bergamot plantations were recorded during this period, and only the area between Reggio and Palizzi was preserved. In fact, during these years the income from the essential oil trade was sufficient only to cover the expenses for the harvesting and processing of the fruits.

Obviously the commerce of bergamot essential oil was even worse during World War II, with the resulting accumulation of large stocks of unsold essential oil. When the situation normalized, new producers arose in the area between Palizzi and Reggio, replacing the cultivated areas which were destroyed by the expansion of the city of Reggio. In the 1960s, a new impulse to the growth of the production of bergamot was recorded in Calabria, but it did not last very long.

Nesci (1994), based on the data provided by the Consorzio del Bergamotto and by the Experimental Station for the Industry of Essential Oils and Citrus Products, investigated the evolution of the surface area relative to the cultivation of bergamot. The author highlighted that, after the growth registered until the beginning of the 1970s, a slow contraction of the total land dedicated to bergamot cultivation began, with a consequent decrease in the total production of fresh fruit and of the essential oil.

The reduction of the cultivated area was due to variation of the purpose of use of the land, with the spread of urbanization and of residential areas; reconversion of the industrial plants due to the recurrent economic crises; and reduction of the water sources for irrigation. Presently, the surface used for bergamot cultivation is estimated to be around 1500 hectares.

1.5   PRESENCE OF BERGAMOT OUTSIDE CALABRIA

The first attempts to introduce bergamot in the United States were made before 1815, with the purpose to study its acclimatization capabilities in Florida, Louisiana and

California; these attempts failed, as did most of the others made to cultivate bergamot outside Calabria.

During World War II, it was not possible to use the essential oil from Calabria
in the United States, but it was possible to find different types of so-called bergamot oils on the market, with a different chemical composition and odor properties
compared to the original bergamot oil produced in Calabria (Guenther 1949). It was
determined that the bergamot oil produced in Brazil was actually a mixture of acety-
lated oils (bois de rose, sweet orange, lemon, etc.). The oil produced in Mexico was
obtained by cold pressure of sweet lime.

The articles by Schwob (1953, 1955) and Huet (1970) report the results obtained
through the analyses performed on samples of bergamot oil obtained from small
industrial productions in Western Africa (Guinea), Northwestern Africa (Mali),
and Northern Africa (Algeria, Morocco, and Tunisia). There is no record of further
industrial productions of bergamot essential oil in these areas.

Bergamot, obtained with material from Calabria, has been present since
the 1960s in different cultivated areas in Argentina (Misiones, Entre Rios, and
Tucumàn) (Drescher et al. 1984). A modest industrial production is found in the
area of Misiones and, mainly, in Tucumàn, close to Monte Grande, department of
Famaillà. In this location the fruits used for transformation are of a seeding variety,
grafted on Cleopatra mandarin; the production of the essential oil in 1980 was 3.4
tons, with an average yield of 530 grams per quintal (g/q). The oil presents appreciated analytical and odor properties.

Brazil and Uruguay are also producers of small amounts of bergamot essential
oil, and represent a good potential for the future development of the international
trade. The only significant competitor with Calabria today is the Ivory Coast. In fact,
up until 1959, with the exception of a few bergamot trees in the Botanical Garden
of Bingerville close to Abidjan, there was no record of plants in this region. It was
in that year that M.M. Lavalade, the pioneer of the development of citrus essential
oil production in this country, decided to start a new plantation of bergamot for the
production of essential oil. He was supported by V. Furon of the IFAC, today IFRA
(French Institute for Research in Fruits and Citrus Overseas), located in Guinea and
Ivory Coast, who selected 1000 plants of bergamot grafted on bitter orange, probably
originating from Morocco or Algeria.

The plants were transferred to Soubrè, 150 km north of Sassandra, protected by
jute bags. Unfortunately most of the plants were affected by molds and only 50 plants
survived. With these plants Lavalade created a nursery, and later he was able to plant
bergamots on a surface of about 40 hectares. The first production started during the
season 1961–1962. For the extraction of the oil the Indelicato MK machine was used.
Following this example, many other farmers of Sassandra started to devote their
lands to the growth of bergamot trees.

In 1969, in the Ivory Coast, 640 hectares (of which 505 was in the Sassandra
area) were cultivated for bergamot oil production. According to Duclos (1997), from
1970 onwards, an extraordinary growth of the bergamot oil production led to an
actual cultivation surface of 1300 hectares. The bergamot plantations considered as
"old" are progressively replaced by new ones. However, the potential production of
essential oil in Ivory Coast is declared to be approximately 50 to 70 tons, an amount

which appears to be much higher than the real production capabilities. This opinion is based on official statistical data provided by the Republic of Ivory Coast (Nesù 1994) reporting a total area of bergamot cultivation in 1989 equal to 260 hectares, a total fruit production of 2800 tons, and that of essential oil equal to 10.7 tons.

Bergamot trees have been cultivated in the South of Turkey mainly between Bodrum and Antalya (Kirbaslar et al. 2000). According to the World Map showing the production centers of essential oils published by Treatt, attached to the recent book by Baser and Buchbauer (2009), today the countries producing for industrial transformation are Italy, the Ivory Coast, Brazil, and China.

REFERENCES

Alberti, L. 1550. Descrittione di tutta Italia nella quale si contiene il sito di essa, l'origine et la signoria delle città et de' Castelli, ecc. in-folio, Bologna.

Baldini, E., Rossi, F., Bassi, V., and Giannini, E. 1982. Agrumi, in Agrumi, frutta e uve nella Firenze di Bartolomeo Bimbi, pittore mediceo, 17–44, Firenze: Consiglio Nazionale delle Ricerche.

Baser K. H. C., and Buchbauer G. 2009. *Handbook of essential oils.* London and New York: Taylor and Francis.

Bomboletti, A. 1879. *Notizie sul bergamotto.* Roma: Artero.

Calabrò Anzalone, F. 1804. *Della balsamica virtù dell'essenza di bergamotta nelle ferite.* Messina, Italy: Letterio Fiumara e Giuseppe Nobolo Soci.

Calvarano, M., Ravanelli, G., and Di Giacomo, A. 1984. Sulle caratteristiche di un'essenza di bergamotto ottenuta da frutti prodotti in Toscana. *Essenz. Deriv. Agrum.* 51: 220–227.

Clarici, P. B. 1726. Istoria e coltura delle piante che sono pel fiore ragguardevoli e più distinte per ornare un giardino in tutto il tempo dell' anno, con un copioso trattato sugli agrumi. 703, Venezia: Andrea Poletti.

Del Riccio, A. 1595. Trattato di Agricoltura (copiato da Antonino Sangallo MS, presso Bibl. Medicea, Università di Firenze).

De Nava, G. 1910. Sull'industria delle essenze in Calabria, Reggio Calabria.

Di Giacomo, A. 1970. Cenni storici sullo sviluppo della coltivazione degli agrumi e dei metodi di estrazione degli olii essenziali. *EPPOS* 52: 140–145, Milano.

Drescher, R. W., Beñatena, H. N., Rossini, G. G. T., Ubiergo, G. O., and Retamar, J. A. 1984. Evaluaciòn de la calidad perfumistica del aceite esencial de *Citrus bergamia* Risso (bergamotta) cultivada en la E.E.A. INTA de Concordia (Entre Rios). *Essenz. Deriv. Agrum.* 54: 192–199.

Duclos, T. 1997. *Bergamot and other citrus from Ivory Coast.* Bois-Colombes, France: Duclos Trading.

Engler, A. 1931. Rutaceae, in *Die natürlichen Pflanzenfamilien.* 2nd. ed. 344, eds. A. Engler and K. Prantl. Leipzig: Engelmann.

Ferrari, B. 1646. Hesperides, sive de malorum aureorum coltura et usu. Libri quatuor, Sumptibus Hermanni Scheus, Roma.

Gallesio, G. 1811. *Traitè du Citrus.* Paris: Louis Fantin.

Giuffrè, A. 1967. Coltivazione-Economia-Tecnologia del Bergamotto, Conferenza tenuta presso il Rotary Club di Reggio Calabria, il 9, 25, 1967.

Griso, L. 1905. Bollettino della Arboricoltura Italiana, 1, I e II trimestre, Napoli: Francesco Giannini & Figli.

Guenther, E. 1949. *The essential oils*, vol. 4, 281. New York: D. Van Nostrand Co.

Hagerty, M. J. (trans.). 1923. *Monograph on the oranges of Wen-Chou, Chekiang*, Hau Yeu-Chihs Chu Lu (originally published 1178). Leiden, the Nederlands: Brill.

Huet, R. 1970. Essai d'introduction d'une industrie de la Bergamotte ou Mali. *Fruits* 25: 709–715.

Kirbaslar, S. I., Kirbaslar, F. G., and Dramur, U. 2000. Volatile constituents of Turkish Bergamot Oil. *J. Essent. Oil Res.* 12: 216–220.

Kunkar, A., and Kunkar, E. 1997. *Il Bergamotto e le sue essenze*. Reggio Calabria: Edizioni "AZ."

Lanza, D. 1949. *Agrumi*, in *Enciclopedia Italiana*, vol. 2, Roma.

Micheli, P. A. Enumeratio quarundam Plantarum sibi per Italiam, et Germaniam observatorum in acta Tomefortii methodum dispositarum, Tomo IX, MS, presso Bibl. Dip. Bot. Univ., Firenze.

Mottareale, G. 1928. Il Bergamotto. *Italia Agricola* 65: 1042–1052.

Nesci, F. S. 1994. Il bergamotto: dall'azienda agraria al mercato. Problemi attuali e prospettive future. *Essenz. Deriv. Agrum.* 64: 312–357.

Risso, A., and Poiteau, A. 1818–1822. *Histoire naturelle des orangers*, 109–116. Paris: Audot.

Schwob, R. 1953. L'essence de bergamotte de la Guinée Francaise. *Ind. Parfum.* 8: 406–409.

Schwob, R. 1955. Etude sur l'essence de bergamotte d'Afrique du Nord. *Ind. Parfum.* 10: 46–50.

Spanò Bolani, D. 1857. *Storia di Reggio*, Reggio Calabria.

Spinelli, G. 1968. *Studio sul bergamotto*, 11–15, Reggio Calabria: Grafiche Sgroi.

Spinelli, P. 1951. Il "Bergamotto di Reggio Calabria" (*Citrus bergamia* Risso). *Rivista di Agricoltura Sub-tropicale e Tropicale* 45: 1–3, 4–6, 7–9.

Strocchi, M. L. 1982. Bartolomeo Bimbi pittore naturalista alla corte di Cosimo dei Medici, in: *Agrumi, Frutta e Uve nella Firenze di Bartolomeo Bimbi pittore mediceo*, 17–44, Firenze: Consiglio Nazionale delle Ricerche.

Targioni Tozzetti, G. 1780. Notizie degli aggrandimenti delle scienze fisiche accaduti in Toscana nel corso degli anni LX del secolo 17, vol. 3, 10. Firenze: G. Bouchard in Mercato Nuovo.

Tolkowsky, S. 1938. *Hesperides: A history of the culture and use of citrus fruits*. London: John Bale Sons & Cumow.

Volckamer, J. C. 1708–1711. Nürnbergische Hesperides, oder Grümdliche Beschreibung der edlen Citronat-, Citronen-und Pomeranzen-Früchte, 2:155, Nürnberg.

Webber, H. J. 1948. History and development of the citrus industry, in *The citrus industry*, eds. H. J. Webber and L. D. Batchelor, vol. 1, 488–489. Los Angeles: University of California Press.

2 Citrus × bergamia Risso & Poiteau

Botanical Classification, Morphology, and Anatomy

Antonio Rapisarda and Maria Paola Germanò

CONTENTS

2.1 SYSTEMATICS, TAXONOMY, AND NOMENCLATURE

Despite its considerable distribution and great economic importance, the origin and evolution of plant species of the genus *Citrus* are still unclear; indeed these species hybridize easily and naturally through cross-pollination. The phylogeny and taxonomy are somewhat controversial; often the differences between species, new botanical variants, or varieties cultivated in a different way can be extremely small. In the case of varieties and hybrids in existence for a long time, only the new methods of molecular investigation are able to distinguish different species, although often this type of response is far from unique (Nicolosi et al. 2000).

2.1.1 *CITRUS* L.

The classification of the citrus affected by botanists in the last 400 years and their taxonomy has been based primarily on morphological and geographical data.

In 1753 Carl von Linné, with his *Species plantarum*, combined all citrus fruits known to him in a genus of two species: *Citrus medica* L., citron, with var. *limon* L., lemon, and *C. aurantium* L., orange with var. *grandis* L., the pomelo, and var. *sinensis* L., the sweet orange (Mabberley 1997). Pehr Osbeck, a follower of Linnaeus, in the diary of his journey to the Far East describes the cultivation of citrus in Canton. In *Species plantarum* Linnaeus used for some species the names of Osbeck's manuscript (Hansen and Fox Maule 1973); conversely, Osbeck in 1755 was one of the first to use Linnaean nomenclature in his *Dagbok Öfwer en Ostindsk Resa*. Although Linnaeus was very interested in the latest news from Osbeck, he did not include this in his work; therefore, several new plants were first described by Osbeck (Merrill 1916).

The genus *Citrus* L., reported in *Species plantarum*, was inserted by Linnaeus and Osbeck in the family of Rutaceae. According to the most reliable classifications of the last century (Hutchinson 1973; Dahlgren 1989; Thorne 1992; Cronquist 1993; Takhtajan 1997) the genus *Citrus* L. belongs to the family Rutaceae (Mabberley 2008), order Sapindales, subclass Rosidae, class Magnoliopsida, division Magnoliophyta, subkingdom Tracheobionta, kingdom Plantae.

The most recent recognized classification of the Rutaceae is due to Thorne (2000), who identified three subfamilies: Rutoideae (120 genera), Aurantioideae (30 genera), and Spathelioideae (5 genera). The subfamily Aurantioideae (which is the correct name for Citroideae or Limonoideae) includes tropical genera and species; within this subfamily, Engler (1931) identified a single tribe (Aurantieae, i.e., Citreae), Tanaka (1936) eight (Micromeleae, Clauseneae, Aegleae, Lavangeae [i.e., Luvungeae] Aurantieae, Meropeae, Atalantieae, and Microcitreae), Swingle (1943) and Swingle and Reece (1967) two (Clauseneae and Aurantieae, i.e., Citreae).

The tribe Aurantieae is composed of three subtribes: Triphasiinae, Citrinae and Balsamocitrinae. The subtribe Citrinae includes three groups; primitive citrus fruit, near citrus fruit, and the true citrus fruit trees. True citrus fruit trees include six genera: *Clymenia, Eremocitrus, Microcitrus, Poncirus, Fortunella,* and *Citrus* (Swingle 1967).

Of the many classification systems of the genus *Citrus* L. formulated in the last century, those proposed by Swingle and Reece (1967) and Tanaka (1977) have been widely accepted. The number of identified species is the major difference between the two systems (Uzun et al. 2009). Swingle (1967) recognized 16 species in the genus *Citrus* L. subdividing in the subgenus *Citrus* with edible fruits and in the subgenus *Papeda* (Hassk) with inedible fruits, while Tanaka (1977) has recorded 162 species. The classification of Tanaka is so detailed that it is not functional, but that of Swingle is certainly more functional for ease of setup even if sometimes it is unsatisfactory.

Recently in the botanical classifications, in addition to morphological data, molecular markers are considered to provide substantial information that is insensitive to variations arising from environmental factors. Several studies tried to examine and investigate the phylogenetic relationships within the genus *Citrus* L., and/ or between this and related genera, based on markers obtained by the use of modern molecular biology techniques: isozymes (Herrero et al. 1996), RFLPs (restriction fragment length polymorphisms) (Federici et al. 1998), SSRs (sample sequence repeat) (Barkley et al. 2006), ISSRs (inter-sample sequence repeat) (Fang et al. 1998; Gulsen and Roose 2001a, b), RAPD (random amplification of polymorphic DNA)

(Federici et al. 1998; Nicolosi et al. 2000), cpDNA sequence (Morton et al. 2003), and AFLP (amplified fragment length polymorphism) (Pang et al. 2007).

In recent years, the hybrid origin of many "species" has been demonstrated within the genus *Citrus* L., also ascertaining what might be the presumed parent species (Nicolosi et al. 2000; Gulsen and Roose 2001b; Pang et al. 2007). This clarifies the differences between the various classifications concerning the number of species in the genus *Citrus* L., whose taxonomy is complicated by means of apomixy and hybridization, with many lines of stable hybrids that could be erroneously granted the status of species. However, the variations that arise by means of spontaneous mutation are often difficult to distinguish even using biochemical or genetic markers (Barkley et al. 2006).

Scora (1975) and Barrett (1976) claim that there are only three real ancestral species within the subgenus *Citrus*: citron (*C. medica* L.), mandarin (*C. reticulata* Blanco) and pomelo (*C. maxima* [L.] Osbeck). Next, Scora (Scora and Kumamoto 1983; Scora 1988) added *C. halimi* B. C. Stone. Therefore, the other cultivated species of the genus *Citrus* that are derived from hybridization between these true species followed, mainly, from natural mutations. Recently, this view has been confirmed by several biochemical and molecular studies (Federici et al. 1998; Nicolosi et al. 2000; Barkley et al. 2006).

2.1.2 *Citrus × bergamia* Risso & Poiteau

Unlike other species of the genus *Citrus* L., literature concerning the phylogeny and taxonomy of bergamot are few and often contradictory. Bergamot is named for the first time in one of the most famous historical classifications of citrus fruits: the *Histoire Naturelle des Orangers* published in 1818 by Pierre Antoine Poiteau (1766–1854) and Joseph Antoine Risso (1777–1845).

Bergamot has been widespread in the Mediterranean for centuries, although the characteristics of the essential oil were not known and appreciated before 1750. The botanical and geographical origins of this plant are still uncertain; it is possible that bergamot is native of Calabria, deriving by mutation from other species even if other probable places of origin may be the Antilles, Greece, and the Canary Islands, where Columbus would bring the plant in Europe, arriving at Calabria from the Spanish city of Berga, hence the name of "bergamot."

In addition to the geographical-origin uncertainty, even the nomenclature of this plant is rather confusing because the taxonomic data in the literature are not unique in this regard. According to the Integrated Taxonomic Information System (ITIS 2011) *Citrus aurantium* L. comprises two subspecies: *Citrus aurantium* ssp. *aurantium* L. or sour orange and *Citrus aurantium* ssp. *bergamia* (Risso & Poiteau) Wight & Arn. Ex Engler or orange bergamot, and therefore *Citrus bergamia* (Risso & Poiteau) Wight & Arn. Ex Engler must be considered synonymous with *Citrus aurantium* ssp. *bergamia* (Risso & Poiteau) Wight & Arn. Ex Engler (ITIS 2011).

Flora Europaea, published by Royal Botanic Garden Edinburgh (Flora Europaea 2011), and the International Plant Name Index (IPNI 2011) reports as the currently accepted name *Citrus bergamia* Risso & Poiteau. According to the International Organization for Plant Information (IOPI 2011) database sponsored by International

Association for Plant Taxonomy (IAPT 2011), founded at the Seventh International Botanical Congress in Stockholm on July 18, 1950, *Citrus aurantium* ssp. *bergamia* (Risso & Poiteau) Wight & Arn. ex Engler. (Engler 1896, 1931) is the valid name while the synonym is: *Citrus bergamia* Risso & Poiteau—that is the basionym for *Citrus aurantium* ssp. *bergamia* (Risso & Poiteau) Wight & Arn. ex Engler.

The online database Tropicos of the Missouri Botanical Garden, USA (TROPICOS 2011) reports *Citrus* × *bergamia* Risso is considered a nothospecies, including a subspecies *Citrus* × *bergamia* ssp. *mellarosa* (Risso) Rivera et al., and two varieties: *Citrus* × *bergamia* var. *unguentaria* M. Roem. and *Citrus* × *bergamia* var. *ventricosa* (Michel) M. Roem. *Citrus* × *bergamia* Risso is considered a basionym of which combinations are: *Citrus* × *aurantium* ssp. *bergamia* (Risso) Engl.; *Citrus* × *aurantium* var. *bergamia* (Risso) Brandis; and *Citrus aurantium* ssp. *bergamia* (Risso) Wight & Arn. ex Engler. These should be considered synonyms: *Citrus limetta* Risso var. *bergamia* Risso in Ann. Mus. Aris 20 (1813) 197; *Citrus bergamota* Raf. In Fl. Tellur. (1838) 141; *Citrus aurantium* var. *bergamia* Wight & Arn. in Prodr. (1834) 98; *Citrus decumana* ssp. *bergamia* Asch. & Graebn. in Syn. mitteleur. Fl. 7 (1915) 291.

In accordance with the International Code of Botanical Nomenclature, the first (in order of time) correct classification of a plant is one that can be used in scientific studies accompanied by the name of the author (Vienna Code 2006, unmodified in Melbourne Code 2011, Chapter II, Section III, Article 11.3: "For any taxon from family to genus inclusive, the correct name is the earliest legitimate one with the same rank."). A fair description of the plant, written in Latin, and the keeping of a dry sample of the plant (called a holotype) in a botanical library (or herbarium), so that in this example we can refer in case of a subsequent dispute or in doubt, are the basic requirements in order for a botanical name to be valid. As of recently, photographs on the Internet with the proper documentation are also valid. Nowadays only the International Botanical Congress, which has been held every five years since 1900, can change the name of a plant according to rules set out in the International Code of Botanical Nomenclature. The currently valid code is the Melbourne Code of 2011, which is available via the Internet. Records of all botanical conferences since 1950 are kept at the International Association for Plant Taxonomy in Vienna, headed by D. J. Mabberley, president since 2005, who is the author of the most recent taxonomic revision of the genus *Citrus* L. based on molecular markers.

Mabberley's classification, clarifying the phylogenetic relationships among the three true species and their involvement in different hybrids, states that in the group of edible citrus fruits, there are only three wild species in *Citrus* genus: *Citrus medica* L. citron, *Citrus maxima* (Burm.) Merril pomelo, and *Citrus reticulata* Blanco mandarin. Each of these is involved in several hybrids. From a pragmatic point of view, as it was decided for other species with similar taxonomic problems, it may be preferable to give up a Linnaean classification and simply refer to cultivars as, for example, *Citrus* "Valencia" (an orange) and *Citrus* "Dancy" (a tangerine). These can then be arranged in groups as required by the International Code of Nomenclature for Cultivated Plants, 1995 (art. 4.1); for example, "*Citrus* Valencia (Group Sweet Orange)." Bergamot is only known from cultivation and consists of a limited and well-defined number of cultivars.

The bergamot is of hybrid origin. It has been suggested that it is a hybrid between sour orange (*C. aurantium* L.) and lemon (*C. limon* [L.] Burm.f.), or a mutation of the latter. Others considered it a hybrid between sour orange (*C. aurantium* L.) and lime (*C. aurantiifolia* [Christm. & Panzer] Swingle) (Federici et al. 2000).

Four groups of bergamot types are recognized: Common group, Melarosa group (fruit rather flattened), Torulosa group (fruit ridged), and Piccola group (dwarf cultivars). Only cultivars in the Common bergamot group are commercially cultivated for the essential oil and three cultivars are grown: "Castagnaro," "Femminello," and "Fantastico" (a.k.a., "Inserto"). Formerly, Femminello and Castagnaro constituted virtually all commercial plantings in the world, but they have largely been replaced by Fantastico, a hybrid of Femminello and Castagnaro.

Art. H.1.1 of Appendix I to the International Code of Botanical Nomenclature of Vienna, unmodified in Melbourne Code 2011, states that "Hybridity is indicated by the use of the multiplication sign × or by the addition of the prefix 'notho-' to the term denoting the rank of the taxon," while Art. H.3.1. establishes that "Hybrids between representatives of two or more taxa may receive a name. For nomenclatural purposes, the hybrid nature of a taxon is indicated by placing the multiplication sign × before the name of an intergeneric hybrid or before the epithet in the name of an interspecific hybrid, or by prefixing the term 'notho-' (optionally abbreviated 'n-') to the term denoting the rank of the taxon (see Art. 3.2 and Art. 4.4). All such taxa are designated nothotaxa."

However, where there is certainty, it is preferable to use a Linnaean system, applying the formula provided for in Art. H.2.1. of Vienna Code, unmodified in Melbourne Code 2011, through which can be derived for the hybrid species their alleged relationship with wild plants. In fact, according to current rules of nomenclature, a hybrid of some taxa may be indicated by placing the multiplication sign between the names of taxa; it is also preferable, based on the recommendation H.2A.1., to place the names or epithets in a formula in alphabetical order. Hybrids between representatives of two or more taxa may receive a name (Art. H.3.1.).

In the case of bergamot, keep in mind that the name the first time assigned to this species hybrid is *Citrus bergamia* Risso & Poiteau (Vienna Code 2006, Chapter II, Section III, Article 11.3, unmodified in Melbourne Code 2011).

A reliable Linnaean classification scheme related to the genus *Citrus* can be the one proposed by Mabberley (1997), which includes three species in each hybrid taxa:

1. *Citrus medica* L., Sp. Pl. 2:782 (1753), involved in the following hybrid taxa:
   a. *Citrus* × *limon* (L.) Osbeck, Reise Ostind. China:250 (1765) as "limonia", pro sp.; Burm.f., Fl. Indica:173 (1768), pro sp.
   [1. *Citrus medica* ×?]
   C. medica L. var. *limon* L., Sp. Pl. 2:782 (1753).
   b. *Citrus* × *jambhiri* Lush., Ind. Forester 36:342 (1910), pro sp.
   [1. *Citrus medica* × 3. *Citrus reticulata* or (1. × ?, i.e. *Citrus* × *limon*) ×3.]
2. *Citrus maxima* (Burm.) Merr., Interp. Herb. Amb.:46, 296 (1917), *Aurantium maximum* Burm., Herb. Amb. Actuar. Ind. Univ.: [16](1755), involved in the following hybrid taxa:
   a. *Citrus* × *aurantiifolia* (Christm.) Swingle, J. Washington Acad. Sci. 3:465 (1913) pro sp., as "aurantifolia."

    b.   *Citrus* × *aurantium* L., Sp. Pl. 2:782 (1753), pro sp.
       [2. *Citrus maxima* × 3. *Citrus reticulata*]
    3. *Citrus reticulata* Blanco, Fl. Filip.:610 (1837).

The previous classification reflects the results obtained at the molecular level by Gulsen and Roose (2001a), which established that the three taxa of *Citrus* L. sp. proposed as ancestral, *C. medical, C. maxima*, and *C. reticulata*, have different patterns of chloroplastidial DNA (cpDNA), further providing that the pomelo has contributed to the chloroplast genome of bergamot.

Bergamot, which has a very high percentage of heterozygosity indicative of an original interspecific hybridization (Herrero et al. 1996), according to research by Federici et al. shows an RFLP profile compatible with a hybridization *C. aurantium* × *C. limetta* (Federici et al. 2000).

The bergamot is considered by Scora (1988) a cross between citron and orange, which is itself a cross between pomelo and mandarin, the formula for which, on the basis of the previous scheme proposed by Mabberley (1997) is "1. × (2. × 3.)," and in accordance with Articles H.1.1, H.2.1, and H.3.1. The Recommendation H.2A.1., Appendix I, Vienna Code 2006, unmodified in Melbourne Code 2011, becomes *C. medica* L. × [*C. maxima* (Burm.) Merr. × *C. reticulata* Blanco], and hence the botanical name of bergamot under the Vienna Code of 2006, Chapter II, Section III, Article 11.3 is *Citrus* × *bergamia* Risso & Poiteau (Mabberley 1997).

2.2   MORPHOLOGY AND ANATOMY

Citrus × *bergamia* is an evergreen small tree that can grow up to 12 m tall, with an erect, cylindrical, dark grayish-brown stem and very thin irregular branches, with or without spines, depending on the variety. In cultivation, trees are pruned up to 4–5 m in height.

The bergamot is propagated by grafting, and the best results are obtained by using the bitter orange, which produces sturdy long-lived trees that are particularly resistant to the inclement weather. The plant comes into flower in April and May and the fruits ripen from November to March.

2.2.1   ROOT SYSTEM

The root system is made up of a tap-root, from which arises plagiotropic secondary roots; in sandy soils it may penetrate down to 5–6 meters but in clay soil the penetration is shallower.

The root hair region occurs a short distance above the region of elongation. The root hairs come from a subepidermal layer. The mature region is situated above the root hair region. Here the root becomes thicker and secondary or lateral roots are developed. The roots in this region are covered by a protective cork layer.

In the root hair region the cortex usually consists of thin-walled parenchyma cells with numerous intercellular spaces. The innermost layer of the cortex constitutes the endodermis. The root cortex is usually wider than the stem cortex, and therefore plays a larger role in storage. Since the plant is not deciduous, carbohydrates are stored in the roots during the winter period. The vascular cylinder comprises all the tissues enclosed by the endodermis. It includes the pericycle, a single layer of thick-walled, tightly packed cells without intercellular spaces from which lateral roots and vascular tissue arise. This conducting tissue consists of xylem and phloem, which are separated from each other by parenchyma. Only primary xylem is present in young roots; it is differentiated into protoxylem, which lies against the pericycle, and the metaxylem lying towards the inside. The primary xylem consists of nonliving, thick-walled xylem vessels and tracheids. The phloem alternates between the arms of the xylem and consists of living thin-walled cells (sieve tubes, companion cells, and phloem parenchyma). The xylem transports water and dissolved substances from the roots to the stem and leaves. The xylem is the main strengthening tissue of the root. The phloem transports organic substances such as carbohydrates from the leaves to the root.

2.2.2 STEM

The young stems have an external epidermis; it consists of a single layer of closely packed living cells. The walls are thickened and covered with a thin waterproof layer called the cuticle. Stomata with guard cells are found in the epidermis. The cortex region is situated inside the epidermis and consists of the collenchymas (three to four layers of cells with thickened cell walls), parenchyma (layers of thin-walled cells, parenchyma, with intercellular spaces), and starch sheath (a single layer of tightly packed rectangular cells bordering the stele of the stem). The vascular cylinder or stele includes the pericycle, vascular bundles, and pith (medulla). The pericycle strengthens the stem. It provides protection for the vascular bundles. The vascular bundles are situated in a ring on the inside of the pericycle of the plant. This distinct ring of vascular bundles is a distinguishing characteristic of dicotyledonous stems. A mature vascular bundle consists of three main tissues—xylem, phloem, and cambium. The phloem is located towards the outside of the bundle and the xylem towards the center. The cambium separates the xylem and phloem, which brings about secondary thickening. The xylem provides a passage for water and dissolved ions from the root system to the leaves. The xylem also strengthens and supports the stem. The phloem transports synthesized organic food from the leaves to other parts of the plant. The cambium produces new xylem and phloem cells, making secondary thickening possible. The pith occupies the large central part of the stem. It consists of thin-walled parenchyma cells with intercellular air spaces. Between each vascular bundle is a band of parenchyma, the medullary rays, continuous with the cortex and the pith. The cells of the pith store water and starch. They allow for the exchange of gases through the intercellular air spaces. The medullary rays transport substances from the xylem and phloem to the inner and outer parts of the stem. The cortex of a young stem may develop lysigenous oil cavities (Esau 1965).

In the region of transition to secondary structure, mature parenchyma cells in the medullary rays, which lie between the adjacent vascular bundles, become meristematic and form the fascicular cambium. The fascicular cambium forms a continuous ring of cambium as it joins up with the inter-fascicular cambium. This cambium ring undergoes division to form secondary phloem to the outside and secondary xylem to the inside. The secondary xylem and secondary phloem are laid down in the form of concentric cylinders on either side of the cambium ring. At certain points the cambium forms parenchyma, which radiates from the middle of the stem through the secondary xylem and secondary phloem to form vascular rays. As a result of these changes the stem's thickness increases. The primary xylem and phloem are pushed further and further apart. The pith remains alive. Some of the parenchyma cells between the vascular bundles continue to exist to form radially directed vascular rays. Annual rings develop in the secondary xylem, each consisting of a layer of spring wood and a layer of autumn wood. A cylindrical meristem develops in the cortex, the cork cambium (phellogen). The cork cambium gives rise to the cork cells (phellem) on the outside and the secondary cortex (phelloderm) on the inside. Opposite the stomata the cork cells (phellem) give rise to lenticells for gaseous exchange. The formation of periderm in the stem of this plant may be considerably delayed, as compared with that of the secondary vascular tissues, or it may never occur, despite the obvious increase in thickness of the stem. In this genus the development of the periderm sometimes commences only after the production of the secondary vascular tissues has reached considerable dimensions.

2.2.3 LEAVES

The leaves are persistent, simple, alternate, and aromatic when bruised; the blade is up to 12 × 6 cm, ovate-oblong or lanceolate, with the apex usually sharp and margin crenate or sometimes just slightly wavy, hairless, leathery, with a dark-green upper surface and light-green lower surface. The leaves present a short petiole, slightly winged, about 13 mm long, articulated near the blade. A midrib is visible and prominent at least in the lower surface of the blade. Many glands are visible as small rounded cavities; they are more abundant near the blade margin. The leaves are renewed every 14–18 months; the abscission is more intense in the months of April and May.

The leaves present a dorsiventral symmetry. In a thin transverse section of the blade three regions are distinguishable, namely the epidermis, mesophyll, and vascular bundles or veins. The mesophyll is differentiated into palisade parenchyma and spongy parenchyma. The palisade parenchyma consists of thin-walled cells which are usually cylindrical in shape. These cells contain large numbers of chloroplasts. The spongy mesophyll is usually ball-shaped with large intercellular spaces, but usually contains fewer chloroplasts than the palisade cells. There is a system of air spaces that communicates with the air chambers behind the stomata. The vein system of the leaf consists of branched vascular bundles. A vein contains the vascular tissue, which consists of xylem and phloem. The lignified xylem cells are situated towards the upper epidermis and the phloem towards the lower epidermis. In the large veins the vascular bundles are usually surrounded by a bundle sheath. Stomata are confined to the lower side.

2.2.4 FLOWERS

Inflorescence in *Citrus × bergamia* are terminal, racemose, and many-flowered with pedicel up to 8 mm long. The actinomorphic flowers are pentamerous, bisexual, and fragrant; terminal flowers are often sterile. The calyx (cup-shaped with short lobes, yellow-green) tube has five pale green sepals, welded to the base, green with oil glands; they persist in the ripened fruit. The corolla presents most often five narrow-elongate, pure-white (pearly white) petals inserted alternately with the sepals. Oil glands are present in the petals. Polyandry androecium with polyadelphous stamens ((13-)21(-28), in (2-)4(-6) groups, sometimes petaloid) have more than twice the number of petals, arranged in two whorls free or welded to the base for a stretch, almost to form a tube around the ovary, which is super, obovate in shape, and divided into 10–15 lodges. The gynoecium presents a syncarpous ovary, a white cylindrical fleshy style, and a yellow-green stigma. In syncarpous gynoecium of this genus each component style may have its own canal (Esau 1965).

The nectary in *Citrus × bergamia* forms a ring around the base of the ovary. Stomata with wide apertures are present on raised portions of the ring. In tangential section of the nectary the stomata are seen to be strikingly rounded in shape, and in cross-section it is seen that the substomatal chambers are fairly deep and the cells below the epidermis are small and compact. The epidermis itself has small, cubical, thick-walled cells covered by a relatively thin cuticle. All the cells of the nectary, including those of the epidermis, have granular colorless content. Sometimes these cells may also contain crystals, as is a common feature in all other tissues of *Citrus × bergamia*. According to their location the floral nectaries of this genus are toral nectaries (those developing on the receptacle), anular (the nactary forms a ring or a part of one on the surface of receptacle) with a prominent ring around the ovary base (Fahn 1972). Within this plant genus a correlation appears to exist between the intensity of nectar secretion and the size of the nectary (Fahn 1979).

2.2.5 FRUITS

Fruit is a slightly flattened subglobose to pyriform berry (hesperidium), 8.5–15 cm × 7–12.5 cm, often with a small navel and a persistent style. The peel is 3–6 mm thick, with numerous glands, though, smooth to rough, sometimes rigged, adherent, shiny green turning to pale yellow when the fruit is ripe. The mexocarp is white, the endocarp is divided into 10–15 lodges containing a greenish-yellow juice, sour and bitter, with few seeds.

The citrus fruit, the hesperidium, is closely related to a berry. It develops from a syncarpous gynoecium with axile placentation. With the development of the fruit the number of cells throughout the ovary increases and, finally, three strata can be distinguished. The exocarp (flavedo) consists of small, dense collenchyma cells which contain chromoplasts. This tissue contains essential oil cavities. The epidermis consists of very small, thick-walled cells. In surface view it resembles a cobbled surface; the cells contain chromoplast and oil droplets. A few scattered stomata can be found in the epidermis. The mexocarp (albedo) consists of loosely connected colorless cells; this tissue has a spongy nature and it is white because of the numerous air spaces in it. The endocarp is relatively thin and consists of very elongated, thick-walled cells that form a compact tissue. The stalked, spindle-shaped juice vesicles, which fill the locules when the fruit ripens, develop from the cells of the inner epidermis and subepidermal layers. Each juice vesicle is covered externally by a layer of elongated cells, which enclose very large, extremely thin-walled juice cells (Fahn 1972).

The juice sacs develop as multicellular hairs and resemble to some extent the "spouting glands" of *Dictamus*. In their center they contain cells with very thin folded walls. These cells produce and store relatively large amounts of osmiophilic droplets, apparently essential oils that originate in the plastids. The cells surrounding these oil-producing cells are arranged similarly to those surrounding the oil cavities in the pericarp. The juice sacs represent trichomes occurring on the adaxial side of the carpels, and their inner structure resembles incompletely developed oil cavities occurring on the abaxial side of the carpels, inside the exocarp (Fahn 1979).

The epidermis of the fruit is composed of polygonal, almost isodiametric cells with more or less slightly wavy thickened anticlinal walls whereas the outer tangential walls are covered with a very thick cuticle due to the presence of a continuous waxy layer (Figure 2.1a).

Anomocytic stomata are evidenced. On a cross-section, the lower region of parenchyma is composed of irregular polygonal cells with thickened walls; these cells increase in their size inward.

2.2.6 SEEDS

Seed (0-)3(-13) per fruit are flattened, 11 mm × 6 mm × 4.4 mm, pale yellow, and usually monoembryonic. The outer integument of the seed or *testa* is coarse, yellowish in color; the inner integument or *tegmen* is membranous and very thin. Inside the seed there is the embryo with more or less semispherical cotyledons. The cotyledons are ivory yellow. Sometimes there is the appearance of two or more embryos in a single seed. The process of apomixes is often accompanied by the formation of a few embryos from the same ovule. Sometimes a normal embryo may develop together with those produced by apomixix. In a single ovule of certain species of *Citrus*, 3 to 12 apomixial adventitious embryos may develop alongside the normal embryo (Fahn 1972).

2.2.7 SECRETORY STRUCTURES

Lysigenous (from the Greek "lysis," or "loosening") intercellular spaces are present in the peel of the fruit of the *Citrus* (Fahn 1972). In these secretory cavities the cells that break down release their secretion into the space and remain themselves in partly collapsed or disintegrated state around the periphery of the cavity (Esau 1965). Essential oil cavities occur characteristically also in the mesophyll of the leaves and in the petiole (Fahn 1972).

On the basis of microscopic light observations, the lacunae of the cavities in *Citrus* have been interpreted by some authors as developing lysigenously and by others as developing schizogenously. On the basis of studies with the aid of the scanning electron microscope using a mixture of methyl- and hydroxyl-propyl-methacrylate (Rapisarda et al. 1992, 1996a–c; De Domenico et al. 2004) it was suggested that the oil cavities of fruits of *C. bergamia* and *C. limon* are formed schizogenously. While the lysogeny does occur during later stages of cavity development, the schizogenesis consists of the separation of the cell walls that may not only be a result of high pectinase activity causing the dissolution of the middle lamella; the essential oil if eliminated from the cell may also cause wall separation and formation of intercellular

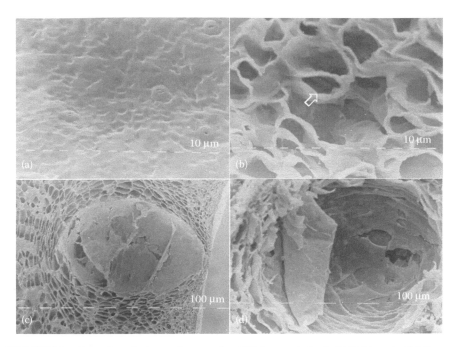

FIGURE 2.1   Scanning electron micrographs of *C. bergamia* fruit. (a) Epidermis. (b) Cross-sectioned fruit peel showing the inside of a secretory cavity developing schizogenously (the arrow indicates the beginning of the separation of adjacent cells). (c) Cross-sectioned fruit peel showing a fully developed secretory cavity surrounded by from four to six layers of secretory cells. (d) Cross-sectioned fruit peel showing the disintegration of the cells surrounding the secretory cavity.

spaces (Figure 2.1b). In this genus the essential oil mainly remains in the plastids inside the cell until its disintegration (Hansen and Fox Moule 1973).

The ultrastructural study of *C. bergamia* fruit peel by means of scanning electron microscopy has clarified the schizolysigenous nature of the secretory tissues, where the essential oil is produced and stored (Rapisarda et al. 1996a). The secretory cavities consist of a central cavity surrounded by four to six layers of secretory cells (Figure 2.1c) having more developed tangential walls respect to the radial ones. This phenomenon begins in the mexocarp, then during the following stages of gland development, due to the disintegration of cells surrounding the central cavity (Figure 2.1d), the exocarp is also affected. These observations confirm that the secretory cavities of *Citrus bergamia* fruits develop mainly schizolysigenously.

2.2.8   ONTHOGENESIS

Germination of citrus seeds is hypogeous; the rootlet emerges through the seed coats. The rootlet will develop into a root system and the tip apex will start the formation of the aerial part of the plant. After formation of the new plantlet, a long period of growth will follow until the tree reaches the flowering period. This will occur after two years at least.

Flower induction consists of the evolution of a vegetative bud into a flowering. Flowering buds are different in aspect, being dome shaped. There is one blooming period, about 2–4 weeks after the winter dormancy. Before opening, flowers are visible and the flower bottoms start to grow and develop while internal meiosis and formation of the gametes takes place. The time of total opening of the flowers is called the anthesis. The stigma of each flower is receptive to the pollen grains for a few days because of the production of a sticky secretion, the stigmatic fluid, where they can adhere.

Pollination is the transfer of the pollen grains from the anthers to the stigma. In *Citrus* it is usually carried by insects, or occurs by either self- or cross-pollination. Fertilization implies the growth of the pollen tube plus the union of the spermatic nucleus with the oosphere with the formation of the zygote. At this time the petals and stamen fall, and the ovary starts to divide at a very high rate of mitosis: fruit set has taken place. Where fruit set has not been completed, flowers will be brown and fall off in a very short period of time. Only the set fruits will start the following physiological period. Pollination, growth of the pollen tube, fertilization, and subsequent seed development seem to stimulate the production of plant hormones that prevent the ovary drop. Although in many cases fertilization is needed for fruit production, production of seedless fruits or parthenocarpy is possible, at least in some varieties. Apparently a sufficient production of hormones in the ovarian tissues makes fruit development possible. In some instances the stimulation of the pollen tube growth or even the fertilization and subsequent seed abortion are needed in order to develop the fruit. Fruits are considered as set when developed to the point that can be expected to remain until maturity, unless physiological or mechanical stress later occurs.

Sexual incompatibility occurs in *Citrus* when fertilization is not possible although male and female components are functional. This incompatibility is due to slow pollen tube growth, apparently caused by the presence of inhibitors in the style.

During the fruit growth period cell division and cell enlargement are the two main physiological processes. Fruit growth includes three stages. The first stage is characterized by slow growth in volume but intense cell division and cell enlargement in all tissues. Oil glands that were present at full bloom enlarge themselves and new ones appear. Juice sacs also present at that time and continue to be formed. Most of the growth in this stage is due to the peel. The second stage is characterized by very rapid fruit growth. Cell enlargement and differentiation predominate. The peel grows in thickness mainly due to the enlargement of the albedo cells. The spongy tissue is formed. At the end of the period, the increase in fruit size causes a decrease in rind thickness. The endocarp becomes the main constituent of the fruit.

The third stage includes all changes affecting the size of the fruit, up to reaching full maturation.

Changes in rind color are mainly due to degradation of the chlorophyll and increase of the carotenoids. Carotenes and xanthophylls are the main pigments responsible for the yellow predominant content in flavonoids. Changes in texture are mainly due to the chemical changes in the albedo. Most changes occur within the juice vesicles. The main changes correspond to a parallel increase in sugar content and organic acid content. Together with this change, pigmentation of the juice also varies. Aromatic compounds provide the final aroma of the pulp and juice.

2.2.9 THE CULTIVARS

There are currently three distinct varieties of *Citrus bergamia* Risso & Poiteau: Femminello, Castagnaro, and Fantastico.

The Femminello cultivar is a fast-growing plant that is not very developed, with lanceolate leaves and the spherical fruit and a thin exocarp rich in essential oils; it is more aromatic and is therefore preferred to the other cultivars. This cultivar is presumed to be the best variety. It is also the earliest cultivar, since collection begins in early October.

The Castagnaro cultivar is a rustic plant, resistent to strong winds and long-lived, with large and lanceolate leaves and globular fruit with exocarp of medium thickness; it has an average content of essential oils. The harvest begins in November.

The Fantastico cultivar is rustic and highly productive, but has larger leaves than other cultivars, with globular, pear-shaped fruit, and oil with excellent aroma. Maturation takes place in November or December. Although it spread more recently than the other two cultivars, it constitutes the largest percentage of fruit production.

The fruit of *Citrus bergamia* Risso & Poiteau, a typically Mediterranean species, has been examined in order to obtain phytognostic markers able to differentiate the cultivars—Castagnaro, Fantastico, and Femminello—that represent the vegetable source for bergamot essential oil production (De Domenico 2004).

Peel fragments of *C. bergamia* fruits, picked from the main branches of five plants for every cultivar that grows in Reggio Calabria province, were examined by means of scanning electron microscopy (SEM Philips mod. 500) and the images of the glandular cavities were analyzed with a Zeiss VIDAS 2.1 Image Analyzer (Rapisarda et al. 1996b, c; Rapisarda et al. 2003).

The micromorphometric analysis of cross-sectioned fruit peels of *C. bergamia* cultivars Castagnaro, Fantastico, and Femminello has yielded size parameters of the sagittal section of the glandular cavities such as the area (a) by which the equivalent circular diameter (dc) was calculated ($\sqrt{4a/\pi}$), the longest (da) and the shortest (db) of the Feret diameters, the linear perimeter (pl), the convex perimeter (pc), and the volume ($v = [\pi/6][db\ da2]$), in addition to shape parameters such as the aspect ratio ($fs = db/da$) and shape factor ($fc = 4\pi a/pl2$). All the conventional statistical data, such as mean, median, variance, standard deviation, skewness, and kurtosis, were provided by the Image Analyzer. The Kolmogorov-Smirnov test was applied to the curves of distribution of the main shape and size parameters investigated. The results have permitted a comparative study of the three cultivars examined (cv. Castagnaro [Figure 2.2a], cv. Femminello [Figure 2.2b], cv. Fantastico [Figure 2.2c and 2.2d]), providing a key factor for their differentiation and for the possibility to obtain useful information about productivity of each cultivar.

The glands number/cm^2 (Table 2.1) in unripe and ripe fruits is constant, even if the sizes of oil cavities change. As the stage of fruit development, position on the plant, pedological characteristic of the ground, climate, light, humidity, latitude, altitude, and so on may influence the quality and quantity of essential oil, these investigation methods, along with chemical analysis, can be useful to follow, in the course of ontogenesis, the development of the secretory tissues in the peel of *C. bergamia* fruits; indeed, the mean value of the volume of the glandular cavities/number of glandular

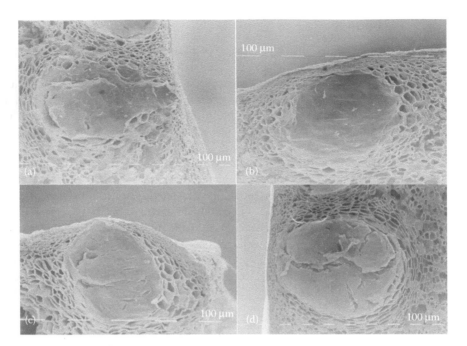

FIGURE 2.2 Scanning electron micrographs of a cross-sectioned fruit peel of *C. bergamia* showing fully developed secretory cavities. (a) Cultivar Castagnaro. (b) Cultivar Femminello. (c, d) Cultivar Fantastico.

TABLE 2.1

Glandula Cavities (number/cm²) Present in the Epicarp of *C. bergamia* Fruit

Cultivar	Glandular Cavities (number/cm²)
Castagnaro	106.27 ± 5.13
Fantastico	132.56 ± 8.61
Femminello	123.58 ± 7.66

cavities per cm² ratio evidences that the essential oil productivity for the three cultivars of *C. bergamia* fruits is: fantastico > castagnaro (−13%) > femminello (−21%).

REFERENCES

Barkley, N. A., Roose, M. L., Krueger, R. R., and Federici, C. T. 2006. Assessing genetic diversity and population structure in a citrus germplasm collection utilizing simple sequence repeat markers (SSRs). *Theor. Appl. Genet.* 112:1519–1531.

Barrett, H. C., and Rhodes, A. M. 1976. A numerical taxonomic study of affinity relationships in cultivated Citrus and its close relatives. *Syst. Bot.* 1:105–136.

Cronquist, A. 1993. *An integrated system of classification of flowering plants.* New York: Columbia University Press.

Dahlgren, G. 1989. The last Dahlgrenogram system of classification of the dicotyledons. In *Plant taxonomy, phytogeography, and related subjects*, eds. K. Tan, R. R. Mill, and T. S. Elias, 249–260. Edinburgh, UK: Edinburgh University Press.

De Domenico, C., Cotroneo A., Trozzi A., and Rapisarda A. 2004. Ultrastructural analysis of the secretory tissues during ontogenesis of *Citrus bergamia* Risso & Poiteau fruit. 35th International Symposium on Essential Oils. Giardini Naxos, Messina. September 29 to October 2, 2004.

Engler, A. 1931. Rutaceae. In *Die natürlichen Pflanzenfamilien*. 2nd ed., eds. A. Engler and K. Prantl, Vol. 19a, 187–359. Leipzig, Germany: Engelmann.

Engler, A. 1986. Rutaceae. In *Die natürlichen Pflanzenfamilien*, eds. A. Engler and K. Prantl, Vol. 3(4), 95–201. Leipzig, Germany: Engelmann.

Esau, K. 1965. *Plant anatomy*. New York, London, Sidney: John Wiley & Sons.

Fahn, A.1972. *Plant anatomy*. Oxford, New York, Toronto, Sydney, Braunschweig: Pergamon Press.

Fahn, A. 1979. *Secretory tissues in plants*. London, New York, San Francisco: Academic Press.

Fang, D. Q., Krueger, R. R., and Roose, M. L. 1998. Phylogenetic relationships among selected *Citrus* germplasm accessions revealed by inter-simple sequence repeat (ISSR) markers. *J. Am. Soc. Hort. Sci.* 123:612–617.

Federici, C. T., Fang, D. Q., Scora, R. W., and Roose, M. L. 1998. Phylogenetic relationships within the genus Citrus (Rutaceae) and related genera as revealed by RFLP and RAPD analysis. *Theor. Appl. Genet.* 96:812–822.

Federici, C. T., Roose, M. L., and Scora, R. W. 2000. RFLP analysis of the origin of *Citrus bergamia, Citrus jambhiri*, and *Citrus limonia Acta Hort. ISHS* 535:55–64.

Flora Europaea. 2011. Published by Royal Botanic Garden Edinburgh (UK): http://rbg-web2. rbge.org.uk/FE/fe.html.

Gulsen, O., and Roose, M. L. 2001a. Chloroplast and nuclear genome analysis of the parentage of lemons. *J. Am. Soc. Hort. Sci.* 126:210–215.

Gulsen, O., and Roose, M. L. 2001b. Lemons: Diversity and relationships with selected citrus genotypes as measured with nuclear genome markers. *J. Am. Soc. Hort. Sci.* 126:309–317.

Hansen, C., and Fox Maule, A. 1973. Pehr Osbeck's collections and Linnaeus's Species plantarum. *Bot. J. Lin. Soc.* 67:189–212.

Herrero, R., Asíns, M. J., Carbonell, E. A., and Navarro, L. 1996. Genetic diversity in the orange subfamily Aurantioideae. I. Intraspecies and intragenus genetic variability. *Theor. Appl. Genet.* 92(5):599–609.

Hutchinson, J. 1973. *The families of flowering plants: arranged according to a new system based on their probable phylogeny*. 3rd ed. Oxford, UK: Clarendon Press.

IAPT. 2011. International Association for Plant Taxonomy, http://www.iapt-taxon.org/index_ layer.php.

IOPI. 2011. International Organization for Plant Information—with IAPT contribution to the prototyping of the IOPI GPC Database, http://www.bgbm.org/IOPI/GPC/default.asp.

IPNI. 2011. International Plant Name Index by Royal Botanic Gardens, Kew, London, UK, http://www.ipni.org/ik_blurb.html.

ITIS. 2011. Integrated Taxonomic Information System—USDA United States Department of Agriculture, (USA): http://www.itis.gov/.

Mabberley, D. J. 1997. A classification for edible. *Citrus* (Rutaceae). *Telopea* 7(2):167–172.

Mabberley, D. J. 2008. *Mabberley's plant-book: A portable dictionary of plants*. 3rd ed. Avon, UK: Cambridge University Press.

Merrill, E. D. 1916. Osbeck's Dagbok Öfwer en Ostindsk Resa. *Am. J. Bot.* 10:571–588.

Morton, C. M., Grant, M., and Blackmore, S. 2003. Phylogenic relationships of the Aurantioideae inferred from chloroplast DNA sequence data. *Am. J. Bot.* 90:1463–1469.

Nicolosi E., Deng Z. N., Gentile A., et al. 2000. *Citrus* phylogeny and genetic origin of important species as investigated by molecular markers. *Theor. Appl. Genet.* 100(8):1155–1166.

Pang, X. M., Hu, C. G., and Deng, X. X. 2007. Phylogenetic relationship within Citrus and related genera as inferred from AFLP markers. *Genet. Res. Crop. Evol.* 54:429–436.

Rapisarda A., Iauk L., and Ragusa S. 1992. Impiego di una miscela di metil-e idrossipropil-metacrilato nello studio di tessuti vegetali in microscopia elettronica a scansione. *Riv. It. EPPOS* 7:56–58.

Rapisarda A., Caruso C., and Ragusa S. 1995. Indagini strutturali sulle sacche secretrici dei frutti di *Citrus bergamia* Risso et Poiteau e *Citrus medica* L. *Acta Technol. Legis Medic.* 6(3):382–387.

Rapisarda A., Caruso C., Iauk L., and Ragusa S. 1996a. Applicazione dell'analisi di immagine nello studio delle ghiandole oleifere dei frutti di alcune specie di *Citrus*. *Essenze e Derivati Agrumari* 56(1):5–12.

Rapisarda A., Pancaro R., and Ragusa S. 1996b. Image analysis: A tool for plant drugs investigations. *Microscopy and Analysis* 44:15–16.

Rapisarda A., Pancaro R., and Ragusa S. 1996c. Image analysis for micromorphometric characterization of vegetable drugs obtained by cogeneric species. *Phytother. Res.* 10:S172–174.

Rapisarda A., Iauk L., and Ragusa S. 2003. A quantitative morphological analysis of some *Hypericum* species. *Pharm. Biol.* 41:1–6.

Scora, R. W. 1975. On the history and origin of *Citrus*. *Bull. Torr. Bot. Club* 102:369–375.

Scora, R. W., and Kumamoto, J. 1983. Chemotaxonomy of the genus *Citrus*, 343–351 in *Chemistry and chemical taxonomy of the Rutales*, eds. P. G. Waterman and M. F. Grundon. London: Academic Press.

Scora, R. W. 1988. Biochemistry, taxonomy and evolution of modern cultivated Citrus. In *Proc. 6th Int. Citrus Cong.*, eds. R. K. Goren and K. Mendel, 1:277–289. Weikersheim, Germany: Margraf Publishers.

Swingle, W. T. 1943. The botany of *Citrus* and its wild relatives of the orange subfamily. *In The citrus industry*, 1st ed., vol. 1, *History, world distribution, botany, and varieties*, 129–474. Berkeley: University of California.

Swingle, W. T., and Reece, P. C. 1967. The botany of citrus and its wild relatives of the orange subfamily. In W. Reuther, H. J. Webber, and L. D. Batchelor, *The citrus industry*, revised 2nd ed., vol. 1, *History, world distribution, botany, and varieties*, 190–430. Berkeley: University of California.

Takhtajan, A. 1997. *Diversity and classification of flowering plants*. New York: Columbia University Press.

Tanaka, T. 1936. The taxonomy and nomenclature of Rutaceae-Aurantioideae. *Blumea* 2:101–110.

Tanaka, T., 1977. Fundamental discussion of *Citrus* classification. *Stud. Citrol.* 14:1–6.

Thorne, R. F. 1992. An updated phylogenetic classification of the flowering plants. *Aliso* 13:365–389.

Thorne, R. F. 2000. The classification and geography of the flowering plants: Dicotyledons of the class Angiospermae (subclasses Magnoliidae, Ranunculidae, Caryophyllidae, Dilleniidae, Rosidae, Asteridae, and Lamiidae). *Bot. Rev.* 66:441–647.

TROPICOS. 2011. Tropicos.org. Missouri Botanical Garden, USA,: http://www.tropicos.org-Home.aspx.

Uzun A., Yesiloglu T., Aka-Kacar Y., Tuzcu O., and Gulsen O. 2009. Genetic diversity and relationships within Citrus and related genera based on sequence related amplified polymorphism markers (SRAPs). *Sci. Hort.* 121:306–312.

3 Cultural Practices

Florin Gazea

CONTENTS

3.1 HISTORY NOTES

Bergamot is cultivated almost exclusively in Calabria and has the same cultivation practices and needs as other citrus species. The presence of bergamot in Calabria is related to its pedoclimatic conditions and mainly to the properties of its soil, which is of alluvial origin, a mixture of sand, clay, and limestone. Bergamot is a middle-sized plant. It generally develops slower than other citrus species, with regular and symmetrical foliage. It is traditionally kept low, approximately 1 m above the ground, mainly as a defense against the wind, which is extremely feared by the Calabrian farmers.

The main cultivars are Castagnaro, Femminello, and Fantastico. The Castagnaro is hardy and long-lived with good development, lance-shaped leaves, and a fruit peel that is a little thick, the collection of which begins in November. The Femminello is fast growing with reduced development, early but not long-lived, with lance-shaped leaves of medium size and thin-skinned fruit. The Fantastico is a rustic plant, with good development and high production. Its large leaves and fruits are harvested between November and February. These plants have significantly higher fruit yield and quality than the other two cultivars.

Typically the traditional implants were realized in a square, usually 5 × 5 m or sometimes 4.5 × 4.5 m. The branches become procumbent, fructifying in large quantity and allowing the fruit to be harvested without the use of a ladder. During the fructification it was thus necessary to sustain them, so the plant was encircled by biforked puncheons (forcine).

Bergamot plants generally propagate by grafting, since those generated by seeds or cuttings often grow weak and do not survive for long. Traditionally bergamot was grafted on lime cutting, obtaining very productive subjects which are quite longevous, while grafting on lemon was not practiced since the resulting plants did grow quickly and robustly but did not last long, and the resulting fruits were scantly aromatic.

3.2 CULTURAL PRACTICES IN BERGAMOT CULTIVATION

Certain cultural practices are required to improve plant development, preserve them from diseases, and increase their lifespan and productivity. These practices are usually carried out in autumn, after the fruit harvest, and in spring during the vegetative weakening. Additional on-field practices are carried out regularly during the year; these are aimed to normalize behavior of the plant. In particular, cultivation practices are related to the two distinct phases: nursery to the seedling production and orchard.

The practices in the nursery include:

- Propagation operations, sowing to prepare the grafts and rootstocks
- Preparation of the soil for the new plants and oxygenation of the soil

The preparation practices in the orchard include:

- Nutrition and fertilization
- Water management to reintegrate the water loss due to the plant transpiration and facilitate the ionic exchange
- Care of the plant, pruning, removal of the dried branches, etc.

3.2.1 PROPAGATION

Propagation is the preliminary activity to begin citrus cultivation. It is performed in nursery and is aimed at genetic improvement or at the production of new plants. Bergamot is propagated by the seeds (gamic way) or through the vegetation (agamic way) using the scions, cuttings, and grafts.

Virus-free citrus cultivars can be prepared through different methods:

- Constitution of nucellar clones
- Clone selection
- Thermotherapy
- Micro-grafts

The most ancient method, simple and more specific, used for the propagation of citrus is seed propagation. This method is effective because unlike other fructiferous plants, citrus possesses the polyembryo property, thus obtaining from a single seed more than one plant with identical properties to the parent plant. However, the plants so obtained also have undesired properties such as spines, are vigorous but start production slowly, are difficult to distinguish between gendered and nucellar plants, present total or partial apireny, and have scant resistance to diseases such as the canker, even if generated from seed free of any virus, disease, or mycoplasma. The seed widely used are those of sweet orange, bitter orange, lemon, trifoliate orange, and Citrange Troyer.

The plants selected to produce the seed must be vigorous, resistant to diseases, tolerate nematodes, be compatible with the graft, be resistant to cold and dryness, and tolerate salinity, and the fruits must result in good quality and shape. Only select plants can be used to produce the seed, as not all possess the desired requirements.

3.2.1.1   Seed Extraction

The seeds are extracted mechanically from the fruits and washed under current water. If they must be stocked for a long time before they are planted, they are washed again with water at 51.5°C and treated with fungicides. They are stored in plastic bags refrigerated at 4–7°C (Scuderi and Terranova 1970). The dry seeds, which float on the water surface, usually contain only one embryo capable of producing a more vigorous plant than those obtained by larger seeds with more than one embryo, but smaller.

3.2.1.2   Sowing and Planting

The location of the sowing must be selected with care. It must be not exposed to strong winds or frequent temperature drops; it should preferably be warm with a constant humidity. In addition, light covers can be used to protect the young plants from the intense solar light, avoid the excessive evaporation from the soil, and provide protection from freeze, factors that greatly affect the safety of the plants. Usually, the best time for citrus sowing is spring or summer. The germination of the seeds starts when the soil reaches 14–15°C; the optimal temperature is 14–18°C. After two years, when the small plants present a diameter at the base of the stem of 0.5 cm, they can be transferred in a seminary and successively put in place for the implant. When they reach the appropriate dimensions (1.5 m height) they can be grafted. The plants are transferred in spring and are placed in rows in single hollows or furrows with 70–90 cm distance and 25 cm depth, with care taken to press the soil around the stem and immediately irrigate. To defend the plant's trunk against the sun's rays it is possible to cover it with aluminium foil and treat it with herbicide (paraquat).

3.2.1.3   Grafting

Grafting the young plants is a multiplication system based on the union of vegetative parts derived from different individuals (Hoepli 1997a) with the following scopes:

* Multiply a selected cultivar (cv), maintaining the original characters.
* Use a cv under particular pedoclimatic conditions.
* Prevent or heal pest attacks.
* Substitute a cv in a citrus field that is already productive with a more appreciated one.

From the grafting will derive an individual that is a bimembral being formed by one subject or rootstock which consists of the root apparatus, and by an object or scions from which will originate the aerial part of the plant, the foliage. The rootstock can derive from a seed (franc) or it can be obtained by self rooting (cloned). The scion can be an embryonic shoot (bud) with a small section of the stem, or it can be a vegetative shoot obtained from the parent plant.

The conditions necessary for a successful graft are:

- Affinity between the scion and the rootstock
- Correct selection of the season (timing)
- Perfect adhesion between the vascular cambium of the stock and the scion
- Use of only healthy buds free of viruses

The graftage is practiced on plants with a trunk having a diameter of 1–1.5 cm that is 30 cm tall above the collar. The best period to graft is spring or autumn. The grafting is practiced 40–60 cm above the ground to avoid the effects of soil moisture and pest attacks.

Seldom is it preferred to proceed directly in the orchard, generating the citrus field. The bud grafting is carried out with the most common bergamot cultivars (the most commonly used is the cv Fantastico) on stocks obtained by the seeds.

3.2.1.4 Propagation by Cutting

Bergamot can be propagated not only through bud graftage using the scions, but also through the cuttings. The multiplication by cuttings is one of the most common techniques; it consists of the removal of a small amount of vegetative part from the parent plant. This removed piece, called the *cutting*, is then encouraged to grow and take root as an independent plant. More commonly, stem cuttings containing at least one bud (woody cuttings) can be used, or more rarely stem cuttings of young green stems (semi-woody or herbaceous cuttings) or from the leaves. The cuttings are put in an appropriate rooting medium for the time necessary to allow root formation.

The selection of the rootstock for the bud propagation is a very important step, and it must satisfy the following conditions:

- High nucellar embryony
- Correct period of practice (April or May)
- Good adaptation to different physical conditions
- Resistance to pest attacks, pathogens, fungi, viruses, and mycoplasma
- Resistance to dryness, salinity, and limestone
- Good adaptability to the nursery technique
- Dwarfing effect on the cv to be used, facilitating the field operations and reducing the production costs
- Improving the quality and quantity of the productivity

The most applied rootstock in the cultivation of bergamot is bitter orange, which satisfies most of the necessary conditions, possesses a very good affinity with most of the grafting materials, and also provides strong resistance to canker. Some drawbacks of using bitter orange as rootstock include scant resistance to *Citrus tristeza* virus (CTV), mal secco disease, and some common diseases such as the *Citrus psorosis* virus, exocortis, and impietratura.

Studies carried out at the "Sezione Operativa Periferica dell'Istituto Sperimentale per l'Agrumicultura di Acireale" on the qualitative effect determined on bergamot propagated by grafting on different rootstocks, highlighted that some stock such as *Citrus volkameriana*, *Citrus macrophylla*, Citrange Troyer, and trifoliate orange demonstrated good properties. The plants obtained from these rootstocks were more productive and more resistant to diseases. The essential oil obtained was of good quality (Caruso et al. 2000; Spina and De Martino 1991; Starrantino et al. 1979). The yield of oil obtained from bergamot grafted on trifoliate orange was comparable to that obtained on the rootstock of bitter orange (2.29% vs. 2.21%, respectively). Moreover, the analysis of productivity in the seven years considered in this study showed higher production from the plants obtained on rootstocks of *Citrus volkameriana* (201 kg) and *Citrus macrophylla* (171 kg) versus 122 kg obtained on bitter orange rootstocks. The best fruit quality (acidity, total solubles, ripening rate, yield of oil) was determined on bergamot plant grafted on *Citrus volkameriana*. These results show that *Citrus volkameriana*, *Citrus macrophylla*, and especially trifoliate orange can represent the best substitute to bitter orange for successful propagation of bergamot.

The grafting approach mainly used in the cultivation of bergamot is called "the triangle" because of the shape given at the base of the scion containing two buds. This is introduced in the stock by a cut under the bark. This step must be done before the cambial activity.

The budding approach is often used with a bud grafting (dormant of vegetative) consisting of a bud with a small portion of tissue, including bark and occasionally wood. The rootstock is prepared by a cut, a T-shaped slit made in the stock plant. The shield is placed under the cut. In spring when the cambium is fully active, it is possible to practice "crown grafting," which consists of inserting one or more cuttings, still dormant, appropriately shaped by an oblique cut or two opposed converging surfaces, inside the rootstock between the bark and the wood.

Fifteen to 20 days after the grafting, it is possible to proceed with the control of the buds. These are alive if still green and the petiole easily detaches. After one to two years the selected plants are placed in a new orchard; this procedure usually occurs in March to April, avoiding the cold season.

3.2.2 Soil and Implant

3.2.2.1 Soil

The climatic conditions optimal for the cultivation of bergamot are not common. In fact, this citrus species has developed almost exclusively in the Ionian Coast of the Reggio Calabria province. In this area the temperature variations and excursions during the year, rainfall, humidity, and soil structure create the optimal habitat for bergamot growth.

The main soil property to be selected for the cultivation of bergamot is one that allows the best rooting development, generating vigorous roots deep enough for the best intake of available water and oxygen, and absorbing nutrients available in the soil as well as those provided by fertilization. Another peculiarity of the soil is its particle size and structure, including the capacity of the particles to agglomerate and flocculate in the presence of calcium ions and organic colloids.

For the cultivation of bergamot the most appropriate soils are alluvial, calcareous clay, characterized by pH values between 6.5 and 7.5. The value of pH is an important parameter to control the growth of microorganisms responsible for the transformation of organic matter into humus, a colloidal matter with high moisture retention capability with higher softness than every other inorganic colloid found in the soil.

Bergamot plant also grows well in sandy soils, if often fertilized and if the loss of water can be overcome through appropriate irrigation procedures. In these soils more deep roots develop, increasing the epigeous apparatus and fructification. In clay soils the rooting process is contrasted by the tenacious asphyctic soil that opposes resistance and affects the flourishing of the aerial part. It is important to avoid soils with clay amounts higher than 35%, silt 20%–25%, calcium carbonate 30%, and soils containing a total amount of sodium and magnesium carbonates, sulphates, and chlorides at 40% of the total salts.

3.2.2.2 Implant

Planting density indicates the distribution of the plants in the cultivation field. The distance between plants depends on the pedoclimatic environment, the rootstock, and the cultivars. Row distance must be adequate to allow the mechanization of cultural operations.

If the procedure is carried out correctly, the plants start fructification from the third or fourth year. However, it is recommended to prune the canopies after the seventh or eighth year. The best period for orchard implanting is March or April, to allow the plant to make new roots and buds with the advent of the good season (Spina and Russo 1985).

3.2.3 NUTRIENT MANAGEMENT

The soil is the environment which, through its physical, chemical, and biological properties, offers to the plant the possibility to live and produce. Nutrient management includes the source of nutrient inputs for crop production. Nutrient inputs can be chemical inorganic fertilizers, manure, green manure, compost, and mined minerals. The elements necessary for the survival of a plant to allow its functions and physiological processes necessary to a regular development are 12, divided based on their concentration and functions in macro (N, P, K), meso (Ca, Mg, S), and micro (Fe, Zn, B, Cu, Mn, Mo) elements. To provide a balanced nutrition the plant requires three elements: nitrogen, phosphorus, and potassium.

A rational and cheap dressing aimed to enhance chemical fertility must take into account the presence in the soil of fertilizing agents, the highest possible productivity, the amount of necessary fertilizers, and the evaluation of costs. The necessary amount of each nutrient is determined by diagnosis of a sample of leaves (5–7 months of age) collected from terminal young healthy branches without fruits (Embleton et al. 1973a, b; Spina et al. 1991).

Bergamot is an evergreen plant; it adsorbs nutrients all year long, with its maximum intake in spring, when the vegetation starts to lengthen the braches and blooms, the second most immediately after the fruits drop, and the third most at the beginning of autumn to allow the development and ripening of the fruits.

The addition of nitrogen to the soil is necessary for the survival of the plant before and during blooming, for the preservation of fruits, and to avoid fruit drops. The best time to administer this compound is the end of winter. The dose per hectare of bergamots is on the order of 55–70 kg of nitrogen to produce 350–400 q of fruit. In practice, the fertilizer is distributed, based on its concentration, around the trunk at a distance of 10–15 cm, or along the rows. Phosphorus participates to the metabolic activity of the plant and to the vegetative development. Its plastic function is well known, since it is part of the protein structure (nucleic and phosphorated proteins), of reserve compounds (phyitins, lecithins, phosphatides, etc.), and is a fundamental part of important enzymes (co-enzyme I and II, ATP, carboxylase, etc.). It is located in vegetable tissues and in reserve organs. It also activates the metabolism during the vegetative cycle; favors blooming, fructification, and lignifications of tissues; and participates in the chlorophyllian phenomena. In the soil it distributes in the absorbable form as orthophosphate if the pH of the soil is below 6.0–6.6 units. Fertilization with phosphorus is performed in function of the plant's needs based on the foliage analysis and is usually administered before the blooming and during the development of the small fruits.

The ash analysis demonstrates that potassium is present as carbonate with a percentage higher than every other element. This element is absorbed by the plant eagerly for its role in the photosynthesis and confers to the plant resistance to dryness, cold, and disease. It also reduces water intake and affects the color of the flowers and fruit. When there is a lack of potassium, the buds do not grow and the quality of the fruits is poor. For a correct fertilization that must be carried out in winter during the vegetative dormancy, and the amount administered must not exceed the concentration of 0.8%–1.2% of dry matter.

The nitrogen fertilizing effect on the bergamot plant growth, productivity, and on the productive parameters (kg of fruit/plant, yield of juice, yield of oil, acidity, total solubles, etc.) have been subjects of different studies carried out at the Agency in Reggio Calabria of the Istituto Sperimentale per l'Agrumicultura in Acrireale (Intrigliolo et al. 1997, 1998, 1999). Contemporaneously were studied the effects on the fruits parameters of other nutritional elements: P, K, Mg, Fe, Zn, and Mn.

From the results attained, it is possible to deduce that the fruit dimensions and the yield of oil are significantly influenced by these elements—with the exception of Ca and Mg, which perform different roles for the vegetative development of bergamot plants. Among all, nitrogen and manganese demonstrate to be fundamental for the production if they are administered to the soil at amounts higher than 200 K/ha, corresponding to 2.15% of N on the dry matter.

The thickness of the skin and the yield of oil appeared to be related to the doses of N, K, and Mn, although the ideal ratio between these elements is not clear due to their reciprocal antagonism. The positive effect exerted by Mn suggests the necessity of particular care for this element during the program of fertilization, mainly for its scant occurrence in the Calabrian bergamot orchards. In addition, it has been proven that soil texture exerts a pronounced effect on the composition and quality of bergamot oils. Interference caused by the physical soil components was related to the yield of oil, which was negatively influenced by the high sand component, in contrast with a positive effect exerted by the presence of silt and clay (Intrigliolo et al. 1999). Positive effects were observed on the composition of the essential oil using increasing amounts of nitrogen fertilizers. Embleton et al. (1996) and Khan et al. (1992) reported that only with optimal fertilization procedures such as administering nitrogen and potassium, were positive effects attained on almost all the parameters of the productivity of bergamot, without taking into account the biochemical antagonism between these two elements.

From what has been reported in the literature, it is evident that it is not an easy task to find the optimal fertilization formulas if the plant needs are not fully determined. This can be done by the foliar analysis, the determination of the availability of nutrients in the soil by physical, mechanical, and chemical analyses. The evaluation of the vegetative and health conditions of the plant must also be considered (Coïc 1966; Hauser 1966).

3.2.4 TILLAGE

Tillage is also the agricultural preparation of the soil by mechanical operations carried out in order to loosen and aerate the top layer of soil, which can facilitate the planting of the crop; to help in the mixing of residue from the harvest, organic matter (humus) and nutrients should be distributed evenly throughout the soil. Tillage may improve productivity by warming the soil, incorporating fertilizer and controlling weeds, but also renders soil more prone to erosion, triggers the decomposition of organic matter releasing CO_2, and reduces the abundance and diversity of soil organisms.

In a bergamot orchard, these procedures are usually carried out:

* Deep soil aeration (max depth 20–25 cm).
* Weeding at the end of winter/beginning of spring (March/April).
* Comminution of the clods to ease watering.
* Light weeding every two or three irrigations from March to September, eliminating the weeds using a mill and a lister.

It is easy to process the soil if the plant patterns are broad, using specific mechanical tools such as mills, listers, and harrows. With this approach it is possible to reduce the time for processing and workforce with beneficial economic effects for the farm. During winter, particularly if there is a freeze, the field operations are adjourned so as not to affect the vegetative activity of the plants. In the practice of bergamot cultivation, and of citrus in general, seldom is it preferred not to process the soil, but to clean it from the weeds by chemical treatments using different classes of compounds, such as paraquat, diquat, glyfosate, and simazine. When using this approach the depth of the freatic zone must be evaluated, and the toxic effect on the operator must be minimized. The use of pesticides can provide advantages such as:

* Increase of the superficial rooting
* Reduction of the amount of water for irrigation, with the consequent increase of the amount of fertilizer available for the plant
* Preservation of the physical properties of the soil, with consequent reduction of erosions.

But it also had the following drawbacks:

- Increasing the cost for the high commercial price of pesticides
- Occurrence of toxic effects (rarely)
- Destruction of field animals (rats, moles, crawlers, etc.)
- Pollution of the freatic layers and marine waters
- Decreasing the microbiological activity

This approach is applied if the pattern does not allow the access to the mechanical tools; the use of herbicide and the biological approach are, however, the best choices for resistant infestants.

3.2.5 IRRIGATION

Irrigation is surely one of the most important practices in cultivation of every kind of vegetable. Water is a very important substance for the soil texture and for the health of plants; it makes the cell turgid, helps the circulation of nutrients and their absorption, and participates in various metabolisms during the different physiological states of the vegetative cycle. In addition, plants easily lose water through evapo-transpiration due to the heat of the sun, requiring replacement with regular and controlled watering; if this is not performed the plant goes through hydro-stress and can fade.

Irrigation practice is dependent on the water capacity of the soil, climate, and properties of the water available (presence of bacteria, microorganisms, salinity, solved gas, pollutants, temperature, and pH).

Water with good cultivation properties must have a salinity index around 1.5; this index is determined by the ratio of the saline residue (%) and the water hardness (°f). The concentration of the sodium ion is critical since it can exert toxic effects on plants; it is an antagonist of Ca and Mg, blocking the absorption of nutrients, the soil's structure degradation, and infiltration and permeability.

Alkalinity is due to the presence of sodium and is determined by the sodium absorption ratio (SAR) formula: $SAR = Na^+/1/2(Ca^{++} + Mg^{++})^{1/2}$, where the concentration of Na^+, Ca^{++}, and Mg^{++} is expressed as milliequivalents/L (Hoepli 1997b). The higher the SAR, the less adequate the water for irrigation; for example, water with SAR equal to 8 or more cannot be used to irrigate bergamots. The same is true for water with total salinity higher than 3.5–5 g/L and a subalkaline or alkaline reaction of the soil with pH > 7.8–8.0. Only water with Na at 60% of the sum of Ca and Mg can be used for applying agrarian chalk in proportions from 4 to 40 tons/ha.

Different methods are available to deliver water to the soil necessary for the correct development of the plant (Marsh 1973; Hoepli 1997b):

- Superficial expansion, requiring an accurate preparation of the soil
- Aspersion methods (above- or under-crown), requiring modest preparation of the soil
- Sub-irrigation, providing the water under the ground surface

It is up to the farmer which to choose based on economic and productive evaluations, for the most adequate method for the environment, the water and nutritive needs of the plant, the physiological vegetative state and, most importantly, the plant pattern of the field, whether or not it can allow the use of mechanical tools.

Environmental factors (rain, wind, solar exposition, relative humidity, and quality of the water available) are fundamental for the choice of irrigation. To cultivate bergamot in Calabria, the following methods are proposed:

- Irrigation by infiltration, distributing the water in superficial furrows and letting it penetrate vertically and horizontally on the soil surface. This system is cheap and requires small amount of water (2–15 L/s). It is possible to use dirty and/or cold water. The water dissipation is minimum and the furrows are fed singularly using plastic pipes (fixed or mobile) with plugholes or siphuncles.
- Irrigation by aspersion (above- or under-crown) when the water is delivered as artificial rain by simple apparatus. This approach does not require the preparation of the soil and can be performed efficiently also in declivity. The water dissipation by percolation and flowage is however restrained, with water saving.
- Micro-irrigation, or rain irrigation, is a method based on the fact that the efficacy of water delivered on the surface where roots are developed is about the same as that delivered on the entire surface; water is distributed in extremely small amounts, through drops from the pipes lying on the ground or suspended above along the tree rows. This method saves a good amount of water, it requires little handling, and it can be used also with salted waters.

3.2.6 PRUNING

Pruning is applied in the cultivation of bergamot, as for every citrus cultivation, with the aim to improve the plant productivity and regulate a harmonious development of the plant by cutting, interfering with the morphology and the physiological phenomena of the vegetative cycle. During the productive vegetative activity it has been proven that pruning is strictly related and complementary to fertilization, irrigation, soil manipulation, and phytopathic defense. This consists of the removal by cuts of the dry and damaged branches, also applied to rejuvenate the plant, making it acquire vigor and vitality in the remaining branches to contrast infections. The tools are generally metallic (scissors, knives, hacksaws); they must be bleached before and after their use to preserve them from rust.

The methods are dependent on the implant's pattern: in narrow patterns (3–4 m between plants) it must be carried out manually and frequently to avoid competition of aerial development, while in ampler patterns, it is possible to use mechanical tools.

Pruning is performed starting at the rootstock generation and continuing for the entire biological cycle. For example, the first pruning is carried out on the franc taproot during the transfer from the sowing bed to the nursery. Thus, following the plant's natural growth, the pruning times are:

- Pruning in nursery, to promote the formation of healthy buds, which will form the crown of the plant
- Pruning the roots (broken or damaged) and vegetative apparatus of the plant before the transfer to the final location

- Pruning of the crown, with vase or globe shapes
- Fructification pruning, performed after harvest in spring, before the vegetative awakening

For correct pruning, the dry, damaged, and infested parts of the plant must be removed, to improve the health of the plant which will live longer.

The three most common bergamot cultivars do not present a particular need for pruning, with the exception of cv Femminello, which develops less and has a shorter life.

Due to the climate change, summers are excessively hot with scant rain, frequent winds, and irregular thermal excursions, so the frequency of pruning is increasing, and it must be combined with more sustained irrigations. It is, however, suggested to avoid pruning of bergamot with mechanical means, even if the field pattern allows them, because of the damage that can be done to this fragile plant.

REFERENCES

Caruso, A., Gazea, F., Terranova, G., Russo G., and Cicciarello, G. 2000. Performance of bergamot on several rootstocks. *Proc. 9th Congress of International Society of Citriculture.* Dec. 2–7, 2000, Orlando, Florida.

Coïc, J. 1966. Aspects chimiques, physiques et biologiques de la fertilization, Chemical fertilizers. *Proc. 17th International Congress, "Chemistry Days 1966,"* 13–29, ed. G. Fauser. Oxford: Pergamon Press.

Embleton, T. W., Reitz, H. J., and Jones, W. W. 1973a. Citrus fertilization, in *The Citrus Industry,* ed. W. Reuther, 3: 122–156. University of California Riverside, California: Berkeley Press.

Embleton, T. W., Jones, W. W., Labanauskas, C. K., and Reuther, W. 1973b. Leaf analysis as a diagnostic tool and guide to fertilization, in *The Citrus Industry,* ed. W. Reuther, 3: 183–198. University of California Riverside, California: Berkeley Press.

Embleton, T. W., Coggis, C. W. Jr., and Witney, G. W. 1996. What is the most profitable use of Citrus leaf analysis? *Proc. Int. Soc. Citriculture* 1: 1261–1264.

Hauser, G. 1966. Food and fertilizer in developing countries, chemical fertilizers. *Proc. 17th International Congress "Chemistry Days 1966,"* 1–12, ed. G. Fauser. Oxford: Pergamon Press.

Hoepli, U. 1997a. *Propagazione delle piante, moltiplicazione; Coltivazione arboree, impianto dell'arboreto.* Milano: Ulrico Hoepli.

Hoepli, U. 1997b. *Fertilizzazione, elementi nutritivi; Acqua, terreno e colture; Capacità idrica; Qualità delle acque irrigue e dei sali disciolti (SAR).* Milano: Ulrico Hoepli.

Intrigliolo, F., Caruso, A., and Giuffrida, A., et al. 1997. Effetto dei parametri produttivi e nutrizionali sulla resa e sulla qualità dell'essenza di bergamotto. *Essenz. Deriv. Agrum.* 67: 42–61.

Intrigliolo, F., Caruso, A., Russo, G., Giuffrida, A., and Gazea, F. 1998. Effetti della fertilizzazione azotata sui parametri vegeto-produttivi e nutrizionali del bergamotto. *Italus Hortus* 5(4): 27–35.

Intrigliolo, F., Caruso, A., and Russo G., et al. 1999. Pedologic parameters related to yield and quality of bergamot oil. *Commun. Soil Soc. Plant Anal.* 30: 2035–2044.

Khan, I. A., Embleton, T. W., Matusumara, M., and Atkin, D. R. 1992. Leaf sampling methods: A problem in the exchange of fertilizer management information in the Citrus world. *Proc. Int. Soc. Citriculture* 2: 564–569.

Marsh, A. W. 1973. Methods of irrigation, in *The Citrus Industry,* ed. W. Reuther, 3: 259–277. University of California Riverside, California: Berkeley Press.

Scuderi, A., and Terranova, G. 1970. Propagazione e semi per soggetti, I vivai, Convegno "Tutto sugli Agrumi, situazione attuale, prospettive future e tecniche colturali," 1–8, Roma: Istituto di Tecnica e Propaganda Agrumaria.

Spina, P., and De Martino, E. 1991. *Gli Agrumi*, 120–127; 133–191. Bologna: Edagricole.

Spina, P., and Russo F. 1985. *Trattato di Agrumicultura*, 186–194. Bologna: Edagricole.,

Starrantino, A., Russo F., and Spina, P. 1979. La micropropagazione in agrumicoltura, *Atti Convegno Tecniche di coltura in vitro*, 219–228, Pistoia, 6 Ott, Italy.

4 Bergamot Diseases

*Giovanni Enrico Agosteo, Santa Olga Cacciola,
Pasquale Cavallaro, Gaetano Magnano
di San Lio, and Mariateresa Russo*

CONTENTS

4.1   INTRODUCTION

Many comprehensive reviews and compendia dealing with citrus diseases have been published (Fawcett 1936; Klotz 1973, 1978; Wallace 1978; Timmer et al. 1999, 2000). However, specific references to diseases of bergamot (*Citrus bergamia*) are rarely found as cultivation of this citrus species is economically relevant only in very restricted areas. As far as the phytosanitary status of bergamot industry in Calabria (southern Italy), the major bergamot producing area in the world, is concerned, three reports are available in the scientific literature of the last 40 years. The first two (Terranova and Cutuli 1975; Terranova et al. 1984) cite several virus and virus-like diseases, including exocortis, cachexia-xyloporosis, concave gum, cristacortis, and impietratura, that were diagnosed on the basis of visual symptoms on naturally infected trees and on artificially inoculated indicator test plants, while the third one (Grasso and Polizzi 1988) is a scant list of fungal and bacterial diseases extrapolated from previously published papers concerning citrus diseases in general. Advances in our knowledge of citrus diseases have since occurred, and some diseases such as Alternaria brown spot had not yet emerged at that time (Faedda et al. 2011), while others such as citrus tristeza virus have assumed much greater importance (Davino et al. 2011). Despite the lack of detailed information, it can be inferred that bergamot is susceptible to numerous diseases affecting other commercially important citrus species. Like other plant diseases, citrus diseases can be classified as infectious (biotic) or noninfectious (abiotic). Biotic diseases are caused by various agents including viroids, virus, phytoplasmas, bacteria, fungi and oomycetes (pseudofungi). Conversely, abiotic diseases are caused by adverse environmental conditions, injuries caused by improper application of chemicals, nutritional disorders, and genetic alterations. A third group includes diseases whose etiology has not been yet determined. In this chapter only major diseases of bergamot occurring in Calabria or threatening the bergamot industry in this region of southern Italy are reported, and management strategy is briefly discussed.

4.2   PHYTOPHTHORA GUMMOSIS, FOOT ROT, ROOT ROT, AND FRUIT BROWN ROT

This complex disease is caused by soil-borne species of *Phytophthora* (Oomycota) and is recognized as a major disease of citrus worldwide (Timmer et al. 2000; Mariau 2001). *Phytophthora* spp. attack citrus plants at all stages and may infect all parts of the tree, roots, stem, branches, twigs, leaves, and fruit, inducing different *facies* of the disease, including twig and leaf dieback, indicated collectively as canopy blight, foot rot, gummosis, root rot, and fruit brown rot. The most common species of *Phytophthora* in bergamot groves in Calabria are *P. citrophthora* and *P. nicotianae*. *P. citrophthora* is the main causal agent of trunk gummosis and fruit brown rot while *P. nicotianae* is the main causal agent of fibrous root rot. The temperature is a major ecologic factor conditioning the seasonal fluctuations and distribution of these two species. *P. nicotianae* is more common in subtropical areas of the world and causes foot and root rot while in these areas *P. citrophthora* is restricted to cooler weather sites and coastal areas. In the Mediterranean region, *P. nicotianae* is not active in winter while *P. citrophthora* is not inhibited by winter temperatures. Conversely, the activity of the latter species is dramatically reduced in hot summer months with the exception of short periods following irrigation. There is evidence that fluctuations of the inoculum of *Phytophthora* are strongly influenced by the physiological conditions of the host plant. Seasonal variations of the physiology of citrus trees, for example, are a major factor in determining the susceptibility of fibrous roots to the rot incited by *P. nicotianae*. Summer activity of this species is directly correlated with both the production of root exudates and the concentration of sugars in the roots but is inversely correlated with starch concentration (Cacciola and Magnano di San Lio 2008). Also, the susceptibility of the bark to the infection of *P. citrophthora* in subtropical and Mediterranean climates varies throughout the year; it is higher in spring and autumn and very low in winter and summer. Like lemon, limes, sweet orange, and grapefruit, bergamot is very susceptible to bark infection. However, bud union at least 40 cm above the ground avoids contact with the soil, which is the major source of inoculum. When, during the 1860s, Phytophthora gummosis almost completely destroyed the lemon cultivations in the area of Messina, bergamot trees of the province of Reggio Calabria survived, mainly due to the habit of the farmers of leaving the base of the tree trunk free from soil, in contrast to the common practice of stacking the soil along the trunk (Alfonso 1875).

The fruit is susceptible to brown rot infections from the ripening phase. The incubation period of this *facies* of the disease varies from 3 to 14 days, according to the temperature, and asymptomatic infected fruit can infect healthy fruit even after harvesting. Severe fruit brown rot epidemics seldom occur in bergamot orchards in Calabria, probably because the maritime climate of the coastal areas, where this citrus is grown, is not conducive to the infection. Moreover, the structure of bergamot trees resulting from conventional pruning is characterized by rising branches and the lower part of the canopy at a height of more than 1 meter from the soil; this canopy shape reduces the risk of fruit infections by rain splashes from the soil. By contrast, Phytophthora gummosis of the trunk and the major limbs above the bud union is very common in bergamot orchards and is favored by sprinkler irrigation that wets the bark. Phytophthora foot and root rot also occur frequently in bearing bergamot orchards planted in heavy or poorly drained soils. Vigor of the tree and fruit production are reduced because water and mineral nutrient uptake are impaired and carbohydrate reserves in the roots are depleted. As a consequence, fruit size and yield are reduced. Other symptoms of Phytophthora foot and root rot include leaf chlorosis, defoliation, and twig dieback. Since the destructive epidemic outbreaks of citrus gummosis and foot rot caused by *P. citrophthora* during the 1860s in the citrus groves of the Mediterranean area, sour orange (*C. aurantium*), which is resistant to this disease, has been the only rootstock used in bergamot groves. After the introduction and the spread of citrus Tristeza virus (CTV) in Italy, sour orange is being replaced by CTV-tolerant citrus rootstocks; however it should be verified if these new rootstocks, universally used for sweet orange, grapefruit, and mandarins, are fit as bergamot rootstocks (see Section 4.8). Moreover, resistance to Phytophthora foot and root rot is just one of the characteristics to be taken into account when selecting a rootstock. The performance of a rootstock depends also on its response to abiotic stresses, including calcareous soil, soil and water salinity, root asphyxiation, and cold hardness or tolerance (Table 4.1). Control of diseases incited by *Phytophthora* spp. by chemical means is a widespread practice especially since very effective systemic fungicides like metalaxyl (trade name Ridomil) and aluminium ethyl-phosphyte or phosetyl-Al (trade name Aliette) have come onto the market. More recently, metalaxyl has been substituted with its enantiomer metalaxyl-M or mefenoxam (trade name Ridomil Gold), which is effective at a lower dosage. Another derivate of phosphorous acid, potassium phosphonate, which is on the market as fertilizer, acts in the same way as phosetyl-Al. Metalaxyl and phosetyl-Al are systemics that are absorbed by the roots, bark, and leaves and translocated within the plant. After absorption by citrus, both fungicides will effectively limit *Phytophthora* activity for a period of 3–4 months. Metalaxyl acts directly on *Phytophthora* to inhibit its growth and sporulation. However, fosethyl-Al and other phosphorous acid derivatives have only weak direct activity against *Phytophthora*, and much of their activity stems from their ability to trigger citrus defense mechanisms.

TABLE 4.1
Susceptibility of Citrus Rootstocks to Biotic and Abiotic Stresses

Rootstock	Tristeza	*Phytophthora*	*Armillaria*	Exocortis	Cold	Salinity	Root Asphyxiation
Trifoliate orange[a]	O	O	●	●	O	●	O
Carrizo and Troyer citranges[a]	⊙	⊙[c]	●	◑	⊙	●	●
Alemow	●[b]	O	—	⊙	●	O	●
Rough lemon	⊙	●●[d]	●	⊙	●	●	●
Cleopatra mandarin	⊙	●	●	⊙	⊙	O	●
Common mandarin	⊙	●●	—	⊙	⊙	◑	—
Sour orange	●●[b]	O	⊙	⊙	⊙	⊙	⊙
Sweet orange	⊙	●	●	⊙	⊙	●	●
Swingle citrumelo[a]	⊙	O	—	⊙	⊙	◑	●
C-35 citrange	⊙	⊙	—	—	—	—	—
Volkamer lemon	⊙	●	—	⊙	●	◑	⊙
Forner-Alcaide N. 5			—	—	—	—	—
Forner-Alcaide N. 2418	O	O	—	—	—	—	—
Yuma Ponderosa lemon	⊙	⊙[e]	—	—	—	—	—

O	*Resistant*	[a]*Mandarin trees are short-lived on these rootstocks*
⊙	*Tolerant*	[b]*Tolerant combinations when scion is lemon*
◑	*Moderately susceptible*	[c]*Carrizo and Troyer citranges are more resistant to Phytophthora gummosis but more susceptible to root rot than sour orange*
●	*Susceptible*	[d]*Schaub rough lemon is susceptible to* Phytophthora citrophthora *but tolerates infections of* P. nicotianae
●●	*Very susceptible*	[e]*Yuma Ponderosa lemon is susceptible to Phytophthora gummosis but tolerates root rot*

4.3 BERGAMOT GUMMOSIS

Bergamot gummosis is a disease whose etiology has not been determined. It is widespread in bergamot groves of the province of Reggio Calabria. The most typical symptom is a gum exudate oozing from the trunk and the limbs from spring to autumn (Magnano di San Lio et al. 1978). Gum, which is water soluble, is washed away by heavy rains. The sour orange rootstock is not affected even when symptoms

FIGURE 4.1 Gum oozing from the trunk and the limbs of a bergamot tree in summer.

on the scion are severe (Figure 4.1). A viral etiology of this disease was hypothesized (Terranova et al. 1978–79, 1984; Catara et al. 1980, 1984), but it has never been demonstrated that bergamot gummosis is graft-transmissible. As a matter of fact, bergamot trees produce gum exudate easily and profusely as a consequence of biotic and abiotic stresses (Matarese Palmieri et al. 1979), and it is likely the syndrome referred to as bergamot gummosis can be imputed to diverse causes, including aerial infections of *Phytophthora*, water stress, concave gum, or severe attacks of the California red scale (*Aonidiella aurantii*).

4.4 MAL SECCO DISEASE

Mal secco disease of citrus is a highly destructive vascular wilt disease confined to the Mediterranean and Black sea areas. It is caused by the mitosporic fungus *Phoma tracheiphila* (syn. *Deuterophoma tracheiphila*). The principal host species is lemon (*C. limon*). Two comprehensive and up-to-date reviews on mal secco disease of citrus have recently been published (Migheli et al. 2009; Nigro et al. 2011). Bergamot has been reported to be very susceptible to mal secco (Perrotta and Graniti 1988; Solel and Salerno 2000; Migheli et al. 2009; Nigro et al. 2011); however, "it is rare to see bergamot plants affected by the disease" (Spinelli 1978) and no destructive effects, comparable to those observed in lemon orchards, have been observed in bergamot groves in Calabria. The most severe infections may be observed when bergamot trees are exposed to marine winds.

4.5 FRUIT DISEASES

4.5.1 STYLAR-END ROT

Stylar-end rot, also named stylar-end breakdown, is a physiological disorder first described on Tahiti lime (*Fortunella* sp. × *C. aurantifolia*), which like bergamot has a persistent stylar-end (Fawcett 1936). It appears as a water-soaked, tan lesion at the stylar end of the fruit and extends rapidly up to half of the fruit surface. The affected area dries, contracts, and becomes sunken (Timmer et al. 2000). Although stylar-end rot is a common disease on bergamot, usually it does not cause serious damage. Femminello is the most susceptible among bergamot cultivars (Terranova and Cutuli 1975; Cutuli and Salerno 1998). Stylar-end rot occasionally appears in the field, particularly if rainy weather is followed by high temperature. Generally, however, it is a postharvest disease. It develops as a consequence of the rupture of juice vesicles that is more likely to occur if the fruit is picked in the early morning when the turgor pressure is high. High temperature at harvest and during transport favors this disorder. Careful handling of fruit helps reduce its incidence.

4.5.2 SEPTORIA SPOT

Septoria spot is a fungal disease caused by species of *Septoria*, including *S. citri*, *S. limonum*, and *S. depressa*. Very probably, however, all these epithets are synonyms. The causal agent of septoria spot of bergamot fruit in Calabria has been identified as *S. limonum* (Agosteo 2002). Leaf and fruit spots caused by *Septoria* species have been reported from most citrus-growing areas in the world. Although they are generally considered of minor significance, infections of fruit rind may become a problem when fresh fruit is sold or when fruit rind is utilized for the extraction of the essential oils, as in the case of bergamot. On bergamot fruit still attached to the tree, the lesions consist of small depressions or pits that extend no deeper than the flavedo. Infections occur when the fruit is green and symptoms become more conspicuous when the fruit ripens. The pits are light tan or buff with narrow greenish margins that become reddish brown as the fruit matures. Symptoms sometimes appear as tear stains. Small black pycnidia may be produced on the fruit lesions. During postharvest storage lesions may coalesce into larger, brown to black sunken blotches that are several centimeters in diameter and may extend into the fruit segment. Mature fruit are more susceptible.

The disease appears most commonly after cool or frosty weather and is more severe when rainfall levels are high. Low temperatures predispose citrus fruit to infections. The disease may be controlled by treatments with copper fungicides before winter rainfall. Moreover, as *S. citri* has saprophytic capability and pycnidia form profusely on dead twigs and leaves, pruning may reduce the incidence of the disease.

4.5.3 ALTERNARIA BROWN SPOT

Alternaria brown spot is caused by the pathotype of the fungus *Alternaria alternata* producing the host-specific toxin ACT. It causes serious defoliation, fruit drop, and fruit blemishes on susceptible cultivars. On young leaves, lesions first appear as small brown to black spots, surrounded by a yellow halo. Lesions expand and appear as circular or irregular blighted areas. Necrotic areas may be colonized by *C. gloeosporioides*; thus symptoms of this disease are sometimes confused with those of anthracnose. Shoot infection and abscission of young leaves produce dieback of twigs. Fruitlets may be infected soon after petal fall and even a single small lesion causes fruitlet drop. On larger fruit, symptoms vary from dark pinpoints to large black lesions. Usually the infections are walled off by periderm but sometimes the fungus continues to invade the tissues, forming large lesions surrounded by a yellow halo, and fruit fall before maturity. Brown spot is an emerging disease in Italy (Faedda et al. 2011) where it was first reported in Sicily and Calabria on Fortune mandarin, which is a very susceptible cultivar. Some other mandarin and mandarin-like cultivars, including Dancy, Minneola, Orlando, Nova, Page, Lee, Sunburst, Encore, Murcott, Michal, Winola, Ponkan, Emperor, Tangfang, and Primosole, are also very susceptible. In contrast, sweet orange and clementine cultivars are resistant. The susceptibility of citrus to Alternaria brown spot depends on a single gene with two alleles, one dominant (*S*) and the other one recessive (*r*), that transmit susceptibility and resistance, respectively. Sweet orange and clementine are homozygous (*rr*), while Fortune and Nova are heterozygous (*Sr*). The susceptibility of bergamot to brown spot has not been determined.

4.5.4 ANTHRACNOSE

The prevalent causal agent of citrus anthracnose is the fungus *Colletotrichum gloeosporioides* (teleomorph, *Glomerella cingulata*). *C. gloeosporioides* is a colonizer of injured and senescent tissues both in the field and after harvest. This fungus is a common symptomless invader of the rind of citrus fruit but is not able to cause disease on healthy tissues. To prevent anthracnose, fruit should be handled carefully to avoid injury and should not be held too long on the tree after ripening and in storage.

4.5.5 SOOTY MOLD

Sooty mold is a black superficial incrustation constituted by fungal growth that appears on leaves, stems, and fruit after trees become infested by honeydew-excreting insects, including aphids, soft scales, mealybugs, and whiteflies. *Capnodium citri* is the prevalent component of sooty mold. Sooty mold may affect tree performance and fruit development by interfering with photosynthesis. It may delay fruit ripening and can be difficult to remove in the packinghouse on fruit with a rough rind. The presence of sooty mold on bergamot fruit may interfere with the extraction of the oil from the rind and may negatively affect its quality. Sooty mold formation is prevented by controlling the insects responsible for the honeydew.

4.6 CITRUS TRISTEZA VIRUS

Citrus tristeza virus (CTV) is the most economically important viral pathogen of citrus worldwide (Lee and Bar-Joseph 2000). Millions of trees on sour orange rootstock have been killed by CTV in several citrus-growing countries, including Argentina, Brazil, Israel, Peru, the United States, Spain, and Venezuela. Severe stem-pitting strains of CTV reduce the productivity and affect the quality of fruit even of citrus trees on CTV-tolerant rootstocks. The tristeza virus probably comes from Asia, which is also the area where the genus *Citrus* originated. It is readily graft-transmissible. Moreover, it is transmitted by aphids in a persistent manner. Long-distance spread occurs by infected plant material. Most species of *Citrus* and some species in other genera of the family Rutaceae are infected by this closterovirus. The susceptibility of the tree to CTV depends on the scion/rootstock combination (Carrero 1981); trees of sweet orange on sour orange rootstock are extremely susceptible to CTV-induced decline while lemon on sour orange rootstock tolerates the infection. CTV was reported for the first time in Italy many decades ago but only recently has it spread epidemically (Davino et al. 2011). As a consequence of the epidemic spread of tristeza in Italy, sour orange rootstock is being substituted for with CTV-tolerant rootstocks, such as Troyer and Carrizo citranges (*C. sinensis* × *Poncirus trifoliata*) or citrumelo (*P. trifoliata* × *C. paradisi*). As far as bergamot is concerned, care must be taken in utilizing these new rootstocks as the quality of the oil may be significantly affected by the rootstock. Moreover, although no direct evidence has been provided, very probably bergamot on sour orange rootstock, like lemon, tolerates CTV-induced decline.

4.7 OTHER VIRUS AND VIRUS-LIKE DISEASES

Several virus and virus-like diseases, including psorosis, exocortis, cachexia-xyloporosis, concave gum, cristacortis, and impietratura, have been reported on bergamot (Terranova and Cutuli 1975; Terranova et al. 1978–79, 1984). In addition to these common diseases, two other virus-like diseases of bergamot have been described—bud knot, which was supposed to be of genetic origin (La Rosa et al. 1984), and bergamot vein yellowing (Protopapadakis et al. 2002), which was proved to be graft-transmissible.

Virus-free bergamot accessions, obtained by shoot-tip grafting from old clones of the cultivars Castagnaro, Fantastico, and Femminello in the national citrus certification scheme, have been available since 1977 at CRA-ACM Acireale, Italy (Caruso et al. 2009).

4.8 DISEASES CAUSED BY PROKARYOTES

Huanglongbing (HLB), previously called greening, is an emerging, highly destructive disease of citrus caused by the phloem-limited gram-negative bacteria *Candidatus* Liberibacter spp. (Garnier and Bovè 2000). Young trees infected by HLB do not produce while mature trees become unproductive. HLB can infect all citrus cultivars, species, and hybrids and some citrus relatives. Sweet oranges, mandarins, and mandarin hybrids are severely affected. Grapefruit, Rangpur lime, lemons, calamondins, and pummelos are less severely affected. Mexican lime, trifoliate orange, and trifoliate orange hybrids are tolerant and show only leaf mottle symptoms. An early symptom is the appearance of yellow shoots, hence the name huanglongbing, which means yellow dragon disease. The fruits on HLB-affected trees are small, lopsided, and poorly colored (the name "greening" stems from this last symptom). Three species of *Candidatus* Liberibacter have been recognized so far: *Candidatus* Liberibacter asiaticus, *Candidatus* Liberibacter africanus, and *Candidatus* Liberibacter americanus. These species are transmitted by psyllids

(*Trioza erytreae* and *Diaphorina citri*) and differ in geographical distribution and some ecological characteristics (Bovè 2006). *Candidatus* Liberibacter asiaticus is the most widespread and dangerous species. HLB is a potential threat to the citrus industry in Italy (Davino et al. 2011) and its causal agents are included in the A1 list of quarantine pathogens of the European and Mediterranean Plant Protection Organization (EPPO). Another disease, called Australian citrus dieback, with symptoms similar to those of huanglongbing, has been reported in citrus-growing areas of eastern Australia. It is most severe and common in grapefruit, but occurs in other citrus species, including bergamot (Broadbent 2000). The causal agent of this disease has not been identified.

REFERENCES

Agosteo, G. E., 2002. First report of Septoria spot on bergamot. *Plant Disease* 86: 71.

Alfonso, F., 1875. *Trattato sulla coltivazione degli agrumi.* Palermo: Luigi Pedone Lauriel.

Bovè, J. M., 2006. Huanglong bing: A destructive, newly-emerging, century-old disease of citrus. *Journal of Plant Pathology* 88: 7–37.

Broadbent, P., 2000. Australian citrus deback. In *Compendium of citrus diseases.* 2nd ed., eds. L. Timmer, S. M. Garnsey, and J. H. W. Graham, 46. St. Paul, MN: The American Phytopathological Society Press.

Cacciola, S. O., and Magnano di San Lio, G., 2008. Management of citrus diseases caused by *Phytophthora* spp. In *Integrated management of diseases caused by fungi, phytoplasma and bacteria*, eds. A. Ciancio and K. G. Mukerji, 61–84. Heidelberg, Germany: Springer Sciences+Business Media B. V.

Carrero, J. M., 1981. *Virosis y enfermedades afines de los cítricos,* 2nd ed. Madrid, Spain: Publicaciones de extension agraria.

Caruso, A., Recupero, S., Reforgiato Recupero, G., and Russo, G., 2009. *Le accessioni di agrumi registrate dal Centro di Ricerca per l'Agrumicoltura e le Colture Mediterranee nel servizio nazionale di certificazione volontaria. Consiglio per la Ricerca e la Sperimentazione in Agricoltura.* Acireale (Catania), Italy: Centro di Ricerca per l'Agrumicoltura e le Colture Mediterranee.

Catara, A., Davino M., and Magnano di San Lio, G., 1980. Investigations on a bergamot gummosis in Calabria (Italy). In *Proceedings Eighth IOCV Conference,* eds. E. C. Calavan, S. M. Garnsey, and L. W. Timmer, 81–85. Riverside, CA: IOCV.

Catara, A., Davino, M., and Magnano di San Lio, G., 1984. Further researches on a virus-like gummosis of bergamot. In *Proceedings Ninth IOCV Conference,* eds. S. M., Garnsey, L. W., Timmer, and J. A., Dodds, 247–249. Riverside, CA: IOCV.

Cutuli, G., and Salerno, M., 1998. *Guida illustrata alle alterazioni dei frutti di agrumi. s.r.l.* Bologna, Italy: Edagricole-Edizioni Agricole della Caldironi.

Davino, S., Saponari, M., Albanese, G. et al., 2011. Il virus della "tristeza" degli agrumi (CTV) mette a rischio l'agrumicoltura italiana. *Protezione delle Colture* 4: 16–23.

Davino, S., Lo Giudice, V., Caruso, A., and Davino, M., 2011 Huanglongbing (greening): Una possibile minaccia per l'agrumicoltura italiana. *Protezione delle Colture* 5: 41–48.

Faedda, R., Pane, A., Cacciola, S. O. et al., 2011. Malattie fungine degli agrumi. *Protezione delle Colture* 4: 24–29.

Fawcett, H. S., 1936. *Citrus diseases and their control,* 2nd ed. New York: McGraw-Hill.

Garnier, M., and Bovè, J. M., 2000. Huanglongbing (greening). In *Compendium of citrus diseases,* 2nd ed, eds. L. Timmer, S. M. Garnsey, and J. H. W. Graham, 46–48. St. Paul, MN: APS Press.

Grasso, S., and Polizzi, G., 1988. Cryptogamic diseases of bergamot in Calabria (Italy). In *Proceedings of the Sixth International Citrus Congress*, eds. R. Goren and K. Mendel, 833–837. Philadelphia/Rehovot: Balaban Publishers.

Klotz, L. J., 1973. *Color handbook of citrus diseases*. Berkeley: University of California, Division of Agricultural Sciences.

Klotz, L. J., 1978. Fungal, bacterial, and nonparasitic diseases and injuries originating in the seedbed, nursery, and orchard. In *The citrus industry. Vol. 4. Crop protection*, eds. W. Reuther, E. C. Calavan, and G. E. Carman, 1–66. Berkeley: University of California, Division of Agricultural Sciences.

La Rosa, R., Napoli, M., Di Silvestro, I., and Catara, A., 1984. Bud knot: A disorder of bergamot. In *Proceedings Ninth IOCV Conference*, eds. S. M. Garnsey, L. W. Timmer, and J. A. Dodds, 241–246. Riverside, CA: IOCV.

Lee, R. F., and Bar-Joseph, M., 2000. Tristeza. In *Compendium of citrus diseases*. 2nd ed., eds. L. Timmer, S. M. Garnsey, and J. H. W. Graham, 61–63. St. Paul, MN: The American Phytopathological Society Press.

Magnano di San Lio, G., Davino M., and Catara, A., 1978. Osservazioni istologiche su una gommosi: del bergamotto. *Riv. Pat. Veg. S. IV*, 14: 99–109.

Mariau, D., ed., 2001. *Diseases of tropical fruit crops*. Montpellier, France: CIRAD and Science Publishers, Inc.

Matarese Palmieri, R., Tomasello, D., and Magnano di San Lio, G., 1979. Origine e caratterizzazione istochimica delle gomme in piante di agrumi con sindromi a diversa eziologia. *Riv. Pat. Veg. S. IV*, 15: 43–49.

Migheli, Q., Cacciola, S. O., Balmas, V., et al., 2009. Mal secco disease caused by *Phoma tracheiphila*: A potential threat to lemon production worldwide. *Plant Disease* 93: 852–867.

Nigro, F., Ippolito, A., and Salerno M. G., 2011. Mal secco disease of citrus: A journey through a century of research. *Journal of Plant Pathology* 93: 523–560.

Perrotta, G., and Graniti, A., 1988. *Phoma tracheiphila* (Petri) Kanchaveli & Gikashvili. In *European handbook of plant diseases*, eds. I. M. Smith, J. Dunez, R. Lelliott, and D. H. Phillips, 396–398. Oxford: Blackwell Scientific Publications.

Protopapadakis, E. E., Tzortzakaki, S., Kasapakis, J., and Tsagris, M., 2002. A new graft-transmissible disease of bergamot in Greece. In *Proceedings Fifteenth IOCV Conference*, eds. N. Duran-Vila, R. G. Milne, and J. V. da Graçia, 382–383. Riverside, CA: IOCV.

Solel, Z., and Salerno, M., 2000. Mal secco. In *Compendium of citrus diseases*. 2nd ed., eds. eds. L. Timmer, S. M. Garnsey, and J. H. W. Graham. St. Paul, MN: The American Phytopathological Society Press.

Spinelli G., 1978. *Studio sul bergamotto*. Reggio Calabria: Grafiche Sgroi.

Terranova, G., and Cutuli, G., 1975. Lo stato fitosanitario del bergamotto in Calabria. *Annali Ist. Sper. Agrumicoltura Acireale* 7/8: 175–189.

Terranova, G., and Reforgiato Recupero, G., 1978–1979. Ulteriori ricerche sulla concave gum del bergamotto (*Citrus bergamia* Risso). *Annali Ist. Sperim. Agrumicoltura Acireale* 11/12: 111–119.

Terranova, G., Starrantino, A., and Russo, F., 1984. Phytosanitary status of bergamot in Italy: In *Proceedings Ninth IOCV Conference*, eds. S. M. Garnsey, L. W. Timmer, and J. A. Dodds, 256–258. Riverside, CA: IOCV.

Timmer, L. W., and Duncan L. W., eds., 1999. *Citrus health management*. St. Paul, MN: The American Phytopathological Society Press.

Timmer, L. W., Garnsey, S. M., and Graham, J. H., eds., 2000. *Compendium of citrus diseases*. 2nd ed. St. Paul, MN: The American Phytopathological Society Press.

Wallace, J. M., 1978. Virus and viruslike diseases. In *The citrus industry, Vol. IV*, eds. W. Reuther, E. C. Calavan, and G. E. Carman, 67–184. St. Paul, MN: The American Phytopathological Society Press.

5 Bergamot in the Calabrian Economy and the Role of Consorzio del Bergamotto

Francesco Crispo

CONTENTS

5.1 PRODUCTION, TRANSFORMATION, COMMERCE

Bergamot was defined by its growers as the "green gold" because in the past even a small piece of land with Bergamot cultivation was sufficient to feed a family. "Small is beautiful" could be the motto for this Italian excellency, with an average plantation not exceeding 1.5 hectares, much smaller than extensive citrus plantations in the rest of the world, and bergamot fruits are still processed in relatively small extraction factories.

Bergamot cultivation occurs on about 1000 hectares along the Reggio Calabria coast, precisely from Scilla to Monasterace. Of this area about 800 hectares are used by farmers who are part of the organization "Organizzazione Produttori Unionberg" (Nesci and Sapone 2011).

The present picture is the result of a slow and continuous decrease, due to the expansion of urban and residential areas, and to the reconversion of the industrial plants. The latter phenomenon was determined by the numerous crises of the market for bergamot essential oil. In 1967, in fact, in Calabria the surface area cultivated with bergamot was about 3800 hectares. At the end of the 1980s the surface shrunk to 1800 hectares (Di Giacomo 1989).

Figure 5.1 provides a snapshot of the evolution of the land dedicated to bergamot cultivation (Crispo and Dugo 2001; Crispo 2011).

Today we count 600 different farmers who are producers of bergamot. Among these farmers, only 348 appear in the registry of the enterprises associated with

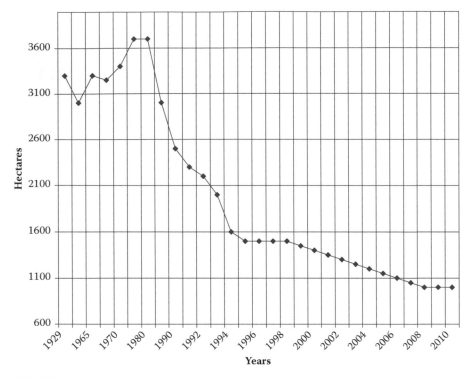

FIGURE 5.1   Evolution of the surface area cultivated with bergamot in Calabria.

the Consorzio del Bergamotto. Not long ago, from a census made in 1973 by the Consorzio del Bergamotto, there were 2700 farmers. These numbers demonstrate, the extreme fragility of the productive division pulverized for a demand structured as an oligopoly (Crispo and Dugo 2001).

Until 1962, when it was mandatory to stock at the Consorzio del Bergamotto, the effective production of oil was under control, and the offer provided bargaining power. The liberalization of trade due to the sentence in law No. 54 in 1962 of the Constitutional Court established the illegitimacy of the mandatory storage, thus reversing the privileged offer onto an oligopoly demand. In fact, the offer is provided by about 20 enterprises of transformation, usually family owned, with few cooperatives and, mainly, by the organization "Organizzazione Produttori Unionberg," which produces an average of 100 tons of oil per year, corresponding to 0.5% of the fruit weight. The essential oil trade is operated by a modest number of export enterprises and/or by transformation/ export enterprises that trade in small amounts inside the country, though mainly with France, Switzerland, and the United States, but also with the United Kingdom, Spain, Japan, and India for its applications in perfumery (Cucuzza and Timpanaro 1996).

Until October 14, 2002, the Consorzio del Bergamotto, a Public Economic Institute delegated by Reggio Calabria for the protection and to increase the value of bergamot and its derivatives, using its transformation plants located in S. Gregorio, represented an important reference in the transformation of the fruit and the trade of the oil. The situation was drastically changed by the law No. 41 issued by the Reggio

Calabria on October 14, 2002. This law revoked the delegation given to consortium for bergamot transformation. Article 12 of the abovementioned law foresaw that, by a public competition, the plants were entrusted to private parties. This law left to the Consorzio the role of promoter, technical assistant, and professional educator.

In the past the Consortium received the fruits by the associated enterprises to be transformed by its plants and consequently located the resulting essential oil on the market, dividing the income among the associated enterprises. Today the row material is subject to single verbal agreements between farmers and transformers based on payments made for amounts of fruits, or more rarely on percentage of oil yield (e.g.. 500 g of oil for one quintal of fruit).

The predominant role in essential oil trade is played by the export enterprises, comprising a union ring between the production and the market, represented by about 10 international companies. They can easily control the price of the oils, which is often quite low, also due to the scant offerings and the presence on the market of heavy amounts of synthetic products.

It is thus possible to understand the slow process which led to the decrease in the amount of land surface cultivated, the abandon of cultural practices, the lack of industrial innovation, and the scant investment in bergamot production. However, during more recent years a new impulse has been registered: natural bergamot oil is more appreciated and consumed; its derivatives are improved with a resulting higher value; new applications of bergamot products are found in the food and pharmaceutical fields.

Recently, bergamot juice has been successfully introduced in gastronomy, confectionery, liqueurs, and other parts of the food industry (Spinelli and Sandicchi 2000). A new and interesting induced activity on the local market is in expansion. The essential oil finds novel applications in pharmaceutical preparations, creating a demand in highly specialized sectors with higher incomes. In fact, although its disinfectant, hygienic, and healing properties have long been known, modern studies prove that bergamot oil also possess anti-inflammatory, antimicrobial, antimycotic, antiviral, neurosedative, and antidepressive properties (Focà 2001). The juice, which was empirically used in the past to reduce hyperlipidemia, today finds new prospective uses due to its antioxidant activity and its well-assessed properties to reduce the hematic level of cholesterol, triglycerides, and glucose (Mollace 2011). Lastly, in addition to the known use of the fruit peel to extract pectines, there are now spreading practices of making decorative objects, flowers, dolls, and boxes through manipulation of the fruit's reversed skin.

It should also be underlined that, in addition to the traditional cold-pressed oil, it is now possible to find on the market oils with certified origin (DOP), commercial oils, furocoumarins-free oils obtained by chemical or physical processes, and other mixed citrus oils, with declared manipulations and physical or chemical transformation performed on natural bergamot cold-pressed oil. These products are usually customized based on particular requisites expressed by the buyers, and created also to contrast with the market of reconstituted oils.

In Table 5.1 are reported the amounts of fruit produced from 2000 to 2010 along with the corresponding price of the essential oil. The values demonstrate how this market division is not constant, but is dependent on variation to the production and to the market tendencies of large companies in the cosmetics sector.

TABLE 5.1
Production of Bergamot Fruits and Market Price of the Essential Oil from 2000 to 2010*

Season	Fruits Produced (tons)*	Market Price (Euro) **
2000–2001	15,100	46.00
2001–2002	25,000	48.00
2002–2003	25,100	37.00
2003–2004	23,900	60.00
2004–2005	29,100	60.00
2005–2006	29,200	40.00
2006–2007	20,000	44.00
2007–2008	12,000	70.00
2008–2009	20,400	72.00
2009–2010	18,000	74.00

* Comité Liaison del'Agrumiculture Méditerraéenne (CLAM), annual reports.
** Nesci F. S., and Sapone N. 2011. Tutela e sviluppo del bergamotto reggino—32nd Congresso Italiano di Scienze Regionali: "Il Ruolo delle Città nell'Economia della Conoscenza." Torino, September 15–17, 2011.

5.2   LAW REGULATIONS AND HISTORICAL ROLE OF THE CONSORZIO DEL BERGAMOTTO

5.2.1   LAW REGULATIONS BEFORE THE CONSORZIO

The market for bergamot oil has often been subject to strong speculative phenomena, leading to recurrent crises in this sector. Disparaging campaigns on the properties of the oil favored the presence of synthetic products on the international market. They contributed to fluctuation in economic value and in production. This led to the intervention of regulations to protect this typical Calabrian product, with the rule of the mandatory stock at the Consorzio and quality control of the oil. In 1819 the first regulation relative to the quality assurance of cold-pressed bergamot oil was issued. The authority of that time (Intendente della Calabria ulteriore I) declared that the use of alembics was forbidden for the production of the oil by distillation. The transgressors were punished by fines and by the confiscation of the instrumentation.

After the economic crisis in 1929 and the perturbations of the market caused by the presence of reconstituted oils in the market, the Italian government with the Royal Decree of Law (R.D.L.) 3/31/1930 No. 438 (converted into law No. 1089 on July 18, 1930) instituted a Public Warehouse in Reggio Calabria, upon request of the local bergamot producers, to deposit the essential oil. In order to export the oil, it was mandatory to possess a certificate of genuineness released by the Royal Laboratory.

On the basis of these regulations, and those determined by the R.D.L. No. 1330 on October 15, 1931, the Consorzio dei Produttori di Bergamotto was founded with Prefect decree No. 3492 on November 11, 1931. The aims of the Consortium were the mandatory stocking of the oil and consequent protection of its quality.

Successively, the Ministry of Agriculture instituted with its decree on September 14, 1934 the Consorzio Provinciale dell'Agrumicoltura, which is the guardian of production and at the same time the research center for better discipline over commerce and its development. The Ministry Decrees on June 5, 1936 and August 26, 1936 authorized the constitution of the public warehouse and bergamot section to handle the bergamot stocks.

Law 829 was passed on April 23, 1936; it converted the R.D.L. No. 278 from February 3, 1936, finalizing the mandatory stock of the essential oil, forbidding its being sold in Italy or abroad if not through the bergamot sector of the Consorzio Provinciale dell'Agrumicoltura controlled by the Ministry of Agriculture. This was necessary to normalize the production, avoid adulteration, limit the competition between producers, and control speculation.

The warrants (amounts of bergamot oil stock in the Consortium warehouse) were useful financial tools used by the Consortium to draw the money, necessary for advance payments, from the bank accounts of the owners of the product.

After the fall of fascism, the stock was handled independently by the Commissary appointed with a Prefect decree on December 11, 1943 and later by an Advisory Commission until the suppression of the Economical Agency with the D.L. 4/26/1945 No. 367.

5.2.2 THE ROLE OF CONSORTIUM

In 1946 the Italian government, stimulated by the requests of bergamot producers to rule the market during the crisis, founded with the D.M. 5/29/1946 published in the official gazette G.U. 6/21/1946, the Consorzio del Bergamotto in Reggio Calabria to protect, promote, and locate on the market the product derivatives of bergamot, and therefore limit the adulteration phenomena and the competition between the producers.

To accomplish these duties, the Consortium had to:

(a) Promote all the possible initiatives to protect Bergamot
(b) Promote basic principles for the good practice in production and use of Bergamot oil
(c) Stock the essential oil as requested by law 829 issued on April 23, 1936

Therefore, the Consortium built the extraction plants in S. Gregorio and managed the stock and commerce of the oil, with an optimal equilibrium between production and the amount of oil sold. Producers were satisfied through being appropriately remunerated, and consumers were happy with the high quality of the product.

This situation was upset by the sentence of the Constitutional Court No. 54 on June 14, 1962, which found the law not sufficiently precise in terms of defense of the market freedom, although finding useful the general asset of the Consortium's aims.

To limit the consequences of this sentence, the advisory board of the Consortium, with a plebiscitary vote, decided to continue the stock under a voluntary approach. It was also requested to the government to propose a new law of the mandatory stock in agreement with the Constitution. Meanwhile the State Council, with decision

No. 218 on July 7, 1964, recognized the Consortium as a public institute without lucrative purposes but of general interest. The Consortium would be under the supervision of the Ministry of Agriculture, who would control the budget, can break up the Administration Board, and nominate a Commissioner.

Following the suppression of the mandatory stock of the oil, the Consortium had to face the international market with the competition of numerous irregular sellers and large quantities of nongenuine essential oils. As a consequence, the demand of pure genuine essential oil drastically decreased, and therefore the price dropped.

This situation reached its apex in the crisis of 1967–1968 when the warehouse of the Consortium stocked an incredible amount of unsold essential oil (100,000 kg) with debits for £1,500,000,000 with the banks financing the conferred production. To deal with the sales drop, the producers founded on March 2, 1969 the new Cooperative "Il Bergamotto" to continue the activity of stocking new oil and at the same time petition the Italian government for help in this matter.

The Italian government thus immediately closed the Administration board of the Consortium and nominated a Commissioner with the decree D.M. del 6/30/1969. Successively on November 29, 1973 was issued law No. 835 establishing the mandatory character of the stocking and providing £500,000,000 to cover the debts of the Consortium with the banks and normalize this sector; the quality control became the duty of the Experimental Station for Citrus Essential Oils and Derivatives (Stazione Sperimentale per le Essenze ed i Derivati Agrumari) in Reggio Calabria (disposition confirmed later by the law No. 170 on May 4, 1983) and established administrative fines for those who did not respect their law.

As a consequence of art. 29 of the above-mentioned law, and of the proxy to agriculture given by the Italian government to the regions, Reggio Calabria issued law No. 7 on February 5, 1977 reintroducing the mandatory stock for the producers, reproposing the regulations on the quality control issued by the National Law, and giving freedom to the producers to directly trade up to 70% of the conferred oil.

This regulation, however, was insufficient for the defense of this sector, since the offer was pulverized by the lack of observance of mandatory stock in contrast with the monopolistic demand, and the consequent drop in price, which was unsatisfactory for the small producers of bergamot.

The nonobservance of the stock could be explained by the lack of financial tools of the Consortium, which left it incapable of handling the loans given to the producers.

A new law was issued by Reggio Calabria (No. 1 on February 14, 2000), facing the crucial problem of the institution of a Rotatory Founding, aimed at warranting the loans to the associated producers and to cover general expenses. The regulations on stocking and quality control on the entire production line are confirmed in concordance to the European Community regulations; the negotiations with transformer and exporter enterprises are regulated by the new Sector's Committee.

According to this last action, the protection and enhancement of the value of bergamot will be performed through:

• Giving the entire fruit production to the Consortium
• Concentrating the offer through the mandatory stock of the oil and every other derivative

- Certification of quality of the entire production line according to the actual regulations
- Issuing circulars regarding the production requirements for the improvements of the production of raw material (bergamot fruits) and the extraction processes
- The transformation of all the fruits by the Consortium's plants
- Trading the product through professional agreements
- Promoting the use of essential oil and other bergamot products through appropriate marketing policies
- Retraining the Consortium's staff
- Organizing professional courses for specialized operators
- Providing technical assistance to the enterprises
- Promoting studies and research on innovation of the technological processes for the preparation of traditional and innovative products
- Promoting intra- and inter-sector projects for the development, using every grant provided by public corporations from the government and by local authorities, along with every financial tool granted by the European Community (EC)
- Sustaining and participating in every enterprise aimed at the recovery of the added value from the use of the essential oil and of every other derivative of the transformation industry (perfumery, cosmetic, pharmaceutical food, etc.)
- Promoting and assisting associated companies in the use of every financial tool from the EC

To accomplish these purposes, the Consortium will count on the technical assistance, the promotion, and the experimentation provided by the Region, and on the technical-scientific collaboration of research facilities (University, Experimental Station for the Citrus Essential Oils and Derivatives of Reggio Calabria, the section in Reggio Calabria of the Experimental Station for the Citrus Culture in Acireale, Regional Research Centers, CNR, etc.) and of experts in the field.

However, the European Union still has not approved this regional Law, because it is in contrast with the EC regulations of free competition, and invites Reggio Calabria not to financially support the Consortium in its economic, industrial, and commercial activity since these are considered government aids.

Reggio Calabria, after long discussions in the Council Committee, between the agriculture professional organization and the industrial ones, finally promulgated the law No. 41 on October 14, 2002 entitled "Regulation for the protection of culture and quality of production of bergamot. Regulation of the Consorzio del Bergamotto." With this law the Consortium is defined as a public corporation that represents the bergamot producers and is dedicated to the promotion, improvement, and valorization of the essential oil and of every other derivative of bergamot. These aims are attained through the professional skills acquired by the operators in this field and through the assistance given to the companies operating in the production line to apply the appropriate financial tools provided at regional, national, and communitarian levels. The regulations also gave the Consortium the role of collaborator for the agricultural program in the sector through the realization of strategic plans.

The true regulatory innovation is reported in subsection 1, art. 14, law 41, literally reporting:

> In order to encourage the survival of bergamot cultivation, particularly in relation to the protection of the landscape, of the environment, of the historical function and of the territory, the Region recognizes, in favor to every bergamot producer associated or not to the Consortium, a financial aid proportional to the surface cultivated, in agreement with the specific regulation approved by the Regional Council, within 4 months from the approval of the present Law.

The rules of the Consortium's activity, according to art. 3 of the above-mentioned law L.R. 41 and to art. 34 of the EU Treaty on March 25, 1957, 57/01/TI, are reported in detail by the Regional regulation No. 10 on April 8, 2009 published on BURC on July 8, 2009, str. Suppl. n.1 BURC part I and II No. 14 on January 8, 2009.

It is undoubted that law No. 41 has deeply changed the aims and activities of the Consortium compared to the past, however giving to this institution a fundamental role for the development of bergamot and its derivatives in the safeguard of the environment and landscape of the territory, one of the most suggestive in Italy.

The role of Consortium for the safeguard and improvement of bergamot was already foreseen in the National Law No. 39 where the Denomination of Controlled Origin (DCO) was immediately given, while the Denomination of Protected Origin (DPO) was proposed by the Italian Agricultural Policy Minister of the EU.

The law mentioned above also foresaw an economic aid of 32 million Italian Lira for the following interventions:

- Expansion of the cultivated fields within the optimal area, in substitution of other Citrus species, to contribute to the limitation of carbon dioxide production in the atmosphere and to the improvement of climatic and environmental conditions.
- Re-grafting, re-grafting with tinning out, simple tinning out of plantations.
- Development of nurseries (in greenhouse cultivation) and company policy.
- Constructing rural buildings.
- Realizing infrastructures of small and medium sizes designed to reduce production costs and favor the restart of cultural practices.
- Building new transformation plants and commercial structures.
- Carrying out studies and research and providing technical assistance.
- Promoting the commercial sector.

Unfortunately the law has not applied since it not compatible with EU regulations.

5.3 FUNDAMENTAL RULES FOR THE COMMERCE OF BERGAMOT OIL

Regarding the quality control and the genuineness assessment of bergamot oil, law No. 170 passed on May 4, 1983 ruled on the export of citrus and their derivatives, authorizing the Experimental Station for the Industry of Citrus Oils and Derivatives located in Reggio Calabria to issue a free certificate of analysis for bergamot oil to be exported. In this case was confirmed the regulations given in law No. 1089 on July 18, 1930, and in R.D.L. No. 278 on February 3, 1936 converted into law No. 829 on April 23, 1936.

These regulations in practice require that only the essential oil of bergamot exported in communitarian countries must be certified pure.

A very useful regulation to relaunch and bring up the value of bergamot is the No. 509/2001 issued on March 15, 2001 in which the European Union established the DPO of bergamot essential oil of Reggio Calabria. This regulation also foresees the constitution of the Consortium for the safeguard and of an Institute for the control, connected to the historic Consorzio del Bergamotto, to protect and respect rigorous rules of production so as to offer on the market a genuine and high-quality essential oil. For this reason the Ministry of the Agricultural Policy with the decree 11/15/2005 (G.U. 11/29/2005, year 146, No. 278, p. 37) and the decree 12/21/2005 (G.U. 1/2/2006, year 147, No. 1, p. 22) indicates the Experimental Station for Citrus Oils and Derivatives as the public authority for the official control on the DPO "olio essenziale di bergamotto di Reggio Calabria" registered by the EU according to the Regulation CEE No. 2081/92.

The final enhancement of the value of every product obtained from bergamot is given to the Higher International Institute of Cosmetic Perfumery and Food Flavors, funded by art. 3 of the Italian State Law No. 246/89, known as the Reggio Decree. This Institute, at this time under construction in the city of Reggio Calabria, will be realized in buildings owned by the Consorzio del Bergamotto, and will be used as a higher education school for perfumers and other professionals in the field. It will also have a pilot center for the production of innovative fragrances and a museum.

REFERENCES

SCIENTIFIC PAPERS AND BOOKS

Crispo F. 2011. Personal communication.
Crispo F., and Dugo G. 2001. Il bergamotto: coltivazione, tecnologie di estrazione, aspetti storici, economici e legislativi. *Atti Congresso Internazionale Bergamotto 98,* Reggio Calabria: Laruffa Editore, 170–174.
Cucuzza G., and Timpanaro G. 1996. L'industria di trasformazione degli agrumi minori in Italia. Catania: Università degli studi di Catania.
Di Giacomo A. 1989. Il bergamotto di Reggio Calabria. Reggio Calabria: Laruffa Editore.
Focà A. 2001. Sull'azione antimicrobica dell'essenza di bergamotto. Catanzaro: Università degli Studi di Catanzaro.
Les exportations d'agrumes du bassin meditarrene, Comité de Liaison del' Agrumiculture Méditerranéenne. *Annual reports,* 2000–2010.
Mollace V. 2011. Hypolipemic and hypoglycemic effect of bergamot juice and bergamot polyphenolic fraction (BPF) in patient with metabolic syndrome. *First International Conference Method of Antioxidant Derivatives in Health and Diseases: Focus on Bergamot.* Reggio Calabria. September 24, 2011.
Nesci F. S., and Sapone N. 2011. *Tutela e sviluppo del bergamotto reggino—32nd Congresso Italiano di Scienze Regionali: "Il Ruolo delle Città nell'Economia della Conoscenza."* Torino, September 15–17, 2011.
Spinelli R., and Sandicchi M. 2000. Il bergamotto ed altri agrumi in gastronomia. Reggio Calabria: Laruffa Editore.

RULES AND LAWS

Decisione Consiglio di Stato No. 218 on July 7, 1964. In Tripodi F. 1993. Il Consorzio del Bergamotto di Reggio Calabria ed il suo Archivio. *Rassegna degli Archivi di Stato* 53 (2,3): 305.

D.L. No. 367 on April 26, 1945. In Tripodi F. 1993. Il Consorzio del Bergamotto di Reggio Calabria ed il suo Archivio. *Rassegna degli Archivi di Stato* 53 (2,3): 304.

D.M. September 14, 1934, D.M. June 5, 1936, D.M. August 26, 1936 (Ministero dell'Agricoltura). In Tripodi F. 1993. Il Consorzio del Bergamotto di Reggio Calabria ed il suo Archivio. *Rassegna degli Archivi di Stato* 53 (2,3): 301.

D.M. May 29, 1946 (Ministero dell'Agricoltura). G.U. No. 135, June 21, 1946.

D.M. June 30, 1969 (Ministero dell'Agricoltura). In Tripodi F. 1993. Il Consorzio del Bergamotto di Reggio Calabria ed il suo Archivio. *Rassegna degli Archivi di Stato* 53 (2,3): 305.

D.M. November 15, 2005 (Ministero delle Politiche Agricole e Forestali). G.U. No. 278 on November 29, 2005.

D.M. December 21, 2005 (Ministero delle Politiche Agricole e Forestali). G.U. No. 1, 1 February 2006. Rectifying D.M. November 15, 2005 on "designazione della SSEA quale autorità pubblica incaricata di effettuare sulla DOP olio essenziale di bergamotto di Reggio Calabria registrata in ambito UE ai sensi del Regolamento CEE No. 2081/92."

Decreto Prefettizio No. 3942 on November 11, 1931. In Tripodi F. 1993. Il Consorzio del Bergamotto di Reggio Calabria ed il suo Archivio. *Rassegna degli Archivi di Stato* 53 (2,3): 300.

Decreto Prefettizio No. 496 on December 11, 1943. Nomina Commissario prefettizio dell'avv. Basilio Catanoso. In Tripodi F. 1993. Il Consorzio del Bergamotto di Reggio Calabria ed il suo Archivio. *Rassegna degli Archivi di Stato* 53 (2,3): 304.

Law No. 1089 on July 18, 1930 converting the R.D.L. No. 438, March 31, 1930. In Tripodi F. 1993. Il Consorzio del Bergamotto di Reggio Calabria ed il suo Archivio. *Rassegna degli Archivi di Stato* 53 (2,3): 300.

Law No. 829 on April 23, 1936 converting the R.D.L. No. 278, February 3, 1936. In Tripodi F. 1993. Il Consorzio del Bergamotto di Reggio Calabria ed il suo Archivio. *Rassegna degli Archivi di Stato* 53 (2,3): 304.

Law No. 835 on November 29, 1973. In Tripodi F. 1993. Il Consorzio del Bergamotto di Reggio Calabria ed il suo Archivio. *Rassegna degli Archivi di Stato* 53 (2,3): 307.

Law No. 170 on May 4, 1983. G.U. No. 135 on May 11, 1983.

Law No. 246 on July 5, 1989. G.U. No. 175 on July 7, 1989.

Law No. 39 on February 25, 2000. G.U. No. 52 on March 3, 2000.

Law Regione Calabria No. 7 on February 5, 1977. In Tripodi F. 1993. Il Consorzio del Bergamotto di Reggio Calabria ed il suo Archivio. *Rassegna degli Archivi di Stato* 53 (2,3): 311.

Law Regione Calabria No. 1 on February 14, 2000. *BURC* No. 10 on February 21, 2000. Supplement Extraordinary, parts I and II.

Law Regione Calabria No. 41 on October 14, 2002. *BURC* No. 19 on October 17, 2002. Supplement Extraordinary No. 1 parts I and II.

R.D.L. No. 1330 on October 15, 1931. In Tripodi F. 1993. Il Consorzio del Bergamotto di Reggio Calabria ed il suo Archivio. *Rassegna degli Archivi di Stato* 53 (2,3): 300.

Regulation CE No. 509 on March 15, 2001. GUCE, March 16, 2001.

Regional Regulation of the Regione Calabria No. 10 on August 4, 2009. BURC, August 7, 2009. Supplement extraordinary No. 1 parts I and II.

Sentence Constitutional Court No. 54 on June 14, 1962. In Tripodi F. 1993. Il Consorzio del Bergamotto di Reggio Calabria ed il suo Archivio. *Rassegna degli Archivi di Stato* 53 (2,3): 305.

ABBREVIATIONS

D. M.: Ministry Decree
Legge: National Italian Law
R.D.L.: Royal Decree-Law
BURC: Bollettino Ufficiale Regione Calabria (Official Bullettin Calabria Region)
G.U.: Gazzetta Ufficiale Italiana (Italian Official Gazette)
GUCE: Gazzetta Ufficiale della Comunità Europea (European Union Official Gazette)

6 Transformation Industry

Angelo Di Giacomo and Vilfredo Raymo

CONTENTS

6.1   BERGAMOT FRUIT

Bergamot fruit is quite similar to other *Citrus* fruits. Specifically, from the outside to
the inside, the following layers are present (Di Giacomo 1974, Figure 6.1):

- A deep layer of epidermal cells.
- The epicarp or flavedo, consisting of a tissue rich in pigments (chlorophyll and carotenoids); in the epicarp, more precisely in the layer immediately under the surface of the fruit, oil glands containing the essential oil
  are irregularly distributed. The glands, with a diameter between 0.4 and
  0.6 mm, are surrounded by cells containing a water solution rich in salts,
  sugars and colloids. When the peels come into contact with water, the higher
  osmotic pressure of the liquid inside the cells causes an additional entrance
  of water, leading to an increase in pressure on the glands. In such conditions, a mechanical action on the peel, *viz.*, pressure or grinding, causes the
  rupture of the glands and the release of the essential oil. Additionally, the
  latter is not only located in the epicarp of the fruit; essential oil can also be

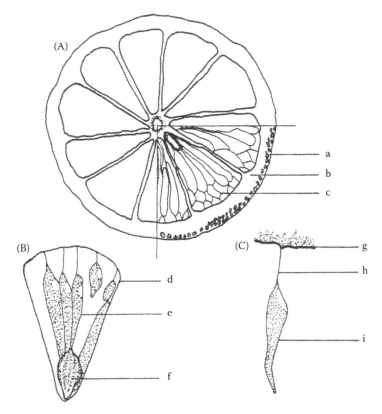

FIGURE 6.1   Section of bergamot fruit. (A) Transverse section. (a) epicarp; (b) mesocarp
[albedo]; (c) endocarp, formed by several sectors. (B) Segment. (d) layer; (e) glandular hair;
(f) seed. (C) Single glandular hair. (g) external layer; (h) peduncle; (i) vesicle containing the juice.

found in the flowers, in the leaves, and in the small branches, as well as in glands than can be found both outside and inside the vegetable cells.

- The albedo is formed of cells of irregular shape, spongy, and of white color, with large inter-cell spaces; the albedo contains mainly cellulose (6.6%–8.2%), pectin (4%–5%), and sugars (8.8%–11%).
- The endocarp is formed of segments distributed around a central axis, the latter having a similar composition to the albedo. The segments, which are enclosed in a thin membrane, contain small juice vesicles, themselves isolated by very thin walls. The seeds or pips are also located inside the endocarp, and can be found often close to the central axe. There are also some small oil glands containing a volatile oil known as recovery essence.

6.2 PRELIMINARY HANDLING OF THE FRUITS

6.2.1 HARVEST

The picking of *Citrus* fruits and, more specifically, those like bergamot for which the production of essential oil is of major importance, must be done very carefully, since the processing will be performed on the peel. Therefore, picking by hand is the best way to preserve the peel for optimum processing and yield. It is also highly important that fruit handling and transportation to the processing industry are carried out by expert operators.

6.2.2 TRANSPORT AND UNLOADING

Transportation to the industrial plant must be carried out in the best conditions to minimize peel damage that would cause breakage of the oil cells. Specifically, wooden or plastic boxes with a 20-kg capacity, or larger containers with a 400-kg capacity, are preferred over the direct use of the bulk truck volume. The bergamot fruits, once in the plant, are first visually checked in order to eliminate those fruits that are unsuitable for processing (for instance, overripe fruits).

6.2.3 STORAGE

It is usually advised to limit the period of storage prior to *Citrus* processing. If the storage period is excessive, the peel will become drier, the release of oil from the cells would become harder, and the oil yield would be reduced.

With regards to bergamot, advantages have been observed in storing the fruit between 24 and 48 hours at a temperature of 15°C before processing. Under such conditions, the oil is recovered more easily, the emulsion is of higher fluidity, and centrifugation is easier. In bergamot processing, storage is normally carried out using the same boxes employed for picking and transport, thus considerably reducing the damage that occurs when the fruit is stored in large containers. In any case, processing should not be performed more than four or five days after picking.

6.2.4 SAMPLING

For each batch of fruit delivered to the industrial plant a representative sample is taken, with the aim of checking a series of important parameters necessary to measure both the oil and juice content of the fruit. The easiest way to measure the oil content is to steam-distill the fruit, previously homogenated, with added water (Clevenger method).

6.3 PROCESSING DIAGRAM

As a first step, the fruit is washed to remove leaves and dust. The washing is normally carried out for some minutes, in a tank and under stirring. Fresh water is constantly added.

Usually for bergamot processing, washing with plain water is not followed by brushing, as for other citrus fruits; moreover, sanitizing solutions are not employed (for instance, Cl-based ones). Both the washing and the rasping are carried out using water sprays. Sometimes, the rasped fruit is further washed with water prior to juice extraction in order to reduce the oil content and the bitterness. The recycled oil emulsion could also increase the microbiological total count of the juice. A general flow chart of bergamot processing (oil extraction, juice extraction, and use of the exhausted peel) is shown in Figure 6.2. The various processing steps are described hereafter.

6.4 OIL EXTRACTION

In order to recover the citrus oils through cold processing (cold-pressed oils), three steps are involved (Rodanò 1930):

- Rupture of the epidermis and of the oil glands that contain the oil
- Creation in the peel of compressed areas, surrounded by areas which are under less pressure, through which the oil can be expressed
- Abrasion of the peel, producing small pieces of debris

The cold-pressed oil can be extracted from both the halved-fruit peel (after juice extraction), or from the entire fruit. When extracting oil from the halved-fruit peel, the first two steps in the list are performed, while when the entire fruit is used, all three steps occur. The extraction system can be applied to the entire fruit, or to the peel, and be classified as manual or mechanical (Spinelli 1967). Today, for obvious economic reasons, manual methods have been totally abandoned. Among the mechanical systems, those that act on the peel (sfumatura and torchiatura) are not used for bergamot oil. In fact, today only extractors that treat the entire fruit are used, specifically the "Speciale" and "Moscato" extractors. Apart from these, mention will be made of extractors that have been used in the past ("Calabrese" machine, Avena, Indelicato MK), along with others recently tested on an industrial scale (FMC In-Line and BOE) with interesting results and that might be used in the future.

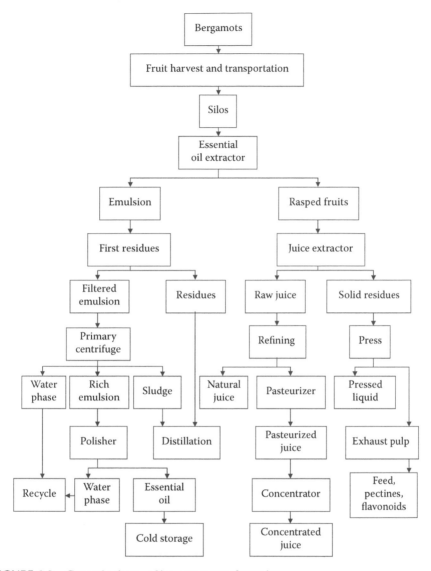

FIGURE 6.2 General scheme of bergamot transformation.

6.4.1 HAND OIL RECOVERY

When Rodanò edited his famous text on citrus derivatives in 1930, bergamot was already mechanically processed. As previously stated, today hand extraction techniques are no longer used to extract the oil from the fruit. Nevertheless, apart the historical interest, hand-derived oils still represent an ideal product. Consequently, manual methods will be herein briefly described. In the "spugna" process (natural sponge), the fruits are cut into halves. The pulp is then removed with a special spoon, and the obtained half is washed with water containing lime, then left for up to 24 hours on grates to drain the water. Such a treatment both neutralizes the acidity in the peel and makes the peel harder, in order to facilitate the release of the oil. The spugna process consists of hand compression, with a rotating movement, of the peel against the natural sponge in a "concolina," which is a terracotta or copper bucket. The oil drops into the bucket from the sponge and is separated, by decantation, from the solid residue. The latter is formed by breakage of the skin tissue.

Another method was defined as "scodella" and was carried out on the entire fruit. Originally, the scodella was made of wood with nails inside. Later the shape was modified into a copper funnel, closed at the bottom, with the internal walls covered by brass needles. The operator kept the scodella between his legs, pressing the fruit delicately and rotating it against the nails. The oil dropped, along with the solid and liquid residues of the peels, to the bottom of the funnel and was separated by decantation.

6.4.2 "Calabrese" Extractor

The "Calabrese" machine was first built in 1844 by Luigi Auteri and Nicola Barillà, and then improved in 1875 by Domenico Carbone (Catanea 1928) (Figure 6.3). In 1931, La Face reported a different model, defined as "modern," also known as the "Polimeni" machine. This extractor consists of two cups (Figure 6.4) with a 30-cm diameter. Fruits are pressed and rasped between the cups, thus allowing the oil to sprinkle. The oil, together with water and peel parts, flows through the lower cup into a bucket. The lower cup is fixed to the machine and presents nails, while the upper one rotates around a vertical axis, pressing and rasping the fruit. To feed the extractor, the upper cup is lifted and the right number of perfect fruits is put onto the lower cup. The upper cup is then closed. At the end of processing, the upper cup is reopened and the fruits are taken out, dried with a sponge to absorb the residual oil, and then sent to the screw press for juice extraction. The number of fruit introduced per extraction ranges from 6 to 10, depending on their size. The loading, processing, and unloading normally takes 15 seconds; therefore the machine can process about 65 kg of fruit per hour. The fact that the extractor works without water is a key factor in the high quality of the oil obtained when compared with other techniques. However, it cannot be denied that problems did arise that at the time could not be overcome, such as the reduced machine capacity and the need to sort the fruit prior to processing. In fact, if fruits with different sizes are introduced into the extractor, the little ones are not rasped; on the other hand, if the pressure is increased, the bigger fruits are damaged, causing an important yield loss of yield. An additional issue is that the capacity of the machine is very low when processing ripe fruits or fruits with irregular shapes.

In the following years, new extractors were introduced that differed from the Calabrese system in terms of improved mechanics, a larger capacity, and easier loading and unloading processes. In 1926, Ugo Giovannelli patented a machine with a moving lower cup. Under the rasping plate three cups linked to each other move from one point to the next. In this manner, it was possible to load and unload the fruit while processing without opening the extractor. The machine could therefore process 50% more fruit than the standard Calabrese machine.

The machine named "Franco" is characterized by bigger cups with a 45-cm diameter, thus allowing 30% more fruit into the extractor compared to the standard machine. In the "Martorana" machine, the chains move (rotation) two extraction

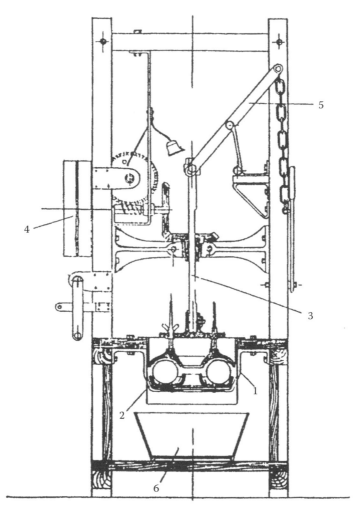

FIGURE 6.3 Section of the "Calabrese" machine. (1–2) cups; (3) vertical axis; (4) shaft; (5) lever; (6) accumulation basin.

chambers. Fruits are introduced in the side and then automatically discharged. This machine can process double the quantity of the standard one, with much lower power consumption.

Of course, all of these machines are no longer employed.

6.4.3 MACHINE "A SBATTIMENTO"

These machines, or dry "Pelatrici" (La Face 1955, 1956), were introduced in Calabria during the 1944–45 season and were used by producers that had a limited quantity of fruit available. The machine can be considered somewhere between the Calabrese and the Pelatrice machines. In fact, the Pelatrici machines could process two or three times the quantity of fruit of the Calabrese machine, with half the

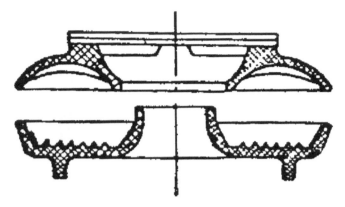

FIGURE 6.4 Detail of the cups of the "Calabrese" machine.

hand-work requirements. The most interesting examples of these machines were built by Romanò (Pellaro, RC), Leonardo and Foti (RC), and Laganà (RC). Such machines were based on the principle of fruit rasping, without a sizer, in a vertical cylindrical chamber characterized by nails on the bottom and on the sides. During processing, the peel is rasped and the oil sprinkles, together with small parts of peel, and is recovered underneath. The obtained raw material is put into woolen bags and then pressed to recover the oil. Such extractors are no longer employed.

6.4.4 "Speciale" Machine

The automatic "Pelatrice" built by Speciale (Figure 6.5) allows oil extraction from the entire fruit (La Face 1958). These are the main models:

- Model DS10   700–1200 kg/hr
- Model DS20   1200–1600 kg/hr
- Model DS40   2500–3500 kg/hr

The machine consists of a certain number of rotating cylinders and an internal cochlea. Both the rollers and the cochlea present stainless steel nails. The rotating rollers are fixed on the bottom of the machine, in a half circle, and the fruit is pushed ahead by the cochlea. The fruits are fed inside the chamber under a spray of water. In the extractor, both the speed of the rollers and of the cochlea can be varied, thus modifying the processing time of the fruit. The quantity of fruit each extractor can process also depends on the number of rotating rollers. All the parts directly in contact with the oil/water emulsion are made of stainless steel. The emulsion is filtered through a net in three steps and then sent to the centrifuge. The "Pelatrice Speciale" is the most popular extractor for bergamot in Calabria.

6.4.5 "Indelicato" Machine

The "Sfumatrice Indelicato MK," built by the Indelicato brothers in Giarre (Sicily), processes the entire fruit. In this extractor the fruits are pricked by nails, and the peel remains

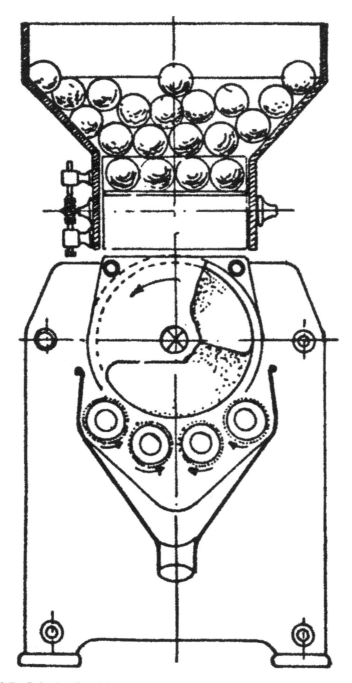

FIGURE 6.5 Pelatrice Speciale.

almost undamaged. The extraction of the oil takes place between two vibrating layers, characterized by many nails. The fruits are pushed by small mechanical hands through the two layers for a time between 15 and 200 seconds. The vibration action pushes the fruits against the nails, thus allowing the oil to sprinkle out under sprays of water. As usual, after filtration, the rather clear emulsion is subjected to centrifugation. This extractor, owing to its poor yield, and after a series of industrial tests, has been abandoned.

6.4.6 PELATRICE AVENA

This extractor, subjected to a patent, was built in 1924 by Giuseppe and Placido Avena in Pistunina (Messina). It is characterized by a round chamber, with two rotating plaits, and has a shape similar to that of a centrifuge (Figure 6.6). The rotating plates are covered with pyramidal stainless steel nails; additionally, the inside of the chamber is covered up to a height of 22 cm with glass nails. The extractor works as follows: the fruits are loaded into the chamber, which is divided into two parts and

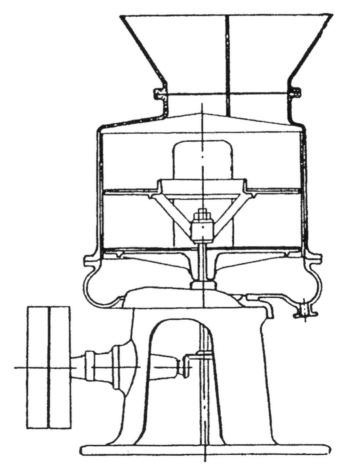

FIGURE 6.6   Pelatrice Avena: section.

has a bottom part that can be opened. Each chamber contains a quantity of fruit corresponding to one plate. Once the lower holes are opened, the fruits are distributed on the plates. As an effect of the centrifugal force they are thrown against the nails. The rasping effect depends on the speed, on the time the fruits are kept inside the chamber, and even on the length of the nails. In the extractor, a spray of water is employed to recover the oil and the peel residues. Once the extraction is over, the rasped fruits are taken out and the machine is loaded again (La Face 1958).

6.4.7 MACHINE MP-C3 (C.M.A., COSTRUZIONI MECCANICHE AGRUMARIE, PELLARO, RC)

This extractor, an automatic centrifugal Pelatrice with three elements, is similar to the Pelatrice Avena (Di Giacomo and Mincione 1994). It is made of stainless steel (apart from the mechanical devices) and has a capacity of 1.9–2.4 tons/hr of bergamot. It is normally fed with an elevator, which transports the fruits from a warmed water tank in which the temperature can be adjusted. The speed of the extraction process is regulated through specific instrumental variations. The three rotating elements are located one after the other, and allow the rasping of the fruits, which must go through different rotating directions from the first element to the third one. The filtration unit is characterized by three elements rotating on the same axis equipped with stainless steel nets of decreasing pore sizes.

6.4.8 FMC IN-LINE EXTRACTOR

This machine can extract the juice and the oil at the same time (Figure 6.7). The machine presents upper and lower cups. The upper ones are moved upwards and downwards, and are characterized by stainless steel fingers that maintain the fruit

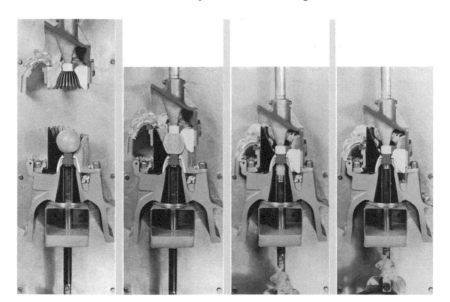

FIGURE 6.7 (**See color insert.**) Extractor FMC In-Line: scheme.

inside the cups during processing. The fruits are first sorted on a sizer and then reach the lower cup through a feeder which is synchronized with the movement of the upper cup. The extraction begins with the fruit compressed between the cups; the lower cup has a tube that is a knife on the upper side, followed by a net; the tube penetrates the fruits, allowing the juice to flow through the net as the upper cup presses the fruit onto the lower one. The stainless-steel fingers break the oil cells, causing the oil to sprinkle out. At the same time, water sprayed from a dedicated upper ringer covers the oil, forming an emulsion. The oil content in the emulsion varies from 0.9% to 1.1%, depending on the kind of fruit, the oil content, and size.

6.4.9 Brown Extractor (Brown International Corporation, Covina, CA)

This type of extractor has been used for bergamot oil extraction only during tests in Sicily. Since such tests have been successful, both in terms of yield and of the quality of the oil recovered, the characteristics of such an extraction system are worthy of description (Di Giacomo and Di Giacomo 2002). The Brown oil extractor (BOE) (Figure 6.8) is exploited for oil recovery. The fruits are first washed, and then after the BOE step, are further processed to extract the juice. The BOE receives fruit through an elevator that insures a constant quantity of fruit. The extractor is characterized by approximately 3,000,000 stainless steel nails, fixed on rotating rollers, which prick the peel. A speed differential, applied to the rollers, allows a variable intensity of extraction. The rollers, in order to perforate the entire surface of the fruit, also move in a horizontal manner. The process of extraction occurs in a few centimeters of water bath to avoid oil losses. Once processed, the fruit is passed on to a series of smooth rollers to recover the residual water-oil emulsion. The emulsion is then filtered and sent to the centrifuge to recover the cold-pressed oil.

6.5 OIL SEPARATION

When the oil is extracted by the Calabrese machine, the oil-water emulsion and the solid residues are similar to those obtained by manual operation ("scodella, spugna").

FIGURE 6.8   (See color insert.) Brown oil extractor (BOE).

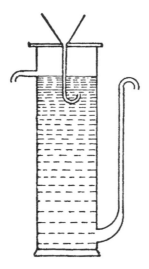

FIGURE 6.9 Florentine vase.

Therefore, the different phases can be separated easily and rapidly through decantation (Figure 6.9): the upper layer is the oil that can be recovered through a simple filtration on paper; the middle layer is an emulsion of water and oil and contains suspended colloids; the lower layer does not contain any oil and therefore is of no interest. To recover the oil contained in the middle layer, the emulsion is absorbed on a natural material (sponge or wool) and manually pressed in a screw press. The liquid obtained is then decanted to recover the oil. Such a procedure has been abandoned, due to the development of more modern extractors, primarily the pelatrice system. The latter requires large volumes of water, and above all, the use of a centrifuge to recover the oil. The water-oil emulsion, containing solid parts, undergoes the following steps:

- Separation of the solid parts in a rotating filter (sieve)
- Primary centrifugation (disc centrifuge; Figure 6.10) to eliminate the solid parts and most of the water; an enriched emulsion containing at least 30% of oil is attained, along with a water phase that is recycled and directed to the extractor
- Secondary centrifugation in a polisher to recover clear oil, with no water

Some processing systems recover the oil with just one centrifugation, but since the water phase can still contain a certain amount of oil, it is nonetheless passed through a second centrifuge, with the water phase then recycled. The recycled water can solubilize the oxygenated components (above all, alcohols and esters) that are very important for the "bouquet." In fact, these components contain hydrophilic groups whereas the terpene hydrocarbons are strongly hydrophobic.

6.6 STORAGE

In most cases, the shelf life of bergamot oil is 24 months if the oil is kept between 8°C and 10°C and in full airtight containers. It is in fact mandatory that the oil be

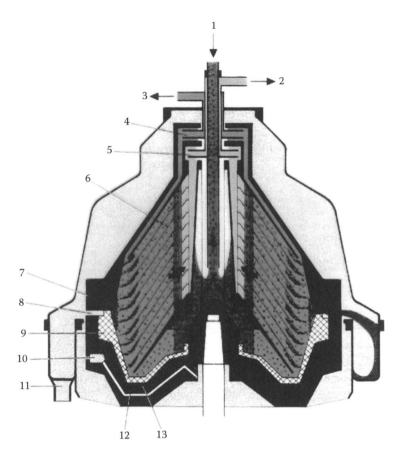

FIGURE 6.10 Round chamber with two rotating plaits. (1) feed; (2) discharge, oil; (3) discharge, water; (4) centripetal pump for water; (5) centripetal pump for oil; (6) discs; (7) sediment holding space; (8) sediment ejection ports; (9) sliding piston; (10) piston valve; (11) solids discharge; (12) opening water duct; (13) closing chamber.

protected from the action of oxygen, as well as the light and high temperatures. It is also important to eliminate traces of water and juice and to avoid contact with iron and copper that catalyze oxidation and color change. The quality of the oil is altered through rather complex processes, enhanced by the temperature, and through reactions of oxidation, reduction and polymerization, generates new components. After centrifugation, traces of water and juice should be eliminated with sodium sulphate anhydrous or sodium bicarbonate. Stainless steel tanks have long ago replaced copper tanks. Until the tanks are kept full, the action of oxygen is negligible. Once some oil is withdrawn from the tanks, it is mandatory to fill the empty space with inert gas (nitrogen, carbon dioxide).

To avoid the oil becoming hazy and/or the precipitation of waxes, it is advisable to apply a de-waxing process in advance. Such a procedure can be carried out in specific jacket tanks, cooled down to the temperature of $-20°C$ for at least seven days. After decanting, the oil is centrifuged and put back into the tank. The containers for the final users normally consist of iron drums or tins, either lacquered or tinned. Even aluminium or stainless-steel drums are sometimes used.

6.7 DISTILLATION

Distillation is one of the most popular ways to recover essential oil. Nevertheless, for citrus fruits this technology is used only in specific cases (e.g., lime), since cold pressing is the most common production procedure. Distilled oils, even if obtained at a very low temperature, are more fragile than cold-pressed oils. This is due to the lack of nonvolatile components, some of which have an odor impact, whereas others are characterized by antioxidant properties that guarantee a longer shelf life. For bergamot distilled oils different names are used, on the basis of the starting raw material:

- Bergamottella distilled oil, obtained from the unripe little fruits that fall down during the summer
- Feccia oils, obtained by distillation from the solid and semifluids residues of the cold extraction by the Calabrese and Pelatrice machines
- Peratoner oil, obtained by distillation at reduced pressure from the residues and the water phase obtained by the Pelatrice extraction at the end of each day of production
- Distilled juice note or recovery essence obtained by distillation from the single-strength juice

The distillation unit (Figure 6.11) is characterized by three main parts: the still, the condenser, and the separation unit of the water-oil emulsion. The still and the

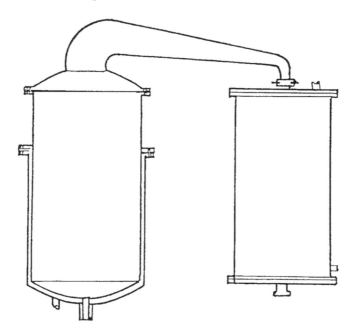

FIGURE 6.11   Distillation apparatus.

condenser are linked through the so-called swing neck. The units exist in various shapes and sizes, with fixed or movable stills. Steam bath distillation has been often used to recover the oil from the flowers and the leaves. The distilled oils obtained are known as "neroli," from the flowers, and "petitgrain" from little branches, leaves, and unripe little fruits. To produce petitgrain oils, in most cases the little branches are obtained during the pruning. Many factors influence the yield and the quality of the petitgrain oil (La Face 1942). Important losses arise when distillation is applied for an excessively long time. Even the time of the year is important to attain the highest yield. Among the other factors that influence the yield and the quality of the petitgrain oil are the age of the tree, and the time between the leaf collection and the beginning of the distillation. Lower yields are typical for old trees and dry leaves.

6.8   TERPENELESS OILS

The main reasons terpeneless oils are produced are:

- The quantity of oxygenated compounds responsible for the characteristic odor of the oil is increased.
- The solubility of the oil in alcohol and other solvents used for food and fragrance production is increased.
- The stability of the product is increased, since the unstable terpene hydrocarbons are deleted.

The distillation unit commonly used to produce terpeneless oils is made of a stainless-steel pot still jacked for the steam circulation; a serpentine for cooling; a fractionating column with Rashing rings; a reflux system with water cooling; a steam tube; a condenser; and two collectors to recover the distilled oil.

During processing, the entire system operates under a vacuum. The processes that separate the fractions rich in esters and alcohols are:

- Distillation of the monoterpene hydrocarbons to obtain the folded terpeneless oils
- Steam bath distillation of the folded oil to eliminate the waxes and the nonvolatile residues
- Final concentration of the terpeneless oils to the required specification

In truth, the production of terpeneless bergamot oil is rather small compared with other essential oils, since bergamot oil is already very rich in oxygenated compounds. In fact, the yield in terpeneless oil is 33%–44% of the starting bergamot oil, whereas it is 3.5%–6% and 1%–3%, respectively, for lemon and orange oil (Di Giacomo 1974). The terpeneless bergamot oil still containing the waxes and nonvolatile residues is normally referred to as "folded."

In order to avoid or reduce the time and the exposure to heating, in more recent years new technologies have been developed to produce citrus terpeneless oils, such as:

- Extraction with alcohol
- Counter-current extraction with two nonmiscible solvents

- Use of two solvents, one polar and the other apolar
- Chromatographic separation on a specific phase, with the use of solvents to recover the oil

To the best of our knowledge, none of these approaches has had an industrial application.

6.9   EXTRACTION WITH SUPERCRITICAL LIQUIDS

Processes using solvents under supercritical conditions allow the extraction of a series of components in conditions of low temperature and are therefore extremely interesting compared to the traditional methods (Moyler 1993; Poiana et al. 1997). These are described in detail in Chapter 10, this volume.

6.10   BERGAPTENE-FREE BERGAMOT OIL

Bergamot oil with a reduced content of bergaptene, also referred to as bergaptene-free, has been widely requested on the international market for the well-known problem of phototoxicity. Some final users themselves apply a process to eliminate bergaptene, prior to use of the oil in perfumes. The process used in Italy is based on the treatment of the oil with an alkaline solution. The concentration of the alkaline solution, the stirring system, and the contact time between the two liquids are the parameters that, if correctly set, can reduce considerably the bergaptene content without deeply modifying the quality of the oil. A high-quality, bergaptene-free oil should not contain more than 30 ppm of bergaptene, even though such a limit depends on the requirements of the final users.

There are three main characteristics of bergaptene-free oil, in order for it to be used in the fragrance field (Di Giacomo and Calvarano 1978):

- The odor must be as close as possible to the crude oil.
- The shelf life must be maintained.
- The analytical chemical and physical characteristics and parameters should not be modified compared to the crude oil, apart from the bergaptene content.

A typical process can be described as follows: to one volume of cold-pressed oil, the same volume of a water solution of potassium hydrate is added. The emulsion is stirred for at least 12 hours at room temperature and the oil is then recovered by centrifugation. The recovered oil is then washed again with plain water and then centrifuged. A final washing step with water and citric acid can be carried out.

6.11   FUROCOUMARIN-FREE BERGAMOT OIL

For specific uses (colorless/light perfumes), it is necessary to totally eliminate the furocoumarins by physical means (distillation under vacuum). The modern units employed today operate at very low temperatures. With such units it is possible to obtain an oil that has almost the same characteristics of a cold-pressed oil. Nevertheless, the lack of natural antioxidant components, eliminated during distillation, makes such oils much more fragile when compared with their cold-pressed counterparts and, hence, they have a shorter shelf life.

6.12 JUICE EXTRACTION

There are different methods to extract the juice from the endocarp. The "Birillatrice" extractors reproduce hand squeezing. The quality of the juice obtained is very high, but the yield is very low. Some processors also employ screw pressing on the entire fruit, after oil extraction, but the quality of the juice recovered is very poor. At the time when the Calabrese machine was used, the juice was recovered by a wooden screw press after oil extraction and was employed to produce calcium citrate (La Face 1931). In a primary step, the fruit was chopped in a mill and the obtained liquid was subjected to the action of a wooden screw press. The screw press operates per batch. For each batch, the material obtained from roughly 200 kg of fruit was loaded, and it took an average of three hours to obtain the juice. Therefore, the throughput was roughly 70 kg per hour. Today, stainless steel screw presses are used to squeeze the fruit coming out from the Pelatrice machine. They consist of two screws, turning in opposite directions, pushing and pressing the fruit against the outlet. The juice is passed through a sieve and is then refined. Some processors pasteurize, and even concentrate, the juice. A typical finisher for citrus juice (FF50 model, from the Speciale company) is characterized by a screw that obliges the juice, under pressure, to pass through 0.5-mm holes (95 holes per cm^2).

Postorino et al. (2001) compared different extraction techniques applied to bergamot, in order to highlight the influence of the technology (Citrostar Model OM2 Bertuzzi and Speciale RS) on the quality of the juice. The Citrostar extractor is similar to the in-line FMC system and allows the extraction of oil and juice at the same time, with no contact occurring between the two products. In the extractor Speciale RS (Polycitrus style) the whole fruit coming out from the Pelatrice machine is sent into the feeder, where two counter-rotating cylinders spread it with nails, push the fruit against a fixed knife, and cut it into halves. The halves are further pushed from the cylinders against two fixed nets with 3-mm holes. The juice is released for the squeezing effect caused by the decrease of the distance between the cylinders and the net; the juice is collected in a tank underneath and then refined.

The bitter taste of bergamot juice is related to the content in naringin and can be reduced by debittering. For debittering, a Varind unit has been used (Calvarano et al. 1995). The plant is characterized by an ultrafiltration unit (cutoff: 10,000 Dalton), and then a stainless steel column containing Amberlite XAD16 resin, whose characteristics meet the FDA directive CFR 173.65. Through such a process it is possible to reduce the naringin content by 99.8%, if the volume of the juice treated is a maximum of 10 times the volume of the resin, and by 85.61% if the volume of the juice is 33 times that of the resin. Since the product is subjected to ultrafiltration, the recovered juice is crystal clear.

6.13 BY-PRODUCT

6.13.1 Exhausted Peel ("Pastazzo")

The exhausted bergamot peel amounts to roughly 50% of the fruit. It is often used fresh, for cattle feeding, and the analytical composition is similar to that of orange and lemon peel, though the fiber content is higher (Maymone and Dattilo 1958). A certain amount of exhausted peel is dried, both to produce pectine and for cattle feeding. Of course, volume is reduced and storage is possible. Prior to peel drying, it is necessary to use a screw press to reduce the water content. To carry out this step successfully, it is necessary to reduce the pectine content of the peel. Such an operation is performed by passing the peel through a mill, and then by adding lime. After a short contact, between 6 and 12 minutes, the semifluid material goes through a screw press that reduces the humidity to 70%–75%, and then is directed to a rotating oven. Often the peel is directly dried, through heating with gas or diesel oil. The liquid from the screw press is filtered and then concentrated up to 72°Bx in a multiple-effect concentration unit that, in some cases, also recovers the residual oil (stripper oil). The molasses obtained is added to the dried peel. The chemical composition of bergamot peel shows that it contains many interesting components, such as pectines and flavonoids, which could possibly be recovered.

6.13.2 Extraction of Naringin

Calvarano et al. (1996) described a process to extract and recover naringin from exhausted bergamot peel. The method allows the extraction of naringin and the separation of the other components of the peel that are soluble in hot water (sugar, organic acids, etc.). The exhausted peel can still be used to recover pectine. The peel is first passed in a mill, and then four volumes of water are added. After filtration, under vacuum, four more volumes of water are added, boiled for 16 minutes, and then filtered under vacuum. The liquid obtained in the two filtration steps is mixed together, concentrated under vacuum, and then purified by ultrafiltration and through a column containing aqueous resins. Column elution is performed with 30% ethanol (three to four times the volume of the resin). The solution obtained is evaporated under vacuum and left to allow the formation of the naringin crystals.

6.14 MODERN TECHNOLOGIES

Bergamot oil is a magnificent natural product, with a unique flavor and fragrance. Its composition, as compared to other citrus oil, is the secret of its properties. It contains only 50% monoterpene hydrocarbons, whereas most citrus oils have from 90% up to 98%. The remaining 50% is made up of alcohols, aldehydes, ketons, esters, sesquiterpens, and waxes. One of the main components is linalool, whose content can vary from 28% at the beginning of the season down to 3% in April, when the fruit is overripe. The early crop oil, rich in linalool, has a very intense green note, which has been much appreciated in the last decade by perfumers for fine fragrance applications and even as a flavor for the finest quality of Earl Grey tea. The other important component is linalyl acetate. This ester can start at only 20% at the beginning of the season and smoothly increase to as much as 34% by the end of the season. This component gives the classic sweet note that is characteristic of the ripe fruit.

In the 6% solid residue we find furocoumarins, coumarins, antioxidants, and many other unique components that make this oil a long-lasting raw material both from the shelf-life point of view and the dry-out of certain fragrances, which among citrus fruits only bergamot oil can guarantee.

One negative about these components is that in some people they can cause spots on the skin or skin irritation. Bergaptene content has been regulated by the IFRA for many years. Some allergens are considered as potentially dangerous, among them limonene, linalool, and citral. Nowadays furocoumarins and other allergens (such as limonene oxides) are under investigation. To continue to use safely bergamot oil in fine fragrances, new products are required that have a reduced amount of furocoumarins and allergens. At the same time, more soluble oils are required in flavors (tea, soft drinks, etc.). Even the demand for colorless oils has grown in the last decade, for the so-called "waterclear" perfumes.

The industry has been obliged to develop new technologies to eliminate begaptene, furocoumarins, residues, color, pesticides, and phthalates, and sometime to make the oil more soluble, reducing to various extents the hydrocarbon content.

The most important and modern technology applied is fractionation under deep vacuum using columns with 10 to 20 Sulzer plates. These units can fractionate the oil through an automatic flux/reflux device. To produce concentrated, folded, or even terpeneless oils, the monoterpene hydrocarbons are distilled at a temperature between 40°C and 50°C. The oil remains in the unit for 1 to 2 hours and is stirred for the entire process, thus obtaining the folded oil. As a second step the oil can be further distilled to eliminate the waxes in order to obtain the terpeneless oil. The qualities demanded by the flavor and fragrance industries can contain from 1% to 20% of total monoterpene hydrocarbons and from a trace to 10% of waxes. To produce colorless oils, a steam distillation can be applied, or even a thin layer distillation under vacuum. Some very modern plants also apply molecular distillation, a technology that produces a colorless oil that is, quality-wise, most similar to the cold-pressed oil.

Fractionation under deep vacuum has recently been introduced to reduce to traces the furocoumarins present in the oil. Even most pesticides and contaminants can be eliminated through fractionation. This technology can also be used to produce high-quality natural isolates with purity up to 98%, such as natural linalool and linalyl acetate, which in the past were obtained by chemical reaction.

In recent years, samples of "low allergen" bergamot oil have become readily available on the market, containing very reduced quantities of limonene, linalool, and citral, but perfumers have found the quality to be very poor and far from the magnificent scent of natural, cold-pressed bergamot oil.

The applications of bergamot oil in food products (tea, ice cream, cakes, spirits) and soft drinks (sodas) are discussed in Chapter 19, this volume.

REFERENCES

Calvarano, M., Postorino, E., Gionfriddo, F., and Calvarano, I. 1995. Sulla deamarizzazione del succo di bergamotto. *Essenz. Deriv. Agrum.* 45:384–398.

Calvarano, M., Postorino, E., Gionfriddo, F., Calvarano, I, and Bovalo, F. 1996. Naringin extraction from exhausted Bergamot peels. *Essenz. Deriv. Agrum.* 66:126–135.

Catanea, A. 1928. *Il bergamotto*. 2nd ed. Catania: Battiato.

Di Giacomo, A. 1974. *Gli oli essenziali degli agrumi*. Milano: EPPOS.

Di Giacomo, A., and Calvarano, M. 1978. Il contenuto di bergaptene nell'essenza di bergamotto estratta a freddo. *Essenz. Deriv. Agrum.* 48:51–83.

Di Giacomo A., and Di Giacomo, G. 2002. Essential oil production. In *Citrus: The genus citrus*, eds. G. Dugo, and A. Di Giacomo. London and New York: Taylor and Francis.

Di Giacomo, A., and Mincione, B. 1994. *Gli olii essenziali agrumari in Italia*. Reggio Calabria: Laruffa Editore.

La Face, D. 1956. L'estrazione dell'essenza di bergamotto. *Essenz. Deriv. Agrum.* 26:219–227.

La Face, D. 1958. Le macchine "pelatrici ad acqua" nella estrazione dell'essenza di bergamotto. *Essenz. Deriv. Agrum.* 28:3–15.

La Face, F. 1931. L'industria del bergamotto. Condizioni attuali e possibilità di miglioramento. *Boll. Uff. Staz. Sper. Ind. Ess. e Der. Agr.* 6:46–67.

La Face, F. 1942. Le essenze di petit-grain. Osservazioni sui fattori che influenzano la resa ed i caratteri di composizione. *Boll. Uff. Staz. Sper. Ind. Ess. e Der. Agr.* 17:27–35.

La Face, F. 1955. Nuove macchine per l'industria agrumaria. *Essenz. Deriv. Agrum.* 25:7–16.

Maymone, B., and Dattilo, M. 1958. Digeribilità e valore nutritivo dei sottoprodotti dell'industria agrumaria italiana. *Alimentazione Animale* 2:5–18.

Moyler, D. A. 1993. Extraction of essential oils with carbon dioxide. *Flavour Fragr. J.* 8:235–237.

Poiana, M., Sicari, V., Mincione, B., and Crispo, F. 1997. Prospettive di utilizzazione della tecnica estrattiva con fluidi supercritici applicata ai derivati della coltura del bergamotto. *Essenz. Deriv. Agrum.* 47:27–41.

Postorino, E., Poiana, M., Pirrello, A., and Castaldo, D. 2001. Studio dell'influenza della tecnologia di estrazione sulla composizione del succo di bergamotto. *Essenz. Deriv. Agrum.* 71:57–66.

Rodanò, C. 1930. *Industria e commercio dei derivati agrumari*. Milano: Hoepli.

Spinelli, G. 1967. *Studio sul bergamotto*. Reggio Calabria: Grafiche Sgroi.

7 Current Day Chromatography-Based Strategies and Possibilities in Bergamot Essential Oil Analysis

Peter Quinto Tranchida, Paola Dugo,
Luigi Mondello, and Giovanni Dugo

CONTENTS

7.1 INTRODUCTION

Prior to describing and discussing current-day analytical possibilities related to the analysis of bergamot (*Citrus bergamia*) essential oil, it is important to briefly define its compositional degree of complexity and the reasons why this sample-type needs to be studied in detail.

In general, any chromatography-based technique must match the degree of complexity of the sample subjected to analysis (e.g., in fingerprinting approaches) or fill a specific and well-defined objective (e.g., target analysis). Hence, considering bergamot essential oil, one could be interested in the qualitative or quantitative analysis of the entire chemical composition (volatiles and/or non-volatiles), or be interested in specific analytes, such as enantiomers, or specific odor constituents. Additionally, the analytical task must be done with minimum time expenditure. It is obvious that an optimized weak method will never produce the best result, while an excessively powerful one will certainly do the job, but at higher economic costs.

The definition of the degree of analytical complexity related to any sample-type is a subjective matter. First, one must differentiate between the different chromatography techniques, which for bergamot essential oil are usually gas chromatography (GC) and high-performance liquid chromatography (HPLC). Considering the former technique, we use "low-complexity" for samples containing up to 50 volatile compounds (e.g., olive oil fatty acids), while the term "medium-complexity" refers to samples containing 50–150 constituents (e.g., bergamot essential oil); any sample containing more than 150–200 compounds is to be defined as "highly complex" (e.g., coffee aroma). It is obvious that, in using such definitions, a direct relation is being made to a standard capillary column (30 m × 0.25 mm ID); the latter, even if of high selectivity, is nearly always an adequate analytical tool for the analysis of low-complexity samples, and often for medium-complexity ones.

Bergamot essential oil is characterized by a volatile fraction, consisting of about 100 compounds, and by differing polarities. The volatile fraction is a group of monoterpene (43%–65%) and sesquiterpene (under 2%) hydrocarbons, and oxygenated derivatives (aldehydes, ketones, alcohols, esters, and oxides), along with smaller, less abundant compounds, such as aliphatic aldehydes, ketones, alcohols, and esters. The bergamot oil oxygenated fraction is characteristic, as it is present in considerable amounts, higher than in other *Citrus* oils. The main oxygen-containing compounds are linalool (4%–26%) and linalyl acetate (24%–35%), with enantiomeric excesses in both cases by far in favor of the *R* isomer. The main hydrocarbon is represented by limonene (29%–46%); β-pinene (5%–7%) and γ-terpinene (6%–8%) are also present in significant amounts. The percentage values provided here (derived from GC analysis, with a flame ionization detector [FID]), relate to the most valued cold-extracted bergamot oil, namely that derived from Southern Calabria (Italy) (G. Dugo et al. 2011). Apart from the specific sample fingerprint, of high interest are the enantiomeric composition, the isotope-ratio profile of specific constituents, the aroma-generating volatiles, and the presence/absence of contaminants. All such parameters can be of great help in studies devoted to the determination of quality, genuineness, and the degree of potential toxicity.

Oxygen heterocyclic compounds, which form a fundamental part of the bergamot oil non-volatile fraction, consist of coumarins, psoralens (furocoumarins), and polymethoxylated flavones. Bergamottin (5-geranoxypsoralen), in particular, is the main oxygen heterocyclic constituent, followed by other less abundant compounds such as citropten (5,7-dimethoxycoumarin), bergapten (5-methoxypsoralen), and

5-geranyloxy-7-methoxycoumarin, on to trace-amount components such as sinense-tin (3',4',5,6,7-pentamethoxy flavone) and tetra-O-methyl-scutellarein (a polyme-thoxy flavone) (G. Dugo et al. 2011). The qualitative and quantitative profiles of such constituents are typical for each *Citrus* essential oil and are parameters which can be exploited to form an opinion on authenticity. Additionally, oxygen heterocyclic compounds can present both pharmacologic and toxicological activities (Frérot and Decorzant 2004; Wang et al. 2006; Wu et al. 2007). In particular, a European law (a modification of the Directive 76/768/EEC) regulates the amounts of photosensi-tizing psoralens in some cosmetics to a maximum level of 1 ppm. For this reason, there is a need for bergapten-free bergamot oils (alkali treated or distilled) for the perfumery and cosmetic industries. Oxygen heterocyclic constituents present strong absorption in the UV region (λ_{max} *circa* 315 nm) and, hence, are commonly detected exploiting such a characteristic, after a normal-phase (NP) or reversed-phase (RP) LC separation.

Gas chromatography is the prime choice for the analysis of bergamot essential oil volatiles, while liquid chromatography is the obvious option for the separation of the high MW constituents. Heart-cutting multidimensional gas chromatography (MDGC) has a well-specified location in the field, while other approaches, such as multidimensional liquid-gas chromatography (LC-GC), can be useful, though are probably not absolutely necessary. Comprehensive two-dimensional chromatogra-phy methods have also been used for the compositional elucidation of bergamot oil, though their real need remains doubtful. As in practically all research fields, the mass spectrometer plays a central role in both GC and LC bergamot-oil applica-tions. It is diminishing to define an MS system as a simple detector because it not only provides precious structural information, but also has the inherent capability to achieve the separation of different analytes, which enter the ion source at the same time. Apart from various forms of mass spectrometry, a series of less information-rich detectors have been exploited.

The objective of the present contribution is to describe and critically discuss the current state-of-the-art in the field of bergamot essential oil analysis, as well as future perspectives.

7.2 GC-BASED METHODS

7.2.1 CONVENTIONAL GC

The relationship between GC and the bergamot oil volatile fraction started in the mid-1950s, when Liberti exploited the novel technique and showed his results at an international meeting (Liberti and Conti 1956). The first bergamot oil chromatogram showed not more than four or five peaks. Though such work was a milestone in the field, it is clear now that the packed column used by the Italian scientist achieved only a group-type separation of what is considered the most appreciated and valuable *Citrus* essential oil. Column technology has evolved considerably, and nowadays it appears little remains to be discovered of the 100 or so compounds forming the volatile fraction of bergamot oil. The term "discovery" refers to molecular structure, quantity, olfactometric potency, enantiomer composition, and isotopic ratios. The presence or absence of contamination is also a highly important analytical issue. The detailed knowledge of such parameters can provide a clear view on the state of a bergamot oil.

7.2.1.1 Conventional Stationary Phases

When talking about the use of a capillary column, it is common to refer to purely theoretical terms such as "separation space" and "peak capacity." However, rather than reference to a column, it is more correct to use the term "separation space (or peak capacity) generated by a capillary-column GC method." The reason for such a specification is due to the dependence of the two theoretical parameters on the operational conditions (injection, pressure, temperature program, detection). Separation space is essentially the (time) space available for the chromatographic separation, and it extends from the dead time (e.g., 5 min) to the end of the temperature program (e.g., 60 min), or to the pre-defined end of an isothermal analysis. It is theoretically possible to generate an infinite separation space in the case of an isothermal analysis (if no analysis stoppage occurs), and an extremely high amount in the case of a temperature program with a very slow gradient (e.g., 0.001°C/min). Though the concepts of separation and peak capacity are different, both parameters are closely related: peak capacity is "the number of peaks that can be positioned side by side (with a specific R_S value), along the separation space generated by a chromatography column." A GC method, using a conventional capillary column (i.e., 30 m × 0.25 mm ID × 0.25 μm d_f), is characterized by a peak capacity usually in the 300–500 range.

Fundamental theoretical dictations have established that no more than 37% of the peak capacity can be exploited to generate peak separation, while many of those resolved peaks are the result of two or more overlapping compounds. Additionally, the final number of totally isolated compounds does not exceed 18% of the peak capacity (Giddings 1990). For example, if a GC method generates a peak capacity value of 400, then about 70 compounds can be fully resolved, which is slightly less than the number of volatiles contained in a bergamot essential oil. It is obvious that such a calculation is an approximation, not considering stationary-phase selectivity.

Bergamot oil constituents are usually analyzed using 30 m × 0.25 mm ID × 0.25 μm d_f columns, with apolar (e.g., polydimethylsiloxane, 5% diphenyl + 95% polydimethylsiloxane) and/or polar (e.g., polyethylene glycol) stationary phases. Peak assignment is generally carried out through MS-database spectral matching, the calculation of linear retention indices (LRI), and the analysis of commercial standards; quantification is most often achieved using a flame ionization detector (FID), considering the relative peak areas generated by the specific GC-FID software.

A great amount of information, often helpful for the assessment of quality and authenticity, can be attained from conventional GC analyses. Even so, co-elutions do occur using any classical phase type. For example, 5% diphenyl apolar phases, which achieve separation mainly on a boiling-point basis, suffer overlapping between (among others) limonene and (Z)-β-ocimene, along with nerol and citronellol. On polyethylene glycol phases, component overlapping can occur between some oxygenated monoterpenes and sesquiterpene hydrocarbons. Such a drawback can be circumvented by using apolar and polar columns in separate applications, and then combining the data attained. A further route to increase resolution could be to use a longer column (e.g., 100 m). In relation to the former option, not all have the availability of two GC systems, and the continuous changing of columns in a single instrument is a tedious process. With regards to the latter choice, the price to pay for a resolution increase (by a factor of 1.7) would be a sensitivity loss and an extended GC run time. Consequently, the most common option for the analysis of bergamot oil volatiles is the "stand-alone" 5% diphenyl one.

It is the authors' opinion that the GC-FID/GC-MS analysis of bergamot essential oil enables a good initial, though not conclusive, idea of the specific state of the oil (this can be defined as the primary screening level). In fact, the quantitative profile of genuine bergamot oil volatiles can undergo variations throughout the season and is characterized by wide concentration ranges. For example, the monoterpene hydrocarbon fraction and the main components of this class tend to increase during the production season; the same trend is observed for linalyl acetate, while linalool presents the opposite behavior. The main consequence of such seasonal variability is that it is rather easy to produce adulterated oils altogether similar to the natural counterparts. With regards to MS detection, this is carried out, more often than not, in the full scan mode, usually with a quadrupole mass analyzer (unit-mass resolution). As mentioned, peak assignment is then achieved by matching the experimental MS spectra with those contained in spectral databases (often commercial). If the latter contain reference LRI values, then such data can be used as a filter during spectra matching, if the native GC-MS software enables such a comparative process. The procedure works as follows: the GC-MS software automatically calculates LRI values relative to each bergamot oil volatile (a mixture of alkanes is subjected to analysis prior to the main application). During the MS database search process, the same software deletes matches with a reference LRI, with respect to that of the "unknown," outside a predefined LRI range (i.e., +/– 5 LRI units).

The final part of this section is devoted to the target analysis of bergamot-oil contaminants, namely pesticides (organophosphorous and organochlorine) and plasticizers. In general, a clean-up step is necessary when analyzing contaminants in a complex and often "dirty" vegetable matrix. However, in a way bergamot essential oil has already been subjected to a pre-separation step, namely the extraction process. Consequently, the sample preparation step can consist of dilution with an appropriate solvent prior to GC analysis.

The preliminary analysis of phytosanitary products (and plasticizers) is normally carried out by means of a single GC column in combination with an element selective detector, such as an electron capture (ECD), flame photometric (FPD), or a nitrogen-phosphorous (NPD) system. In order to achieve positive identification, mass spectrometric structural elucidation is required. It is common to use a quadrupole MS system in the full-scan mode, combined with database searching, for such a purpose (if the target analyte is present in sufficient concentrations).

If positive identification was not performed through full-scan MS data, then qualitative and quantitative analysis can be carried out by operating in the selected ion monitoring (SIM) mode with a specific number of diagnostic ions (generally one quantifier and two qualifiers). Calibration curves are generated using appropriate internal standards. The SIM mode is very sensitive, reaching limits of detection at the ppb level and lower. Extracted ion chromatograms (EIC) can be used not only to evaluate the selectivity of the GC separation, but also for quantification. However, the EIC approach cannot match the sensitivity of the SIM mode.

7.2.1.2 Alternative Stationary Phases

In general, insufficient resolution can be improved by increasing column efficiency and/or selectivity. Taking for granted that a specific capillary is being employed under ideal conditions, and apart from increasing length, then selectivity is another fundamental parameter that can generate improved separations. In fact, selectivity can have a much greater impact on a separation than efficiency. Such a consideration is valid for low to medium complexity samples, while for highly complex ones, peak capacity (or efficiency) is the most important parameter.

The selectivity of all the traditional phases is well known in the bergamot field, and in the *Citrus* oil field in general. On the other hand, much less is known about novel phases. Recently, an ionic liquid (IL) stationary phase (SLB-IL59) was used for the analysis of a lemon essential oil in a comparative study with a classical nonpolar and polar column (Ragonese et al. 2011). Room temperature ionic liquids are a class of solvents, generally formed of an organic cation containing N or P (i.e., alkyl imidazolium, phosphonium), counterbalanced by an anion of organic or inorganic nature. Properties such as low volatility, thermal stability, and excellent selectivity towards a variety of chemical classes make ionic liquids suitable as GC stationary phases. In the IL study, bergamot oil was not analyzed, though the published results are worthy of description for a possible future employment of the IL59 column for bergamot oil analysis. In fact, the IL phase employed demonstrated good efficiency, and a polarity similar to that of a "wax" column, but with a higher thermal stability (300°C vs. 280°C) and less chemical bleed. Considering the specific application, almost all lemon oil volatiles were separated using the IL column; however, a series of co-elutions occurred on the apolar phase, such as limonene and 1,8-cineole, along with neral and carvone. The performance of the IL capillary was also superior to that of the polar column; for example, neryl acetate overlapped with β-bisabolene on the "wax" phase, while they were well resolved on the IL one. It must be emphasized, however, that we are still talking about the primary screening level.

An additional valuable contribution toward the reliable assessment of the specific state of a bergamot essential oil can be attained using enantio-GC. In general, the determination of enantiomer compositions is of high importance in the essential oil field, with derivatized cyclodextrins the most popular stationary-phase choice for direct enantio-GC analysis. The enantiomeric profiles relative to α-thujene, limonene, linalool, linalyl acetate, sabinene, and β-pinene in bergamot essential oil do not vary during the year of production. In particular, typical chiral markers are linalyl acetate and linalool, with the R-(−) isomer present in excess of 99%. Due to the stability and specificity of enantiomer profiles, such an analysis-type can be defined as a secondary screening level. The aforementioned enantiomers can be well-separated using a diethyl-tert-butyl-silyl β-cyclodextrin phase. However, it is obvious that enantioselective phases do increase the number of compounds requiring separation, and thus there is an increased possibility of overlapping. A good choice is to carry out a GC (typically apolar) pre-separation on a bergamot oil, and then direct target enantiomer pairs to a chiral selector using a heart-cutting multidimensional GC system. An alternative is to use an MS detector and extracted-ion chromatograms to eliminate the contribution of interfering compounds.

Although the elution sequence of specific enantiomers in bergamot oil is well known, this assumption may not be valid for all analysts, especially newcomers to the field. In fact, it is well known that MS systems can produce a name for two enantiomers, but cannot discriminate between them. To circumvent such a lack of knowledge, Liberto et al. (2008) constructed an enantiomer MS database, a tool potentially of great help for (general) essential oil analysis. The database was created using four chirally selective columns, including not only pure mass spectra but also LRI data. Because a universally selective chiral capillary is not available, the columns were chosen to cover a wide range of enantiomer separations. The LRI data was used as a filter during library matching in the manner mentioned at the end of Section 7.2.1.1.

7.2.1.3 Isotopic Analysis

Apart from enantioselectivity, isotope discrimination during plant biosynthesis can be exploited to evaluate geographic origin and adulteration of natural bergamot oil. For such scopes, isotope ratio mass spectrometry (IRMS), combined with gas chromatography, is another useful analytical tool. Isotope ratio mass spectrometry is an analytical method which enables the measurement of deviations of isotope abundance ratios, from an agreed standard, by only a few parts per thousand for C, as well as for other elements such as H, N, O, and S. Prior to entrance into the gas source, each element must be converted into a gaseous species. In the field of *Citrus* essential oils, the GC-IRMS determination of the $^{13}C/^{12}C$ ratio is now rather well established, and is obtained by converting the C atoms of a specific analyte into CO_2. Such a transformation is achieved using a combustion chamber, and then by comparing the C isotope ratio of that constituent to that of a known reference. Hence, GC-IRMS provides no clue on the structure of a given analyte. A dimensionless quantity (δ) is used to express the isotope ratio value of a specific analyte, in relation to the reference, and is expressed in ‰ (Brenna et al. 1997).

In single-column GC-IRMS and enantio-GC experiments, the possible occurrence of co-elution could provide erroneous results in terms of both $\delta^{13}C$ values and enantiomer ratios. Moreover, the series of connections and instrumental parts from the column outlet to the IRMS inlet (combustion reactor, transfer lines, water trap, and open split) can cause a considerable degree of band broadening. To overcome such disadvantages, Juchelka et al. (1998) described a Deans switch MDGC-IRMS system for the analysis of "flavor components": a 5% phenyl column was employed in the first dimension, while a cyclodextrin-based one was used in the second (independent ovens were employed). Although such an approach has not been applied to bergamot oil, such instrumentation could provide reliable information on quality and authenticity.

In conclusion, though isotopic data is a valuable tool to evaluate the specific state of a bergamot oil, it must be accompanied by chiral information. The reason is related to the fact that the enantiomer ranges, for specific compounds, are very narrow, while the $^{13}C/^{12}C$ authenticity ranges are a little wider. Furthermore, in the

Citrus oil field, authenticity ranges are constructed using the $^{13}C/^{12}C$ values of 10–15 volatiles. If only 1 or 2 compounds out of 15 deviate slightly from an authenticity range, then there is no real basis to define that oil as "adulterated" or as "derived from a different geographical origin." On the other hand, in the presence of a deviation of the enantiomer composition of linalool and/or linalyl acetate, then a more firm opinion can be formed.

7.2.1.4   GC-Olfactometry

The technique known as GC-olfactometry (GC-O) enables the assessment of odor-active components contained in mixtures of volatile compounds (essential oils, perfumes, foods, etc.) on a one-to-one basis. There is often no relationship between analyte concentration and its contribution to the overall odor of a sample. It is worthy of note that even enantiomers, in many cases, are characterized by an entirely different odor. Consequently, GC-O can be exploited to distinguish between optical isomers, something that cannot be done using MS information.

In a GC-O process, the compounds eluting from a GC capillary are normally split between the human olfactory system and a common GC detector (usually an FID or MS). A GC-O experiment provides both a chromatogram and an "aromagram," thus enabling the identification, quantification, and assessment of odor type/intensity of specific constituents (d'Acampora Zellner et al. 2008). To the "pure" GC-O analyst, a chromatogram does not represent a flavor or fragrance.

Recently, Sawamura et al. (2006) subjected bergamot essential oil to GC-O, GC-FID, and GC-MS analyses, and then "constructed" a bergamot oil aroma mixture. Aroma extraction dilution analysis (AEDA) was used to elucidate the most important aroma-contributing compounds. The bergamot oil sample was diluted and then analyzed in a stepwise manner to determine the flavor dilution (FD) factors and relative flavor activities (RFA). The FD factor of an analyte corresponds to the number of times a sample can be diluted before its odor is not perceived. RFA values are calculated by relating the FD factor and the concentration (in % weight) of each odorant. If an abundant compound is not recognized after a few dilutions, then its RFA value will be low.

Fifty-eight sensorial descriptions were given for the same number of bergamot oil constituents; of these, 55 were identified and quantified. Four volatiles, namely (Z)-limonene oxide, decanal, linalyl acetate, and geraniol, were described as being bergamot-like. High FD factors were associated with several compounds, such as β-pinene (8), geraniol (8), (Z)-β-ocimene (7), neral (7), linalool (7), and geranial (7) (Figure 7.1); the highest RFA value was given to (Z)-β-ocimene (25.9), meaning it was the most active compound (Figure 7.2), followed by octanal (24.0), perillyl acetate (21.6), and decanal (19.5). The authors affirmed that, apart from the odors that form the bergamot-oil background aroma, the contribution of other compounds with high FD and/or RFA values is equally important. For example, (Z)-β-ocimene, which was quantified at the 0.02% level and was defined as being green/spicy, was used in the constructed bergamot oil mixture.

In general, GC-O is a valuable tool not only for the elucidation of the most important contributors toward the generation of bergamot oil aroma, but also for an evaluation of quality. It would appear that conventional GC columns can separate

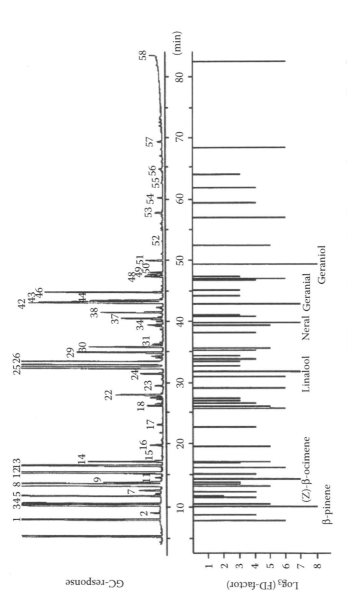

FIGURE 7.1 GC chromatogram and aromagram constructed with RFA values. For peak identification refer to source. (From Sawamura et al., *Flavour Fragr. J.* 21, 609–615, 2006. Reproduced with permission from John Wiley & Sons.)

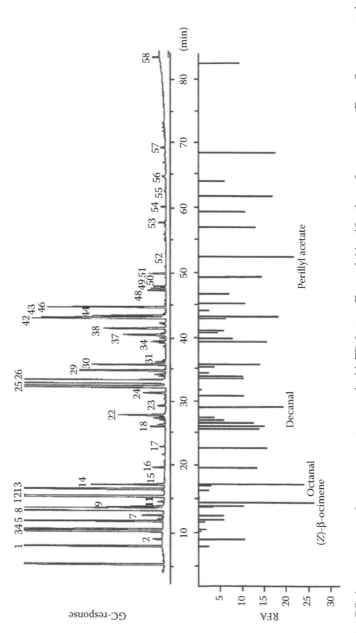

FIGURE 7.2 GC chromatogram and aromagram constructed with FD factors. For peak identification refer to source. (From Sawamura et al., *Flavour Fragr. J.* 21, 609–615, 2006. Reproduced with permission from John Wiley & Sons.)

almost all such odor generators in bergamot essential oil. Consequently, the use of multidimensional GC-O approaches would probably not be necessary. The situation is different for highly complex fragrances (i.e., perfumes), where the use of heart-cutting MDGC-O systems is often required (Eyres et al. 2007).

7.2.2 Fast and Very Fast GC

A conventional GC analysis of a bergamot oil can take up to an hour (using a 30-m column), a run time probably acceptable for most laboratories, though not for all. It is obvious that analysis times can be reduced by increasing the gas flow and/or the temperature gradient. There is no problem with such a choice if the initial analytical objectives are maintained. If, on the contrary, an analyst requires a run-time reduction with no cost in terms of column efficiency, then the use of a micro-bore column is certainly the most advisable choice (Korytár et al. 2002). It has only been in the last decade that fast GC instrumental requirements have been entirely satisfied, enabling the expansion of rapid methods.

A reduction of the column ID and film thickness leads to an increase in plate number. In general, for capillary columns with a high phase ratio, the minimum plate height (H_{min}) is equivalent approximately to the column ID; therefore, a 10 m × 0.10 mm ID column can potentially generate the same theoretical plate number, namely 100,000, as a 25 m × 0.25 mm ID column.

In terms of mobile phase, H_2 is the best choice for high-speed GC analysis, though safety precautions must be considered. Hydrogen is characterized by higher optimum linear velocities (approximately 55–60 cm/s for a 10 m × 0.10 mm ID capillary) compared to He; furthermore, if higher-than-ideal gas velocities are employed, resolution losses are limited.

Temperatures programs also require careful tuning: it has been demonstrated that a gradient of 10°C/dead time is a good compromise (Blumberg and Klee 1998). For high-speed GC bergamot oil applications, an instrumental capability of 80°C–100°C/min is more than sufficient. If a 10 m × 0.10 mm ID capillary is employed under ideal gas flow conditions, then a 10°C–20°C/min gradient can be considered as optimum.

Sample capacity is proportional to ID cubed and, hence, micro-bore capillaries can accommodate lower sample amounts. However, compared to conventional GC methods, sensitivity is more or less maintained because the injection of lower sample amounts is counterbalanced, in part, by the generation of narrower peaks.

The accurate reconstruction of narrow GC peaks requires high acquisition frequencies; thus detector characteristics are another important issue. Furthermore, the detector must be characterized by low dead volumes and a rapid rise time. A non-sufficient acquisition frequency can hinder correct quantification, while an excessively high frequency can cause a noise increase and, consequently, a decrease in signal-to-noise ratios. It is generally accepted that 10 data points per peak are enough for correct peak reconstruction. In this respect, a 20-Hz sampling frequency is sufficient for fast GC experiments (3–15 min run time), while in very fast GC analyses (1–3 min analysis time), higher frequencies are needed (25–50 Hz).

The use of a mass spectrometer in micro-bore column fast GC experiments is obviously desirable. In rapid GC applications, two MS systems are generally employed; namely, low-resolution time-of-flight (ToF) and quadrupole (q) instruments. The former is a non-scanning device characterized by very high acquisition frequency (up to 500 Hz) under wide mass range conditions; such instruments generate consistent spectral profiles at each data point (no skewing), a feature which is exploited for the deconvolution of co-eluting compounds, thus reducing the need of high-resolution separations. ToF-MS systems are suited for very fast and, in particular, for ultra-fast GC experiments (less than 1 min). For fast and very fast GC purposes less-expensive scanning systems such as single quadrupoles can be employed, even for quantitative analysis. Ultimate generation quadrupole mass spectrometers, operated in the full scan mode, can generate up to 50 spectra/s using a "normal" mass range (Purcaro et al. 2010). An additional advantage is that qMS systems are employed for the construction of commercial MS spectral databases.

If a 10 m × 0.10 mm ID column is employed instead of a 25 m × 0.25 mm ID column, then a bergamot oil conventional GC method can be fully translated, and a speed enhancement factor in the 4–6 range can be expected. If very fast GC conditions are applied, then a price must be paid in terms of separation quality, which is entirely acceptable if target analytes remain separated. Both types of rapid approaches have been employed for the analysis of bergamot essential oil.

Tranchida et al. (2006) translated a conventional GC-FID method to a fast GC one in an automated headspace solid-phase microextraction (HS-SPME) investigation on bergamot oil. The conventional HS-SPME-GC method (a 30-μm polydimethylsiloxane fiber and 0.25-mm ID apolar capillary were employed) achieved analyte extraction and separation in "50 + 50" minutes. In the fast GC application, a reduced-volume (7-μm) fiber enabled rapid equilibration (15 min), while a 10 m × 0.1 mm ID apolar column enabled fast separations (12.5 min). A cryo-trap was installed just below the injector and was exploited to reduce band broadening during the desorption step. Cooling had a positive focusing effect on the more volatile bergamot oil components.

The investigation highlighted the advantages of using a micro-bore column in an SPME-GC application—the reduction of both the equilibration and chromatography times. The former benefit was related to the obliged use of a low-volume fiber due to the limited column sample capacity. The chromatograms relative to the conventional and fast GC applications are shown in Figure 7.3. As can be observed, resolution was altogether comparable in the two applications. The numbered peaks were used to make a direct comparison between the two traces. Though a slight sensitivity reduction was observed in the fast analysis, this did not hinder the detection of all peaks of interest.

A very fast direct-injection GC-FID analysis on bergamot oil has been carried out by Mondello et al. (2004) using a 10 m × 0.10 mm ID apolar column under drastic experimental conditions (70°C/min temperature gradient; H_2: 81.5 cm/sec). The rapid application was achieved in 3.3 minutes, with a speed gain of nearly 14 times that of a conventional approach. The attainment of such a low GC run-time had a moderate cost in terms of resolution: the quantification of 56 compounds was reported, only 6 less than the conventional GC-FID analysis.

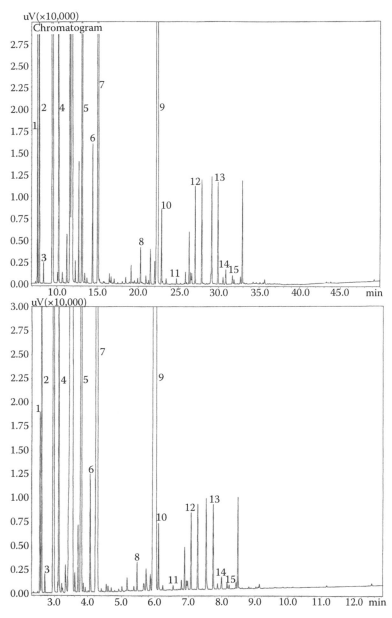

FIGURE 7.3 Upper chromatogram: A conventional HS-SPME-GC application on bergamot essential oil. Peaks: (1) α-thujene; (2) α-pinene; (3) camphene; (4) myrcene; (5) γ-terpinene; (6) terpinolene; (7) linalool; (8) octyl acetate; (9) linalyl acetate; (10) geranial; (11) nonyl acetate; (12) neryl acetate; (13) *trans*-α-bergamoptene; (14) *cis*-β-farnesene; (15) germacrene D. Lower chromatogram: A cryo-trap fast HS-SPME-GC analysis on bergamot essential oil (peak identification as indicated before). (From Tranchida et al., *J. Chromatogr. A* 1103, 162–165, 2006. Reproduced with permission from Elsevier.)

A ToF-MS instrument has been used as a detector in the very fast GC analysis of bergamot oil (Veriotti and Sacks 2002). The chromatographic separation was finished in less than 140 seconds (a 50°C/min temperature gradient was applied) and was carried out using a dual-capillary stop-flow approach: a 7 m × 0.18 mm ID trifluoropropylmethylpolysiloxane column was connected to a "5% phenyl" one of the same dimensions. A pressure pulse, applied to the column conjunction point, stopped the primary-column flow for specific time periods, improving the overall GC performance. A spectral generation frequency of 25 Hz was applied, enabling spectral deconvolution; for such a purpose, at least two complete mass spectra between the peak apexes of co-eluting compounds (corresponding to a minimum peak apex separation of 120 msec) were required for the deconvolution software to work. The very fast application proposed, which can be considered somewhere between multi-column GC and MDGC, is to be evaluated for its analytical potential rather than for the (low) number of bergamot oil volatiles separated and identified.

Recently, a rapid-scanning qMS instrument has been used in fast GC applications on a variety of *Citrus* oils (Scandinaro et al. 2010). The fast GC-qMS method developed was characterized by a 10-minute run-time (10°C/min temperature gradient; H_2: 50 cm/sec). One of the main aims of the research was to compare experimental LRI values to conventional GC LRIs contained in an MS database. After several GC-MS experiments on six *Citrus* oils (including bergamot), it was found that an LRI range of +/− 15 units was found necessary in the fast experiment.

The fast GC-qMS approach was also employed for the construction of a novel mass spectral database named EI-MS F&F (electron-impact-MS flavor and fragrance). In brief, the authors subjected 200 essential oils (*Citrus* and non-*Citrus*) to fast GC-qMS analysis. A single spectrum was extracted from each chromatogram by averaging the spectra relative to all peaks in the retention time window, apart from the solvent, and was then added to the EI-MS F&F database. Hence, each spectrum attained can be considered in the same way as a direct injection into an ion source, without the formation of solvent ions.

The EI-MS F&F database was found to be an effective tool to give a reliable name to an unknown essential oil, while it was obviously of no use for the identification of unknowns. With regards to *Citrus* oils, the database was capable of recognizing different oil types (e.g., sweet orange, bergamot, lemon, etc.) but it could not distinguish between oils produced in different periods (e.g., green, yellow, and red mandarin oils).

Fast enantioselective applications on bergamot oil have also been carried out: 10 m, 5 m, and 2 m × 0.1 mm ID cyclodextrin capillaries were used by Bicchi et al. (2008) in the GC-MS analysis of bergamot oil chiral compounds (β-pinene, limonene, linalool, linalyl acetate, and α-terpineol). Surprisingly good results were attained using the shortest column that were not exclusively related to the short analysis time. In fact, it was found that the low column efficiency was counterbalanced by a higher selectivity due to the lower elution temperatures. The chiral analysis of a bergamot oil using a reference cyclodextrin (2,3-di-O-ethyl-6-O-tert-butyldimethylsilyl-β-CD) column (25 m × 0.25 mm ID × 0.25 μm), as well as the 5 and 2 m × 0.1 mm ID counterparts, are shown in Figure 7.4. The extracted ion chromatograms for limonene are also reported, demonstrating the similar resolution values attained.

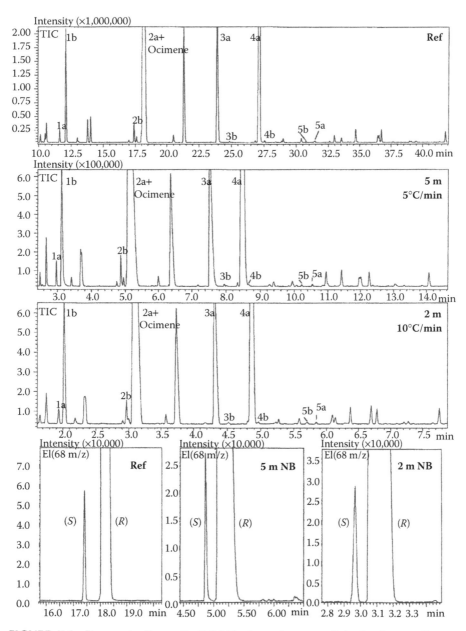

FIGURE 7.4 Bergamot oil enantio-GC-MS chromatograms attained using a reference (Ref) cyclodextrin column (25 m × 0.25 mm ID × 0.25 μm) and micro-bore capillaries (5 and 2 m). The bottom extracted ion chromatograms refer to the separation of limonene enantiomers. Peak identification: (1) β-pinene; (2) limonene; (3) linalool; (4) linalyl acetate; (5) α-terpineol. (From Bicchi et al., *J. Chromatogr. A* 1212, 114–123, 2008. Reproduced with permission from Elsevier.)

The exploitation of extracted ions enabled the elimination of interferences, exalting the selectivity of the MS process.

In conclusion, it can be affirmed that with regards to the potential of such approaches to provide conclusive information on the specific state of a bergamot essential oil, the previous observations made on conventional GC methods are also valid for rapid ones.

7.2.3   Heart-Cutting Multidimensional Methods

An effective way to increase the selectivity and separation power of a chromatography system is by using a multidimensional chromatography (MDC) instrument. MDC approaches can be divided in (conventional) heart-cutting and comprehensive techniques; the former approach allows the transfer of selected chromatography bands, from a primary to a secondary column. It must be emphasized that the number of chromatography bands subjected to analysis in the second dimension is limited because too many transfers could cause the loss of resolution already achieved on the primary column. The peak capacity of an MDC method is equal to the sum of that of the first and second column, the latter multiplied by the transfer number (x) $[n_{c1} + (n_{c2} \times x)]$: for example, if both columns are characterized by a peak capacity of 500 and three transfers are carried out, then a final value of 2000 is obtained.

7.2.3.1   Heart-Cutting Multidimensional GC

Over the last 20 years, the number of bench-top GC-MS instruments sold worldwide has increased greatly. Moreover, and in relation to the use of a two-dimensional technique such as GC-MS, there has been wide debate on the necessity (or not) of MDGC. It is our opinion that both conventional MDGC and mass spectrometric approaches are necessary, and would ideally be combined.

Heart-cutting MDGC experiments, in the *Citrus* oil field in general and in that of bergamot in particular, mainly refer to the elucidation of the enantiomeric profiles of a series of target isomers. The one-dimensional enantio-GC analysis of a bergamot oil can fail in providing reliable measurements due to a lack of peak capacity. Thus, it is advisable to pre-separate the essential oil on an apolar or polar column, then to transfer target isomer pairs on a chiral column, and possibly then to an MS system. The MDGC chiral analysis of a bergamot essential oil enables a conclusive idea on the specific state of an oil (secondary screening level), because it is also possible to attain primary screening data (from the first-dimension analysis).

The last 10 years have witnessed the development and commercial availability of effective MDGC instruments. Such systems are characterized by accurate electronic pressure control, are not of such elaborate construction as what they were in the past, are supplied with user-friendly software enabling complete automatic instrumental control, and give highly repeatable results, even under multiple-cut conditions. In terms of transfer device, the Deans switch has conquered the analytical scene. If one desires to use a switching-valve MDGC instrument, then the required hardware (a switching valve, connections, an auxiliary pressure source, etc.) must be purchased and then installed in the GC. Two examples of present-day MDGC systems will be herein described.

A conventional MDGC system, with a microfluidic transfer device (Agilent Technologies) has been recently reported (Quimby et al. 2007). The interface, of the Deans-switch type, is characterized by high thermal stability and chemical inertness, low dead volumes, and leak-free connections, and is constructed using capillary flow technology (CFT): through holes and flow channels are etched into stainless steel plate halves, which are then folded and heated to a very high temperature (greater than 1000°C); afterwards, high pressure is applied to generate a metal sandwich. Conditions for stand-by, cutting, and backflushing processes are created by using auxiliary electronic pressure control. The transfer device contains five ports: two entrances are fixed, being linked to a two-way solenoid valve, with the valve connected to the auxiliary pressure unit. Two ports are connected to the primary and secondary analytical columns, while the last is linked to a restrictor, with the same flow resistance as the second capillary. Such a requisite is highly important because the pressure drop across the primary column must not alter, switching between the two operational modes. If differences in the pressure drop occur, then there will be a mismatch between the programmed chromatography-band-cut windows, which are set on the basis of a preliminary stand-by separation, with those which occur during the MDGC analysis. The restrictor is usually connected to a detector (usually an FID) to monitor the first-dimension separation. Figure 7.5 shows how the fluidic device achieves the bypass (stand-by) and inject (cut) states. The first-dimension flow, which is always lower than the auxiliary flow, enters the transfer device through the central port. In the stand-by configuration, the solenoid valve directs the auxiliary flow to the top-left part of the Deans switch, which is connected, through an internal channel, to the port linked to the second column. The additional gas flow is divided in two parts once inside the interface—one crosses the internal vertical channel ending up in the restrictor, while the other part is directed to the second column. Before entrance to the restrictor, the auxiliary flow is mixed with that of the first dimension. When the solenoid valve is switched to the cutting configuration, the auxiliary flow is directed to the bottom-left part of the Deans switch, which is connected through an internal channel to the port linked to the restrictor. Inside the interface, the auxiliary gas flow is divided between the restrictor and the second column. As in the stand-by mode, the additional flow is mixed with that of the primary column.

An effective Deans-switch MDGC system has been developed and commercialized by Shimadzu Corporation. The MDGC instrument is equipped with a quadrupole mass spectrometer and two independent GC ovens. The stainless steel transfer device (circa 3 cm long) is located in the first oven and is characterized by low dead volumes, high thermal stability, and chemical inertness. There are five interface connections: two fixed branches derive from an auxiliary pressure source, two ports are connected to the primary and secondary analytical columns, and the remaining port is for a stand-by detector. A fused-silica restrictor (R_1) is fixed inside and crosses the interface. Figure 7.6 illustrates two schemes of the entire Shimadzu transfer system in the "stand-by" (Figure 7.6a) and "cut" (Figure 7.6b) positions. The five-port metallic interface requires a web of external connections to create the classic MDGC conditions. In both configurations, an advanced pressure control unit (APC) supplies a gas flow at constant pressure to an external fused-silica restrictor (R_3) and to a two-way electrovalve (V). The electrovalve is linked to two metal branches, one with

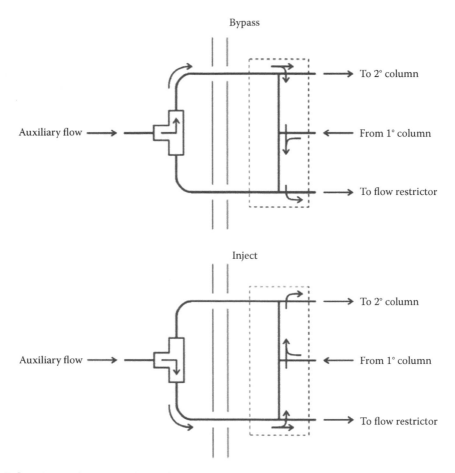

FIGURE 7.5 Scheme of the "Agilent" Deans switch in the bypass (stand-by) and inject (cut) modes. (From Seeley et al., *Anal. Chem.* 79, 1840–1847, 2007. Reproduced with permission from American Chemical Society.)

an in-line fused-silica restrictor (R_2) and the other without: R_2 produces a slightly higher pressure drop than that generated by R_3 ($\Delta P_2 > \Delta P_3$). During stand-by analysis (Figure 7.6a), the ACP pressure is reduced on the side of the first dimension (e.g., 200 kPa − ΔP_3), while it reaches the second dimension branch unaltered. As a consequence, analytes eluting from the first column are directed to the "monitor" detector. The transfer device is converted to the cutting mode upon activation of the electrovalve (Figure 7.6b): the pressure on the first-dimension side of the interface remains unaltered, while the pressure on the second-dimension side becomes 200 kPa − ΔP_2 (a pressure lower than 200 kPa − ΔP_3). It is obvious that, under such pressure conditions, primary-column chromatography bands are free to reach the second capillary. The instrument is controlled by using dedicated software, which also enables the calculation of fundamental GC parameters, such as gas flows, linear velocities, and percentage of analyte transfer.

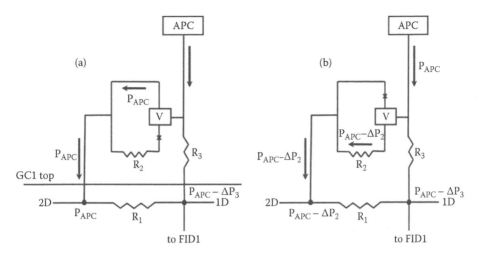

FIGURE 7.6 Scheme of the "Shimadzu" Deans switch in the (a) stand-by and (b) cut configurations. Abbreviation definitions are reported in the text.

With respect to the Agilent system, the main differences are that the external design of the (three-restrictor) transfer system is a little more elaborate; the capillary linked to the stand-by detector does not need to be characterized by the same flow resistance as that of the secondary column (meaning that if one wants to change the second column, then the restrictor does not also need to be replaced). However, both commercial instruments work in an effective manner. The Shimadzu MDGC instrument has been used for the apolar-chiral analysis of bergamot oil (G. Dugo et al. 2011), as well as in other *Citrus* oil applications (Sciarrone et al. 2010).

The MDGC-MS system is a sufficient analytical tool to attain reliable information on the specific state of a bergamot essential oil. In fact, such an instrument can provide reliable qualitative and quantitative data relative to the volatile fraction, and to the enantiomer composition. It is also obvious that the availability of a GC-IRMS instrument can give further strength to a specific evaluation.

7.2.3.2 Heart-Cutting Multidimensional LC-GC

Heart-cutting multidimensional LC-GC will be described among the GC-based methodologies because the LC step is characterized by a low peak capacity and can be considered as a pre-separation, or sample clean-up step. The GC process, on the other hand, provides much higher resolution, and can be considered as the main analytical step.

A series of issues must be considered when using an LC-GC instrument. First, any compound that is transferred from the LC system must be suited to GC analysis; consequently, practically all LC-GC methods derive from a previous GC approach. An additional factor is that the LC process aims to isolate groups of analytes on a polarity basis; afterwards, the homogenous and ideally simple mixtures are then subjected to GC analysis. Samples containing a high number of chemical classes and an interfering matrix are most suited to LC-GC analysis. A further aspect is related to the elimination of high amounts of liquid mobile phase, which is the most difficult obstacle from a technological viewpoint. Additionally, the transfer of large sample amounts, from the LC to the GC unit, means that sensitivity is enhanced. Finally, the combination of a mass spectrometer to an LC-GC system is highly convenient because the mass spectra attained are normally free of interferences, enabling a much easier interpretation if compared to conventional GC-MS spectra. In the LC-GC field, various transfer systems have been developed to enable the passage of heart-cuts from the first to the second dimension. The reader is directed to the literature for detailed information (Grob 1991).

The majority of *Citrus* essential oil LC-GC studies, including those related to bergamot oil, were carried out in the 1990s (Mondello 1996a, 1996b); in recent years things have been much quieter. Now, considering that the chemical classes of volatiles contained in bergamot oil can be nicely pre-separated using NP-LC, the GC step (e.g., a 5% diphenyl or "wax" column) could then be used for the complete isolation of the compounds contained in each fraction. Hence, a couple of LC-GC processes could easily succeed in the full elucidation of the volatile profile of a bergamot oil. However, is such a powerful approach necessary? For any type of analytical scope related to the volatile fraction contained in bergamot oil, this can be successfully achieved using an MDGC instrument. In short, if you can get the same results using either LC-GC or MDGC, then use the latter! It could be argued that an advantage of LC-GC over classical MDGC is its enhanced sensitivity. However, as will been seen, highly sensitive methods such as comprehensive two-dimensional GC (GC × GC) have not provided further insight into the composition of bergamot oil. Perhaps the sensitivity of LC-GC, or even better, GC × GC, could be of help for the analysis of target trace-amount compounds such as contaminants.

7.3   LC-BASED METHODS

7.3.1   CONVENTIONAL LC

The main percentage of bergamot essential oil is formed of volatile compounds, so it is not surprising that gas chromatography methods play a major analytical role. However, the analysis of the non-volatile fraction is also of interest, not only for scopes of characterization, but also from a human metabolic viewpoint.

As mentioned previously, the oxygen heterocyclic constituents are commonly detected exploiting the strong UV absorbance, by using either an NP or RP-HPLC column. It is worth noting that several oxygen heterocyclic compounds are not available as pure standards, so the use of an "information-rich" detector, such as the mass spectrometer, is also highly advisable. MS systems can also be of great support in the case of co-eluting oxygen heterocyclics, though this is not a major problem when analyzing bergamot oil, which is of rather low complexity.

The development of atmospheric pressure ionization (API) methods has led to considerable expansion of MS detection in the HPLC field. Among the API interfaces, atmospheric pressure chemical ionization (APCI) and electrospray (ESI) complement each other with regards to analyte molecular weight, polarity, and chromatography conditions. In particular, ESI is useful for polar and ionic solutes ranging from 100 to 150,000 Da in MW, while APCI is applicable to non- and medium-polarity molecules with MW in the 100–2000 Da range (Niessen 1999).

Polymethoxylated flavones are more polar than psoralens and coumarins, and can be subjected to analysis using both ESI and APCI in the positive mode; psoralens and coumarins ionize only when APCI in the positive mode is employed (P. Dugo et al. 1999; P. Dugo et al. 2000). API techniques are soft ionization ones, with very little or no fragmentation and the formation of a pseudomolecular ion.

Frérot and Dercorzant (2004) employed RP-HPLC (C18, 150 × 2.1 mm ID × 3 µm) in combination with UV, APCI-MS, and fluorescence detection for the analysis of furocoumarins in six *Citrus* oils, including bergamot. Considering the latter sample, bergamottin accounted for more than 99% of the total furocoumarin content. The alkaline treatment of the oil explained the low concentration of bergapten found (approximately 8 ppm). On the basis of the data provided by the UV detector, oxypeucedanin was quantified at the 53 ppm level; such an outcome was not confirmed by the MS or the fluorescence detectors. Additionally, the co-elution of epoxybergamottin with a compound characterized by a molecular weight of 274 resulted in an overestimated UV result. Because of the high selectivity of MS detection, it was found to be the best option for quantification purposes.

In a recent study focused on the multidisciplinary analysis of a series (50) of bergamot essential oils, herniarin (7-methoxycoumarin) was reported in cold-pressed samples for the first time (G. Dugo et al. 2011). The authors used UV/vis and MS detection, along with pure standard compounds, for identification/quantification, and a C18 fused-core column (150 × 4.6 mm ID × 2.7 µm) for separation. In recent years, fused-core particles have been gathering an increasing number of followers due to their high efficiency. Fused-core particles, with diameters of ≈ 3 µm, are characterized by comparable efficiency to <2 µm fully porous ones (Carr et al. 2011). An HPLC-UV chromatogram relative to the cold-pressed bergamot oil oxygen heterocyclic fraction, showing "GC"-like efficiency, is illustrated in Figure 7.7.

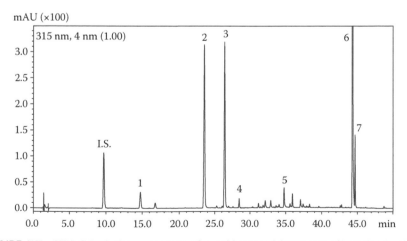

FIGURE 7.7 HPLC-UV chromatogram of a cold-pressed bergamot oil attained using a fused-core particle column. For peak identification refer to source. (From Dugo, G. et al., *J. Essent. Oil Res.* 24, 93–117, 2011. Reproduced with permission from Allured.)

In cold-pressed oils, the authors confirmed the presence of four main compounds, namely two coumarins (citropten and 5-geranyloxy-7-methoxycoumarin) and two psoralens (bergapten and bergamottin); two polymethoxyflavones, sinensetin and tetra-O-methylscutellarein, were quantified for the first time, while herniarin had only been previously reported in concentrated oil (Costa et al. 2010). The presence of herniarin, a coumarin found predominantly in lime oils (Dugo and Russo 2011), was confirmed by high-resolution mass spectrometry.

7.3.2 FAST HPLC

Recently, there have been a series of impressive developments with the objective of increasing speed (plate generation/unit of time) in HPLC applications. Such fast approaches include the use of small non-porous, fully porous, and core-shell particles, as well as monolithic columns. The employment of higher pressures and/or temperatures can be used to reduce analysis times. Evolution in column technologies has acted as a stimulant toward the development of improved LC instrumentation, with higher pressure capabilities, reduced extra-column sources of band broadening, better fittings, and faster detectors (Carr et al. 2011).

One of the most popular approaches toward faster analysis and higher column efficiency in HPLC is that related to a reduction in particle diameter. Short-packed columns with microparticles can maintain essentially the same separation power as a conventionally packed one, if methods are properly tuned. The main disadvantage of using smaller particles is the increased pressure requirements. Bonaccorsi et al. (1999) employed a reversed-phase 30 × 4.6 mm ID column, packed with 3 μm particles, for the rapid analysis of oxygenated heterocyclic compounds in five *Citrus* oils, including bergamot. The separation of the target analytes was achieved in less than seven minutes, while resolution was maintained at a satisfactory level.

The employment of monolithic columns is another interesting route toward faster HPLC analysis (Carr et al. 2011). Such columns are made of one single piece of adsorbent material, filling the entire column, and are characterized by high permeability due to their increased porosity. Furthermore, time losses between consecutive applications are greatly reduced due to fast re-equilibration. An additional advantage is that efficiency is not significantly affected at high flow rates. Currently, substantial improvements are being made in monolithic-column technology, as demonstrated by the excellent performance of second-generation silica monoliths (Carr et al. 2011).

The effectiveness of C18 monoliths in the fast analysis of *Citrus* oil oxygenated heterocyclic compounds (1-min separations) was evaluated in preliminary research, prior to the development of a comprehensive 2D LC system (P. Dugo et al. 2004).

7.4 COMPREHENSIVE MULTIDIMENSIONAL METHODS

The main difference between a heart-cutting and a comprehensive chromatography (CC) method is related to the number of transferred fractions—in a CC approach, the entire initial sample is analyzed in each dimension through continuous sampling. In

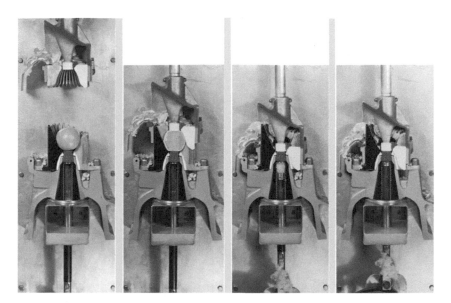

FIGURE 6.7   Extractor FMC In-Line: scheme.

FIGURE 6.8   Brown oil extractor (BOE).

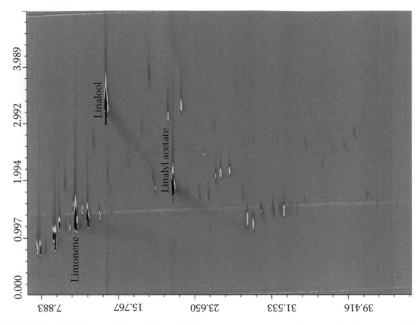

FIGURE 7.8 Full-scan GC × GC-MS chromatogram of a cold-pressed bergamot oil with the peaks relative to limonene, linalool, and linalyl acetate indicated.

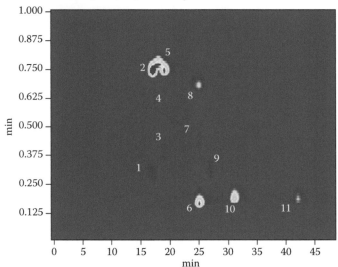

FIGURE 7.12 Comprehensive 2D normal-phase—reversed-phase LC separation of the oxygen heterocyclic fraction of a lemon essential oil sample. Peak identification: (1) unknown; (2) 5-geranyloxypsoralen (bergamottin); (3) 5-isopentenyloxypsoralen (isoimperatorin); (4) 5-geranyloxy-8-methoxypsoralen; (5) 5-geranyloxy-7-methoxycoumarin; (6) 5,7-dimethoxycoumarin (citropten); (7) 5-methoxy-8-isopentenyloxypsoralen (phellopterin); (8) 8-geranyloxypsoralen; (9) 5-isopentenyloxy-8-epoxyisopentyloxypsoralen; (10) 5-epoxyisopentyloxypsoralen (oxypeucedanin); (11) 5-methoxy-8-(2,3-epoxyisopentyloxy)psoralen. (From Dugo et al., *Anal. Chem.* 76, 2525–2530, 2004. Reproduced with permission from American Chemical Society.)

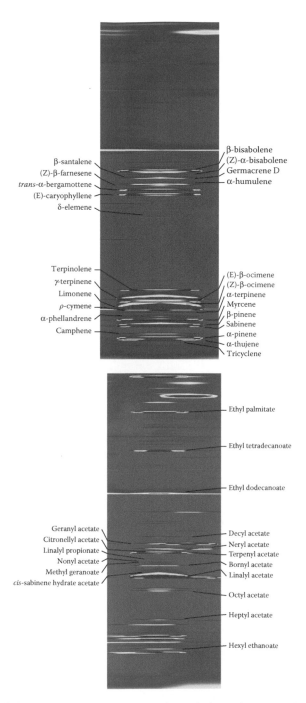

FIGURE 7.10 LC × GC chromatogram expansions relative to bergamot oil. (a) Mono- and sesquiterpene hydrocarbons; (b) Esters. The *y*-axis goes from 0 to 60 min; the *x*-axis (1.5-min width) corresponds to three cuts.

FIGURE 18.3 The bottle of "Eau de Cologne Imperiale" created by Pierre-François-Pascal Guerlain (1853).

an ideal CC system, the total peak capacity becomes that of the first column multiplied by that of the second ($n_{c1} \times n_{c2}$).

Comprehensive chromatography methods are generally characterized by the following general characteristics: (1) two columns, with distinct selectivities, are connected in series; (2) a transfer device is situated between the two dimensions; (3) the function of the interface is, in all cases, to cut and release continuous fractions from the first to the second column.

The methods that have been employed in the *Citrus* oil field have been comprehensive 2D GC (GC × GC), LC-GC (LC × GC), and LC (LC × LC). Bergamot essential oil has been analyzed using GC × GC and LC × GC. To the best of the authors' knowledge, bergamot oil has not been subjected to LC × LC analysis, even though the most interesting and fruitful *Citrus* oil applications have been, without a doubt, the LC × LC ones.

7.4.1 Comprehensive 2D GC

Comprehensive 2D GC was first described in 1991, and can be certainly considered one of the most revolutionary evolutions in the GC field (Liu and Phillips 1991). In GC × GC instruments, the transfer device is defined as a modulator, is normally located at the head of the second column, and functions in a continuous manner throughout the application. Any modulator enables the isolation, and (often) compression and re-injection (modulation process) of chromatography bands from a conventional column (e.g., 30 × 0.25 mm ID) onto a short segment of micro-bore capillary (e.g., 1–2 m × 0.10 mm ID). The modulation period is the time between a transfer process and the following one, and corresponds to the analysis time in the second dimension (typically 4–8 s). Each peak eluting from the first dimension must be subjected to three or four modulations to maintain primary-column resolution. During the fast second-dimension analysis, modulation is carried out on the subsequent fraction. An apolar phase is often employed in the first dimension; then, packets of isovolatile solutes are subjected to a rapid 2D separation, generally on a medium- or high-polarity column. Such a column combination guarantees slow and fast peak production in the first and second dimension, respectively.

Comprehensive 2D GC experiments are achieved through a variety of modulators, with cryogenic systems being the most common devices. The reader is directed to the literature for detailed information (Tranchida et al. 2011). Cryogenic modulation generates very narrow chromatography bands, and therefore fast detectors are mandatory; furthermore, entrapment has a beneficial effect on sensitivity: signal-to-noise ratios increase to the 10–50 factor range, depending on the experimental conditions and the cryogen employed.

In all comprehensive chromatography experiments, dedicated software is needed for visualization and quantification. For example, if a one-hour GC × GC application is considered, with a six-second modulation period, then 600 sequential six-second mini-chromatograms will form a "raw" (monodimensional) GC × GC chromatogram. The latter is transformed in a bidimensional chromatogram as follows: every single second-dimension analysis (or mini-chromatogram) is positioned at a 90-degree angle to an *x*-axis, and is characterized by a first-dimension retention time normally expressed in minutes. Every compound separated on the second column is aligned along a *y*-axis and is characterized by an ellipse form and by a retention time value expressed in seconds. The color intensity and dimension of each bidimensional peak are directly related to the detector response. With regards to quantification issues, peak areas relative to the same compound in each fast 2D chromatogram are summed.

The advantages of comprehensive 2D GC, over conventional GC, include increased sensitivity, selectivity, and separation power; more simple identification, due to the formation of group-type patterns (for homologous series of compounds); and speed (comparable to very fast GC experiments, considering the number of peaks resolved per unit of time). If a mass spectrometer is hyphenated to a GC × GC system, then the most powerful approach for the analysis of volatile compounds is generated (GC × GC-MS).

The number of published GC × GC investigations in the bergamot oil field is very low; such a statement is valid for all *Citrus* oils. The reason is most probably due to the simple fact that the need for such a powerful method for *Citrus* oil analysis has yet to be demonstrated.

Shellie and Marriott (2002) extended the concept of enantio-MDGC to the field of GC × GC—a GC × enantio-GC-MS method was developed for analysis of bergamot oil. The research was more an interesting method-optimization exercise than a feasible analytical route for chiral analysis. The authors exploited the vacuum-outlet conditions for the separation of enantiomers on a short chiral capillary column (1 m × 0.25 mm ID). A micro-bore apolar column (10 m × 0.1 mm ID) was used in the first dimension, and was exploited not only for separation purposes but also as a restriction; the 0.25-mm ID column was suited for low-pressure rapid chiral separations because subatmospheric conditions extended across the entire second-dimension, producing an increase in the optimum linear velocity. Though wrap-around (analytes with a second-dimension retention time exceeding the modulation period) was rather evident, some nice second-dimension enantiomer separations were illustrated.

A GC × GC-MS chromatogram (unpublished data), relative to the volatile fraction of a cold-pressed bergamot oil, is shown in Figure 7.8. The main compounds (limonene, linalool, and linalyl acetate) are indicated in the bidimensional plot. Not more than 80–90 compounds appear on the 2D plane, highlighting the medium-complexity nature of bergamot oil.

7.4.2 Comprehensive 2D LC-GC

In the GC field, it is well known that no matter which stationary phase is used, retention is always dependent to a certain extent on analyte boiling points. For this reason, it is extremely hard to find a compound which elutes early from the first GC × GC dimension and late from the second, or vice-versa. On the contrary, entirely different separation mechanisms can be attained in the field of multidimensional LC-GC. A highly orthogonal choice is a polarity-type LC separation, combined with a boiling-point GC one.

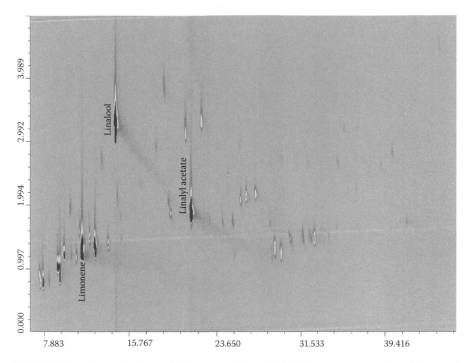

FIGURE 7.8 **(See color insert.)** Full-scan GC × GC-MS chromatogram of a cold-pressed bergamot oil with the peaks relative to limonene, linalool, and linalyl acetate indicated.

The technological leap from heart-cutting to comprehensive LC-GC is rather large, and probably for such a reason, LC × GC instruments have only been rather recently developed. The first description of an LC × GC instrument was in 2000, and was only applicable to volatile compounds (Quigley et al. 2000). Later, an entirely automated LC × GC system was exploited in lipid analysis (de Koning et al. 2004). The instrument was operated using stop-flow conditions and combined with ToF mass spectrometry. Two transfer devices were evaluated; a dual side-port syringe and a six-port switching valve, both providing satisfactory results.

Only a single LC × GC application has been carried out in the *Citrus* oil field, specifically on bergamot oil (Mondello et al. 2008): the LC analysis was carried out under gradient NP conditions (300 × 1 mm ID, silica) and enabled a polarity-based separation. The first fraction was composed of mono- and sesquiterpene hydrocarbons, followed by esters, aliphatic and monoterpene aldehydes, and mono-/sesquiterpene alcohols. LC fractions, 30 seconds wide (corresponding to 160 µL), were transferred to the GC system using a dual-port syringe (Figure 7.9). The latter was linked, through two transfer lines, to the LC detector outlet (entrance A) and to waste (exit B). A rubber plug was located at the end of the syringe piston. Throughout the waste mode, the plug is located below line A, while during the transfer mode the plug is situated between the lines. The multidimensional system was capable of both heart-cutting LC-GC and LC × GC analysis. Solvent elimination

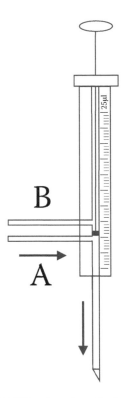

FIGURE 7.9   Scheme of the LC × GC syringe transfer device in the transfer configuration.

was achieved with a programmed-temperature vaporizer injector, while GC separation was carried out on a 25 m × 0.25 mm ID apolar column (the LC pumps were
stopped during each GC analysis). LC × GC chromatogram expansions, relative to
the hydrocarbon and ester fractions of bergamot oil, are shown in Figure 7.10. The
LC × GC application on bergamot oil was a successful one, also demonstrating the
effectiveness of both hardware and software.

7.4.3   Comprehensive 2D LC

Low degrees of cross-correlation can be achieved in the LC × LC field because of
the existence of methodologies characterized by entirely different selectivities (e.g.,
NP and RP). It must be noted that such a benefit is not always exploited because the
combination of a series of LC methodologies can be a rather complicated issue; in
fact, problems related to the precipitation of buffer salts, solvent immiscibility, or 1D
mobile phase-2D stationary phase incompatibility can arise (Cortes and Rothman
1990).

  Comprehensive two-dimensional liquid chromatography was introduced by Erni
and Frei in 1978 (Erni and Frei 1978), followed by Bushey and Jorgenson at the end
of the 1980s (Bushey and Jorgenson 1990). Since that time, the LC × LC systems

(a)

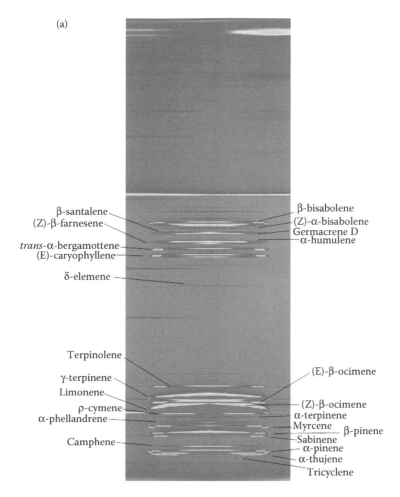

FIGURE 7.10   (See color insert.) LC × GC chromatogram expansions relative to bergamot oil. (a) Mono- and sesquiterpene hydrocarbons; (b) Esters. The *y*-axis goes from 0 to 60 min; the *x*-axis (1.5-min width) corresponds to three cuts.

developed can be basically divided into three groups: in one, a switching valve with two sample loops interfaces the first and second dimensions; in another approach, a switching valve links the primary column to two parallel secondary ones; the remaining group is characterized by methods based on the use of a switching valve, connecting the first and second dimension. Instead of sample loops, the valve is equipped with two trapping columns. When using two parallel columns, these must be characterized by identical retention properties, while peak sampling frequencies can be increased. Additionally, the instrumentation is of higher complexity, and two detectors are needed. Peak focusing is enhanced when using trapping columns, though modulation periods are generally higher. With regards to the use of two loops, it is the most popular and simple methodology, though peak focusing is less pronounced.

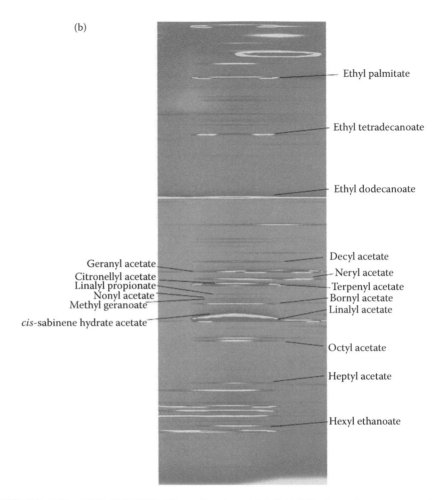

(b)

Ethyl palmitate

Ethyl tetradecanoate

Ethyl dodecanoate

Decyl acetate
Geranyl acetate
Citronellyl acetate
Linalyl propionate
Nonyl acetate
Methyl geranoate
cis-sabinene hydrate acetate
Neryl acetate
Terpenyl acetate
Bornyl acetate
Linalyl acetate

Octyl acetate

Heptyl acetate

Hexyl ethanoate

FIGURE 7.10 **(CONTINUED) (See color insert.)** LC × GC chromatogram expansions relative to bergamot oil. (a) Mono- and sesquiterpene hydrocarbons; (b) Esters. The *y*-axis goes from 0 to 60 min; the *x*-axis (1.5-min width) corresponds to three cuts.

Comprehensive two-dimensional liquid chromatography has not been used for the analysis of non-volatiles contained in bergamot oil, though such an application will most probably be achieved in the future. Worthy of note is that the first-ever combination of normal-phase and reversed-phase modes, in an LC × LC system, was applied to the analysis of the oxygen heterocyclic components contained in lemon oil (P. Dugo et al. 2004). For this experiment, which also coincided with the first *Citrus* oil analysis in the LC × LC field, a micro-bore silica column was used in the first dimension (isocratic conditions) while a short C18 monolithic column was employed in the second (gradient elution). Monolithic columns are a good choice in the second dimension because they enable both high flows (with little efficiency losses) and rapid re-equilibration times. Considering the first dimension, the employment of a micro-bore column is advisable for the following reasons: the small column

ID avoids excessive dilution, generating flows that are compatible with transfer volumes; a preconcentration step at the head of the second column is not required; and solvent incompatibility, using different separation modes (such as in this case), can be avoided.

A gradient was applied in the second dimension every 60 seconds, and was necessary to elute the oxygenated compounds within 48 seconds (plus 12 s re-equilibration time). The columns were linked to a 10-port valve, equipped with two 20-μL loops, which were alternately filled with first-dimension 60-second chromatography bands. While one loop was in the cut position, the content of the other loop was flushed onto the monolithic column; at the end of the transfer period, the valve was switched, inverting the loop positions.

It is worthy to note that no degradation of the chromatography performance was observed in the transfer of small volumes of immiscible solvent onto the second column. Furthermore, the first-dimension mobile phase composition was stronger than that at the head of the monolithic column, generating a peak focusing effect during each second-dimension injection. The LC × LC instrumentation used is shown in Figure 7.11, while the lemon oil bidimensional chromatogram is illustrated in Figure 7.12. Peak identification (eight psoralens and two coumarins) was achieved through the comparison of retention times with those of standard compounds (when available) and UV spectra.

In the specific case of lemon oil, it can be observed that a single-column application would have failed in the resolution of the heterocyclic compounds. Considering

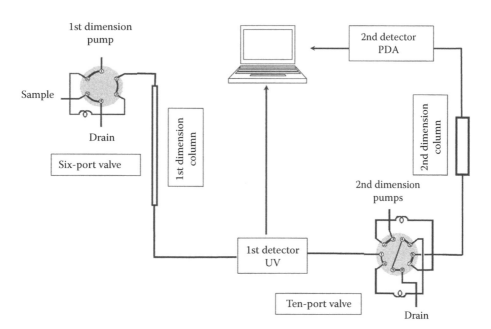

FIGURE 7.11 Schematic of the LC × LC instrument used for the analysis of lemon essential oil. (From Dugo et al., *Anal. Chem.* 76, 2525–2530, 2004. Reproduced with permission from American Chemical Society.)

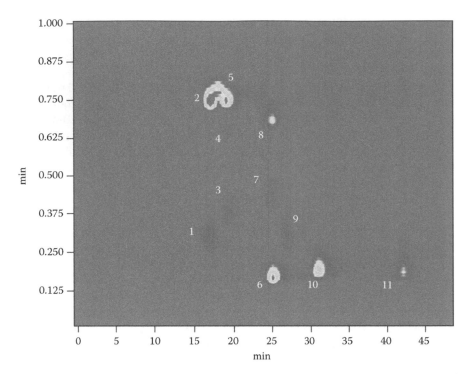

FIGURE 7.12 (See color insert.) Comprehensive 2D normal-phase—reversed-phase LC separation of the oxygen heterocyclic fraction of a lemon essential oil sample. Peak identification: (1) unknown; (2) 5-geranyloxypsoralen (bergamottin); (3) 5-isopentenyloxypsoralen (isoimperatorin); (4) 5-geranyloxy-8-methoxypsoralen; (5) 5-geranyloxy-7-methoxycoumarin; (6) 5,7-dimethoxycoumarin (citropten); (7) 5-methoxy-8-isopentenyloxypsoralen (phellopterin); (8) 8-geranyloxypsoralen; (9) 5-isopentenyloxy-8-epoxyisopentyloxypsoralen; (10) 5-epoxyisopentyloxypsoralen (oxypeucedanin); (11) 5-methoxy-8-(2,3-epoxyisopentyloxy) psoralen. (From Dugo et al., *Anal. Chem.* 76, 2525–2530, 2004. Reproduced with permission from American Chemical Society.)

the first dimension, in accordance with the NP separation mechanism, isomers are well-resolved, as can be seen for bergamottin (peak 2) and 8-geranyloxypsoralen (peak 8), while homologues, on the contrary, are not very well separated because hydrocarbon substituents contribute little to analyte retention. For example, bergamottin and isoimperatorin (peak 3), which differ in the alkyl chain length, elute very closely in the first dimension. Obviously, compounds with differing polarities are easily separated under NP conditions. The NP separation was useful in the resolution of a series of analytes that would have co-eluted under RP conditions (e.g., peaks 6, 10, and 11).

Considering the RP dimension, hydrophobicity dominates retention. In fact, components with a geranyloxy side chain are more retained (peaks 2, 4, 5, and 8), compared to those with a shorter alkyl chain (peaks 3, 7, and 9), with each group located in a distinct chromatogram zone. The position of the geranyloxy side chain is less discriminating compared to the NP mode, and components such as bergamottin and

8-geranyloxypsoralen have similar RP retention times. The RP mode was useful in the resolution of a series of compounds that otherwise would have overlapped in the first dimension (e.g., peaks 3, 4, 5 and 6, 7, 8).

7.5 CONCLUSIONS AND FUTURE PERSPECTIVES

Hopefully, a broad overview of the current chromatography-based strategies and possibilities in the field of bergamot essential oil has been herein given. The number of instrumental options is high, and the present contribution can direct the bergamot oil analyst to the most appropriate choice on the basis of the initial analytical objective.

With regards to conventional GC applications, it can be concluded that even if an MS dimension is employed, the overall resolving power of the system is often lower than that required. The GC analysis of bergamot oil chiral constituents is a typical example. It must be added, however, that novel GC stationary phases, such as ionic liquids, hold promise for the future.

Single-column LC methods appear to be a valid tool for the elucidation of the non-volatile fraction, especially if MS data are available. Furthermore, high-resolution stationary phases, such as fused-core particles, have given some very good results.

In GC applications, the availability of a heart-cutting MDGC-MS system is an ideal solution for practically all that is demanded from the analysis of the bergamot oil volatile fraction. There appears to be little room for future improvement in the MDGC instrumental field. If the MDGC-MS results attained can be confirmed by using a GC-IRMS instrument, then all the better.

The need for GC × GC or LC × GC is yet to be demonstrated, with possible room for use in the determination of trace-amount natural constituents and/or contaminants. The superior resolving power of LC × LC has given some excellent results in the *Citrus* field, but has not found yet application for bergamot oil. However, comprehensive chromatography methods will most probably be evaluated in future investigations in the bergamot oil field.

In conclusion, the analytical tools currently available, if used in an appropriate manner, can reveal practically all economically convenient bergamot oil adulterations. However, such illegal practices tend to evolve, becoming more sophisticated, in attempts to remain invisible to the analytical community. Bergamot oil analysts must always be aware of such a tendency and remain attached to well-known practices but also be open to innovation.

REFERENCES

Bicchi, C., Liberto, E., Cagliero, C., et al. 2008. Conventional and narrow bore short capillary columns with cyclodextrin derivatives as chiral selectors to speed-up enantioselective gas chromatography and enantioselective gas chromatography–mass spectrometry analyses. *J. Chromatogr. A* 1212:114–123.

Blumberg, L. M., and Klee, M. S. 1998. Method translation and retention time locking in partition GC. *Anal. Chem.* 70:3828–3839.

Bonaccorsi, I. L., McNair, H. M., Brunner, L. A., Dugo, P., and Dugo, G. 1999. Fast HPLC for the analysis of oxygen heterocyclic compounds of Citrus essential oils. *J. Agric. Food Chem.* 47:4237–4239.

Brenna, J. T., Corso, T. N., Tobias, H. J., and Caimi, R. J. 1997. High-precision continuous-flow isotope ratio mass spectrometry. *Mass Spectrom. Rev.* 16:227–258.

Bushey, M. M., and Jorgenson, J. W. 1990. Automated instrumentation for comprehensive two-dimensional high-performance liquid chromatography of proteins. *Anal. Chem.* 62:161–167.

Carr, P. W., Stoll, D. R., and Wang, X. 2011. Perspectives on recent advances in the speed of high-performance liquid chromatography. *Anal. Chem.* 83:1890–1900.

Cortes, H. J., and Rothman D. L. 1990. Multidimensional high-performance liquid chromatography. In *Multidimensional chromatography: Techniques and applications*, ed. H. J. Cortes, 219–250. New York: Marcel Dekker.

Costa, R., Dugo, P., Navarra, M., et al. 2010. Study on the chemical composition variability of some processed bergamot (*Citrus bergamia*) essential oils. *Flavour Fragr. J.* 25:4–12.

d'Acampora Zellner, B., Dugo, P., Dugo, G., and Mondello, L. 2008. Gas chromatography–olfactometry in food flavour analysis. *J. Chromatogr. A* 1186:123–143.

de Koning, S., Janssen, H-G, van Deursen, M., and Brinkman, U. A. Th. 2004. Automated on-line comprehensive two-dimensional LC × GC and LC × GC-ToF MS: Instrument design and application to edible oil and fat analysis. *J. Sep. Sci.* 27:397–409.

Dugo, G., Bonaccorsi, I., Sciarrone, D., et al. 2011. Characterization of cold-pressed and processed bergamot oils by GC-FID, GC-MS, GC-C-IRMS, enantio-GC, MDGC, HPLC, HPLC-MS-IT-TOF. *J. Essent. Oil Res.* 24:93–117.

Dugo, P., Favoino, O., Luppino, R., Dugo, G., and Mondello, L. 2004. Comprehensive two-dimensional normal-phase (adsorption)–reversed-phase liquid chromatography. *Anal. Chem.* 76:2525–2530.

Dugo, P., Mondello, L., Dugo, L. Stancanelli, R., and Dugo, G. 2000. LC-MS for the identification of oxygen heterocyclic compounds in citrus essential oils. *J. Pharm. Biomed. Anal.* 24:147–154.

Dugo, P., Mondello, L., Sebastiani, E., et al. 1999. Identification of minor oxygen heterocyclic compounds of Citrus essential oils by liquid chromatography-atmospheric pressure chemical ionisation mass spectrometry. *J. Liq. Chrom. & Rel. Technol.* 22:2991–3005.

Dugo, P., and Russo, M. 2011. The oxygen heterocyclic components of Citrus essential oils. In *Citrus oils*, eds. G. Dugo and L. Mondello, 405–443. Boca Raton, FL: CRC Press.

Erni, F., and Frei, R. W. 1978. Two-dimensional column liquid chromatographic technique for resolution of complex mixtures. *J. Chromatogr. A* 149:561–569.

Eyres, G., Marriott, P. J., and Dufour, J-P. 2007. The combination of gas chromatography–olfactometry and multidimensional gas chromatography for the characterisation of essential oils. *J. Chromatogr. A* 1150:70–77.

Frérot, E., and Decorzant, E. 2004. Quantification of total furocoumarins in Citrus oils by HPLC coupled with UV, fluorescence, and mass detection. *J. Agric. Food Chem.* 52:6879–6886.

Giddings, J. C. 1990. Use of multiple dimensions in analytical separations in multidimensional chromatography. In *Multidimensional chromatography: Techniques and applications*, ed. H. J. Cortes, 1–27. New York: Marcel Dekker.

Grob, K. 1991. *On-line coupled LC-GC*. Heidelberg: Hüthig.

Juchelka, D., Beck, T., Hener, U., Dettmar, F., and Mosandl, A. 1998. Multidimensional gas chromatography coupled on-line with isotope ratio mass spectrometry (MDGC-IRMS): Progress in the analytical authentication of genuine flavor components. *J. High Resol. Chromatogr.* 21:145–151.

Korytár, P., Janssen, H-G, Matisová, E., and Brinkman, U. A. Th. 2002. Practical fast gas chromatography: Methods, instrumentation and applications. *Trends Anal. Chem.* 21:558–572.

Liberti, A., and Conti, G. 1956. Possibilità di applicazione della cromatografia in fase gassosa allo studio delle essenze. In *Atti I Convegno internazionale di studi e ricerche sulle essenze*. Reggio Calabria, Italy.

Liberto, E., Cagliero, C., Sgorbini, B., et al. 2008. Enantiomer identification in the flavour and fragrance fields by "interactive" combination of linear retention indices from enantioselective gas chromatography and mass spectrometry. *J. Chromatogr. A* 1195:117–126.

Liu, Z., and Phillips, J. B. 1991. Comprehensive two-dimensional gas chromatography using an on-column thermal modulator interface. *J. Chromatogr. Sci.* 29:227–231.

Mondello, L., Casilli, A., Tranchida, P. Q., et al. 2004. Fast GC for the analysis of Citrus oils. *J. Chromatogr. Sci.* 42:410–416.

Mondello, L., Dugo, G., and Bartle, K. D. 1996a. Coupled HPLC-HRGC-MS: A new method for the on-line analysis of real samples. *Am. Lab.* December: 41–49.

Mondello, L., Dugo, G., Dugo, P., and Bartle, K. D. 1996b. On-line HPLC-HRGC in the analytical chemistry of Citrus essential oils. *Perfum. Flav.* 21:25–49.

Mondello, L., Dugo, P., and Dugo, G. 2008. Multidimensional and comprehensive chromatography (MDGC, GC × GC, LC × LC, LC × GC) for the analysis of complex matrices. In *Proceedings of the Tenth International Symposium on Hyphenated Techniques in Chromatography and Hyphenated Chromatographic Analyzers*, Bruges, Belgium.

Niessen, W. M. A. 1999. *Liquid chromatography–mass spectrometry.* 2nd ed. Chromatographic Science Series, vol. 79. New York: Marcel Dekker.

Purcaro, G., Tranchida, P. Q., Ragonese, C., et al. 2010. Evaluation of a rapid-scanning quadrupole mass spectrometer in an apolar × ionic-liquid comprehensive two-dimensional gas chromatography system. *Anal. Chem.* 82:8583–8590.

Quigley, W. W. C., Fraga, C. G., and Synovec, R. E. 2000. Comprehensive LC × GC for enhanced headspace analysis. *J. Microcol. Sep.* 12:160–166.

Quimby, B., McCurry, J., and Norman, W. 2007. Capillary flow technique for gas chromatography: Reinvigorating a mature analytical discipline. *LC GC The Peak* April: 7–15.

Ragonese, C., Sciarrone, D., Tranchida, P. Q., et al. 2011. Evaluation of a medium-polarity ionic liquid stationary phase in the analysis of flavor and fragrance compounds. *Anal. Chem.* 83:7947–7954.

Sawamura, M., Onishi, Y., Ikemoto, J., Thi Minh Tu, N., and Thi Lan Phi, N. 2006. Characteristic odour components of bergamot (*Citrus bergamia* Risso) essential oil. *Flavour Fragr. J.* 21:609–615.

Scandinaro, M., Tranchida, P. Q., Costa, R., et al. 2010. Rapid quality control of flavours and fragrances by using fast gas chromatography-mass spectrometry and multi-MS library search procedures. *LC GC Eur.* 23:456–464.

Sciarrone, D., Schillipiti, L., Ragonese, C. et al. 2010. Thorough evaluation of the validity of conventional enantio-gas chromatography in the analysis of volatile chiral compounds in mandarin essential oil: A comparative investigation with multidimensional gas chromatography. *J. Chromatogr. A* 1217:1101–1105.

Seeley, J. V., Micyus, N. J., Bandurski, S. V., Seeley, S. K., and McCurry, J. D. 2007. Microfluidic Deans switch for comprehensive two-dimensional gas chromatography. *Anal. Chem.* 79:1840–1847.

Shellie, R., and Marriott, P. J. 2002. Comprehensive two-dimensional gas chromatography with fast enantioseparation. *Anal. Chem.* 74:5426–5430.

Tranchida, P. Q., Lo Presti, M., Costa, R., et al. 2006. High-throughput analysis of bergamot essential oil by fast solid-phase microextraction–capillary gas chromatography–flame ionization detection. *J. Chromatogr. A* 1103:162–165.

Tranchida, P. Q., Purcaro, G., Dugo, P., and Mondello, L. 2011. Modulators for comprehensive two-dimensional gas chromatography. *Trends Anal. Chem.* 30:1437–1461.

Veriotti, T., and Sacks, R. 2002. High-speed characterization and analysis of orange oils with tandem-column stop-flow GC and time-of-flight MS. *Anal. Chem.* 74:5635–5640.

Wang, X., Nakagawa-Goto, K., Kozuka, M., et al. 2006. Cancer preventive agents. Part 6: Chemopreventive potential of furanocoumarins and related compounds. *Pharmaceut. Biol.* 44:116–120.

Wu, C-S., Lan, C-C. E., Wang, L-F., et al. 2007. Effects of psoralen plus ultraviolet A irradiation on cultured epidermal cells in vitro and patients with vitiligo *in vivo. Br. J. Dermatol.* 156:122–129.

8 The Composition of the Volatile Fraction of Peel Oils

Giovanni Dugo, Antonella Cotroneo,
Ivana Bonaccorsi, and Maria Restuccia

CONTENTS

8.1 INTRODUCTION

The volatile fraction of bergamot oil represents about 95% of the whole oil. It is composed of monoterpene and sesquiterpene hydrocarbons and their oxygenated derivatives (alcohols, esters, aldehydes, and oxides). Small amounts of aliphatic alcohols, aldehydes, and esters are also present. The main components are limonene, linalyl acetate, and linalool (Dugo et al. 2002). The composition of this fraction of bergamot oil greatly differs from that of other citrus peel oils because of the large amount of oxygenated compounds, mainly linalyl acetate and linalool, which can be more than 50% of the entire volatile fraction. In all the other citrus peel oils, the oxygenated volatile compounds range from 1% in sweet orange, grapefruit, and clementine, to 6% in lime oils (Di Giacomo and Mincione 1994).

The analytical technique which most contributed to the acquisition of knowledge on the composition of the volatile fraction, as for all the other citrus oils, is gas chromatography with conventional detectors or hyphenated to mass spectrometry. The very first application of gas chromatography to the analysis of bergamot oil, which also represents the first application of GC to citrus essential oils in general, was made by Prof. Arnaldo Liberti. At the beginning of the 1950s in Messina he obtained the first chromatogram of bergamot oil presented at the 1st International Symposium on Essential Oils, held in Reggio Calabria in 1956 (Liberti and Conte 1956). This chromatogram, obtained on a column packed with triacetyl phosphate on celite, is reported in Figure 8.1, which is relative to monoterpene hydrocarbon. In this figure are only five peaks, probably α-pinene, β-pinene, sabinene, limonene, and γ-terpinene. Two years later the same research group (Liberti and Cartoni 1958), taking into consideration the complexity of the volatile fraction of citrus oils and the limited separation power at the time achievable by gas chromatography, which did

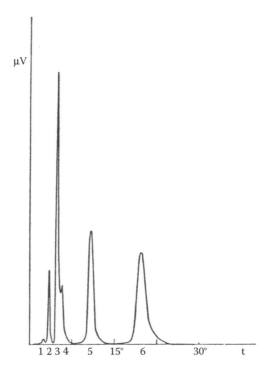

FIGURE 8.1 Gas chromatogram of a bergamot essential oil. Column: trycresylphosphate on celite; column temperature: 100°C. (From Liberti, A., and Conte, G. 1956. *Atti I Congresso Internazionale di Studi e Ricerche sulle Essenze*, Reggio Calabria, Italy, March 1956.)

not allow sufficient resolution of all the components, proposed a method to fractionate bergamot oil by physical and chemical means, and successive GC analyses of the single fractions. The scheme of this method is reported in Figure 8.2, while the chromatograms obtained for the fractions are reported in Figure 8.3.

Since then, gas chromatographic techniques evolved along with the development of knowledge on the composition of bergamot oil volatile fraction as well as that of other citrus oils. First, packed columns were used, then capillary columns made of steel, then glass, and finally fused silica; GC coupled with mass spectrometry (GC/MS) and recently multidimensional techniques allowed for the highlighting of several hundreds of components in bergamot oil, most of which are confidently identified. To better see this evolution, it is helpful to compare Figure 8.1 with the chromatogram obtained with a fused silica capillary column coated with DB-5 reported in Figure 8.4. In the latter chromatogram more than 100 peaks are evident and well resolved; most of them are also fully identified (Dugo et al. 2011).

Most of the studies on the composition of bergamot oil are devoted to the quantitative determination of the components identified. These were carried out on oils industrially produced in Calabria (Calvarano 1963, 1965; Di Giacomo et al. 1963; Calvarano and Calvarano 1964; Liberti and Goretti 1974; Huet 1981; Mazza 1986; Dugo et al. 1987, 1991, 1999, 2012; Dugo 1994; Lamonica et al. 1990; Chouchi et al. 1995; Verzera et al. 1996, 1998; Ferrini et al. 1998; Sawamura et al. 1999b, 2006; Gionfriddo et al. 2000, Russo et al. 2001; Poiana et al. 2003; Tranchida et al. 2006; Belsito et al. 2007; Sciarrone 2009; Costa et al. 2010) and in other regions such as

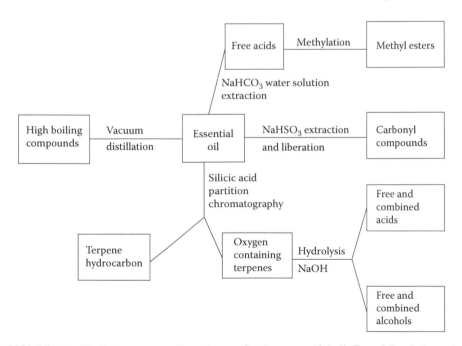

FIGURE 8.2 Preliminary separation scheme of a citrus essential oil (From Liberti, A., and Cartoni, G. P. In *Gas chromatography*, ed. D. H. Desty, 321–329. London: Butterworth's Scientific Publications, 1958.)

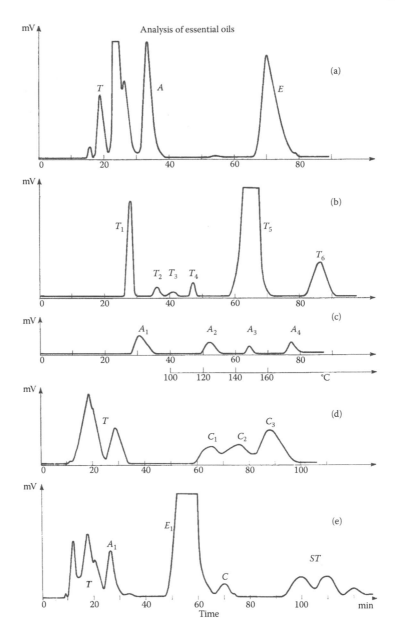

FIGURE 8.3 Bergamot oil: Chromatograms of different fractions. (a) Whole bergamot oil: (T) Monoterpene hydrocarbons, (A) alcohols, (E) esters; (b) monoterpene fraction: (T₁) α-pinene, (T₂) camphene, (T₃) unknown, (T₄) β-pinene, (T₅) limonene, (T₆) terpinene; (c) methyl esters of free fatty acids: (A₁) octanoic acid, (A₂) geranic acid, (A₃) decanoic acid, (A₄) dodecanoic acid; (d) carbonyl compounds: (C₁) unknown, (C₂) geranial, (C₃) neral; (e) high boiling compounds: (A₁) linalool, (E₁) linalyl acetate, (ST) sesquiterpene hydrocarbons. (From Liberti, A., and Cartoni, G. P. In *Gas chromatography*, ed. D. H. Desty, 321–329. London: Butterworth's Scientific Publications, 1958.)

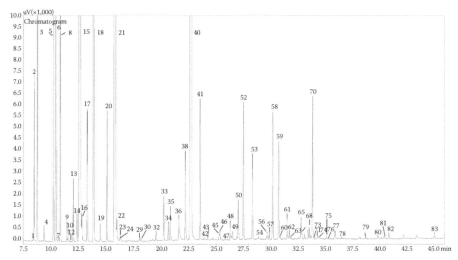

FIGURE 8.4 GC-FID chromatogram of cold-pressed bergamot oil. Peak identification: (1) tricyclene; (2) α-thujene; (3) α-pinene; (4) camphene; (5) sabinene; (6) β-pinene; (7) 6-methyl-5-hepten-2-one; (8) myrcene; (9) octanal; (10) α-phellandrene; (11) δ-3-carene; (12) hexyl acetate; (13) α-terpinene; (14) *p*-cymene; (15) limonene; (16) (Z)-β-ocimene; (17) (E)-β-ocimene; (18) γ-terpinene; (19) *cis*-sabinene hydrate; (20) octanol; (21) terpinolene; (22) linalool; (23) nonanal; (24) heptyl acetate; (25) 4,8-dimethyl-1,3(E),7-nonatriene; (26) fenchol (correct isomer was not identified); (27) *cis*-limonene oxide; (28) *trans*-limonene oxide; (29) isopulegol; (30) camphor; (31) citronellal; (32) terpinen-4-ol; (33) α-terpineol; (34) decanal; (35) octyl acetate; (36) nerol; (37) citronellol; (38) neral; (39) carvone; (40) linalyl acetate; (41) geranial; (42) perillaldehyde; (43) bornyl acetate; (44) ascaridole; (45) undecanal; (46) nonyl acetate; (47) methyl geranate; (48) linalyl propanate; (49) δ-elemene; (50) α-terpinyl acetate; (51) citronellyl acetate; (52) neryl acetate; (53) geranyl acetate; (54) β-elemene; (55) dodecanal; (56) decyl acetate; (57) *cis*-α-bergamotene; (58) β-caryophyllene; (59) *trans*-α-bergamotene; (60) (Z)-β-farnesene; (61) (E)-β-farnesene; (62) α-humulene; (63) β-santalene; (64) γ-curcumene; (65) germacrene D; (66) *trans*-β-bergamotene; (67) bicyclogermacrene; (68) (Z)-α-bisabolene; (69) (E,E)-α-farnesene; (70) β-bisabolene; (71) (Z)-γ-bisabolene; (72) δ-cadinene; (73) β-sesquiphellandrene; (74) (E)-γ-bisabolene; (75) (E)-α-bisabolene; (76) *cis*-sesquisabinene hydrate; (77) (E)-nerolidol; (78) spathulenol; (79) *trans*-sesquisabinene hydrate; (80) 2,3-dimethyl-3-(4-methyl-3-pentenyl)-2-norbonanol; (81) campherenol; (82) α-bisabolol; (83) nootkatone. (From Dugo, G. et al. *J. Essenz. Oil Res.* 24(2), 93–117, 2012.)

Sicily, Italy (Mondello et al. 2003, 2004b), Corsica (Huet 1981), the Ivory Coast (Huet 1981; Sciarrone 2009), Argentina (Ricciardi et al. 1982), Brazil (Koketsu et al. 1983; Franceschi et al. 2004), and Uruguay (Dellacassa et al. 1997); some were limited to the monoterpene hydrocarbons (Calvarano 1963; Di Giacomo et al. 1963). There are also studies on the composition of oils extracted in laboratory from fruits cultivated in Calabria (Calvarano 1968; Sawamura et al. 1999a, 1999b, 2000; Kiwanuka et al. 2000; Gionfriddo et al. 2003; Guerrini et al. 2009) or from other regions, such as Corsica, France (Huet and Dupuis 1969), California (Ortiz et al. 1978), Tuscany, Italy (Calvarano et al. 1984), Argentina (Drescher et al. 1984), China (Huang et al. 1986,

1987), Turkey (Baser et al. 1995; Kirbaslar et al. 2000, 2001), Uruguay (Dellacassa et al. 1997), Japan (Sawamura et al. 1999a, 1999b, 2000), Greece (Melliou et al. 2009) and Sicily (Verzera et al. 2000, 2003); others on commercial oils of unknown geographic origin (Ikeda et al. 1962; Schenk and Lamparsky 1981; Inoma et al. 1989; Oberhofer et al. 1999; Feger et al. 2001; Kubeczka and *Formàček* 2002; Cosimi et al. 2009; Fantin et al. 2010; Wei and Shibamoto 2010); and some on oils evidently adulterated or contaminated (*Formàček* and Kubeczka 1982). Some articles focused on the quantitative determination of the single components (MacLeod and Buigues 1964) or simply on their identification (Sundt et al. 1964; Vernin and Vernin 1966; Wenninger et al. 1967; Dixon et al. 1968; Mookherjee 1969; Ziegler 1971; Calabrò and Currò 1973; Hérisset et al. 1973; Di Corcia et al. 1975; Ehret and Moupetit 1982; Ohloff et al. 1986; Mazza 1987a). Di Giacomo and Calvarano (1970) also listed the components identified up to that time, including those determined by nonchromato-graphic techniques. Bergamot oil has also been used as a real matrix to develop and optimize advanced methods and analytical techniques (Cartoni et al. 1987; Micali et al. 1990; Lanuzza et al. 1991; Mondello et al. 1994, 1995a, 1995b, 1998, 2003, 2004a, 2004b; Veriotti and Sacks 2002). Moreover, Onischi et al. (2003) and Sawamura et al. (2006) studied the characteristics odor components.

The composition of the volatile fraction of bergamot oil has been revised by Dugo et al. (2002, 2011) and by Lawrence in his periodical reviews published in *Parfumer & Flavorist* (Lawrence, 1979–2002).

In this chapter, after a short description of the results limited to qualitative studies and to the determination of single components will be deeply described the composition of industrial oils from Calabria separate from those obtained in other geographic areas; then commercial oils of unknown origin will be reviewed, and finally those extracted in the laboratory. Some information will also be given on the composition of bergapten-free oils and on oils extracted from immature fruits, as well as on the oils recovered from the cold-extraction residues and on results obtained by particular analytical techniques. The parameters which influence the composition of cold-pressed peel oils will also be discussed.

The results reported in the tables, depending on the original source, represent the composition of a single sample, average values, or when possible the variability range. The results are expressed to the second decimal figure if the original source gave this information. If in the tables values are reported with only one decimal place, it means that the results were so expressed in the original article. In each table the components are listed alphabetically grouped by class of substances. Many tables are followed by an appendix containing the following information (if available): geographic origin, cultivar, production technology, number of samples that generated the results, analytical technique, method applied to identify the components, how the quantitative results were expressed, and characteristics of the chromatographic columns (length in meters, internal diameter in millimeters, stationary phase film thickness in micrometers). In the appendices to the tables are also the components identified only in one article, indicating in parentheses the percent amount.

In text and tables the term "tr" (traces) indicates percent amounts less than 0.05% if the results were reported to the first decimal figure, or less than 0.005% if they were reported to the second decimal figure; the symbol (+) indicates the component present but not quantitatively determined; *cis*- and *trans*-linalool oxide must be considered, if not otherwise indicated, in their furanoid form; the asterisk (*) indicates that for that component the correct isomer was not identified in the original article; a (t) indicates that the component was tentatively identified.

At the end of the chapter Table 8.15 lists all the components present, according with literature, in the volatile fraction of bergamot oil.

8.2   QUALITATIVE ANALYSES

Table 8.1 lists the qualitative assays carried out before 1975. In addition to the results attained by the analyses reported in Table 8.1, it should be mentioned that additional studies were carried out in the same period by MacLeod and Buigues (1964), who identified in bergamot oil trace amounts of nootkatone; by Sundt et al. (1964), who identified in bergamot *cis*- and *trans*-octa-5-en-2-one, geraniol, α-terpineol, *cis*- and *trans*-jasmone, *cis*- and *trans*-geranic and dihydrogeranic acids, and the carboxylic acids C_8 to C_{11}, C_{16}, and from C_{22} to C_{26}; by Vernin and Vernin (1966), who found trace mounts of methyl N-methylanthranilate; by Mookherjee (1969), who isolated from bergamot oil by column liquid chromatography and preparative gas chromatography and identified by spectroscopic techniques dehydrocineole and the following bifunctional compounds: 3-methyl-3-acetoxytocta-1,5-diene-7-one; 2,6-dimethylocta-1,7-diene-3,6-diol; 2,6-dimethylocta-7-enyl-acetate; 2,6-dimethyl-4-acetoxyocta-2,7-dienyl-6-ol; 1-hydroxydihydrocarveol; *cis*- and *trans*-2,6-dimethylocta-2,7-dienyl-acetate; 3-hydroxycitronellyl acetate; 6,7-epoxylinalyl acetate; 7-hydroxylinalyl acetate; and 8-hydroxy-*p*-menth-2en-1-yl methyl ether. From information available in literature, Di Giacomo and Calvarano (1970) also indicated the presence in bergamot oil of numerous components among which not yet listed here were octene*, δ-3-carene, α- and β-phellandrene, α-terpinene, α-thujene, heptanal, citronellol, dihydrocuminic alcohol*, nerol, perilla alcohol, terpinen-4-ol, octyl acetate, and pyrrole. In the same review Di Giacomo and Calvarano also listed some nonvolatile components such as long chain carboxylic acids, coumarins, psoralens, and diterpenes. Later the qualitative composition of the volatile fraction of bergamot oil was reviewed by La Face (1972).

More recent studies determined the presence of sesquiterpene alcohols *T*-cadinol, β-eudesmol, and farnesol* (Ehret and Moupetit 1982), (–)-(4S,8R)-8-epi-α-bisabolol and (–)-(4R,8S)-4-epi-β-bisabolol (Ohloff et al. 1986). Mazza et al. (1987a), by fractionation on silica gel columns and GC/MS analysis of each fraction, highlighted the presence in bergamot oil of 193 components. These components, with the exception of 8,9-didehydronootkatone, *cis*- and *trans*-isopiperitenol, *p*-menth-1,8-dien-4-ol, *cis*- and *trans*-3,7-dimethyl-octa-2,7-dien-1-acetoxy-6-ol, *cis*- and *trans*-3,7-dimethyl-octa-1,6-dien-3,8-diol, and *cis*- and *trans*-acetoxylinalool oxide, were already identified by the same author in a previous article (Mazza 1986) on the composition of bergamot oil, which will be discussed in the next paragraphs.

Di Giacomo and Di Giacomo (1990) listed the principal components in bergamot oil and reported the chromatogram obtained on a SE-52 capillary column. A similar chromatogram obtained on SE-52 is also reported by Huet (1991a).

TABLE 8.1

Qualitative Composition of Bergamot Oils (1967–1975)

	1	2	3	4	5	6
Camphene			+		+	+
p-Cymene		+	+		+	
Limonene			+		+	+
Myrcene			+		+	+
Ocimene*		+	+			
α-Pinene		+	+		+	+
β-Pinene			+		+	+
Sabinene			+			+
γ-Terpinene		+	+			+
Terpinolene			+			+
trans-α-Bergamotene	+		+	+		
β-Bisabolene	+		+	+		
Caryophyllene*	+		+	+		
α-Humulene			+	+		
Decanal			+			+
Nonanal			+			+
Geranial			+		+	+
Neral			+		+	+
Geraniol			+			+
Linalool			+		+	+
α-Terpineol			+		+	+
Geranyl acetate			+		+	
Linalyl acetate			+		+	+
Neryl acetate			+			+

Notes: + Present; * Correct isomer not characterized

Appendix to Table 8.1:

1. Wenninger et al. (1967). Column chromatography; GC on packed column; fractionated distillation; derivatization with AgNO$_3$; IR.
2. Dixon et al. (1968). Preparative GC; NMR; IR; MS. Dixon et al. also found linalool oxide*.
3. Ziegler (1971). Ziegler also found β-humulene, dodecanal, octanal, and nootkatone.
4. Calabrò and Currò (1972). Preparative GC; analytical GC on columns packed with five stationary phases; IR. Calabrò and Currò also found β-farnesene*, linear chained hydrocarbons C$_{20}$–C$_{33}$ and correspondent iso- and anteiso-isomers C$_{21}$–C$_{26}$.
5. Hérrisset et al. (1973). GC on packed column. Hérrisset et al. also found citronellal, borneol, bornyl acetate, and cineole*.
6. Di Corcia et al. (1975). GC on packed column.

Mondello et al. (1994, 1995b, 1998) by HPLC-HRGC e HRGC-MS interactively using linear retention indices (LRI) on SE-52 and Carbowax 20M, identified the following components: camphene, δ-3-carene, *p*-cymene, limonene, myrcene, (*E*)-β-ocimene, (*Z*)-β-ocimene, α-phellandrene, α-pinene, β-pinene, sabinene, α-terpinene, γ-terpinene, terpinolene, α-thujene, tricyclene, *cis*-α-bergamotene, *trans*-α-bergamotene, β-bisabolene, *cis*-γ-bisabolene, β-caryophyllene, γ-elemene, (Z)-β-farnesene, germacrene B, germacrene D, α-humulene, β-santalene, decanal, dodecanal, nonanal, octanal, undecanal, citronellal, geranial, perilla aldehyde, neral, 6-methyl-5-hepten-2-one, camphor, nootkatone, dodecanol, octanol, citronellol, geraniol, isopulegol, linalool. nerol, 2,3-dimethyl-3-(4-methyl-3-pentenyl)-2-norbornanol, *trans*-sabinene hydrate, terpienen-4-ol, α-terpineol, α-bisabolol, campherenol, (E)-nerolidol, decyl acetate, heptyl acetate, hexyl acetate, nonyl acetate, octyl acetate, bornyl acetate, citronellyl acetate, geranyl acetate, linalyl acetate, linalyl propanate, methyl geranate, neryl acetate, *cis*-sabinene hydrate acetate, α-terpinyl acetate, 1,8-cineole, *cis*-limonene oxide, *trans*-limonene oxide, *cis*- and *trans*-linalool oxide (furanoid form), citropten, and bergapten. Figure 8.5 reviews one chromatogram obtained by Mondello et al. (1998) on capillary fused silica column coated with Carbowax 20 M. More details on these last articles are given in Chapter 7, this volume.

8.3 QUANTITATIVE COMPOSITION

8.3.1 Industrial Cold-Extracted Oils from Calabria (Italy)

The first studies on the quantitative composition of the volatile fraction of berga-mot oil industrially produced in Calabria, carried out at the Experimental Station in Reggio Calabria (Stazione Sperimentale di Reggio Calabria), relative to 91 samples (Calvarano 1963, 1965; Calvarano and Calvarano 1964), and by Liberti and Goretti (1974), were limited to the dosage of a few components. These results are summa-rized in Table 8.2. The values reported in Table 8.2 were obtained on steel capil-lary columns coated with UCON LB55OX (Calvarano 1963, 1965; Calvarano and Calvarano 1964) or with glass capillary columns coated with FFAP (Liberti and Goretti 1974); they are affected by the limits of the chromatographic technique of those days. In particular the values of δ-3-carene (0.08%–0.23%) are not compatible with genuine bergamot oils, since this component is usually present at trace levels, but should be explained by the probable chromatographic peak overlaps. Calvarano and Calvarano (1964) also analyzed four samples of adulterated bergamot oil, which were characterized by higher limonene amounts, and three of the four samples had lower content of linalol in comparison with genuine ones.

In the same period Di Giacomo et al. (1963) determined the composition of the monoterpene hydrocarbons fraction (*p*-cimene, tr–2.39%; limonene, 61.60%–66.33%; myrcene, 1.25%–2.25%; α-pinene, 2.08%–2.36%; β-pinene, 10.30%–11.61%; sabinene, 4.56%–6.19%; γ-terpinene, 13.51%–14.00%) and Di Giacomo (1972) described the purity parameters applied at the Experimental Station in Reggio Calabria (Stazione Sperimentale di Reggio Calabria) for genuine bergamot oils. Among these were the limits of esters amount (30%–45%) and of free alcohols (15%–30%).

Table 8.3 summarizes the results published on bergamot volatile composition after 1981. The results from Zani et al. (1991) are not reported since they refer to an unusual content of camphene (0.93%), probably due to co-elution or to a wrong identification of this component.

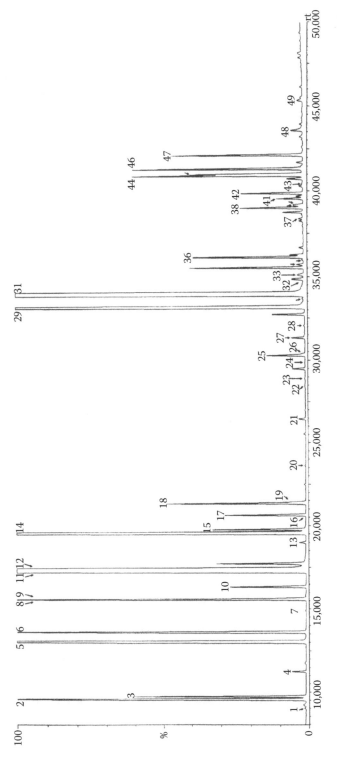

FIGURE 8.5 Total ion current chromatogram of a bergamot essential oil on Carbowax 20 M column. (1) tricyclene; (2) α-pinene; (3) α-thujene; (4) camphene; (5) β-pinene; (6) sabinene; (7) δ-3-carene; (8) myrcene; (9) α-phellandrene; (10) α-terpinene; (11) limonene; (12) 1,8-cineole; (13) (Z)-β-ocimene; (14) γ-terpinene; (15) (E)-β-ocimene; (16) hexyl acetate; (17) p-cymene; (18) terpinolene; (19) octanal; (20) 6-methyl-5-hepten-2-one; (21) nonanal; (22) cis-linalool oxide (furanoid form); (23) cis-limonene oxide; (24) trans-linalool oxide (furanoid form); (25) octyl acetate; (26) citronellal; (27) decanal; (28) camphor; (29) linalool; (30) octanol; (31) linalyl acetate; (32) nonyl acetate; (33) bornyl acetate; (34) terpinen-4-ol; (35) undecanal; (36) β-caryophyllene; (37) citronellyl acetate; (38) neral; (39) α-humulene; (40) decyl acetate; (41) α-terpineol; (42) terpinyl acetate; (43) dodecanol; (44) neryl acetate; (45) geranial; (46) β-bisabolene; (47) geranyl acetate; (48) nerol; (49) geraniol. (From Mondello, L. et al. *EPPOS* 26, 3–27, 1998.)

TABLE 8.2

Percentage Composition of Industrial Cold-Pressed Calabrian Bergamot Oils (1963–1974)

Camphene	0.03–0.10
δ-3-Carene	0.08–0.23
p-Cimene	tr–1.68
α-Phellandrene	0–0.09
Limonene	18.63–48.50
Myrcene	0.40–1.50
α-Pinene	0.61–2.27
β-Pinene + Sabinene	4.45–11.86
Sabinene	1.26
γ-Terpinene	3.48–11.76
Terpinolene	0.22–0.83
α-Thujene	0.20–0.43
Decanal	0.42–0.69
Nonanal	0.07–0.18
Citronellal	0.04–0.16
Geranial	0.26–0.71
Neral	0.16–0.64
Citronellol	0.25–0.61
Linalool	7.07–29.12
Terpinen-4-ol	0.05–0.09
α-Terpineol	0.10–0.43
Octyl acetate	0–0.12
Geranyl acetate	0.27–0.82
Linalyl acetate	23.48–35.62
Neryl acetate	0.44–1.19
Terpinyl acetate*	0.17–0.54

Sources: Calvarano, M., *Essenz. Deriv. Agrum.* 33, 67–101, 1963; Calvarano, M., *Essenz. Deriv. Agrum.* 35, 197–211, 1965; Calvarano, M., and Calvarano, I., *Essenz. Deriv. Agrum.* 34, 71–92, 1964; Liberti, A., and Goretti, G., *Essenz. Deriv. Agrum.* 44, 197–208, 1974.

tr, traces; * correct isomer not characterized

The values in Table 8.3 are in good agreement, although for numerous compounds the range of variability is quite wide. Take notice of the high values of (E)-β-ocimene reported by Couchi et al. (1995) (1.06%) and by Russo et al. (2001) (2.7%); of (Z)-β-ocimene reported by Mazza (1986), Couchi et al. (1995), and Russo et al. (2001) (respectively 0.78%, 0.43%, and 1.5%); and the value of (Z)-β-farnesene (0.45%) reported by Belsito et al. (2007).

Most of the results summarized in Table 8.3 are relative to a single sample or to very few samples, while the values reported in column 3 are relative to 1540 samples, surely genuine, representative of the entire productive seasons, 1984–85, 1986–87,

TABLE 8.3
Percentage Composition of the Volatile Fraction of Industrial Cold-Pressed Calabrian Bergamot Oils

Hydrocarbons	1	2	3	4	5	6	7
Monoterpene							
Camphene	0.04	tr	0.02–0.05	0.11	–	0.03	0.02–0.05
δ-3-Carene	–	–	tr–0.01	–	–	–	tr–0.02
p-Cymene	0.34	0.54	0.01–0.89	1.29	–	0.97	–
Limonene	35.64	38.35	24.07–54.85ᵃ	32.14	29.8–30.4	35.62	37.61–50.48
Myrcene	0.90	2.04	0.63–1.81	2.33	1.2–1.4	0.83	0.74–1.38
allo-Ocimene	–	tr	trᵈ	–	–	–	–
(E)-β-Ocimene	–	0.23	0.02–0.42	1.06	–	0.03	0.12–0.35ᶠ
(Z)-β-Ocimene	–	0.78	0.01–0.07	0.43	–	–	0.12–0.35ᶠ
α-Phellandrene	0.04	–	0.01–0.06	0.18	–	0.03	0.02–0.09
β-Phellandrene	–	–	a	–	–	0.19	–
α-Pinene	1.51	1.38	0.72–1.84	0.99	1.1–1.2	1.45	0.92–1.56
β-Pinene	7.05	5.77	4.81–12.80ᵇ	6.49	6.7–7.5	8.93	4.11–9.55
Sabinene	1.23	1.26	4.81–12.80ᵇ	–	1.2–1.3	1.16	0.50–1.32
α-Terpinene	–	0.21	0.08–0.28	–	–	0.10	0.09–0.19
γ-Terpinene	7.63	7.67	5.27–11.38	7.54	7.2–7.9	6.53	5.73–10.23
Terpinolene	0.17	0.36	0.21–0.47	0.72	0.3–0.4	0.23	0.03–0.40
α-Thujene	–	–	0.19–0.49	0.25	0.3	–	0.21–0.40
Tricyclene	–	–	tr–0.01	–	–	–	–
Sesquiterpene							
α-Bergamotene*	–	–	–	–	–	0.30	–
cis-α-Bergamotene	–	tr	0.02–0.05	–	–	–	–
trans-α-Bergamotene	0.32ᵃ	0.36	0.16–0.44	–	–	–	0.28–0.50

trans-β-Bergamotene	—	—	—	—	—	—	—
Bicyclogermacrene	0.43–0.65	0.45	0.4–0.5	0.52	0.01–0.08	0.46	0.60[t]
β-Bisabolene	—	—	—	—	0.21–0.65	—	—
(E)-α-Bisabolene	—	—	—	—	—	—	—
(Z)-α-Bisabolene	—	—	—	—	0–tr	—	—
(E)-γ-Bisabolene	—	—	—	—	—	—	—
(Z)-γ-Bisabolene	—	—	—	—	0–0.01	tr	—
δ-Cadinene	—	—	—	—	—	tr	—
β-Caryophyllene	0.26–0.48	0.21	0.3–0.4	0.45	0.15–0.55	0.27	—
β-Elemene	—	+	—	—	—	—	—
δ-Elemene	0.01–0.04	—	—	—	0–0.06	tr	—
(E,E)-α-Farnesene	—	—	—	—	0–tr	—	—
(E)-β-Farnesene	—	0.07	—	tr	tr	—	—
(Z)-β-Farnesene	0.03–0.14	—	—	0.08	0.03–0.09	—	—
Germacrene D	—	0.03	—	—	0.03–0.11	tr	—
α-Humulene	—	+	—	0.03	0.01–0.04	tr	—
β-Santalene	—	+	—	—	tr–0.02	tr	—
β-Selinene	—	—	—	0.04	—	tr	—
β-Sesquiphellandrene	—	—	—	—	—	tr	—
Aldehydes							
Aliphatic							
Decanal	0.04–0.08	—	—	0.10	0.04–0.10	—	0.09
(E)-2-Decenal	—	+	—	—	0–0.01	—	—
Dodecanal	—	+	—	—	tr–0.05	—	—
Nonanal	—	—	—	—	0.01–0.08	—	0.03

continued

TABLE 8.3 (CONTINUED)
Percentage Composition of the Volatile Fraction of Industrial Cold-Pressed Calabrian Bergamot Oils

	1	2	3	4	5	6	7
Octanal	–	–	0.02–0.08	–	–	–	0.03–0.09
Undecanal	–	–	tr–0.02	–	–	–	tr–0.04
Monoterpene							
Citronellal	0.02	–	0.01	–	–	–	0.01–0.03
Geranial	0.35	0.31	0.16–0.20[c]	0.43	0.3–0.4	0.47	0.21–0.42
Neral	0.23	0.19	0.11–0.13	0.28	–	0.16	0.12–0.30
Perilla aldehyde	–	tr	0.16–0.20[c]	–	–	–	–
Ketones							
Aliphatic							
6-Methyl-5-hepten-2-one	–	tr	tr–0.01	–	–	–	tr–0.05
Monoterpene							
Camphor	–	–	tr–0.01	–	–	–	–
Carvone	–	–	0–tr	–	–	–	–
Sesquiterpene							
Nootkatone	–	tr	0.01–0.10	–	–	0.07	0.01–0.10
Alcohols							
Aliphatic							
Decanol	–	tr	–	–	–	–	–
Octanol	–	tr	0–0.03	–	–	–	–
Monoterpene							
Citronellol	–	tr	0.01–0.11[d]	–	–	+	–
Geraniol	0.02	tr	0–0.01	–	–	0.11	–

Isopulegol	–	–	tr–0.01	–	–	–	–
Linalool	12.67	9.60	1.58–22.68	11.63	10.1–13.5	8.67	2.23–9.96
Nerol	0.05[f]	tr	0.01–0.11[d]	0.04	–	–	0.01–0.05
cis-Sabinene hydrate	–	–	0.01–0.06	0.07	–	0.04	–
trans-Sabinene hydrate	–	tr	–	–	–	–	0.03–0.09
Terpinen-4-ol	0.29	tr	0.01–0.04	tr	–	0.03	0.02–0.04
α-Terpineol	0.08	0.18	0.03–0.13	0.09	–	0.16	0.02–0.10
Sesquiterpene							
α-Bisabolol	–	–	0.01–0.03	–	–	–	–
Campherenol	–	–	0.01–0.02	–	–	–	–
Nerolidol*	–	–	–	–	–	–	–
(E)-Nerolidol	–	tr	0.01–0.04	–	–	+	–
Norbornanol[e]	–	–	0.01–0.04	–	–	+	–
Spathulenol	–	tr	–	–	–	–	–
Esters							
Aliphatic							
Decyl acetate	–	tr	tr–0.05	–	–	+	–
Heptyl acetate	–	tr	tr–0.02	–	–	+	–
Hexyl acetate	–	tr	0–tr	–	–	–	–
Nonyl acetate	–	tr	0.01–0.05	–	–	+	tr–0.05
Octyl acetate	0.12	tr	0.06–0.22	–	–	0.12	0.08–0.15
Monoterpene							
Bornyl acetate	–	tr	0.01–0.04	–	–	–	tr–0.04
Citronellyl acetate	–	tr	tr–0.06	–	–	–	–
Geranyl acetate	0.26	0.73	0.11–0.84	0.50	0.4–0.5	0.47	0.25–0.87
Linalyl acetate	32.71	26.88	15.09–41.36	30.06	33.7–35.5	29.12	19.99–32.89

continued

TABLE 8.3 (CONTINUED)
Percentage Composition of the Volatile Fraction of Industrial Cold-Pressed Calabrian Bergamot Oils

	1	2	3	4	5	6	7
Linalyl propanate	–	tr	0.01–0.07	–	–	–	0.01–0.04
Methyl geranate	–	–	tr–0.02	–	–	–	tr–0.05
Neryl acetate	–	0.62	0.13–0.67	0.47	0.4–0.5	–	0.25–0.53
trans-Sabinene hydrate acetate	–	–	0.05–0.13[1]	0.07	–	–	–
α-Terpinyl acetate	–	0.10	0.07–0.27	0.18	–	–	0.14–0.29
Ethers and Oxides							
Monoterpene							
1,8-Cineole	–	tr	tr–0.02	–	–	–	–
cis-Limonene oxide	–	tr	tr–0.02	–	–	+	tr–0.02
trans-Limonene oxide	–	tr	tr–0.01	–	–	+	tr–0.02
cis-Linalool oxide	–	tr	0–tr	–	–	–	–
trans-Linalool oxide	–	tr	0–tr	–	–	–	–
Monoterpene							
Caryophyllene oxide*	–	tr	–	–	–	–	–
Others							
Octanoic acid	–	tr	–	–	–	–	–
(*E*)-Solanone[1]	–	–	–	–	–	0.10	–

Hydrocarbons	8	9	10	11	12	13	14	15a	15b	15c
Monoterpene										
Camphene	–	tr	tr	0.02	0.03	0.01–0.03	–	0.03	0.02–0.03	0.02–0.03
δ-3-Carene	–	–	–	tr	–	–	–	tr	tr	tr
p-Cymene	1.0	0.7	1.4	g	0.37	0.07–0.49	–	0.30–0.59	0.08–0.23	0.07–0.13
Limonene	41.3	32.1	37.2	32.52[g]	42.80	37.40–49.10	38.1	41.75–45.78	39.39–44.52	28.71–40.46
Myrcene	5.1	1.0	0.8	1.10	0.91	0.68–0.99	1.1	0.99–1.22	0.92–1.07	0.52–1.00
allo-Ocimene	–	–	–	–	–	–	–	–	–	–
(E)-β-Ocimene	2.7	0.3	–	0.21	0.17	0.16–0.24	0.1	0.23–0.25	0.17–0.25	0.21–0.27
(Z)-β-Ocimene	1.5	tr	tr	0.05	0.04	0–0.03	–	0.06–0.07	0.02–0.08	0.03–0.08
α-Phellandrene	–	tr	tr	0.10[h]	0.02	0.02	–	0.02–0.03	0.02–0.03	0.02–0.03
β-Phellandrene	–	–	0.2	–	–	–	–	–	–	–
α-Pinene	1.2	1.5	1.3	1.01	1.04	0.51–1.15	1.5	1.21–1.27	0.91–1.27	0.91–1.22
β-Pinene	7.3	7.0	6.2	5.20	5.59	3.50–6.88	5.4	5.73–6.35	4.90–7.02	4.92–7.08
Sabinene	1.2	0.9	1.1	0.91	0.89	0.48–0.95	1.1	1.01–1.07	0.88–1.16	0.89–1.20
α-Terpinene	–	0.2	0.1	0.05	0.11	0.04–0.14	–	0.05–0.13	0.09–0.16	0.11–0.16
γ-Terpinene	7.9	8.5	6.8	5.49	6.19	6.36–8.10	7.3	7.12–7.84	6.32–7.53	5.86–7.68
Terpinolene	0.6	0.4	0.3	0.18	0.24	0.18–0.28	0.4	0.23–0.31	0.25–0.31	0.25–0.32
α-Thujene	0.3	0.4	–	0.27	0.27	0.12–0.27	0.4	0.32–0.34	0.23–0.33	0.24–0.32
Tricyclene	–	–	–	–	0.01	0–0.03	–	tr–0.01	tr	tr
Sesquiterpene										
α-Bergamotene*	–	–	tr	–	–	–	0.6	0.02	0.02	0.02
cis-α-Bergamotene	–	–	–	–	–	0.01–0.27	–	0.30–0.35	0.25–0.33	0.22–0.30
trans-α-Bergamotene	–	0.6	–	–	0.25	0.01–0.25	–	0.02	0.01–0.02	0.01–0.02
trans-β-Bergamotene	–	–	–	–	0.01	0–0.01	–	0.02	0.01–0.02	0.01–0.02
Bicyclogermacrene	–	–	–	–	0.01	0–0.01	–	0.01–0.07	0.01–0.03	0.02–0.03

continued

TABLE 8.3 (CONTINUED)
Percentage Composition of the Volatile Fraction of Industrial Cold-Pressed Calabrian Bergamot Oils

	8	9	10	11	12	13	14	15a	15b	15c
β-Bisabolene	0.6	0.8	0.8	0.47	0.36	0.10–0.39	0.8	0.44–0.52	0.36–0.51	0.32–0.44
(E)-α-Bisabolene	–	–	–	–	0.01	0–0.01	–	0.01	0.01	0.01
(Z)-α-Bisabolene	–	–	–	–	0.03	0.01–0.03	–	0.03–0.04	0.03–0.04	0.03–0.04
(E)-γ-Bisabolene	–	–	–	–	tr	0–0.01	–	tr–0.03	tr	tr–0.01
(Z)-γ-Bisabolene	–	–	–	–	tr	0–0.01	–	tr	tr–0.01	tr–0.01
δ-Cadinene	–	–	–	–	tr	0–0.03	–	tr	tr–0.01	tr
β-Caryophyllene	–	0.6	0.3	0.55	0.25	0.17–0.27	0.5	0.33–0.39	0.31–0.38	0.27–0.33
β-Elemene	–	–	–	–	tr	0–0.02	–	tr–0.03	tr–0.01	tr–0.01
δ-Elemene	–	–	–	–	0.01	0–0.01	–	0.01–0.02	0.01–0.02	tr–0.02
(E,E)-α-Farnesene	–	–	–	–	0.01	0–0.01	–	0.01	tr–0.01	0.01–0.02
(E)-β-Farnesene	–	–	–	–	0.04	0–0.04	–	0.05–0.06	0.05–0.06	0.05–0.07
(Z)-β-Farnesene	–	–	–	0.45	0.01	0–0.04	0.1	0.01	0.01–0.02	0.01–0.09
Germacrene D	–	–	tr	–	0.03	0.01–0.04	–	0.03–0.05	0.04–0.05	0.04–0.06
α-Humulene	–	–	tr	–	0.02	0.01–0.02	–	0.02–0.03	0.02–0.03	0.02–0.03
β-Santalene	–	–	–	–	0.01	0–0.01	0.1	tr–0.01	0.01	0.01
β-Selinene	–	–	–	–	–	–	–	–	–	–
β-Sesquiphellandrene	–	–	–	–	tr	–	–	0.01–0.03	tr–0.01	tr–0.01
Aldehydes										
Aliphatic										
Decanal	–	0.1	tr	0.07	0.05	0.01–0.04	0.1	0.04–0.06	0.04–0.06	0.03–0.07
(E)-2-Decenal	–	–	–	–	–	–	–	–	–	0.01
Dodecanal	–	–	+	–	–	0–0.03	–	tr	tr	tr
Nonanal	–	–	tr	tr	0.03	0.01–0.02	0.1	0.02–0.04	0.01–0.09	0.03–0.06

Octanal	0.03–0.06	0.04–0.05	0.04–0.05	–	0.01–0.04	0.05	0.10[h]	tr	–	–
Undecanal	0.01–0.02	tr–0.01	0.02–0.04	–	0–0.02	0.01	–	–	–	–
Monoterpene										
Citronellal	0.01	0.01	0.01–0.02	–	0–0.01	0.01	–	–	–	–
Geranial	0.25–0.36	0.22–0.39	0.24–0.33	0.2	0.14–0.30	0.24	0.69	0.3	0.3	–
Neral	0.16–0.25	0.15–0.27	0.15–0.21	–	0.12–0.21	0.16	0.14	0.2	0.2	–
Perilla aldehyde	tr–0.01	tr–0.01	tr–0.01	–	0–0.01	tr	–	–	–	–
Ketones										
Aliphatic										
6-Methyl-5-hepten-2-one	0.01–0.04	0.01–0.02	0.01–0.08	–	0–0.03	0.02	–	tr	–	–
Monoterpene										
Camphor	tr–0.01	tr	tr–0.01	–	–	–	–	–	–	–
Carvone	tr–0.01	tr–0.01	0.01	–	0–0.01	0.01	–	tr	–	–
Sesquiterpene										
Nootkatone	0.02–0.10	0.01–0.08	0.03–0.06	–	0–0.07	0.08	–	0.1	0.1	–
Alcohols										
Aliphatic										
Decanol	–	–	–	–	–	–	–	tr	–	–
Octanol	0.03–0.05[i]	0.02–0.05[i]	0.03–0.05[i]	–	0.02–0.05[i]	–	–	–	tr	–
Monoterpene										
Citronellol	tr–0.01	tr–0.02	0.01–0.02	–	0–0.02	0.01	–	–	–	–
Geraniol	–	–	–	–	–	–	–	0.1	–	–
Isopulegol	tr	tr	tr	–	0–0.01	–	–	–	–	–
Linalool	5.98–25.99	5.98–15.44	4.28–10.85	6.4	3.69–36.14	5.55	14.1	8.8	12.1	6.9
Nerol	tr–0.14	tr–0.09	0.01–0.04	–	0–0.08	0.02	0.17	0.1	0.1	–

continued

TABLE 8.3 (CONTINUED)

Percentage Composition of the Volatile Fraction of Industrial Cold-Pressed Calabrian Bergamot Oils

	8	9	10	11	12	13	14	15a	15b	15c
cis-Sabinene hydrate	–	0.1	–	0.06	0.02	0.02–0.05[i]	–	0.03–0.05[i]	0.02–0.05[i]	0.03–0.05[i]
trans-Sabinene hydrate	–	0.1	0.1	–	–	–	–	–	–	–
Terpinen-4-ol	–	tr	tr	0.05	0.02	0.02–0.16	0.2	0.03	0.02–0.03	0.02–0.03
α-Terpineol	–	0.1	0.1	0.13	0.06	0.05–0.43	0.1	0.08–0.10	0.06–0.13	0.06–0.13
Sesquiterpene										
α-Bisabolol	–	–	tr	0.01	0.02	0–0.03	–	0.02	0.01–0.03	tr–0.02
Campherenol	–	–	–	–	0.01	0–0.01	–	tr–0.01	tr–0.01	tr–0.01
Nerolidol*	–	–	tr	0.02	–	–	–	–	–	–
(E)-Nerolidol	–	–	–	–	0.02	0–0.02	–	0.02–0.04	0.02–0.06	0.02–0.03
Norbornanol[l]	–	–	–	–	–	0–0.01	–	0.01	0.01–0.02	0.01
Spathulenol	–	–	–	–	0.01	0–0.01	–	0.01	tr–0.01	0.01
Esters										
Aliphatic										
Decyl acetate	–	–	tr	0.03	0.04	0.01–0.03	–	0.03–0.04	0.02–0.04	0.02–0.05
Heptyl acetate	–	–	–	–	tr	–	–	tr–0.01	tr–0.01	0.01
Hexyl acetate	–	–	–	–	0.02	0–0.01	–	tr–0.01	tr–0.01	0.01–0.03
Nonyl acetate	–	–	tr	0.10	0.01	0.01–0.02	–	0.02	0.01–0.02	0.02–0.03
Octyl acetate	–	0.1	–	0.14	0.10	0.05–0.09	–	0.08–0.11	0.07–0.11	0.06–0.14
Monoterpene										
Bornyl acetate	–	–	–	0.12	–	0.01	–	0.02	0.01–0.02	0.01–0.02
Citronellyl acetate	–	–	0.1	–	–	0–0.02	–	0.01–0.02	0.01–0.02	0.01–0.02
Geranyl acetate	2.4	0.4	0.3	0.57	0.31	0.18–0.40	0.7	0.36–0.55	0.24–0.40	0.30–0.34
Linalyl acetate	18.8	29.7	30.1	31.01	27.14	11.80–30.00	28.9	24.44–27.59	24.22–30.16	27.76–34.67
Linalyl propanate	–	0.1	–	–	0.03	0–0.04	–	0.03–0.04	0.02–0.03	0.03–0.06

Methyl geranate	1.2	–	–	–	tr	0–0.01	–	tr–0.01	tr–0.01	tr–0.01
Neryl acetate	0.5	0.5	+	0.28	0.30	0.24–0.36	0.8	0.39–0.44	0.31–0.39	0.25–0.44
trans-Sabinene hydrate acetate	–	–	–	–	–	–	–	–	–	–
α-Terpinyl acetate	0.3	0.3	0.2	–	0.14	0.05–0.16	–	0.14–0.18	0.10–0.19	0.07–0.19
Ethers and Oxides										
Monoterpene										
1,8-Cineole	–	–	tr	–	–	–	–	–	–	–
cis-Limonene oxide	–	–	tr	–	–	0–0.01	–	tr–0.02	tr–0.01	tr
trans-Limonene oxide	–	–	tr	–	tr	0–0.01	–	0.01–0.02	tr–0.01	tr
cis-linalool oxide	–	–	tr	–	–	–	0.1	–	–	–
trans-linalool oxide	–	–	0.1	–	–	–	–	–	–	–
Sesquiterpene										
Caryophyllene oxide	–	–	tr	–	–	–	0.2	–	–	–
Others										
Octanoic acid	–	–	tr	–	–	–	–	–	–	–
(E)-Solanone[i]	–	–	0.1	–	–	–	–	–	–	–

a   limonene + β-phellandrene
b   β-pinene + sabinene
c   geranial + perilla aldehyde
d   citronellol + nerol
e   2,3-dimethyl-3-(4-methyl-3-pentenyl)-2-norbonanol
f   (E)-β-ocimene + (Z)-β-ocimene
g   limonene + p-cymene
h   α-phellandrene + octanal
i   octanol + cis-sabinene hydrate

tr, traces; * correct isomer not characterized; ¹ tentative identification; + identified but not quantitatively determined

continued

TABLE 8.3 (CONTINUED)
Percentage Composition of the Volatile Fraction of Industrial Cold-Pressed Calabrian Bergamot Oils

Appendix to Table 8.3:

1. Huet (1981). One sample; GC on capillary column coated with Carbowax 20M; relative percentage of peak areas.

2. Mazza (1986). Calabria, Italy; tree samples; column chromatography on silica gel; GC on capillary column coated with Carbowax 20 M; GC/MS; relative percentage of peak area. Mazza also found traces amount of *p*-cymenene, γ-acoradiene, α-bisabolene*, α-muurolene, epi-β-santalene, α-selinene, piperitone, decanol, hexanol, (*E*)-hex-2-en-1-ol, (*Z*)-hex-3-en-1-ol, nonanol, 1-pentanol, 2-pentanol, *cis*- and *trans*-carvacrol, *cis*- and *trans*-*p*-menth-2,8-dien-1-ol, 3,7-dimethyl-octa-1,5-dien-3,7-diol, 3,7-dimethyl-3-acetoxy-octa-1,5-dien-7-ol, 3,7-dimethyl-3-acetoxy-octa-1,7-dien-6-ol, *cis*- and *trans*-*p*-menth-2,8-dien-1-ol, 3,7-dimethyl-octa-1,5-dien-3,7-diol, 3,7-dimethyl-octa-1,7-dien-3,6-diol, perilla alcohol, *trans*-pinocarveol, *cis*-hex-3-en-1-yl acetate, 3-(3,4,5-trimethoxyphenyl)-propenyl acetate, geranyl propanate, neryl propanate, *trans*-pinocarvyl acetate, sabinyl acetate, *cis*- and *trans*-linalool oxide (pyranoid form), *p*-menth-1-en-4,5-oxide, *p*-menth-4-en-1,2-oxide, linalyl acetate oxide (two isomers), methyl N-methylanthranilate, acetic acid. In several oxidated oils, Mazza also found cumin aldehyde, *p*-menth-1-en-9-al, myrtenal, carvone, dihydrocarvone, cumin alcohol, 2,6-dimethylocta-1,5,7-trien-3-ol; limonene-10-ol, myrtenol, *cis*- and *trans*-epoxyocimene*, cresol, dihydro-2,3-epoxygeranyl acetate, dihydro-2,3-epoxyneryl acetate, methylacetophenone, perillene, 2-vinyl-2-methyltetrahydrofuran.

3. Dugo et al. (1987, 1991, 1999), Dugo (1994). Lamonica et al. (1990), Verzera et al. (1996, 1998). Calabria, Italy; Pelatrice; Fantastico, Femminello, Castagnaro bergamot; about 1500 samples from 1984 to 1998; column chromatography on neutral alumina; GC on capillary columns coated with SE-52 and DB-5; GC/MS; linear retention indices; relative percentage of peak areas. These authors also found dodecane (0%–0.01%), γ-elemene (0.01%–0.02%), germacrene B (0%–0.04%), tetradecanol (0%–0.01%), dodecanol (0%–0.01%), indole (0%–0.01%) and trace amounts of tridecanal, β-bisabolol, and undecyl acetate.

4. Chouci et al. (1995). Calabria, Italy; one sample of Pelatrice oil; GC on capillary column (30 m × 0.25 mm × 0.25 μm) coated with DB-5; GC/MS; wt%. Chouci et al. also found tridecane (0.05%), δ-2-carene (0.11%), 1,3,8-*p*-menthatriene (0.17%), aromadendrene (0.36%), γ-muurolene (0.07%), isomenthone (0.01%), menthone (0.06%), octen-3-ol (0.08%), dihydrocitronellol (0.05%), menthol (0.16%), neomenthol (0.05%), *(E,E)*-farnesol (0.01%), *(Z)*-β-santalol (0.01%), isomenthyl acetate (0.10%), and trace amounts of caryl acetate, neomenthyl acetate.

5. Ferrini et al. (1998). Calabria, Italy; two samples; GC/FID and GC/MS on capillary column (60 m × 0.20 mm × 0.25 μm) coated with SPB-5; NIST/EPA/MSDC MS library; relative percentage of peak areas. Ferrini et al. also found β-ocimene* (0.4%–0.5%), bergamotene* (0.3%).

6. Sawamura et al. (1999a). Calabria, Italy; one sample; GC/FID and GC/MS on capillary column (50 m × 0.22 mm × 0.25 μm) coated with Thermon 600T; wt% using as internal standards n-heptanol and methyl myristate. Sawamura et al. also found, not quantitatively determined, δ-muurolene, verbenol, (Z)-nerolidol, hydroxylinalool.

7. Gionfriddo et al. (2000). Calabria, Italy; 100 samples extracted with Pelatrice Speciale representative of kg 7.802 of essential oil product from January to March; GC/FID on capillary column (30 m × 0.25 mm × 0.25 μm) coated with DB-5; relative percentage of peak areas.

8. Russo et al. (2001). Calabria, Italy; one sample; GC/FID and GC/MS on capillary column (25 m × 0.25 mm × 0.25 μm) coated with SE-52; relative percentage of peak areas.

9. Poiana et al. (2003). Sicily, Italy; one sample; GC/MS on capillary column (30 m × 0.2 mm × 0.25 μm) coated with HP-5MS; Wiley MS library; relative percentage of peak areas.

10. Sawamura et al. (2006). Calabria, Italy; one sample produced with a Pelatrice-type extractor; GC/FID, GC/MS on capillary column (60 m × 0.25 mm × 0.25 μm) coated with DB-Wax; for GC/MS was also used a DB-1 column; GC/O on capillary column (60 m × 0.53 mm × 1 μm) coated with DB-Wax; wt%; Sawamura et al. also found trace amounts of cedrol. In the work were reported the retention indices on DB-Wax and DB-1 columns of the identified components and their odor description at the GC sniffing port during GC/O. An example of a most probable aroma model of bergamot oil was also indicated. The same results were previously reported by Onishi et al. (2003).

11. Belsito et al. (2007). Calabria, Italy; one sample; GC/MS for qualitative and quantitative analyses on capillary column (30 m × 0.25 mm × 0.25 μm) coated with HP-5MS; wt% using tetradecane as internal standard. GC/FID (for calculation of LRI) on capillary column (30 m × 0.25 mm × 0.25 μm) coated with Equity-5MS.

12. Costa et al. (2010). Calabria, Italy; one sample; GC/FID and GC/MS on capillary column (30 m × 0.25 mm × 0.25 μm) coated with SLB-5MS; MS libraries: Flavour and Fragrance Natural and Synthetic Compounds (FFNSC) homemade; Adams (2007); Hochmuth (2006); NIST05; LRI on SLB-5MS are reported; relative percentage of peak areas. In this paper the composition of the oil is also reported as wt%. Costa et al. also found isobornyl acetate (0.01%) and trace amounts of epi-β-bisabolol.

13. Sciarrone (2009, personal communication). Calabria, Italy; 13 samples; for experimental conditions see point 12 of this appendix. Sciarrone et al. also found 4,8-dimethyl-1,3(E),7-nonatriene (0%–0.01%), ascaridole (0%–0.01%).

14. Menichini et al. (2010). Calabria, Italy; one sample; GC/MS on capillary column (30 m × 0.25 mm × 0.25 μm) coated with SE-30; relative percentage of peak areas of GC/MS analysis.

15. Dugo et al. (2012). Calabria, Italy: (a) 8 samples produced in 2008–09 season; (b) 11 samples produced in 2009–10 season; (c) 23 samples produced in 2010–11 season; for experimental conditions see point 12 of this appendix. Dugo et al. also found 4,8-dimethyl-1,3(E),-nonatriene (tr–0.01%), Y-curcumene (tr–0.01%), cis-sesquisabinene hydrate (tr–0.01%), trans-sesquisabinene hydrate (tr–0.01%), ascaridole (tr–0.01%) and trace amounts of cis-isocitral and fenchol*.

1987–88, 1991–92, 1992–93 and 1996–97. The values in column 7 are relative to 100 samples produced during the 1998–99 season, and those in columns 15a–15c refer to 42 samples representing the 2008–09, 2009–10, and 2010–11 seasons of production; these samples are representative of 42,000 kg. Table 8.3 shows that bergamot oil is characterized by lower average amounts of limonene than every other citrus peel oil, and by high percentages of linalool and linalyl acetate. Among monoterpene hydrocarbons, γ-terpinene, sabinene, and β-pinene are also largely represented. Sesquiterpene compounds are well represented and among these, listed in decreasing amount order, are β-bisabolene, β-caryophyllene, and *trans*-α-bergamotene.

Figure 8.6 (Verzera et al. 1996) shows the average composition in classes of components of bergamot oil. The considerable amount of alcohols and esters are evident, which together represent about 40% of the whole volatiles while carbonyls are not higher than 1%.

More information on the bergamot volatile fraction is available in the review by Ohloff (1994). In this review is reported the percent composition of a bergamot oil (*p*-cymene 0.5%–3.5%; limonene 26%; myrcene 0.6%–1.3%; β-phellandrene 0.1%; α-pinene 1%–2.2%; β-pinene 4%–10%; sabinene 1%; β-terpinene 3.2%; γ-terpinene 5%–11%; terpinolene 0.3%–1%; α-bergamotene* 0.3%; β-bisabolene 0.6%; caryophyllene* 0.2%; farnesens* 0.04%; nonanal 0.08%; octanal 0.08%; citronellal 0.02%; neral + geranial 0.6%; carvone 0.09%; octanol 0.02%; fenchol* 0.01%; geraniol 0.6%; linalool 16%; nerol 0.08%; terpinen-4-ol 0.06%; α-terpineol 0.3%; decyl acetate 0.03%; octyl acetate 0.15%; citronellyl acetate 0.02%; geranyl acetate 0.4%–0.8%; linalyl acetate 34%; neryl acetate 0.4%–1.2%) probably obtained from literature data. Ohloff (1994) also lists the sesquiterpene compounds in bergamot oil, probably obtained from a study carried out by Regula Näf: *cis*-α-bergamotene (tr), *trans*-α-bergamotene (0.36%), β-bisabolene (0.015%), β-farnesene* (0.03%); α-humulene (0.04%); germacrene D (0.015%), bergamotenal, lanceal*, β-sinensal, camphorenone, nootkatone, 8,9-didehydro-nootkatone; α- and β-bisabolol; farnesol*; guajenol; nerolidol*, two santalene hydrates, and spathulenol. The same author cites jasmone, dihydrojasmone, octa-5-en-2-one, coumarin, indole, and methyl N-methylanthranilate as peculiar examples of the compositional aspect of bergamot oil.

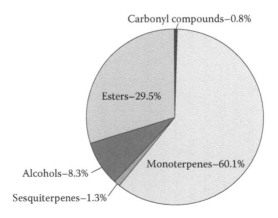

FIGURE 8.6 Average composition for class of compounds in 1082 cold-pressed bergamot oils. (From Verzera A., et al. *Perfum. Flav.* 21(6), 19–34, 1996.)

Ohloff (1994) also reported the presence of β-terpinene, which was an unusual result since this component has never been reported in bergamot oil by other authors.

Huet (1991b) reports the range of variability, limited to the principal components, for Italian bergamot oil, determined from literature data.

In the recent volume edited by Sawamura, Poiana (2010) revised a limited number of articles on the volatile fraction of bergamot oil produced in Calabria and summarized in table only the ranges of variability for the main components.

8.3.2 OILS FROM DIFFERENT GEOGRAPHIC ORIGINS AND COMMERCIAL OILS

Ikeda et al. (1962) reported a commercial oil's composition of the monoterpene hydrocarbons, which in total represented 32.2% of the whole oil. Their results were *p*-cimene, 0.7%, limonene, 22.3%, myrcene, 0.4%, ocimene*, 0.1%; α-pinene, 0.6%; β-pinene, 3.5%; sabinene, 0.5%; γ-terpinene, 0.5%; and α-thujene, 0.1%.

Huet and Dupuis (1968), in a study on oils extracted from bergamot produced in Corsica and from different African countries, reported for the different samples the physicochemical indices and the content of linalool and linalyl acetate. The values of these two compounds are reported in Table 8.4.

These authors tried to find a direct relationship between the amount of alcohols and esters and the temperature and moisture of the cultivated fields. They concluded that humidity above 70% led to an increase of the amount of oxygenated compounds, mainly linalool, and that lower temperatures, mainly when at the last stage of fruit ripening, favored the increase of esters.

The results published after 1981 on bergamot oil not produced in Calabria and on commercial oils are summarized in Table 8.5.

TABLE 8.4
Linalool and Linalyl Acetate Content in Bergamot Oils of Different Geographic Origin

Country of Origin	Linalool	Linalyl Acetate
Guinea[a]	23.0%–35.6%[c]	32.5%–40.4%[c]
Guinea[b]	11.93%–31.79%[c]	32.27%–43.61[c]
Ivory Coast	21.7%–33.0%[c]	32.8%–44.1%[c]
Algeria	9.3%[d]	14.6%[d]
Cameroon	7.3%[d]	34.8%[d]
Morocco	2.4%–9.3%[c]	16.0%–33.7%[c]
Mali	2.9%–20.8%[c]	21.0%–44.6%[c]
Corsica	8.0%–19.0%[c]	59.9%–77.5%[c]

Source: Huet, R., and DuPuis, C., *Fruits* 23:301–311, 1968.

[a] industrial oils.

[b] Schwob (1965).

[c] percent of esters and alcohols determined by classical methods and expressed as linalool and linalyl acetate, respectively.

[d] percent of linalool and linalyl acetate determined by gas chromatography.

TABLE 8.5
Percentage Composition of Industrial Cold-Pressed Bergamot Oils from Other Countries and Commercial Oils

	Corsica			Ivory Coast			Argentina	Brazil	
Hydrocarbons	1a	1b	1c	2	3a	3b	4	5	6
Monoterpene									
Camphene	0.03	0.02	0.03	—	0.01	0.02	0.01	0.04–0.08	—
δ-3-Carene	—	—	—	—	0.01	tr	—	—	—
p-Cymene	0.23	0.08	0.13	0.2–0.5	0.14	0.15	0.16	0.59–0.86	—
Limonene	28.38	26.24	31.16	23.5–24.5	40.90	45.89	43.29	32.85–34.66	38.16
Myrcene	0.77	0.69	0.70	1.0–1.1	0.71	0.83	0.92	0.92–1.32	0.94
(E)-β-Ocimene	—	—	—	—	0.08	0.15	—	—	—
(Z)-β-Ocimene	—	—	—	—	0.02	0.02	—	—	—
α-Phellandrene	0.07	0.09	0.09	—	0.01	0.02	—	0.13–0.15	—
β-Phellandrene	—	—	—	—	—	—	—	0.02–0.04	—
α-Pinene	1.56	0.98	0.76	0.6–0.8	0.57	1.09	0.88	1.62–1.74	0.64
β-Pinene	7.51	4.76	3.47	3.6–4.5	3.58	6.72	5.49	7.63–10.60	3.05
Sabinene	1.26	0.89	0.69	0.7–0.9	0.55	0.93	—	1.00–1.68	0.61
α-Terpinene	—	—	—	—	0.03	0.12	—	—	—
γ-Terpinene	6.90	4.97	4.14	3.5–4.1	3.78	8.03	4.03	3.95–6.73	4.12
Terpinolene	0.21	0.21	0.21	0.2	0.12	0.25	0.25	0.42–0.50[c]	0.17
α-Thujene	—	—	—	0.2	0.11	0.25	—	—	0.15
Tricyclene	—	—	—	—	0.02	0.02	—	—	—
Sesquiterpene									
cis-α-Bergamotene	—	—	—	—	0.01	0.01	—	—	—
trans-α-Bergamotene	0.29[a]	0.29[a]	0.32[a]	—	0.17	0.22	—	0.09–0.15	0.31
Bicyclogermacrene	—	—	—	—	tr	0.01	—	—	—

β-Bisabolene	0.56[t]	0.70[x]	0.77[t]	0.4	0.26	0.34	—	—	0.44
β-Caryophyllene	—	—	—	0.4	0.24	0.22	—	0.16–0.33	0.45
δ-Elemene	—	—	—	—	tr	0.01	—	—	—
(E,E)-α-Farnesene	—	—	—	—	0.01	tr	—	—	—
(Z)-β-Farnesene	—	—	—	—	0.01	0.01	—	—	—
Germacrene D	—	—	—	—	0.02	0.04	—	—	—
α-Humulene	—	—	—	—	0.02	0.01	—	—	—
β-Santalene	—	—	—	—	0.01	tr	—	—	—
β-Sesquiphellandrene	—	—	—	—	—	tr	—	—	—
Aldehydes									
Aliphatic									
Decanal	0.06	0.07	0.06	—	0.02	0.03	—	tr–0.04	—
Dodecanal	—	—	—	—	—	tr	—	tr	—
Nonanal	0.07	0.04	0.03	—	0.01	0.01	—	0–0.06	—
Octanal	—	—	—	—	0.01	0.01	—	0.42–0.50[c]	—
Tetradecanal	—	—	—	—	—	—	—	—	—
Undecanal	—	—	—	—	0.01	tr	—	—	—
Monoterpene									
Citronellal	0.06	0.04	0.04	—	tr	0.01	0.23	tr[d]	—
Geranial	0.58	0.25	0.25	0.3	0.24	0.17	0.05	0.79–1.25	0.27
Neral	0.30	0.14	0.11	—	0.18	0.11	—	0.20–0.21	0.18
Perilla aldehyde	—	—	—	—	0.01	tr	—	—	—
Ketones									
Aliphatic									
6-Methyl-5-hepten-2-one	—	—	—	—	0.01	0.01	—	tr	—

continued

TABLE 8.5 (CONTINUED)
Percentage Composition of Industrial Cold-Pressed Bergamot Oils from Other Countries and Commercial Oils

	Corsica			Ivory Coast			Argentina	Brazil	
Hydrocarbons	1a	1b	1c	2	3a	3b	4	5	6
Sesquiterpene									
Nootkatone	–	–	–	–	tr	0.05	–	–	–
Alcohols									
Aliphatic									
Octanol	–	–	–	–	0.03[h]	0.01[h]	–	tr	–
Monoterpene									
Citronellol	–	–	–	–	0.01	–	0.29	0.17–0.28	–
Geraniol	0.02	0.02	0.05	–	–	–	–	0.03–0.07	–
Isopulegol	–	–	–	–	tr	tr	–	–	–
Linalool	18.94	26.37	22.07	26.4–29.4	19.39	6.81	15.21	16.76–18.22	15.30
Nerol	0.09	0.19[t]	0.33[t]	–	0.10	–	–	0.04–0.08	–
cis-Sabinene hydrate	–	–	–	–	0.03[h]	0.01[h]	–	–	–
trans-Sabinene hydrate	–	–	–	–	–	–	–	–	–
Terpinen-4-ol	0.21	0.25	0.28	–	0.01	0.03	–	tr	–
α-Terpineol	0.11	0.16	0.23	–	0.10	0.16	0.17	0.20–0.27	–
Sesquiterpene									
α-Bisabolol	–	–	–	–	0.01	0.02	–	–	–
Campherenol	–	–	–	–	0.01	tr	–	–	–
(*E*)-Nerolidol	–	–	–	–	0.01	0.02	–	–	–
Norbornanol[f]	–	–	–	–	0.01	0.01	–	–	–

Esters

Aliphatic

Decyl acetate	0.03	—	—	—	0.01	0.02	—	—	—
Heptyl acetate	—	—	—	—	tr	tr	—	—	—
Hexyl acetate	—	—	—	—	tr	0.01	—	—	—
Nonyl acetate	—	—	—	—	tr	0.02	—	—	—
Octyl acetate	—	0.16	0.16	—	0.04	0.08	—	tr[d]	—
Monoterpene									
Bornyl acetate	—	—	—	—	0.01	0.01	—	—	—
Citronellyl acetate	—	—	—	—	0.01	0.02	—	0.04–0.06	—
Geranyl acetate	0.19	0.16	0.17	0.2–0.3	0.18	0.26	0.38	0.15–0.24	—
Linalyl acetate	29.48	31.11	32.66	30.50–35.4	27.53	26.18	27.40	22.68–27.85	34.71
Linalyl propanate	—	—	—	—	0.02	0.02	—	—	—
Methyl geranate	—	—	—	—	0.01	tr	—	—	—
Neryl acetate	—	—	—	0.4–0.8	0.38	0.27	0.23	tr	0.50
α-Terpinyl acetate	—	—	—	—	0.05	0.12	—	—	—

Ethers and Oxides

Monoterpene

1,8-Cineole	—	—	—	—	—	—	—	—	—
cis-Limonene oxide	—	—	—	—	tr	tr	—	—	—
trans-Limonene oxide	—	—	—	—	tr	tr	—	—	—

continued

TABLE 8.5 (CONTINUED)

Percentage Composition of Industrial Cold-Pressed Bergamot Oils from Other Countries and Commercial Oils

Hydrocarbons	Uruguay	Sicily	Commercial						
	7	8	9	10	11	12	13	14	15
Monoterpene									
Camphene	0.03	0.03–0.04	–	tr	–	0.03	tr	–	–
δ-3-Carene	–	tr	–	–	–	–	–	–	–
p-Cymene	0.25–0.32	0.62	2.7	3.61	0.67	0.30	0.7	0.23	0.49
Limonene	38.13–42.54[a]	42.07	33.0	26.71	24.50	33.10	38.4	38.85	38.46
Myrcene	1.02–1.03	0.88–0.98	0.9	0.36	1.78	0.89	1.2	1.48	–
(E)-β-Ocimene	0.18–0.19	0.21	–	0.13	0.82	0.27	0.3	–	–
(Z)-β-Ocimene	0.02–0.03	0.02	–	0.78	0.46	0.04	–	0.37	–
α-Phellandrene	0.04–0.05	0.03	–	–	–	–	tr	–	–
β-Phellandrene	a	–	–	–	–	0.21	–	–	–
α-Pinene	1.08–1.14	1.27–1.43	1.0	0.70	0.46	1.04	1.5	1.60	1.62
β-Pinene	6.55–6.88[b]	7.04	5.7	5.11	2.97	5.87	7.0	6.30	8.07
Sabinene	6.55–6.88[b]	1.16	1.0	0.72	tr	1.02	1.4	1.38	–
α-Terpinene	0.12–0.13	0.15	–	–	–	0.13	0.2	–	0.33
γ-Terpinene	6.19–6.48	6.82–7.84	3.7	1.15	2.54	6.54	9.2	9.13	5.53
Terpinolene	0.26–0.27	0.24–0.32	–	tr	tr	0.28	0.5	0.59	–
α-Thujene	0.27–0.28	0.32–0.33	–	0.15	tr	0.27	0.4	–	–
Tricyclene	0–tr	tr	–	–	–	–	–	–	–
Sesquiterpene									
cis-α-Bergamotene	tr	–	tr	–	–	–	tr	–	–
trans-α-Bergamotene	0.31–0.39	0.22–0.29	0.23	0.20	0.91	0.29	0.3	0.41	–
Bicyclogermacrene	0.04–0.05	0.03	–	–	–	–	tr	–	–

β-Bisabolene	0.43–0.56	0.43	0.57	0.31	0.13	–	0.7	–	–
β-Caryophyllene	0.34–0.50	0.33	0.27	0.18	0.11	0.33	0.5	0.54	0.49
δ-Elemene	tr	tr	–	–	–	–	–	–	–
(E,E)-α-Farnesene	tr	0.03	–	–	–	–	–	–	–
(Z)-β-Farnesene	0.04–0.06	0.03–0.07	–	–	–	–	–	–	–
Germacrene D	0.06–0.10	0.02–0.05	–	–	–	–	0.1	–	–
α-Humulene	0.05–0.07	–	tr	–	–	–	tr	–	–
β-Santalene	0.01–0.02	0.03	tr	–	–	–	tr	–	–
β-Sesquiphellandrene	–	0.01	–	–	–	–	–	–	–
Aldehydes									
Aliphatic									
Decanal	0.06–0.07	0.06	–	tr	tr	0.04	tr	–	–
Dodecanal	0.03–0.04	0.01	–	0.16	–	–	–	–	–
Nonanal	0.04–0.05	0.03	–	tr	–	0.02	–	–	–
Octanal	0.02–0.03	0.05	–	–	–	0.04	tr	–	–
Tetradecanal	tr	0.01	–	–	–	–	–	–	–
Undecanal	0.01–0.02	0.01	–	tr	tr	0.05	–	–	–
Monoterpene									
Citronellal	0.01	0.01	–	–	–	–	tr	–	–
Geranial	0.16–0.20[c]	0.15–0.36	–	0.16	0.36	–	0.2	–	0.30
Neral	0.11–0.13	0.24	0.13	0.13	–	0.21	–	–	0.16
Perilla aldehyde	0.16–0.20[c]	–	0.11	0.11	–	–	–	–	–
Ketones									
Aliphatic									
6-Methyl-5-hepten-2-one	–	0.01	–	–	–	–	tr	–	–

continued

TABLE 8.5   (CONTINUED)
Percentage Composition of Industrial Cold-Pressed Bergamot Oils from Other Countries and Commercial Oils

	Uruguay	Sicily	Commercial						
	7	8	9	10	11	12	13	14	15
Sesquiterpene									
Nootkatone	0.04–0.06	0.05	–	–	–	–	–	–	–
Alcohols									
Aliphatic									
Octanol	tr	tr	–	–	–	–	–	–	–
Monoterpene									
Citronellol	–	–	–	–	–	–	–	–	–
Geraniol	tr	g	0.05	–	–	0.07	–	–	–
Isopulegol	–	–	–	–	–	–	–	–	–
Linalool	8.91–9.52	6.79–7.53	13.45	18.60	22.49	14.63	8.4	16.66	–
Nerol	0.02–0.03	0.05	0.10	–	–	0.08	–	–	–
cis-Sabinene hydrate	0.02	0.04	tr	–	–	–	tr	–	–
trans-Sabinene hydrate	–	–	tr	–	–	0.04	–	–	–
Terpinen-4-ol	0.02	0.03	tr	tr	–	0.03	tr	–	–
α-Terpineol	0.03–0.04	0.09	0.13	0.12	–	0.09	0.1	–	0.29
Sesquiterpene									
α-Bisabolol	0.02–0.03	tr–0.03	tr	–	–	–	–	–	–
Campherenol	0.02–0.03	0.01–0.02	–	–	–	–	–	–	–
(*E*)-Nerolidol	0.01–0.03	0.01–0.02	tr	–	–	–	–	–	–
Norbornanol[f]	0.01–0.02	tr–0.01	–	–	–	–	–	–	–

Esters

Aliphatic

Decyl acetate	0.02–0.03	0.02–0.03	—	—	—	—	—	—	—
Heptyl acetate	0.01	—	—	—	—	—	—	—	—
Hexyl acetate	—	—	—	—	—	—	tr	—	—
Nonyl acetate	0.02–0.03	0.01–0.02	—	—	—	—	—	—	—
Octyl acetate	0.12–0.13	0.07–0.12	—	—	—	0.09	0.2	—	—
Monoterpene									
Bornyl acetate	0.01–0.03	0.05	—	—	—	—	—	—	—
Citronellyl acetate	0.01–0.03	0.03	—	—	—	0.04	—	—	—
Geranyl acetate	0.26–0.32	0.36	0.46	0.64	0.38	0.20	—	0.32	0.59
Linalyl acetate	28.85–31.83	27.32–31.92[g]	31.30	38.13	40.61	32.51	27.2	21.41	23.49
Linalyl propanate	0.05–0.06	0.03	—	—	—	—	—	—	—
Methyl geranate	tr–0.01	0.01	—	—	—	—	—	—	—
Neryl acetate	0.26–0.33	0.18–0.39	0.42	0.47	0.21	0.68	0.5	0.48	—
α-Terpinyl acetate	0.12–0.15	0.17	0.21	—	—	0.14	0.3	0.18	—

Ethers and Oxides

Monoterpene

1,8-Cineole	tr	—	—	—	—	0.04	—	—	0.24
cis-Limonene oxide	—	0–tr	tr	—	—	0.02	—	—	—
trans-Limonene oxide	—	tr	tr	—	—	—	—	—	—

tr, traces; * correct isomer not characterized; † tentative identification; +, identified but not quantitatively determined.

a limonene + β-phellandrene.

b β-pinene + sabinene.

c terpinolene + octanal.

d citronellal + octyl acetate.

continued

TABLE 8.5 (CONTINUED)
Percentage Composition of Industrial Cold-Pressed Bergamot Oils from Other Countries and Commercial Oils

e geranial + perilla aldehyde.

f 2,3-dimethyl-3-(4-methyl-3-pentenyl)-2-norbonanol.

g geraniol + linalyl acetate.

h octanol + *cis*-sabinene hydrate.

Appendix to Table 8.5:

1. Huet (1981). (a) Corsica; (b) Ivory Coast; (c) Ivory Coast (Coci cultivar); GC on capillary column coated with Carbowax 20 M.

2. Ferrini et al. (1998). Ivory Coast; two samples; GC/FID and GC/MS on capillary column (60 m × 0.20 mm × 0.25 μm) coated with SPB-5; NIST/EPA/MSDC MS library; relative percentage of peak areas. Ferrini et al. also found allo-ocimene (0.3%), bergamotene* (0.3%).

3. Sciarrone et al. (2009). Two samples from Ivory Coast; for experimental conditions see point 12 of appendix to Table 8.3. Sciarrone et al. also found *trans*-β-bergamotene (0.01%, 0.01%), (E)-α-bisabolene (0.01%, 0.01%), (Z)-α-bisabolene (0.02%, 0.02%), (Z)-γ-bisabolene (tr, 0.01%), δ-cadinene (0.02%, tr), β-elemene (0.04%, tr), (E)-β-farnesene (0.03%, 0.04%), camphor (0.01%, tr), carvone (0.02%, 0.01%), fenchol* (0. 01%), and trace amounts of 4,8-dimethyl-1,3(Z)-7-nonatriene γ-curcumene, (E)-γ-bisabolene, *cis*- and *trans*-sesquisabinene hydrate, spathulenol, hexyl acetate, ascaridole.

4. Ricciardi et al. (1982). Argentina; FMC. Ricciardi et al. also found trace amounts of nonanol.

5. Koketsu et al. (1983). Brazil; FMC; tree sample; GC on packed columns of Carbowax 20 M and SE-30 and on capillary column (30 m × 0.2 mm) coated with Carbowax 20 M: retention times and Kovats indices; relative percentage of peak areas.

6. Franceschi et al. (2004). Brazil; one sample; GC/MS on capillary column (30 m × 0.25 mm × 0.25 μm); coated with HP-5MS; Wiley MS library; relative percentage of peak areas.

7. Dellacassa et al. (1997). Uruguay; Sfumatrice; tree samples; GC on capillary column coated with SE-52; GC/MS; relative percentage of peak areas.

8. Mondello et al. (2003, 2004b). Sicily, Italy; one sample; conventional GC/FID and GC/MS on capillary column (30 m × 0.25 mm × 0.25 μm) coated with RTX-5MS; Fast GC/FID on capillary column (10 m × 0.1 mm × 0.1 μm) coated with RTX-5MS; MS libraries: Adams, Flavour and Fragrance Natural and Synthetic Compounds (FFNSC) homemade; relative percentage of peak areas. Tranchida et al. (2006). Sicily, Italy; one sample; conventional HS-SPME-conventional GC/FID: 30 μm PDMS fiber and a capillary column (25 m × 0.25 mm × 0.25 μm) coated with SE-52; HS-SPME-Fast GC/FID: 7 μm PDMS fiber and a capillary column (10 m × 0.1 mm × 0.1 μm) coated with SE-52; relative percentage of peak areas. These authors also found sesquithujene (0.02%), camphor (0.01%), (E)-γ-bisabolol (0.01%–0.02%), *trans*-sesquisabinene hydrate (0.01%). In the table are reported the results obtained with conventional method. The results obtained with fast method were very similar. More information is reported in Chapter 7, this volume.

9. Schenk and Lamparsky (1981). GC on capillary column coated with UCON. Schenk and Lamparsky also found in bergamot oil traces amount of ar-curcumene, a-santalene, lilial, 2,6-dimethyl-6-acetoxy-oct-7-en-3-one, 2,6-dimethyl-6-acetoxy-octa-1,7-dien-3-one, 2,6-dimethyl-6-acetoxy-octa-1-en-7-one, cumin alcohol, *cis*- and *trans*-2,6-dimethyl-octa-1,5,7-trien-3-ol, β-photosantalol A, *p*-mentha-1,8-dien-9-ol, *p*-menth-1-en-9-ol, *cis*- and *trans*-*p*-menth-2-en-1-ol, α-cadinol, caryophyllenol I and II, cariophyllene alcohol, hotrienyl acetate, *p*-mentha-1,3-dien-7-yl acetate, *p*-mentha-1,7-dien-4-yl acetate, *p*-mentha-1,7(10)-dien-2-yl acetate, *p*-mentha-1,8-dien-9-yl acetate, *p*-menth-1-en-9-yl acetate, terpinen-4-yl-acetate, *cis*- and *trans*-2,3-ocimene oxide, humulene oxide I, isocaryophyllene oxide, 2,3-epoxygeranyl acetate, 2,3 epoxyneryl acetate.

10. Inoma et al. (1989). One commercial sample; GC on capillary columns (25 m × 0.2 mm × 0.2 μm) coated with OV-101 and Carbowax 20 M; GC/MS; relative percentage of peak areas. Inoma et al. also found trace amounts of *cis*- and *trans*-carveol and *trans*-carveol and of limonene oxide*.

11. Oberhofer et al. (1999). Italy; one commercial sample; GC/FID on capillary columns coated with HP-5 (25 m x 0.32 mm × 0.52 μm), OV-1 (25 m × 0.25 mm × 0.3 μm), Carbowax (25 m × 0.25 mm × 0.3 μm); GC/sniffing technique on capillary column (25 m × 0.53 mm × 0.3 μm) coated with FSOT-RSL-150; GC/IR/MS on capillary columns coated with RSL-200 (30 m × 0.52 mm × 0.25 μm) or with Stabilwax (60 m × 0.32 mm × 0.25 μm); NBS and Wiley MS libraries; EPA-REVA and Robertet IR libraries; relative percentage of peak areas.

12. Kubeczka and Formáček, (2002). Italy; one sample; GC/FID on capillary column (50 m) coated with VG-11; relative percentage of peak areas.

13. Cosimi et al. (2009). One sample. GC/FID on capillary columns (30 m × 0.25 mm × 0.25 μm) coated with HP-WAX or HP-5. GC/MS on capillary column (30 m × 0.25 mm × 0.25 μm) coated with DB-5; MS libraries: NIST 98, Adams; relative percentage of peak areas. Cosimi et al. also found traces amount of (Z)-α-bisabolene, isobornyl acetate.

14. Fantin et al. (2010). One sample; GC/FID and GC/MS on capillary column (30 m × 0.32 mm × 0.15 μm) coated with SE52; relative percentage of peak areas.

15. Wai and Shibamoto (2010). One sample; GC/FID and GC/MS on capillary column (30 m × 0.32 mm × 0.32 μm) coated with DB-wax; relative percentage of peak areas.

It is difficult to provide comments on the results summarized in Table 8.5, mainly because of their large variability, which is due to nontechnological parameters (see Section 8.4). These results are relative to a single sample or a few samples, and therefore not representative of entire productive seasons, during which the composition can vary significantly; nor are they representative of the entire productive areas of the countries of origin, which can affect the composition of the oil. It is possible to notice, however, that many oils from the Ivory Coast and Brazil have similar or lower amounts of β-pinene and of γ-terpinene than genuine oils from Calabria.

In addition to the information summarized in Table 8.5 more data are available in literature on commercial bergamot oils. Formàček and Kubeczka (1982) analyzed two commercial oils originating from Italy, one from Messina, the other one from Reggio Calabria. Both samples, as declared by the authors, were roughly adulterated. Among the numerous anomalous values of their compositions, these oils presented very high amounts of α-terpinyl acetate, 6% and 4%, respectively.

Lis-Balchin et al. (1998) studied the relationship between bioactivity and composition of some essential oils. For commercial bergamot, they reported the following amounts: limonene (31%–38%), α-pinene (2%–4%), linalool (11.0%–21.9%), and linalyl acetate (37.6%–38.5%).

Binder et al. (2001) analyzed 24 commercial oils among which 6 were clearly adulterated, as demonstrated by the high values of the enantiomeric ratios for linalool and linalyl acetate, while 18 samples appeared genuine since the values of the enantiomeric ratios of these two compounds were within the limits of genuine oils. In the oils analyzed, Binder et al. (2001) also determined the percentage of some components:

	a	b
p-Cymene	0.5%–1.6%	0.2%–1.4%
Limonene	34%–43%	32%–46%
β-Pinene	4.0%–6.7%	4.3%–7.4%
γ-Terpinene	3.8%–7.2%	5.0%–8.0%
Geranial	0.2%–0.4%	0.2%–0.4%
Linalool	8.9%–15.0%	3.4%–18.0%
Linalyl acetate	25%–33%	24%–34%

(a) Oils with enantiomeric ratios of linalool and linalyl acetate outside the limits of genuine oils; (b) Oils with enantiomeric ratios of linalool and linalyl acetate within the limits of genuine oils.

Most of the oils analyzed by Binder et al. (2001) also presented high amounts of *p*-cymene indicative of bad storage conditions. Nichols (2008) reported results for a commercial sample of bergamot oil with 14.8% linalool.

8.3.3 BERGAPTEN-FREE OILS

Due to its photosensitizing properties, the amounts of bergapten and related compounds are rigorously regulated in perfumes and cosmetic products. This aspect is treated in detail in Chapter 24, this volume. For this reason the cosmetic and perfumery industries require bergamot oil with a low content of bergapten. As asserted by Di Giacomo (1990), the process to eliminate or reduce bergapten must not compromise the olphactive properties of the oil, must not reduce the shelf-life, and must not modify the physicochemical properties not related to the content of coumarins and psoralens. In the commerce bergamot oil is considered "bergapten-free" when the amount of bergapten is lower than 50 ppm. The most common industrial process used to reduce the amount of bergapten which respects the conditions proposed by Di Giacomo (1990) consists of a treatment of the oil with a water solution of alkali. The concentration of alkali, the agitation system, the duration of the contact between the water and oil phase, and the temperature are all parameters to be optimized since they determine the final concentration of bergapten and the quality of the resulting oil. During this process the alkaline hydrolysis of the lactone ring of bergapten occurs, with the formation of the corresponding water-soluble salt. Similar to this is also hydrolized the lactone ring of citropten which, as bergapten, presents a methoxy group in position 5 of the ring. Under these conditions the lactone ring of 5-geranyloxy-7-methoxycoumarin and of bergamottin is not destroyed due to the steric hindrance of the geranyloxy group present in position 5 of the ring, which protects the molecule against the alkaline attack. Bergamot oil can also be industrially processed by reduced-pressure distillation to decrease the amount of furocoumarins; this process leads to the elimination or reduction of all the psoralens and coumarins present in the nonvolatile residue of the oil.

Laboratory techniques to apply supercritical CO_2 have been described (Poiana et al. 2003). Another recent approach (Figoli et al. 2006) is pervaporation. These approaches have not been yet successfully applied at industrial scale.

The composition of bergapten-free oils obtained by alkaline treatment or by distillation was studied by Verzera et al. 1998; Ferrini et al. 1998; Poiana et al. 2003; Costa et al. 2010; Dugo et al. 2012. Belsito et al. (2007) analyzed the bergapten-free oil obtained in laboratory by vacuum distillation from fruits peels. The composition of these oils is reported in Table 8.6.

The results in Table 8.6 show that the composition of bergapten-free oil is almost identical to the cold-pressed oils. The light alkaline treatment and the optimal distillation conditions do not determine changes of the volatile fraction, with the exception of the slight increase of terpinen-4-ol in some of the distilled oils.

8.3.4 BERGAMOTTELLA OILS

Most of the information given in this section is of historical interest; in fact recent knowledge on production, commerce, and composition of this product is not available; the only available results are those published by Calavarano (1961) in a study using classical methods, and by the same author (1963, 1968) using chromatography.

In summer, before bergamot fruits are mature, depending on the weather conditions, blooming time, and the load of the trees, a large amount of fruit can prematurely fall. These immature fruits are named "bergamottella." They can remain on the ground for a long time, exposed to moisture and heat, and deteriorate, leading to a very dark color. The oil obtained by cold extraction is as dark as these fruits, and is called "nero di bergamoto" (black of bergamot). Those fruits, which stay on the ground for a shorter time, are less exposed to the direct sunlight and maintain a

TABLE 8.6
Percentage Composition of the Volatile Fraction of Industrial Bergapten-Free Bergamot Oils

Hydrocarbons	Distilled						Alkali Treated			
	1a	2	3a	4	5a	6a	1b	3b	5b	6b
Monoterpene										
Camphene	0.04	–	tr	0.03	0.03	0.02–0.03	0.04	tr	0.03	0.03
δ-3-Carene	tr	–	–	tr	–	tr	tr	–	–	tr
p-Cymene	0.09	–	0.5	f	0.98	0.16–0.26	0.13	0.6	0.55	0.11–0.19
Limonene	34.24[a]	32.4	33.1	46.29[f]	39.98	38.45–42.82	34.54[a]	32.5	43.75	40.12–43.23
Myrcene	1.01	1.3	0.9	1.47	0.86	0.88–0.99	1.06	0.9	0.97	0.90–0.99
(E)-β-Ocimene	0.32	–	–	0.26	0.18	0.18–0.22	0.35	–	0.18	0.17–0.18
(Z)-β-Ocimene	0.08	–	tr	0.07	0.04	0.03	0.10	tr	0.05	0.02–0.03
α-Phellandrene	0.03	–	tr	0.06[e]	0.02	0.02	0.03	tr	0.02	0.02–0.03
α-Pinene	1.16	1.2	1.4	1.21	0.99	0.90–1.11	1.16	1.4	1.01	0.90–1.02
β-Pinene	7.55[b]	6.6	6.7	5.54	5.27	4.72–5.70	7.53[b]	6.7	5.41	5.03–5.41
Sabinene	7.55[b]	1.1	1.1	0.91	0.85	0.84–1.00	7.53[b]	0.9	0.89	0.88–0.94
α-Terpinene	0.17	–	0.3	0.13	0.09	0.14–0.15	0.17	0.1	0.11	0.13–0.14
γ-Terpinene	7.15	7.6	8.7	5.95	5.20	6.71–6.87	7.28	8.5	6.15	6.52–6.74
Terpinolene	0.33	0.3	0.5	0.27	0.22	0.30–0.31	0.33	0.4	0.24	0.29
α-Thujene	0.30	0.3	0.4	0.14	0.25	0.24–0.29	0.31	0.4	0.26	0.23–0.26
Tricyclene	tr	–	–	–	0.01	tr	tr	–	0.01	tr
Sesquiterpene										
cis-α-Bergamotene	0.03	–	–	–	0.01	0.02	0.03	–	0.02	0.02
trans-α-Bergamotene	0.24	–	0.5	–	0.22	0.26–0.29	0.29	0.5	0.25	0.29–0.33
trans-β-Bergamotene	–	–	–	–	0.01	0.01–0.02	–	–	0.01	0.02
Bicyclogermacrene	–	–	–	–	0.01	tr–0.01	–	–	0.01	0.01

β-Bisabolene	0.25	0.5	0.5	0.32	0.27	0.30–0.35	0.41	0.7	0.37	0.43–0.51
(E)-α-Bisabolene	–	–	–	–	0.01	tr–0.01	–	–	0.01	0.01
(Z)-α-Bisabolene	tr	–	–	–	0.02	0.02	tr	–	–	0.03–0.04
β-Caryophyllene	0.26	0.4	0.5	0.33	0.24	0.31–0.34	0.30	0.6	0.27	0.36–0.38
δ-Elemene	tr	–	–	–	0.01	tr	tr	–	0.01	tr
(E,E)-α-Farnesene	tr	–	–	–	–	tr	tr	–	–	tr
(E)-β-Farnesene	–	–	–	–	0.04	0.04–0.05	–	–	0.05	0.05–0.06
(Z)-β-Farnesene	0.04	–	–	0.12	0.01	0.01–0.02	0.06	–	0.01	0.01–0.02
Germacrene D	0.02	–	–	–	0.03	0.03	0.03	–	0.03	0.04–0.05
α-Humulene	0.02	–	–	–	0.02	0.02	0.02	–	0.02	0.03
β-Santalene	0.01	–	–	–	0.01	0.01	0.01	–	0.01	0.01
Aldehydes										
Aliphatic										
Decanal	0.06	–	0.1	–	0.04	0.04–0.05	0.06	0.1	0.04	0.05–0.06
Dodecanal	0.02	–	–	tr	–	tr	0.03	–	–	tr
Nonanal	0.03	–	–	tr	0.02	0.01–0.04	0.03	–	0.03	0.03–0.04
Octanal	0.06	–	tr	0.06c	0.04	0.04	0.05	tr	0.05	0.04–0.05
Undecanal	0.01	–	–	0.04	0.01	0.01	0.01	–	0.01	0.01
Monoterpene										
Citronellal	0.01	–	–	0.01	0.01	0.01	0.01	–	0.01	0.01
Geranial	0.29c	0.3	0.3	0.17	0.26	0.28–0.37	0.29c	0.3	0.31	0.28–0.32
Neral	0.21	–	0.2	0.10	0.18	0.18–0.25	0.18	0.2	0.19	0.18–0.21
Perilla aldehyde	0.29c	–	–	–	0.26	tr–0.01	0.29	–	–	tr
Ketones										
Aliphatic										
6-Methyl-5-hepten-2-one	0.01	–	–	–	0.01	tr–0.01	0.02	–	0.03	0.02–0.03

continued

TABLE 8.6 (CONTINUED)
Percentage Composition of the Volatile Fraction of Industrial Bergapten-Free Bergamot Oils

	Distilled						Alkali Treated			
	1a	2	3a	4	5a	6a	1b	3b	5b	6b
Monoterpene										
Camphor	tr	–	–	–	–	tr	tr	–	–	tr
Carvone	tr	–	–	–	0.01	0.01	tr	–	–	tr–0.01
Sesquiterpene										
Nootkatone	0.01	–	tr	–	–	–	0.06	0.1	0.03	0.04–0.06
Alcohols										
Aliphatic										
Octanol	tr	–	tr	–	–	0.03–0.05[j]	tr	tr	–	0.03[j]
Monoterpene										
Isopulegol	tr	–	–	–	tr	tr	tr	–	–	tr
Linalool	13.57	11.3	11.9	7.81	9.96	10.02–13.73	11.44	11.8	10.68	8.53–10.92
Nerol	0.05[d]	–	0.1	tr	0.05	tr–0.02	0.06[d]	0.1	0.05	0.03–0.05
cis-sabinene hydrate	0.03	–	tr	0.08	0.03	0.03–0.05[j]	0.03	tr	0.03	0.03[j]
Terpinen-4-ol	0.03	–	0.1	0.03	0.03	0.03–0.05	0.02	tr	0.02	0.02–0.03
α-Terpineol	0.08	–	0.1	0.03	0.08	0.10–0.15	0.09	0.1	0.08	0.09–0.12
Sesquiterpene										
α-Bisabolol	tr	–	–	–	–	–	0.02	–	0.03	0.02
Campherenol	tr	–	–	–	–	–	0.02	–	0.01	tr
(E)-Nerolidol	tr	–	–	–	0.01	tr	0.02	–	0.02	0.01–0.02
Norbornanol[e]	tr	–	–	–	–	tr	0.01	–	–	0.01

Esters

Aliphatic

	1	2	3	4	5	6	7	8	9	10
Decyl acetate	0.01	–	–	0.01	0.02	0.02	0.02	–	0.03	0.03
Heptyl acetate	0.01	–	–	–	0.01	tr	0.01	–	0.01	tr
Hexyl acetate	tr	–	–	–	0.01	tr	tr	–	–	tr
Nonyl acetate	0.02	–	–	0.04	0.01	0.01	0.02	–	0.02	0.01–0.02
Octyl acetate	0.10	–	0.1	0.13	0.08	0.07–0.08	0.11	0.1	0.08	0.08–0.09

Monoterpene

	1	2	3	4	5	6	7	8	9	10
Bornyl acetate	0.02	–	–	0.05	–	0.01–0.02	0.02	–	–	0.01–0.02
Citronellyl acetate	0.02	–	–	–	0.01	0.01–0.02	0.02	–	0.01	0.02
Geranyl acetate	0.24	0.5	0.3	0.19	0.25	0.21–0.26	0.35	0.4	0.27	0.29–0.32
Linalyl acetate	31.18	33.6	30.6	26.31	25.07	27.61–29.55	31.99	31.3	26.53	28.88–30.83
Linalyl propanate	0.03	–	0.1	–	0.03	0.02–0.04	0.04	0.1	0.03	0.03
Methyl geranate	tr	–	–	–	0.01	tr–0.01	tr	–	0.01	tr–0.01
Neryl acetate	0.26	0.4	0.4	0.14	0.28	0.32–0.34	0.31	0.4	0.32	0.37–0.43
trans-Sabinene hydrate acetate	0.04[g]	–	0.1[h]	–	–	–	0.06[g]	0.1[h]	–	–
α-Terpinyl acetate	0.12	–	0.2	–	0.11	0.12–0.13	0.15	0.3	0.11	0.13–0.14

Ethers and Oxides

Monoterpene

	1	2	3	4	5	6	7	8	9	10
cis-Limonene oxide	tr	–	–	–	0.01	tr	tr	–	–	tr
trans-Limonene oxide	tr	–	–	–	0.01	tr	tr	–	–	tr

tr, traces; *correct isomer not characterized.
a limonene + β-phellandrene.
b β-pinene + sabinene.
c geranial + perilla aldehyde.
d citronellol + nerol.

continued

TABLE 8.6 (CONTINUED)
Percentage Composition of the Volatile Fraction of Industrial Bergapten-Free Bergamot Oils

e α-phellandrene + octanal.

f limonene + *p*-cymene.

g *trans*-sabinene hydrate acetate.

h *cis*-sabinene hydrate acetate.

i octanol + *cis*-sabinene hydrate.

Appendix to Table 8.6:

1. Verzera et al. (1998). Calabria, Italy: (a) average composition of 7 distilled bergapten free oils; (b) 5 alkali treated bergapten free oils; GC on capillary column (30 m × 0.32 mm × 0.40–0.45 μm) coated with SE-52; GC/MS on capillary column (30 m × 0.25 mm × 0.25 μm) coated with DB-5; MS libraries: Adams and FFNSC homemade library; relative percentages of peak areas. Verzera et al. also found germacrene B (0.01%) in sample (b) and a trace amount in sample (a) and trace amounts of dodecane, 2-decenal*, tetradecanal, dodecanol, geraniol, cis- and trans-linalool oxide, and indole.

2. Ferrini et al. (1998). Sicily, Italy; one sample of distilled bergapten-free oil; for experimental conditions see point 5 of appendix to Table 8.3. Ferrini et al. also found β-ocimene* (0.4%), bergamotene* (0.3%).

3. Poiana et al. (2003). Sicily, Italy; (a) one sample distilled and (b) one sample alkali treated of bergapten-free oils; for experimental conditions see point 9 of appendix to Table 8.3. Poiana et al. also found trace amounts of bergapten.

4. Belsito et al. (2007). Calabria, Italy; one sample of bergapten-free oil obtained by vacuum distillation of the peel of bergamot fruits. For experimental conditions see point 11 of appendix to Table 8.3.

5. Costa et al. (2010). Calabria, Italy; (a) one sample distilled and (b) one sample alkali treated of bergapten-free oils; for experimental conditions see point 12 of appendix to Table 8.3. Costa et al. also found, respectively in samples (a) and (b), epi-β-bisabolol (0, 0.02%), isobornyl acetate (0.01%, 0.02%).

6. Dugo et al. (2012). Calabria, Italy; range of the composition 7 samples of distilled bergapten free oils (a), and 3 samples alkali treated bergapten free oils (b); for experimental conditions see point 12 of appendix to Table 8.3; Dugo et al. also found in samples (a) 4,8-dimethyl-1,3(E),7-nonatriene (tr-0.01%), Y-curcumene (0-0.01%) and trace amounts in one or more of analyzed samples of (E)-γ-bisabolene, δ-cadinene, β-elemene, β-sesquiphellandrene, citronellol, fenchol*, ascaridole, and in samples (b) (E)-γ-bisabolene (tr-0.01%), γ-curcumene (0.01%), spathulenol (0.01%) and trace amounts in one or more of analyzed samples of 4,8-dimethyl-1,3(E),7-nonatriene, δ-cadinene, β-elemene, β-sesquiphellandrene, citronellol, cis- and trans-sesquisabinene hydrate

lighter color. From these it is possible to extract by mechanical means "bergamottella oil," which presents better quality than the black of bergamot oil. In each case the color of the oil resembles that of the fruits. Just before ripening, strong winds could cause the fall of numerous green fruits. These are collected and cold-extracted to produce an "emerald-green" oil. The fruits that fall in summer when still too small to be extracted by mechanical means are used to obtain an oil by distillation, or "distilled bergamottella" (Calvarano 1961). The production of these oils is very scant and not regular, as it depends on the fruit load on the trees.

Calvarano (1961) determined the physicochemical parameters and the total content of alcohols, esters, and aldehydes in oils industrially produced from bergamottella, by Pelatrice machine with and without water during the period June to September. The author determined the behavior of alcohols, esters, and aldehydes, which varied during the season (these results are discussed in Section 8.4.1 of this chapter); the differences of the composition of oils extracted in the same period from dark fruits (black of bergamot), yellowish fruits (yellow bergamottella oil), and green fruits (green bergamottella oil); the differences between oils of the same lot of fruits extracted by pelatrice with water, dry pelatrice, and successive screw presses of the solid detritus; and the differences of the composition during the cycle of production. These results are summarized in Table 8.7. In conclusion, in this study the

TABLE 8.7
Percentage of Esters, Alcohols, and Aldehydes in Oils Extracted from "Bergamottella" with Different Pigmentation Grades

	Brown Fruits (Black of Bergamot)	Yellow Fruits (Yellow Bergamottella Oil)	Green Fruits (Green Bergamottella Oil)
Esters	23.2	20.1	36.0
Alcohols	28.2	31.6	33.6
Aldehydes	0.3	0.5	0.7

Oils extracted from the same fruit stock by pelatrice either with water and without (dry) (I, II)

	Pelatrice with Water	Pelatrice Dry		
		a	b	c
Esters	41.1	39.4	36.1	37.0
Alcohols	27.9	30.2	32.3	28.9
Aldehydes	0.5	0.5	0.4	0.4

Oils collected during the process by the pelatrice with water

	Start	Middle	End
Esters	39.4–41.1	37.1	37.0–37.4
Alcohols	27.9–30.2	32.4	32.4–34.2
Aldehydes	0.5	0.5	0.5

[a] oil from the separator
[b] oil from the screw press
[c] = a + b

best quality, expressed in function of esters amount, was obtained from the yellowish fruits, using the pelatrice machine with water in the first part of the process.

Calvarano (1963) analyzed 13 industrial samples of black bergamot oil and 3 samples of distilled bergamottella. The same author (Calvarano 1968) analyzed some samples solvent extracted in laboratory from fruits collected every 10 days from July to January. The oils collected from July to September can be considered bergamottella oils. The ranges of variabilities determined by Calvarano (1963, 1968) for the three types of oil are reported in Table 8.8. These oils are characterized by higher content of p-cymene and a lower content of monoterpene hydrocarbons than traditional bergamot oil. In black of bergamot, the monoterpene hydrocarbons ranged between 17% and 25%, while in traditional bergamot oil usually this fraction ranges between 30% and 48%.

8.3.5 RECOVERED OILS

Small amounts of low-quality oil can be recovered from the residue of cold extraction using different procedures:

- "Ricicli" (recycled oils): Oils recovered by centrifuge of the recycled water at the end of the daily process of extraction. The yield of the recovered oil is about 1% of the total oil extracted mechanically.
- "Torchiati" (screw-pressed oils): Oil recovered by small hydraulic presses from the solid residues in the secondary separator. The yield of the recovered oil is about 3.5% of the total oil extracted mechanically. The name "torchiati" used for citrus oils different from bergamot indicates the cold-pressed oil extracted by screw-pressing of the whole fruits and successive centrifugation.
- "Pulizia dischi" oils: Oils recovered by decanted liquid residues from the secondary separator disk at the end of the daily process. The yield of the recovered oil is about 0.5% of the total oil extracted mechanically.
- Distilled or "fecce" oils: Oils recovered by distillation of the semi-fluid "fecce" wastes automatically ejected by the primary separator. In the past, the distillation was performed at atmospheric pressure, while recently it was carried out at reduced pressure using the Peratoner method.
- Peratoner oils: In present days it is also preferred in numerous industrial plants to use reduced-pressure distillation to recover oil from all the above mentioned residues and from water phase. Using this procedure it is possible to recover oils of better quality than those obtained by traditional recovery processes.

Recovered oils should be sold separately as their addition to cold-pressed oils is not allowed.

8.3.5.1 Oils Recovered by Cold Procedures

The scant information available on the composition of the volatile fraction of these oils is summarized in Table 8.9. These results refer to the study carried out by

TABLE 8.8
Percentage Composition of the Volatile Fraction of Black of Bergamot, Bergamotella, and Bergamotella Distilled Oils

Hydrocarbons	1a	1b	2
Aliphatic			
Monoterpene			
Camphene	–	tr	0.02–0.04
δ-3-Carene	–	–	0.07–0.13
p-Cymene	tr–11.07	tr–1.60	0.26–0.36
Limonene	54.29–67.68	13.21–16.01	24.21–25.61
Myrcene	1.20–1.96	0.56–0.72	0.66–0.85
α-Phellandrene	–	–	0.02–0.06
β-Phellandrene	–	tr	–
α-Pinene	1.59–3.25	0.41–0.77	1.08–1.41
β-Pinene	11.25–21.77	1.83–3.27	4.03–5.40[b]
Sabinene	–	–	4.03–5.40[b]
γ-Terpinene	8.27–15.29	2.20–2.74	5.67–7.83
Terpinolene	tr–9.95	0.42–1.84	0.44–0.58
α-Thujene	0.78–1.40	0.08–0.13	0.03–0.06
Sesquiterpene			
Bisabolene*	–	–	0.21–0.28
Aldehydes	0.41–0.60[a]		
Aliphatic			
Decanal	–	–	0.20–0.26
Heptanal	–	–	0.04–0.07
Nonanal	–	–	0.06–0.09
Octanal	–	–	0.09–0.12
Monoterpene			
Citronellal	–	–	0.01–0.02
Geranial	–	–	0.36–0.48
Neral	–	–	0.30–0.43
Alcohols	28.91–41.86[a]	62.43–67.73[a]	
Monoterpene			
Citronellol	–	–	0.40–0.56[c]
Geraniol	–	–	0.01–0.03[t]
Linalool	–	–	24.22–32.93
Nerol	–	–	0.40–0.56[c]
Terpinen-4-ol	–	–	0.03–0.04
α-Terpineol	–	–	0.10–0.20

continued

TABLE 8.8   (CONTINUED)
Percentage Composition of the Volatile Fraction of Black of Bergamot, Bergamotella, and Bergamotella Distilled Oils

	1a	1b	2
Esters	17.30–39.40[a]	1.82–2.02[a]	
Aliphatic			
Octyl acetate	–	–	0.01–0.02
Monoterpene			
Geranyl acetate	–	–	0.09–0.22
Linalyl acetate	–	–	25.05–30.80
Neryl acetate	–	–	0.19–0.47
α-Terpinyl acetate	–	–	0.06–0.25

tr, traces; t tentative identification; * correct isomer not characterized.

[a] total aldehydes as citral, total alcohols as linalool, total esters as linalyl acetate were determined with classical methods.

[b] β-pinene + sabinene.

[c] citronellol + nerol.

Appendix to Table 8.8:

1. Calvarano (1963). (a) Range of the composition of 13 samples of nero di bergamotto; (b) range of the composition of 3 samples of distilled bergamottella oil; analysis of monoterpene hydrocarbons fraction by GC/FID on stainless steel capillary column (46 m × 0.50 mm) coated with UCON LB55OX.

2. Calvarano (1968). Range of the composition of 14 samples (2 samples for each period) of bergamottella oil laboratory solvent extracted from the peel of fruit picked from July 20 to September 22; GC/FID on stainless steel capillary column (46 m × 0.50 mm) coated with UCON LB55OX.

Verzera et al. (1998) on six samples of "Ricicli," eight samples of "Torchiati," and four samples of "Pulizia dischi" produced during the 1996–97 season. The comparison of recovered oils with cold-pressed ones produced in the same season indicated that the recovered oils presented a total content of linalyl acetate and terpinen-4-ol close to the maximum amount of that determined in cold-pressed oils; the content of octanol (0.023%–0.044%) was more than the highest amount determined in cold-pressed oils (0.010%); α-terpineol, probably due to enzymatic reactions, was much higher in "Torchiati" and in "Ricicli" than in "Pulizia dischi" and cold-pressed oils. Authors concluded that the values of terpinen-4-ol, α-terpineol, and mainly octanol could be used to distinguish the oils recovered by cold procedures from those traditionally cold-extracted.

8.3.5.2   Oils Recovered by Distillation

For many years the only information on distilled oils from "fecce" was that published by Calvarano (1963) on the monoterpene hydrocarbons of two samples. The two oils were characterized by a high amount of *p*-cymene due to the bad storage

TABLE 8.9
Percentage Composition of the Volatile Fraction of Industrially Recovered Bergamot Oils

Hydrocarbons	Cold-Recovered					Distilled			
	1a	1b	1c	1d	1e	2	3a	3b	
Aliphatic									
Dodecane	tr	tr	tr	0.01	tr	—	—	—	—
Monoterpene									
Camphene	0.03	0.03	0.03	0.05	0.02	0.01–0.02	0.02–0.03	0.02	0.04
δ-3-Carene	tr	tr	tr	tr	0.01	—	tr	tr	tr
p-Cymene	0.11	0.11	0.12	2.74	—	0.13–0.16	0.25–0.58	0.033	3.86
Limonene	32.07[b]	30.19[b]	31.39[b]	35.04[b]	23.14[b]	36.21–45.14	31.36–43.23	44.37	50.67
Myrcene	0.81	0.79	0.83	1.32	0.71	0.62–0.82	0.76–1.00	1.00	1.14
(E)-β-Ocimene	0.20	0.21	0.20	0.52	0.28	0.19–0.21	0.21–0.29	0.23	0.13
(Z)-β-Ocimene	0.02	0.02	0.01	0.24	0.12	0.01–0.03	0.06–0.10	0.12	0.11
α-Phellandrene	0.03	0.02	0.02	—	—	0.01–0.02	0.02	0.02	0.01
β-Phellandrene	[b]	[b]	[b]	[b]	[b]	—	—	—	—
α-Pinene	1.12	1.08	1.27	1.38	0.45	0.46–0.72	0.58–1.01	0.64	1.38
β-Pinene	8.15[a]	7.96[a]	8.81[a]	7.92[a]	4.24[a]	3.39–4.82	3.50–5.40	4.56	8.35
Sabinene	8.15[a]	7.96[a]	8.81[a]	7.92[a]	4.24[a]	0.43–0.59	0.56–0.85	0.65	1.27
α-Terpinene	0.14	0.16	0.16	0.09	0.13	0.10–0.13	0.12–0.17	0.18	0.05
γ-Terpinene	7.57	7.58	7.57	4.65	5.99	6.20–7.55	5.77–7.38	8.25	3.68
Terpinolene	0.30	0.32	0.30	0.06	0.31	0.23–0.27	0.30–0.36	0.40	0.19
α-Thujene	0.28	0.28	0.32	0.32	0.12	0.11–0.18	0.15–0.26	0.17	0.33
Tricyclene	tr	tr	tr	tr	tr	0.02	tr	—	tr

continued

TABLE 8.9 (CONTINUED)
Percentage Composition of the Volatile Fraction of Industrially Recovered Bergamot Oils

	Cold-Recovered					Distilled		
	1a	1b	1c	1d	1e	2	3a	3b
Sesquiterpene								
cis-α-Bergamotene	0.03	0.04	0.03	0.03	0.02	0–0.01	0.01	0.01
trans-α-Bergamotene	0.27	0.34	0.25	0.31	0.27	0.07	0.08–0.12	0.14
trans-β-Bergamotene	–	–	–	–	–	–	tr–0.01	0.01
Bicyclogermacrene	0.03	0.04	0.03	0.02	0.01	–	tr–0.01	0.01
β-Bisabolene	0.37	0.51	0.35	0.39	0.36	0.09–0.10	0.11–0.15	0.19
(Z)-α-Bisabolene	tr	tr	tr	tr	tr	0.01	0.01	0.01
(Z)-γ-Bisabolene	tr	0.01	tr	tr	0.01	–	tr	–
β-Caryophyllene	0.26	0.33	0.25	0.21	0.25	0.10–0.11	0.12–0.17	0.13
δ-Elemene	tr	tr	tr	tr	tr	–	tr–0.01	tr
(E,E)-α-Farnesene	tr	tr	tr	tr	tr	–	tr	tr
(E)-β-Farnesene	–	–	–	–	–	–	0.01–0.03	0.02
(Z)-β-Farnesene	0.06	0.07	0.06	0.04	0.04	–	tr–0.01	0.01
Germacrene B	tr	0.01	tr	tr	tr	–	–	–
Germacrene D	0.05	0.07	0.03	0.03	0.03	0.01	0.01–0.02	0.02
α-Humulene	0.02	0.03	0.02	0.02	0.02	0.01	0.01	0.02
β-Santalene	0.01	0.01	0.01	0.01	0.01	–	tr	0.01
Aldehydes								
Aliphatic								
Decanal	0.06	0.07	0.07	0.16	0.17	0.03	0.02–0.04	0.01
(E)-2-Decenal	tr	0.01	0.01	tr	tr	–	–	–
Dodecanal	0.03	0.04	0.02	0.04	0.04	–	tr	–

Nonanal	0.03	0.04	0.04	0.03	0.03	0.01–0.11	0.04–0.13	0.06	0.02
Octanal	0.03	0.04	0.05	0.18	0.02	0.01–0.02	0.01–0.03	tr	tr
Tetradecanal	tr	tr	tr	tr	tr	–	–	–	–
Undecanal	0.01	0.01	0.01	0.02	0.01	–	tr–0.02	tr	tr
Monoterpene									
Citronellal	0.01	0.01	0.01	0.01	tr	–	tr	tr	tr
Geranial	0.26^d	0.32^d	0.28^d	0.15^d	0.12^d	0.13–0.19	0.11–0.27	0.05	0.04
Neral	0.16	0.20	0.18	0.16	0.09	0.12–0.15	0.12–0.22	0.05	0.04
Perilla aldehyde	0.26^d	0.32^d	0.28^d	0.15^d	0.12^d	–	tr–0.01	tr	0.01
Ketones									
Aliphatic									
6-Methyl-5-hepten-2-one	0.05	0.01	0.01	0.04	0.01	0.01–0.03	0.02–0.1	0.08	0.03
Monoterpene									
Camphor	tr	tr	tr	tr	0.01	–	tr–0.01	tr	tr
Carvone	tr	tr	tr	tr	tr	0–0.01	tr–0.01	tr	0.03
Sesquiterpene									
Nootkatone	0.03	0.09	0.08	0.01	tr		tr	–	–
Alcohols									
Aliphatic									
Dodecanol	tr	tr	tr	tr	tr			–	–
Octanol	0.03	0.02	0.04	0.22	0.02	0.01–0.03^e	0.03–0.05^e	0.03^e	0.05^e
Monoterpene									
Citronellol	0.07^c	0.06^c	0.07^c	0.59^c	0.13^c		tr–0.03	tr	0.05
Geraniol	tr	tr	tr	tr	tr			–	–
Isopulegol	tr	tr	tr	tr	tr		tr–0.01	–	–
Linalool	9.46	10.50	8.22	25.40	36.85	18.38–33.14	18.08–33.98	27.52	13.89

continued

TABLE 8.9 (CONTINUED)
Percentage Composition of the Volatile Fraction of Industrially Recovered Bergamot Oils

| | Cold-Recovered | | | | | Distilled | | | |
	1a	1b	1c	1d	1e	2	3a	3a	3b
Nerol	0.07[c]	0.06[c]	0.07[c]	0.59[c]	0.13[c]	0–0.07	0.06–0.13	0.04	0.03
cis-Sabinene hydrate	0.03	0.02	0.04	0.01	0.01	0.01–0.03[e]	0.03–0.05[e]	0.03[e]	0.05[e]
Terpinen-4-ol	0.03	0.04	0.02	0.53	0.35	0.13–0.15	0.15–0.29	0.17	0.12
α-Terpineol	0.33	0.18	0.07	3.98	1.30	0.30–0.42	0.40–0.83	1.67	0.72
Sesquiterpene									
α-Bisabolol	0.01	0.02	0.02	0.01	tr	–	tr	–	–
Campherenol	0.01	0.02	0.01	0.01	tr	–	–	–	–
(E)-Nerolidol	0.02	0.03	0.02	0.02	0.01	–	tr–0.01	tr	0.01
Norbornanol[f]	0.01	0.01	0.01	0.01	tr	–	tr	–	–
Esters									
Aliphatic									
Decyl acetate	0.02	0.02	0.01	0.02	0.02	0.01	0.01–0.03	0.01	0.02
Heptyl acetate	0.01	0.01	0.01	0.02	0.01	–	tr–0.01	tr	tr
Hexyl acetate	tr	tr	tr	0.01	tr	0–0.01	0.01	0.01	0.02
Nonyl acetate	0.02	0.03	0.02	0.05	0.04	0.01	0.01–0.03	0.01	0.09
Octyl acetate	0.09	0.14	0.09	0.05	0.01	0.06–0.09	0.07–0.14	0.08	0.11
Monoterpene									
Bornyl acetate	0.01	0.02	0.01	0.16	0.02	0.01	0.01–0.02	0.01	0.01
Citronellyl acetate	0.02	0.03	0.02	0.04	0.03	0.01	tr–0.02	0.01	–
Geranyl acetate	0.18	0.30	0.15	1.39	0.45	0.31	0.22–0.53	0.10	0.33
Linalyl acetate	36.30	36.50	37.64	8.88	22.76	16.81–18.71	14.26–22.46	8.24	11.66
Linalyl propanate	0.04	0.04	0.04	0.05	0.03	0.02	0.02–0.03	0.01	0.01

Methyl geranate	tr	0.01	tr	tr	tr	–	tr–0.01	tr	0.06
Neryl acetate	0.22	0.31	0.21	0.92	0.40	0.28	0.28–0.47	0.12	0.29
trans-Sabinene hydrate acetate	0.06	0.05	0.07	–	tr	–	–	–	–
α-Terpinyl acetate	0.16	0.19	0.15	0.18	0.14	0.05–0.08	0.07–0.16	0.05	0.11
Ethers and Oxides									
Monoterpene									
1,8-Cineole	tr	tr	tr	–	–	–	–	–	–
cis-Limonene oxide	tr	tr	tr	0.01	0.01	–	tr–0.01	–	0.06
trans-Limonene oxide	tr	tr	tr	0.03	0.01	–	tr–0.01	tr	0.04
cis-Linalool oxide	tr	tr	tr	–	tr	–	–	–	–
trans-Linalool oxide	tr	tr	tr	–	tr	–	–	–	–
Others									
Indole	tr	tr	0.01	0.01	tr	–	–	–	–

tr, trace.

a   β-pinene + sabinene.
b   limonene + β-phellandrene.
c   citronellol + nerol.
d   geranial + perilla aldehyde.
e   octanol + cis sabinene hydrate.
f   2,3-dimethyl-3-(4-methyl-3-pentenyl)-2-norbonanol.

Appendix to Table 8.9:

1. Verzera et al. (1998). Average composition of (a) 8 Torchiati oils; (b) 6 Ricicli oils; (c) 4 Pulizia dischi oils; (d) one distilled oil at reduced pressure; (e) one distilled oil at atmosphere pressure; for the experimental conditions see point 1 of appendix to Table 8.6.

2. Sciarrone et al. (2009, personal communication). Range of the composition of tree samples distilled at reduced pressure; for experimental conditions see point 12 of appendix to Table 8.3.

3. Dugo et al. (2012). Range of the composition of 15 samples (Peratoner) distilled at reduced pressure produced during 2008–09, 2009–10, 2010–11 seasons (a); 2 fecce oils distilled at reduced pressure (b); Dugo et al. also found (E)-α-bisabolene (0–tr), (E)-γ-bisabolene (0–tr), 4,8-dimethyl-1,3(E),7-nonatriene (0%–0.01%), γ-curcumene (0%–tr), β-elemene (0%–tr) fenchol* (0%–0.01%), β-sesquiphellandrene (tr–0.01%), ascaridole* (0%–0.01%), *cis*-sesquisabinene hydrate (0%–tr), ascaridole* (0%–0.01%). For experimental conditions see point 12 of appendix to Table 8.3.

conditions of the raw material used for distillation, and the condition applied to distill the oil. The results reported by Calvarano (1963) are reported below:

	1	2
p-Cymene	1.74%	0.84%
Limonene	19.64%	29.19%
Myrcene	0.85%	1.34%
β-Phellandrene	tr	tr
α-Pinene	0.72%	1.65%
β-Pinene	4.53%	8.04%
γ-Terpinene	4.02%	5.27%
Terpinolene	tr	tr
α-Thujene	0.23%	0.50%

More recently, Verzera et al. (1998) analyzed one oil distilled from fecce at atmospheric pressure; fecce oils obtained at reduced pressure were analyzed by Verzera et al. (1998) and Dugo et al. (2012). Dugo et al. (2012) also analyzed several samples of oils recovered by distillation at reduced pressure simultaneously from all the residues of the cold-extraction process (Peratoner). These results are included in Table 8.9. The composition of these last oils, likely obtained from the residues of the cold extraction by modern vacuum distillation techniques, do not present substantial differences from the cold-pressed oils, with the exception of slight increases in terpinen-4-ol and α-terpineol. In some of the fecce oils analyzed by Verzera et al. (1998) and by Dugo et al. (2012) were high amounts of p-cymene, terpinen-4-ol, and α-terpineol.

8.3.6 CONCENTRATED OILS

The food industry's increased interest in bergamot oil has generated higher demand for concentrated oils obtained by modern procedures of high vacuum distillation. In contrast with what usually occurs for other citrus concentrated oils, for bergamot the concentration does not refer to standard parameters, but the procedure is customized to meet the requirements requested by consumers, requiring different concentration of hydrocarbons, and the presence or not of the waxy residue. The results available in literature on industrially concentrated bergamot oil are limited to six samples analyzed by Dugo et al. (2012) reported in Table 8.10. Authors divided the oils analyzed into two groups, based on the amount of monoterpene hydrocarbons and on the distillation technology. In the first group the oils had monoterpene hydrocarbons ranging between 2.5% and 3%, in the second one (terpene-free and colorless oils) they were between 8% and 19%. Based on the composition of the oils, the first group was concentrated ca. three folds; those of the second one were concentrated ca. two folds. In literature are also available results on samples prepared in laboratory by Fantin et al. (2010), who produced terpene-free oil from a commercial bergamot oil by inclusion of deoxycholic acid. Here is reported the composition, in class of substances, of the original oil and of the terpene-free oil after one cycle and two cycles of inclusions.

TABLE 8.10
Percentage Composition of the Volatile Fraction of Industrial Concentrated Bergamot Oils

Hydrocarbons	1a	1b
Aliphatic		
4,8-dimethyl-1,3(E),7-nonatriene	tr–0.01	0.01–0.02
Monoterpene		
Camphene	–	tr–0.01
δ-3-Carene	–	tr
p-Cymene	0.03–0.15	0.23–0.33
Limonene	0.74–1.00	2.69–11.86
Myrcene	0.02–0.20	0.17–0.39
(*E*)-β-Ocimene	0.04–0.13	0.16–0.20
(*Z*)-β-Ocimene	tr–0.06	0.05–0.06
α-Phellandrene	–	tr–0.01
α-Pinene	0.01	0.01–0.25
β-Pinene	0.04–0.06	0.07–1.39
Sabinene	0.01–0.02	0.01–0.25
α-Terpinene	tr–0.01	0.02–0.05
γ-Terpinene	1.18–1.27	4.01–4.20
Terpinolene	0.15–0.16	0.40–0.59
α-Thujene	tr–0.01	0.01–0.07
Tricyclene	tr	tr
Sesquiterpene		
cis-α-Bergamotene	0.03–0.04	0.02–0.03
trans-α-Bergamotene	0.37–0.64	0.31–0.35
trans-β-Bergamotene	0.02–0.04	0.01
Bicyclogermacrene	0.01–0.04	0.01–0.02
β-Bisabolene	0.36–0.97	0.26–0.31
(*E*)-α-Bisabolene	0.01–0.02	tr
(*Z*)-α-Bisabolene	0.03–0.07	0.02–0.03
(*E*)-γ-Bisabolene	tr–0.01	tr
δ-Cadinene	tr–0.01	tr
β-Caryophyllene	0.54–0.79	0.40–0.46
γ-Curcumene	tr–0.01	tr
β-Elemene	tr–0.01	tr–0.01
δ-Elemene	0.01–0.02	0.01–0.02
(*E,E*)-α-Farnesene	0.01	tr–0.01
(*E*)-β-Farnesene	0.05–0.12	0.04–0.05
(*Z*)-β-Farnesene	0.02–0.03	0.01–0.02
Germacrene D	0.04–0.10	0.04
α-Humulene	0.03–0.06	0.02–0.03

continued

TABLE 8.10   (CONTINUED)
Percentage Composition of the Volatile Fraction of Industrial Concentrated Bergamot Oils

	1a	1b
β-Santalene	0.01–0.03	0.01
β-Sesquiphellandrene	tr–0.01	tr
Aldehydes		
Aliphatic		
Decanal	0.12–1.11	0.10–0.12
Dodecanal	tr	–
Nonanal	0.04–0.05	0.06–0.07
Octanal	tr	tr–0.01
Undecanal	0.01	0.01
Monoterpene		
Citronellal	0.02–0.04	0.02–0.03
Isogeranial	–	tr
Geranial	0.60–0.71	0.50–0.58
Neral	0.44–0.52	0.40–0.45
Perilla aldheyde	0.01	0.01
Ketones		
Aliphatic		
6-Methyl-5-hepten-2-one	tr	tr–0.01
Monoterpene		
Camphor	0.01	0.01
Carvone	0.01–0.02	0.01–0.02
Sesquiterpene		
Nootkatone	tr–0.12	tr
Alcohols		
Aliphatic		
Octanol	0.04–0.08[a]	0.10–0.14[a]
Monoterpene		
Citronellol	tr–0.01	tr–0.01
Fenchol*	tr–0.01	0.01
Isopulegol	tr	tr–0.01
Linalool	19.00–32.17	27.56–29.57
Nerol	0.07–0.11	0.09–0.16
cis-Sabinene-hydrate	0.04–0.08[a]	0.10–0.14[a]
Terpinen-4-ol	0.05–0.08	0.09–0.20
α-Terpineol	0.15–0.30	0.28–0.49

TABLE 8.10 (CONTINUED)
Percentage Composition of the Volatile Fraction of Industrial Concentrated Bergamot Oils

	1a	1b
Sesquiterpene		
α-Bisabolol	tr–0.05	tr
Campherenol	tr–0.01	–
(*E*)-Nerolidol	0.01–0.04	tr–0.01
Norbornanol[b]	tr–0.02	tr
cis-Sesquisabinene hydrate	tr–0.01	tr
trans-Sesquisabinene hydrate	tr–0.01	–
Spathulenol	tr–0.03	tr–0.01
Esters		
Aliphatic		
Decyl acetate	0.04–0.07	0.02–0.03
Heptyl acetate	0.01	0.01–0.02
Hexyl acetate	–	tr
Nonyl acetate	0.02–0.03	0.02–0.03
Octyl acetate	0.20–0.26	0.15–0.27
Monoterpene		
Bornyl acetate	0.03–0.04	0.02–0.03
Citronellyl acetate	0.02–0.04	tr–0.02
Geranyl acetate	0.50–0.72	0.34–0.49
Linalyl acetate	59.79–70.92	47.47–57.73
Linalyl propanate	0.03–0.08	0.03–0.04
Methyl geranate	0.01	tr–0.01
Neryl acetate	0.69–0.94	0.41–0.54
α-Terpinyl acetate	0.19–0.31	0.17–0.24
Ethers and Oxides		
Monoterpene		
cis-Limonene oxide	tr	tr–0.01
trans-Limonene oxide	tr–0.01	0.01

tr, traces; * correct isomer not characterized.

[a] octanol + *cis*-sabinene.

[b] 2,3-dimethyl-3-(4-methyl-3-pentenyl)-2-norbonanol.

Appendix to Table 8.10:

1. Dugo et al. (2012). Calabria, Italy; (a) 3 samples of colored deterpenated oils; (b) 3 samples of colorless deterpenated and wax-free oils. For experimental conditions see point 12 of appendix to Table 8.3.

| | Starting oil | Deterpenated Oils | |
		One inclusion cycle	Two inclusion cycles
Monoterpene hydrocarbons	59.93%	36.99%	19.54%
Sesquiterpene hydrocarbons	0.95%	1.28%	1.69%
Alcohols	16.66%	18.29%	21.66%
Esters	22.39%	43.36%	57.07%

8.3.7 OILS EXTRACTED IN LABORATORY

Calvarano analyzed bergamot oils extracted in laboratory with light petroleum from fruits harvested from July to January of the successive year. The results relative to unripe fruits (bergamottella) were discussed in Section 8.3.4; those relative to the seasonal variation of the composition in function of the ripening stage will be discussed in Section 8.4.1.

Huet and Dupuis (1969) analyzed oils extracted in laboratory with light petroleum or by distillation from fruits harvested in Corsica during ripening; Kirbaslar et al. (2001) analyzed Turkish oils cold-pressed in laboratory from fruits harvested from November to January of the successive year. The results of these two articles, relative to the seasonal variation, will be also discussed in Section 8.4.1. The oils from Corsica (Huet and Dupuis 1969) obtained by distillation, compared to the solvent-extracted ones obtained in the same period, had higher amounts of alcohols and lower amount of linalyl acetate.

Yoshida et al. (1971) analyzed one bergamot oil extracted from fruits cultivated in Japan and asserted its composition was very similar to oils produced in Calabria. The components identified by these authors were limonene, 26.0%; *a*-pinene, 0.6%; β-pinene, 5.5%; linalool, 29.1%; linalyl acetate 29.3%.

In Table 8.11 are results relative to the volatile fraction of bergamot oils extracted in laboratory since 1978. Among the values in Table 8.11 are δ-3-carene, determined by Huang et al. (1986) and by Kirbaslar et al. (2000), which are exceptionally high for bergamot oil. The normal amount of δ-3-carene in bergamot oil never exceeds 0.01% of the volatile fraction. Values of geraniol and α-terpineol reported by Huang et al. (1986) are also quite high. The presence of valencene, identified in some samples analyzed by Melliou et al. (2009), is also unusual.

As expected, distilled oils obtained by Kiwanuka et al. (2000), Kirbaslar et al. (2001), and Melliou et al. (2009), compared to the cold-pressed ones by the same authors show lower amounts of linalyl acetate and higher content of linalool and other monoterpene alcohols, particularly α-terpineol. Even in the distilled oils obtained by Huang et al. (1987), the amount of linalyl acetate is lower than the cold-pressed oils extracted by the same authors, and α-terpineol is present at a considerably higher level. It is, however, surprising that in these distilled oils linalool is present at percentages lower than cold-pressed oils, and at the same time esters different from linalyl acetate are present at higher amounts.

In addition to those reported in Table 8.11, in literature are more results on oils extracted in laboratory. Kiwanuka et al. (2000) not only obtained the results given in Table 8.10, but also analyzed the oil obtained by distillation of the peels at pH 2.5

TABLE 8.11

Percentage Composition of the Volatile Fraction of Laboratory-Extracted Bergamot Oils

Hydrocarbons	1	2a	3	4	Cold-Pressed 5a	5b	6	7a	7b
Monoterpene									
Camphene	0–tr	0.01	0.02	0.02	–	–	tr	0.03	0.03
δ-3-Carene	–	–	–	tr	–	–	0.9	–	–
p-Cymene	0–0.2	0.11	0.35	–	0.3	tr	0.1	0.33	0.02
Limonene	40.2–40.8	35.42	32.28	27.46–41.35[a]	38.8	24.3	23.7[c]	35.66	21.31
Myrcene	0.8	0.95	0.78	0.67–0.95	0.9	0.7	2.0	0.87	0.66
β-Ocimene*	–	–	–	–	–	–	–	–	–
(E)-β-Ocimene	–	–	0.14	0.15–0.19	–	–	–	0.03	0.02
(Z)-β-Ocimene	–	–	0.02	0.02–0.03	–	–	tr	–	–
α-Phellandrene	–	0.01	–	0.05	–	–	–	0.03	0.02
β-Phellandrene	–	–	0.15	a	–	–	c	0.23	0.19
α-Pinene	0.7–1.0	0.52	0.81	0.70–1.00	1.6	1.3	0.5	1.52	1.29
β-Pinene	4.5–5.1	2.90	3.02	3.93–5.78b	8.9	6.8	3.0	10.09	7.47
Sabinene	–	0.58	0.55	3.93–5.78b	–	–	0.5	1.47	1.25
α-Terpinene	–	–	0.07	0.08–0.12	0.2	0.1	0.2	0.16	0.13
γ-Terpinene	4.5–5.0	4.80	4.12	3.80–5.82	8.3	5.6	4.7	7.32	5.30
Terpinolene	0–0.3	0.16	0.18	0.17–0.25	0.3	0.2	0.4	0.28	0.21
α-Thujene	–	–	–	0.17–0.25	–	–	0.2	–	–
Tricyclene	–	–	–	tr	–	–	–	–	–
Sesquiterpene									
α-Bergamotene*	–	0.15	0.28	–	–	–	–	0.25	0.25

continued

TABLE 8.11 (CONTINUED)
Percentage Composition of the Volatile Fraction of Laboratory-Extracted Bergamot Oils

					Cold-Pressed				
	1	2a	3	4	5a	5b	6	7a	7b
cis-α-Bergamotene	—	—	—	0.01–0.03	—	—	tr	—	—
trans-α-Bergamotene	—	—	—	—	—	—	0.9	—	—
trans-β-Bergamotene	—	—	—	0.20–0.30	—	—	—	—	—
Bicyclogermacrene	—	—	—	0.03	—	—	—	—	—
β-Bisabolene	—	0.95	1.43	0.28–0.34	—	—	1.2	0.45	0.76
(Z)-γ-Bisabolene	—	—	—	—	—	—	—	—	—
δ-Cadinene	—	—	0.33	—	—	—	—	—	—
β-Caryophyllene	—	0.53	0.25	0.25–0.31	—	—	0.2	0.15	0.22
δ-Elemene	—	—	—	tr–0.01	—	—	—	tr	tr
(E,E)-α-Farnesene	—	—	—	tr	—	—	—	tr	—
(E)-β-Farnenese	—	0.57	0.07	0.02–0.03	—	—	—	0.06	0.11
(Z)-β-Farnesene	—	—	—	0.03–0.07	—	—	0.1	—	—
Germacrene B	—	—	—	—	—	—	—	0.01	—
Germacrene D	—	—	—	0.03	—	—	0.1	0.03	0.03
α-Humulene	0.8ᵉ	—	—	0.03	—	—	0.1	0.01	0.01
β-Santalene	—	—	—	0.02–0.05	—	—	—	tr	tr
Aldehydes									
Aliphatic									
Decanal	—	0.08	0.04	0.05	0.1	tr	0.1	0.11	0.08
(E)-2-Decenal	—	—	—	—	—	—	—	tr	—
Dodecanal	—	—	—	tr–0.02	—	—	tr	tr	—
Nonanal	—	0.03	0.02	0.03–0.06	—	—	tr	—	—

Octanal	–	0.11	0.03	0.02	–	tr	–	–	–
Tetradecanal	–	–	–	tr	–	–	–	–	–
Undecanal	–	0.08	–	tr–0.02	–	–	–	–	0.01
Monoterpene									
Citronellal	–	0.03	–	0.01	–	–	tr	–	–
Geranial	0–0.1	0.14	–	0.31–0.33[c]	0.3	–	tr[c]	0.39	0.37
Neral	0–0.6	0.04	0.36	0.21	0.2	0.2	0.4	0.25	0.27
Perilla aldehyde	–	–	–	0.31–0.33[c]	–	–	tr[c]	–	–
Ketones									
Aliphatic									
6-Methyl-5-hepten-2-one	–	0.01	tr	tr	–	–	–	–	–
Monoterpene									
Camphor	–	–	–	tr	–	–	–	–	–
Sesquiterpene									
Nootkatone	–	–	–	0.02–0.04	0.1	tr	0.5	0.18	0.07
Alcohols									
Aliphatic									
Dodecanol	–	–	–	–	–	–	–	–	–
Octanol	–	–	–	tr–0.01	–	–	–	–	–
Monoterpene									
Citronellol	0–0.1	0.05	–	–	–	tr	0.3[d]	tr	0.01
Geraniol	–	0.15	0.04	tr	tr	tr	–	0.09	0.02
Isopulegol	–	–	–	–	–	–	–	–	–
Linalool	7.4–15.5	17.38	16.27	12.61–31.30	4.2	18.2	14.7	4.81	20.02
1-Hydroxy linalool	–	–	–	–	–	–	0.1	0.01	0.01

continued

TABLE 8.11 (CONTINUED)

Percentage Composition of the Volatile Fraction of Laboratory-Extracted Bergamot Oils

					Cold-Pressed				
	1	2a	3	4	5a	5b	6	7a	7b
Nerol	–	0.09	0.03	0.05–0.13	tr	0.1	0.3[d]	0.02	0.07
cis-Sabinene-hydrate	–	–	–	0.03–0.05	–	–	–	0.07	0.03
trans-Sabinene hydrate	–	–	–	–	–	–	–	tr	–
Terpinen-4-ol	–	0.05	–	0.01–0.02	0.1	0.1	tr	0.03	0.55
α-Terpineol	0.8[e]	0.07	0.18[f]	0.07–0.09	–	–	0.1	0.19	0.10
Sesquiterpene									
α-Bisabolol	–	–	–	0.01–0.02	–	–	0.1	–	–
Campherenol	–	–	–	0.01	–	–	–	–	–
(E)-Nerolidol	–	–	–	0.01–0.02	–	–	0.1	0.01	0.01
(Z)-Nerolidol	–	–	–	–	–	–	–	0.01	–
Norbornanol[g]	–	–	–	0.01	–	–	–	–	–
Spathulenol	–	–	–	–	–	–	–	0.01	–
Esters									
Aliphatic									
Decyl acetate	–	0.05	–	tr–0.02	–	–	0.1	tr	tr
Heptyl acetate	–	–	–	0.01	–	–	–	tr	tr
Hexyl acetate	–	0.11	–	–	–	–	0.2	–	–
Nonyl acetate	–	0.05	–	0.02–0.03	–	–	tr	tr	tr
Octyl acetate	0–0.2	0.19	0.09	0.06–0.09	–	–	0.2	0.19	0.16
Monoterpene									
Bornyl acetate	–	–	–	0.01	–	–	tr	–	–
Citronellyl acetate	–	0.13	0.05	0.02–0.03	–	–	0.1	–	–

	8			9a	10	11		12a	12b
Geranyl acetate	0-0.1	0.77	0.24	0.09-0.17	0.3	0.2	1.6	0.42	0.24
Linalyl acetate	28.1-30.8	31.15	37.39	24.62-30.89	32.1	39.0	38.7	30.76	36.37
Linalyl propanate	—	—	—	0.04-0.06	—	—	—	—	—
Methyl geranate	—	—	—	tr-0.01	—	—	—	—	—
Neryl acetate	0-0.5	0.91	0.02	0.13-0.21	—	—	1.6	—	0.01
cis-Sabinene hydrate acetate	—	—	—	—	—	—	—	—	—
trans-Sabinene hydrate acetate	—	—	—	—	—	—	—	—	—
α-Terpinyl acetate	0-0.5	0.19*	0.18[r]	0.04-0.11	—	—	—	—	—
Ethers and Oxides									
Monoterpene									
1,8-Cineole	—	0.16	—	tr	—	—	—	—	—
cis-Limonene oxide	—	—	—	tr	—	—	tr	—	—
trans-Limonene oxide	—	—	—	tr	—	—	tr	—	—
cis-Linalool oxide	—	0.02	—	—	—	—	—	—	—
trans-Linalool oxide	—	—	—	—	—	—	tr	—	—
Hydrocarbons									
Monoterpene									
Camphene	0.02-0.03	—	—	tr	0.01-0.04	0-0.04	—	—	—
δ-3-Carene	tr	—	—	—	0-0.01	—	—	—	—
p-Cymene	0.02-0.04	—	—	0.1	0.06-0.15	0.01-0.21	—	—	—
Limonene	38.70-52.49[a]	—	—	36.4-37.2	30.49-47.14	16.95-60.00	—	25.58	10.54-34.88
Myrcene	0.91-1.24	—	—	1.2-1.3	0.52-1.53	0.37-1.89	—	0.58	0.26-0.77
β-Ocimene*	—	—	—	—	—	—	—	—	—
(E)-β-Ocimene	0.13-0.16	—	—	0.3-0.4	0.15-0.29	0.01-0.19	—	0.24	0.12-0.24
(Z)-β-Ocimene	0.01	—	—	tr	0.02-0.06	0-0.04	—	0.20	0.07-0.27

continued

TABLE 8.11 (CONTINUED)
Percentage Composition of the Volatile Fraction of Laboratory-Extracted Bergamot Oils

	8	9a	10	11	12a	12b
α-Phellandrene	0.02–0.04	–	0.01–0.03	0–0.09	0.01	0.01
β-Phellandrene	a	–	–	–	–	–
α-Pinene	0.71–1.18	0.5–0.6	0.23–1.49	0.27–1.52	0.13	0–0.79
β-Pinene	3.70–7.57[b]	2.7–3.9	3.55–6.00	2.30–8.89[b]	0.23	0.08–0.69
Sabinene	3.70–7.57[b]	0.3–0.4	0.26–1.25	2.30–8.89[b]	–	–
α-Terpinene	0.10–0.17	0.2	0.07–0.19	0–0.20	0.16	0–0.23
γ-Terpinene	4.62–7.77	5.1–5.9	4.04–7.77	1.03–9.22	10.04	4.28–10.26
Terpinolene	0.21–0.33	0.3	0.20–0.37	0.07–0.38	0.60	0.28–0.57
α-Thujene	0.18–0.31	0.1–0.2	0.07–0.41	0.02–0.39	0.23	0.15–0.29
Tricyclene	tr	–	–	0–0.01	–	–
Sesquiterpene						
α-Bergamotene*	–	–	–	–	0.63	0.21–0.87
cis-α-Bergamotene	tr	tr	–	tr	–	–
trans-α-Bergamotene	0.28–0.29	0.4	0–0.01	0.09–0.42	–	–
trans-β-Bergamotene	–	–	–	–	–	–
Bicyclogermacrene	0–0.03	–	–	0–0.03	–	–
β-Bisabolene	0.36–0.46	0.6	0.43–0.93	0.12–0.51	–	–
(Z)-γ-Bisabolene	tr	–	–	tr	–	–
δ-Cadinene	–	–	–	–	–	–
β-Caryophyllene	0.22–0.31	0.3	0.30–0.61	0.11–0.80	0.34	0.11–0.59
δ-Elemene	tr	–	0.04–0.12	–	–	–
(E,E)-α-Farnesene	0–0.02	–	0.04–0.08	tr	–	–
(E)-β-Farnesene	–	–	–	–	–	–

(Z)-β-Farnesene	—	—	0–0.11	0.07–0.17	0.1	tr–0.05
Germacrene B	—	—	—	—	—	0–0.01
Germacrene D	—	—	0–0.17	0.09–0.18	tr	0–0.05
α-Humulene	0–0.04	—	0–0.17	0.03–0.07	0.1	0–0.08
β-Santalene	—	—	tr	—	tr	tr
Aldehydes						
Aliphatic						
Decanal	—	—	0.01–0.33	0.03–0.10	tr	0.03–0.08
(E)-2-Decenal	—	—	tr	—	—	—
Dodecanal	—	—	tr	0.02–0.05	tr	0.02–0.05
Nonanal	—	—	0.01–0.05	0–0.05	tr	0.01–0.04
Octanal	—	—	0.01–0.09	0.01–0.03	—	0.02–0.03
Tetradecanal	—	—	tr	0.04–0.06	—	tr–0.01
Undecanal	—	—	0.01–0.04	0.01–0.03	—	tr–0.01
Monoterpene						
Citronellal	—	—	0–0.04	0.02–0.04	tr	0.02
Geranial	0.01–0.83	0.48	0.14–0.52	0.24–0.53	tr	0.15–0.43[c]
Neral	0–0.46	0.35	0.10–0.39	0.17–0.34	0.4	0.10–0.30
Perilla aldehyde	—	—	—	—	—	c
Ketones						
Aliphatic						
6-Methyl-5-hepten-2-one	—	—	tr	—	—	tr
Monoterpene						
Camphor	—	—	tr	0–0.01	—	tr
Sesquiterpene						
Nootkatone	—	—	0–0.07	0.05–0.18	0.1	0.05–0.10

continued

TABLE 8.11 (CONTINUED)
Percentage Composition of the Volatile Fraction of Laboratory-Extracted Bergamot Oils

	8	9a	10	11	12a	12b
Alcohols						
Aliphatic						
Dodecanol	tr	–	–	tr	–	–
Octanol	tr	–	–	tr	–	–
Monoterpene						
Citronellol	$0–0.03^d$	–	–	$0–0.12^d$	–	–
Geraniol	tr	–	–	tr	–	–
Isopulegol	tr	–	–	tr	–	–
Linalool	1.20–10.08	7.9–18.7	2.84–13.57	1.09–29.02	15.33	14.50–20.18
1-Hydroxy linalool	–	0.1	–	–	–	–
Nerol	$0–0.03^d$	0.2	0.02–0.06	$0–0.12^d$	0.01	0.01
cis-Sabinene-hydrate	0.03–0.05	–	–	0–0.06	0.10	0.05–0.10
trans-Sabinene hydrate	–	–	0.02–0.04	0.01–0.09	–	–
Terpinen-4-ol	0.01–0.02	tr	0.02–0.05	0.01–0.05	0.09	0.01–0.09
α-Terpineol	0.02–0.05	0.1	0.03–0.07	0.03–0.17	0.26	0.01–0.33
Sesquiterpene						
α-Bisabolol	0.02	tr	0.01–0.03	0.01–0.03	–	–
Campherenol	0.02	–	0.01–0.02	0.01–0.02	–	–
(E)-Nerolidol	tr–0.01	0.1	0.01–0.05	0–0.16	–	–
(Z)-Nerolidol	–	–	–	–	–	–
Norbornanol^g	0.01–0.02	–	–	0–0.01	–	–
Spathulenol	–	–	–	–	–	–

Esters

Aliphatic

Decyl acetate	0.01–0.02	0.1	0.03–0.06	tr	—	—
Heptyl acetate	tr	—	—	tr	—	—
Hexyl acetate	tr	0.1	—	tr	—	—
Nonyl acetate	0.01–0.04	tr	0.02–0.06	tr	—	—
Octyl acetate	0.04–0.18	0.1–0.2	0.05–0.15	0.01–0.19	—	—

Monoterpene

Bornyl acetate	0.01–0.02	tr	0.01–0.03	0.01–0.02	—	—
Citronellyl acetate	0.02–0.03	0.1	0.01–0.05	0–0.04	—	—
Geranyl acetate	0.15–0.40	0.5–0.7	0.23–0.71	0.02–0.54	—	0–0.21
Linalyl acetate	24.92–37.71	29.5–36.3	28.23–42.10	19.26–41.96	40.51	30.33–40.51
Linalyl propanate	0.03–0.04	—	0–0.01	0.01–0.19	0.16	0.01–0.17
Methyl geranate	tr–0.01	—	0.01	tr	—	—
Neryl acetate	0.21–0.39	0.7–1.1	0.24–0.64	0.05–0.57	—	0.01–0.07
cis-Sabinene hydrate acetate	—	—	0.01–0.14	tr	—	—
trans-Sabinene hydrate acetate	0.05–0.09	—	—	0.01–0.09	—	—
α-Terpinyl acetate	0.10–0.17	—	0.16–0.30	0–0.16	0.35	0.01–0.35

Ethers and Oxides

Monoterpene

1,8 Cineole	tr	tr	—	tr	—	—
cis-Limonene oxide	tr	—	0.01–0.02	tr	—	—
trans-Limonene oxide	tr	—	0.01–0.02	tr	—	—
cis-Linalool oxide	—	—	—	tr	—	—
trans-Linalool oxide	—	—	—	tr	—	—

continued

TABLE 8.11 (CONTINUED)
Percentage Composition of the Volatile Fraction of Laboratory-Extracted Bergamot Oils

	Solvent Extracted						Distilled				
	13	2b	14a	15	16	17	2c	2d	14b	9b	12c
Monoterpene											
Camphene	0.05	0.21	–	–	–	0.03	0.04	0.04	–	tr	–
δ-3-Carene	0.14ⁱ	–	–	–	–	2.04	–	–	–	–	–
p-Cymene	0.57	0.10	–	0.13	0.80	1.56	0.37	0.15	–	tr	–
Limonene	32.37	41.88	40.20	18.78	48.42	45.21	45.11	41.14	41.65	33.2	31.66
Myrcene	1.17	1.34	1.53	2.53	2.10	1.43	1.97	1.92	2.13	2.3	1.33
β-Ocimene*	–	–	0.12	–	–	0.27	–	–	0.46	–	–
(E)-β-Ocimene	–	–	–	0.15	–	–	–	0.74	–	0.3	0.79
(Z)-β-Ocimene	–	–	–	–	–	–	–	–	–	–	0.50
α-Phellandrene	–	0.02	–	–	–	tr	0.04	0.03	–	0.1	0.05
β-Phellandrene	–	–	0.31	–	–	–	–	–	0.35	–	–
α-Pinene	1.88	0.94	0.81	0.56	1.25	0.84	1.30	1.27	0.78	0.9	0.88
β-Pinene	13.34ᵇ	4.35	4.94	0.73	6.15	4.82	5.79	5.45	4.96	3.8	0.78
Sabinene	13.34ᵇ	0.88	1.04	0.60	1.25	0.78	1.00	1.00	0.98	0.5	–
α-Terpinene	–	–	0.13	0.10	–	–	–	–	0.17	0.7	0.30
γ-Terpinene	12.60	5.25	7.27	6.37	8.26	1.35	6.02	5.41	7.73	6.5	10.32
Terpinolene	0.20	0.18	0.29	–	–	0.14	0.43	0.44	0.43	–	0.76
α-Thujene	–	–	–	–	–	0.18	0.03	0.01	–	0.2	0.29
Tricyclene	–	–	–	–	–	–	–	–	–	–	–
Sesquiterpene											
α-Bergamotene*	0.86	0.04	–	–	–	0.02	0.03*	–	–	–	0.20
cis-α-Bergamotene	–	–	–	–	–	–	–	0.07*	–	0.4	–

trans-α-Bergamotene	–	–	–	0.38	–	–	–	–	–	tr	–
trans-β-Bergamotene	–	–	–	–	–	–	–	–	–	–	–
Bicyclogermacrene	–	0.77	–	–	–	–	–	–	–	–	–
β-Bisabolene	1.28	–	–	0.58	–	0.02	0.35	0.35	–	0.8	–
(Z)-γ-Bisabolene	–	–	–	–	–	–	–	–	–	–	–
δ-Cadinene	–	–	–	–	–	–	–	–	–	–	0.20
β-Caryophyllene	0.25	0.39	–	0.37	–	0.11	0.22	0.21	–	0.3	–
δ-Elemene	–	–	–	–	–	–	–	–	–	–	–
(E,E)-α-Farnesene	–	–	–	–	–	0.15	–	–	–	–	–
(E)-β-Farnenese	–	0.45	–	–	–	–	–	–	–	tr	–
(Z)-β-Farnesene	–	–	–	–	–	–	0.22	0.20	–	–	–
Germacrene B	–	–	–	–	–	–	–	–	–	–	–
Germacrene D	–	–	–	–	–	–	–	–	–	–	–
α-Humulene	–	–	–	–	–	tr	–	–	–	–	–
β-Santalene	–	–	–	–	–	–	–	–	–	0.1	–
Aldehydes											
Aliphatic											
Decanal	0.29	0.12	0.10	–	0.04	0.08	0.07	0.11	–	–	–
(E)-2-Decenal	–	–	–	–	–	–	–	–	0.14	–	–
Dodecanal	–	–	–	–	–	–	–	–	–	tr	–
Nonanal	0.09	0.04	–	–	–	–	0.04	0.04	–	–	–
Octanal	–	0.02	–	–	–	–	0.02	0.03	–	–	–
Tetradecanal	–	–	–	–	–	–	–	–	–	–	–
Undecanal	–	0.07	–	–	–	–	0.23	0.21	–	–	–
Monoterpene											
Citronellal	0.02	0.05	–	–	–	0.02	0.03	0.07	–	tr	–

continued

TABLE 8.11 (CONTINUED)
Percentage Composition of the Volatile Fraction of Laboratory-Extracted Bergamot Oils

	Solvent Extracted				Distilled						
	13	2b	14a	15	16	17	2c	2d	14b	9b	12c
Geranial	0.21	0.06	0.22	0.28	2.58	0.48	0.06	0.12	0.22	0.8	0.11
Neral	0.16	0.07	–	–	1.75	0.36	0.10	0.10	–	–	0.15
Perilla aldehyde	–	–	–	–	–	–	–	–	–	–	–
Ketones											
Aliphatic											
6-Methyl-5-hepten-2-one	–	–	0.01	–	–	0.02	0.01	0.01	–	–	–
Monoterpene											
Camphor										0.1	
Sesquiterpene											
Nootkatone	–	–	0.25	–	–	–	–	–	0.11	–	–
Alcohols											
Aliphatic											
Dodecanol	–	–	–	–	–	–	–	–	–	–	–
Octanol	–	–	–	–	–	–	–	–	–	–	–
Monoterpene											
Citronellol	–	0.06	–	–	0.18	tr	0.03	0.03	–	–	–
Geraniol	–	0.20	0.10	–	0.52	5.67	1.35	1.48	1.24	1.7	–
Isopulegol	–	–	–	–	–	0.01	–	–	–	–	–
Linalool	4.39	7.56	5.22	27.35	12.85	17.89	12.12	13.56	10.68	20.1	31.76
1-Hydroxyl linalool	–	–	–	–	–	–	–	–	–	–	–
Nerol	–	0.14	0.06	0.22	0.24	0.52	0.55	0.56	0.48	0.8	0.46

cis-Sabinene-hydrate	–	–	–	–	–	–	–	–	–	–	–
trans-Sabinene hydrate	0.05	–	–	–	–	–	–	–	–	–	–
Terpinen-4-ol	0.03	–	0.02	–	–	0.16	0.04	0.03	0.15	0.5	0.19
α-Terpineol	0.11	–	0.04	0.13	3.07	2.54	2.87	3.11	2.04	3.0	3.85
Sesquiterpene											
α-Bisabolol	–	–	–	–	–	–	–	–	–	–	–
Campherenol	–	–	–	–	–	–	–	–	–	–	–
(*E*)-Nerolidol	tr	–	–	–	–	tr	–	–	–	–	–
(*Z*)-Nerolidol	–	–	–	–	–	tr	–	–	–	–	–
Norbornanol[g]	–	–	–	–	–	–	–	–	–	–	–
Spathulenol	–	–	–	–	–	–	–	–	–	–	–
Esters											
Aliphatic											
Decyl acetate	–	0.05	–	–	–	0.02	0.11	0.12	–	tr	–
Heptyl acetate	–	–	–	–	–	–	–	–	–	–	–
Hexyl acetate	–	0.16	–	–	–	–	0.01	0.02	–	0.1	–
Nonyl acetate	0.03	0.03	–	–	–	0.03	0.03	0.03	–	tr	–
Octyl acetate	0.02	0.17	–	–	–	0.11	0.15	0.17	–	–	–
Monoterpene											
Bornyl acetate	–	–	–	–	0.06	–	–	–	–	0.1	–
Citronellyl acetate	0.15	0.11	–	–	–	0.15	0.09	0.09	–	tr	–
Geranyl acetate	0.58	0.58	0.24	0.19	–	1.31	1.71	1.93	1.27	2.7	1.33
Linalyl acetate	23.00	31.66	30.50	36.12	9.72	11.37	14.67	15.89	21.41	17.3	10.72
Linalyl propanate	–	–	–	–	–	–	–	–	–	–	0.05
Methyl geranate	0.05	–	–	–	–	0.05	–	–	–	–	–
Neryl acetate	0.40	0.48	0.25	0.31	–	0.88	1.13	1.28	0.77	2.1	0.70
cis-Sabinene hydrate acetate	–	–	–	–	–	–	–	–	–	–	–

continued

TABLE 8.11 (CONTINUED)
Percentage Composition of the Volatile Fraction of Laboratory-Extracted Bergamot Oils

	Solvent Extracted						Distilled				
	13	2b	14a	15	16	17	2c	2d	14b	9b	12c
trans-Sabinene hydrate acetate	–	–	–	–	–	–	–	–	–	–	–
α-Terpinyl acetate	0.03	0.22*	–	0.11	–	0.03	0.29*	0.31*	–	–	–
Ethers and Oxides											
Monoterpene											
1,8-Cineole	–	0.25	–	–	–	–	0.28	0.25	–	tr	–
cis-Limonene oxide	–	–	–	–	–	–	–	–	–	–	–
trans-Limonene oxide	–	–	–	–	–	0.07	–	–	–	–	–
cis-Linalool oxide	–	0.02	–	–	–	–	0.02	0.02	–	–	–
trans-Linalool oxide	–	–	–	–	–	–	–	–	–	–	–

tr, traces; * correct isomer not characterized; †, tentative identification.

a limonene + β-phellandrene.
b β-pinene + sabinene.
c geranial + perilla aldehyde.
d citronellol + nerol.
e α-humulene + α-terpineol.
f α-terpineol + α-terpenyl acetate.
g 2,3-dimethyl-3-(4-methyl-3-pentenyl)-2-norbornanol.

Appendix to Table 8.11:

1. Dresher et al. (1984). Entre Rios, Argentina; 2 samples cold-pressed from the cv. Monaco; GC on packed columns of Carbowax 20 M and OV-17; relative percentage of peak areas. These authors also analyzed samples extracted from the cvs. Castagnaro and Femminello; these oils presented high content of monoterpene hydrocarbons (limonene 70%–80%) and trace amounts of oxygenated compounds.

2. Huang et al. (1987). China: (a) one sample cold-pressed; (b) one sample solvent extracted; (c) one sample steam distilled; (d) one sample SDE extracted; GC/FID on capillary columns coated with OV-101 (57 m × 0.3 mm) or PEG-20M (54 m × 0.20 mm); GC/MS on capillary column (30 m × 0.20 mm) coated with SE-54; relative percentage of peak areas.

3. Baser et al. (1995). Turkey; one sample cold-pressed using an expeller-type press; GC on capillary columns coated with Thermon 600T. Baser et al. also found γ-muurolene (0.05%), hexanal (0.02%), and trace amounts of (E)-2-hexenal, 6-methyl-3-heptanol.

4. Dellacassa et al. (1997). Uruguay; 4 samples cold-pressed from fruit picked from May to July; GC/FID on capillary column (25 m × 0.32 mm) coated with SE-52; GC/MS on capillary columns (60 m × 0.32 mm) coated with SE-52 or Carbowax 20 M; relative percentage of peak areas.

5. Sawamura et al. (1999a). (a) One sample of Fantastico bergamot oil from Italy; (b) one sample of Balotin bergamot oil from Japan extracted by hand-pressing of flavedo; GC/FID and GC/MS on capillary column (50 m × 0.22 mm × 0.25 μm) coated with Thermon 600T; wt% using heptanol and methylmiristate as internal standards. Sawamura et al. also found trace amount of decanol in sample (b).

6. Kirbaslar et al. (2000). Turkey; one sample cold-pressed applying manual pressure on the rind of the fruits; GC/FID and GC/MS on capillary column (30 m × 0.25 mm × 0.52 mm) coated with HP-Innowax; relative percentage of peak areas. Kirbaslar et al. also found m-cymene (0.3%), phenyl ethyl alcohol (0.5%), octadienyl formate* (0.1%), and trace amount of p-menth-1-en-8-yl acetate.

7. Sawamura et al. (1999b). Sawamura (2000). (a) One sample hand-pressed from fruits of the cv. Fantastico grown in Calabria, Italy; (b) one sample hand pressed from fruits of the cv. Fantastico grown in Japan; GC/FID and GC/MS on capillary column (50 m × 0.25 mm × 0.25 μm) coated with Thermon 600T; these authors also found trans-carveol (0.01%), perilla alcohol (0.01%), verbenol (0.01%), solanone† (0.12%), and trace amounts of β-elemene, δ-muurolene, α-sinensal, p-cymen-8-ol, 2,7-dimethyl-2,6-octadien-1-ol, and caryophyllene oxide* in sample (a), solanone† (0.08%) and trace amount of verbenol in sample (b).

8. Verzera et al. (2000). Sicily, Italy; range of the composition of two samples each hand-pressed from the cvs. Fantastico, Femminello, and Castagnaro (typical Calabrian cvs.) and from a new clone of Femminello named "PCF"; GC/FID on capillary column (30 m × 0.32 mm × 0.40–0.45 μm) coated with SE-52; GC/MS on capillary columns coated with Mega-5MS (30 m × 0.25 mm × 0.25 μm) or Megawax (30 m × 0.32 mm × 0.40–0.45 μm); Adams MS library; relative percentage of peak areas. Ranges reported in table are obtained from the original data provided by the authors and not from the average values reported in the article.

9. Kirbaslar et al. (2001). Turkey; (a) range of the composition of hand-pressed oils from fruits picked from December 1998 to February 1999; (b) one sample steam distilled; GC/FID and GC/MS on capillary column (30 m × 0.25 mm × 0.52 μm) coated with HP-Innowax. MS libraries: Wiley and NBS; relative percentage of peak areas. Kirbaslar et al. also found p-menth-1-en-9-yl acetate (0.1%–0.2%), 3,7-dimethyl-3-hydroxy-1,6-octadienyl formate (0.1%) in sample (a), 1,3-divinyl benzene (0.2%), α-terpinyl isobutyrate (0.2%), and trace amounts of m-cymene in sample (b).

continued

TABLE 8.11 (CONTINUED)
Percentage Composition of the Volatile Fraction of Laboratory-Extracted Bergamot Oils

10. Gionfriddo et al. (2003). Calabria, Italy; cvs. Femminello, Castagnaro, and Fantastico; range of the composition of 25 oils cold-extracted from fruits picked from December 2000 to February 2001; GC/FID and GC/MS on capillary columns (30 m × 0.25 mm × 0.25 µm) coated with BP-1 or BP-20; NIST 1.7 MS library; LRI on BP-20 and BP-1 are reported. Gionfriddo et al. also found β-sesquiphellandrene (0.04%–0.07%), tridecanal (0.01%–0.06%).

11. Verzera et al. (2003). Sicily, Italy; 203 samples hand-pressed from fruits picked from 1997 to 2000 from plant grafted on different rootstock (sour orange, Carrizzo citrange, trifoliate orange, Alemow, Volkammerian lemon, Troyer citrange); GC/FID on capillary column (30 m × 0.32 mm × 0.25 µm) coated with SE-52; GC/MS on capillary columns (30 m × 0.25 mm × 0.25 µm) coated with MEGA-5MS or Megawax; Adams MS library; relative percentage of peak areas. Ranges reported in table are obtained from the original data provided by the authors and not from the average values reported in the article. Verzera et al. also found trace amounts of dodecane, (Z)-α-bisabolene, and indole.

12. Melliou et al. (2009). Kefalonia, Greece; (a) one sample cold-pressed from fruits collected on January 15; (b) several samples cold-pressed from fruits collected from December to March during two productive seasons; (c) one sample hydrodistilled from fruits collected on January 15; oil extracted from fruits harvested on 15th of January showed the best quality values; GC/FID and GC/MS on capillary column (30 m × 0.25 mm × 0.25 µm) coated with HP 5MS; NIST/NBS Wiley MS library; relative percentage of peak areas. Melliou et al., also found valencene (0.2%; 0%–0.63%; 0.08% respectively in samples (a), (b), and (c).

13. Calvarano et al. (1984). Toscana, Italy; one sample solvent-extracted from fruits of the cv. Femminello; GC/FID on stainless steel capillary column (46 m × 0.50 mm) coated with UCON LB550X; relative percentage of peak areas.

14. Kiwanuka et al. (2000). Calabria, Italy; cv. Castagnaro: (a) one sample solvent extracted, (b) one sample steam distilled; GC/FID on capillary column (50 m × 0.32 mm × 0.5 µm) coated with SPB-5; GC/MS on capillary column (60 m × 0.25 mm ×x 0.25 mm) coated with CP-SIL 8CB; Adams MS library; relative percentage of peak areas.

15. Guerrini et al. (2009). Italy. One sample extracted by sonication of the epicarp of fruits suspended in CHCl₃; GC/FID and GC/MS on capillary column (30 m × 0.25 mm × 0.15 µm) coated with VF-5MS; LRI on VF-5MS are reported; relative percentage of peak areas.

16. Ortiz et al. (1978). California; one sample steam distilled by Cleavenger; GC/FID and GC/MS on packed column of LAC 2R 446.

17. Huang et al. (1986). China; one sample steam distilled; GC/FID on capillary column (50 m × 0.25 mm) coated with OV-101; GC/MS on capillary column (30 m × 0.20 mm) coated with SE-54; LRI on OV-101 are reported; relative percentage of peak areas. Huang et al. also found dodecane (0.037%), pentadecane (0.21%), hexanol (0.01%), (Z)-hex-3-en-1-ol (0.01%), nonanol (0.02%), dihydroxy linalool (0.07%), endofenchol (0.01%), lavandulol (0.02%), sabinene hydrate* (0.06%), citronellyl formate (0.02%), geranyl formate (0.02%), γ-heptalactone (0.04%), and trace amounts of pulegone, butyl acetate, 1,4-cineole.

and that distilled by the homogenized fruits. If compared to the distilled oil reported in Table 8.11 (column 14b), these two oils showed a moderate increase of linalool (ca. 13% in both oils); a decreased amount of linalyl acetate evident in the oil obtained at pH 2.5 (~16%) and extremely evident in the oil distilled from the whole homogenized fruits (~5%); and an increase in monoterpene alcohols different from linalool which in the two oils are 8.4% and 10%, respectively. Statti et al. (2004) determined the biological activity and the amount of linalool, linalyl acetate, and bergapten in solvent-extracted oils from fruits harvested in Calabria at different altitudes and latitudes. Their results raised doubts. From their data, the extracted yields from the whole fruits ranged approximately between 0.1% and 0.8%; the percentage of linalool in the extracts was between 0.003% and 1.20%, and that of linalyl acetate from 0.05% to 3.77% (both exceptionally low); the amount of bergapten ranged from 0.47% to 3.92% (this last value is about 10 times more than what is usually in bergamot). The biological activity does not seem to be related to the amounts of these three compounds determined in the extracts.

8.4 NONTECHNOLOGICAL PARAMETERS INFLUENCING COMPOSITION OF OIL

As previously stated, the composition of the volatile fraction of bergamot oil from Calabria shows a wide range of variability. This can contrast with the short season of production and the restricted areas of cultivation of this fruit, practically limited to some coastal areas of Calabria. Bergamot is in fact cultivated in the low Tyrrhenian and low Ionian coasts, between Villa S. Giovanni and Brancaleone, with a few cultivated fields in the Locri area. The maximum altitude is 200 meters above the sea level (Figure 8.7). The cultivated area is about 1300 ha divided into numerous small firms.

The composition of the essential oil is strongly influenced by the harvest period (Calvarano 1968; Huet and Dupuis 1969; Dugo et al. 1987, 1991, 2012; Lamonica et al. 1990; Verzera et al. 1996, 1998) and by the locations of origin of the fruits (Huet and Dupuis 1968) even if they are located in a small region such as Calabria (Dugo et al. 1987, 1991; Verzera et al. 1996); to a lesser extent the composition of the oil is influenced by the different cultivars of the fruits (Verzera et al. 1996). The result in column 3 of Table 8.3 for 1540 samples indicates that the amount of limonene varies between 24% and 55%, linalool ranges from 1.5% and 23%, and linalyl acetate from 15% to 41%. The width of these ranges, wider than for other citrus oils, often due to a small number of samples as can be seen from the histograms reported in Figure 8.8, does not allow application of the variability of ranges and the seasonal variations of the composition of bergamot oil as fundamental parameters to express the oil genuineness. Dugo et al. (2012) also observed, for the seasons 2009–10 and 2010–11, significant differences in the average composition of Calabrian oils in respect to the oils produced during numerous past seasons.

8.4.1 Changes in the Composition during the Productive Season

The amount of single components and the class of components can vary during the season, with quantitative differences reproducible from one year to the other,

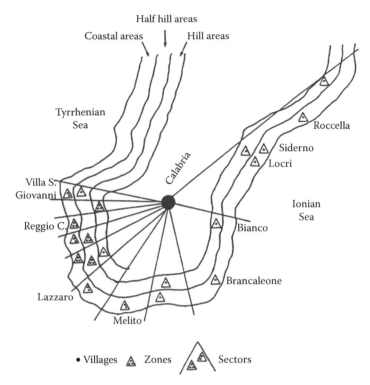

FIGURE 8.7 Occurrence of bergamot cultivation in Calabria, Italy. (From Dugo, G., et al. *Flavour Fragr. J.* 6, 39–56, 1991.)

although some differences in the average values are recorded in the same period of two different years. These differences are probably due to particular climatic conditions that affect the ripening stage of the fruits.

Calvarano (1961) determined the amount of esters, alcohols, and aldehydes in oils extracted industrially from bergamottella collected between June and September. The analyses, carried out by classical means, allowed the determination of the trend of these classes of components: esters increased from June to July, then in September decreased to less than the starting level; alcohols instead showed an almost opposite trend; and aldehydes increased from 0.5% to 0.9%. The results obtained by Calvarano (1961) for esters and alcohols are graphically reported in Figure 8.9. The variation of the composition of bergamot oil as a function of the harvest period was successively studied by Calvarano (1968). In samples obtained from fruits harvested in July 1967 (unripe) until January 1968 (ripe), every 10 days, solvent extracted in laboratory, Calvarano (1968) observed that the amount of linalool decreased in the oil from 33% to 14%; on the contrary, linalyl acetate increased from 25% to 34%; the amount of β-pinene + sabinene and of γ-terpinene increased respectively from 4% to 10% and from 5% to 10%. These behaviors are reported in Figure 8.10 in a graph obtained from the results by Calvarano (1968).

Huet (1970) determined the content of esters and alcohols in oils extracted from fruits of two clones of bergamot, cultivated in Mali, harvested from October 1968 to October 1969. For both the clones, the esters increased from October to April–June

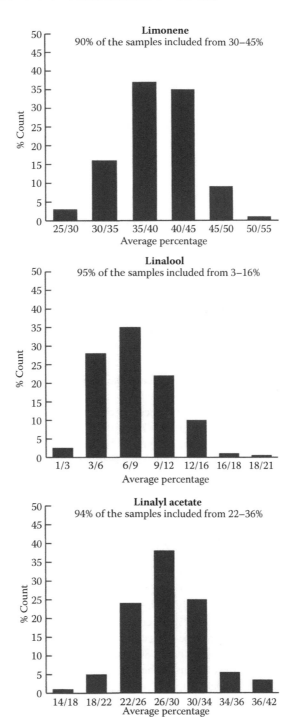

FIGURE 8.8 Average percentage distribution of limonene, linalool, and linalyl acetate in 1082 bergamot oils. (From Verzera A., et al. *Perfum. Flav.* 21(6), 19–34, 1996.)

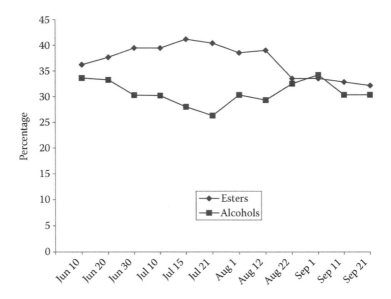

FIGURE 8.9   Average variation from June to September of alcohols and esters in bergamot-tella oils produced in Calabria in 1960. (From the results of Calvarano, I., *Essenz. Deriv. Agrum.* 31, 109–127, 1961.)

of the successive year, then decreased to reach their initial values. Relative to the alcohols, the two clones showed different behavior from October until February of the next year, and then showed a constant pronounced decrease. The same results are reported by the same author in a different article (Huet 1991c). The behavior of alcohols and esters observed by Huet can be seen in Figure 8.11. The author asserted that, based on its composition, the oil of bergamot from Mali could be marketed if produced during February and March: in fact, before February the amount of esters (less of 35%) was too low; after February the content of alcohols (less than 10%) was insufficient.

The seasonal variations of bergamot oil produced in Calabria were investigated by Dugo et al. (1987, 1991, 2012) and by Verzera et al. (1998). In Figures 8.12 through 8.14 are reported the fortnightly behavior for three consecutive seasons (Dugo et al. 1991). As can be seen from the figures, the average content of alcohols (mainly linalool) during the season goes from 12% in the first half of December to about 5% in March. In the different years, the decrease of alcohols is balanced by the correspondent increase of esters (mainly linalyl acetate) or monoterpene hydrocarbons, or both. Carbonyls, never above 1%, decrease during the season. Each component usually shows a trend in accordance to its chemical family. The results used to obtain Figures 8.12 through 8.14 are relative to 841 samples of genuine oils.

Similar behaviors to those discussed here can be observed in the results reported by Verzera et al. (1998) relative to 452 oils produced in the 1996–97 season, as reported in Figure 8.15. They were confirmed by the recent results obtained by Dugo et al. (2012) for samples produced during the 2008–09, 2009–10, and 2010–11 seasons, each representative of 1000 kg of essential oil (Figure 8.16). In Figure 8.10a and Figures 8.12 through 8.16, it is possible to observe that the variations registered

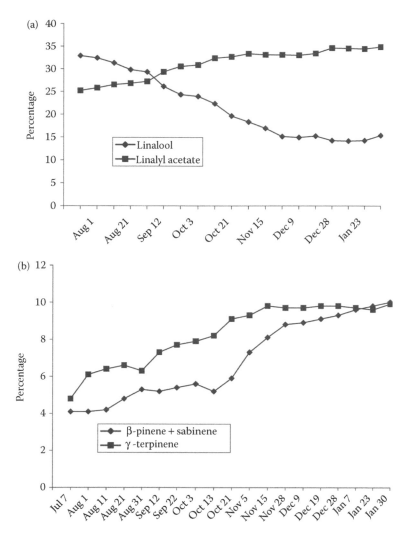

FIGURE 8.10 Variation from July 1967 to January 1968 of linalool and linalyl acetate (a) and sabinene β-pinene and γ-terpinene (b) in bergamot oils produced in Calabria. (From the results of Calvarano, M., *Essenz. Deriv. Agrum.* 35, 197–211, 1968.)

for classes of compounds and for single components in Calabrian bergamot oils during the productive seasons of different years, even with some quantitative differences, are similar and well represent the typical variation of the composition of this oil as a function of the harvest period of the fruits.

Compositional variation similar to the Calabrian oils was recorded by Huet and Dupuis (1969) in oils from Corsica extracted in laboratory with light petroleum from fruits harvested between November and January, and by Kibaslar et al. (2001) in Turkish cold-pressed oils obtained in laboratory from fruits harvested from December to February. Huet and Dupuis observed that, from November to January, linalool decreased from 30% to 16% and linalyl acetate increased from 26% to 44%;

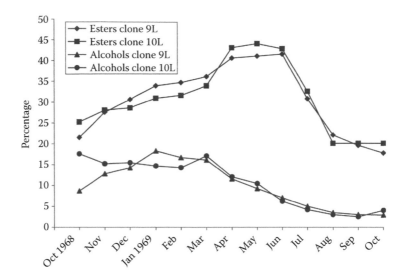

FIGURE 8.11   Variation from October 1968 to October 1969 of alcohol and esters in berga-mot oils from Mali. (From the results of Huet, R., *Fruits* 25(10), 709–715, 1972.)

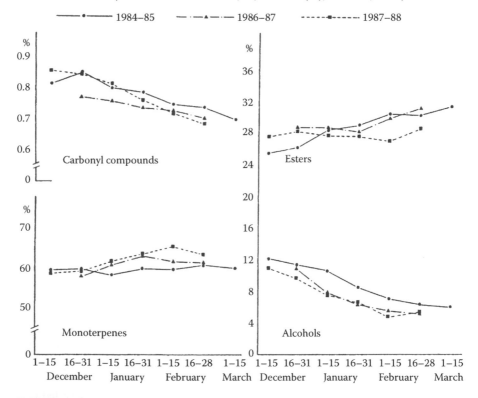

FIGURE 8.12   Average fortnightly variations of the content of monoterpene hydrocarbons, carbonyl compounds, esters, and alcohols in bergamot essential oil produced in Calabria dur-ing three productive seasons. (From Dugo, G., et al. *Flavour Fragr. J.* 6, 39–56, 1991.)

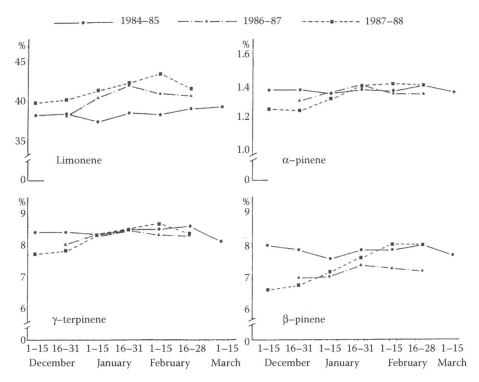

FIGURE 8.13 Average fortnightly variations of the content of limonene, γ-terpinene, α-pinene, and β-pinene in bergamot essential oil produced in Calabria during three productive seasons. (From Dugo, G., et al. *Flavour Fragr. J.* 6, 39–56, 1991.)

monoterpene hydrocarbons varied irregularly during this period. In the same article, the composition of oils obtained in laboratory by distillation of fruits harvested between June and January was reviewed; the composition of these oils was influenced by the extraction technology. The results reported by Kirbaslar et al. (2001) show that during the season linalool decreased from 18.7% to 7.9%; linalyl acetate increased from 29.5% to 36.3%; limonene remained constant at 36% or 37%. β-pinene and γ-terpinene, even if present at smaller concentrations, show trends similar to those reported by Calvarano (1968) in Calabrian oils extracted in laboratory. Some of the variations observed by Kirbaslar et al. are reported in Figure 8.17. The values determined by Dellacassa et al. (1997) in Uruguayan cold-pressed oils obtained in laboratory from fruits harvested from May to July appear irregular, as shown in Figure 8.18. Only linalool behaves in agreement with the results previously described from the beginning of June to the second half of July. Surprisingly, limonene decreases from the end of May to the end of June (from 35.6% to 27.5%), then increases in the second half of July to 41.3%. Melliou et al. (2009) studied the seasonal variations of the main components in Greek bergamot oil (reported in Figure 8.19). They found that the best value of the ratio linalool/linalyl acetate (0.38) and of the sum of linalool + linalyl acetate (55.6%) was in samples extracted in mid-January. None of the trends determined in Greek oils resemble those reported for Calabrian oils.

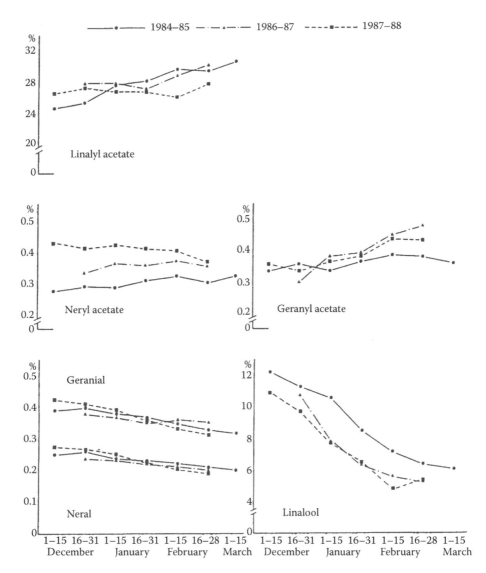

FIGURE 8.14   Average fortnightly variations of the content of linalyl acetate, neryl acetate, geranyl acetate, geranial, neral, and linalool in bergamot essential oil produced in Calabria during three productive seasons. (From Dugo, G., et al. *Flavour Fragr. J.* 6, 39–56, 1991.)

8.4.2   Compositional Variations of Oils Produced in Different Regions of the Same Country

In Figure 8.7 (Dugo et al. 1991) is a schematically traced map of the southern part of Calabria indicating the regions where bergamot is cultivated. Districts are grouped based on their altitudes (coastal, half-hill, and hill areas) from the Tyrrhenian to the Ionian seas. The different areas are divided into 10 sectors; in each sector, independent of its altitude, are included one or more areas at the same latitude overlooking the same

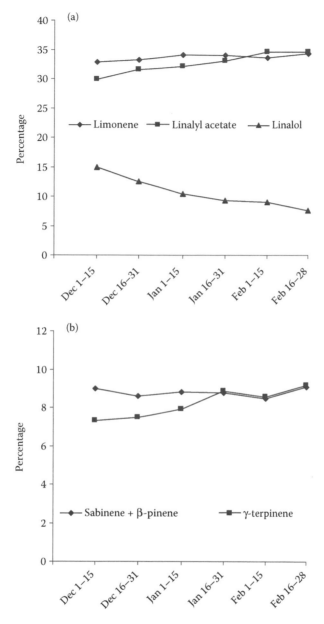

FIGURE 8.15 Average fortnightly variations of the content of limonene, linalool, and linalyl acetate (a), β-pinene + sabinene and γ-terpinene (b), in bergamot oil produced in Calabria during the 1996–97 productive season. (From the results of Verzera, A., et al. *EPPOS* 25, 17–38, 1998.)

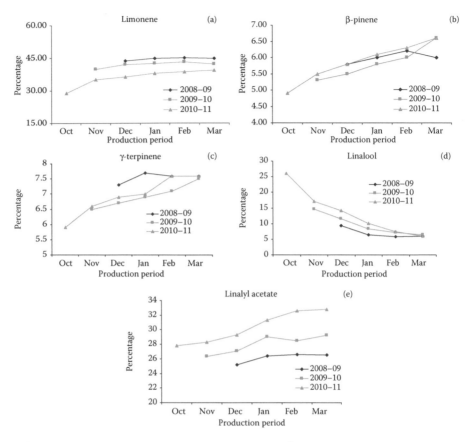

FIGURE 8.16 Variation of the content of limonene (a), β-pinene (b), γ-terpinene (c), linalool (d), and linalyl acetate (e) in bergamot oil produced in Calabria during the 2008–09, 2009–10, 2010–11 productive seasons. (From Dugo, G. et al. *J. Essenz. Oil Res.* 24(2), 93–117, 2012.)

coast. According to the results attained by Dugo et al. (1987, 1991) the composition of the volatile fraction is not sensibly dependent on the altitude of the same zone (each considered a homogeneous productive area) but it is dependent on the latitude and on the overlooking coast. In a graph of the average content of class of compounds as a function of the sector of origin of the fruits (Figures 8.20 through 8.22), it is possible to observe that monoterpene hydrocarbons (as a class of compounds or as single components) show a concave trend, while total alcohols, esters, and single compounds of these classes (e.g., linalool and linalyl acetate) show an opposite behavior than that of hydrocarbons. Some minor components, such as carbonyls and some alcohols, show an irregular behavior. Neryl and geranyl acetate behave similarly to monoterpene hydrocarbons. These results are reproducible for different seasons, as shown in Figures 8.20 through 8.22, which refer to three different years of production. What was observed by Dugo et al. (1987, 1991) has been confirmed by the results obtained by Verzera et al. (1998) for oils produced in 1996–97 (Figure 8.23). The concave or convex trend was observed for all the sectors of production with the exception of sector 9 (see Figure 8.7 and 8.20).

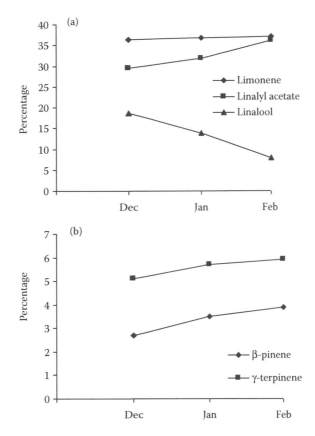

FIGURE 8.17 Variation from December 1998 to January 1999 of the content of limonene, linalool, and linalyl acetate (a), β-pinene and γ-terpinene (b) in bergamot oil from Turkey. (From the results of Kirbaslar, F. G., et al. *J. Essent. Oil Res.* 12, 216–220, 2001.)

Calvarano et al. (1984) analyzed an oil solvent extracted in laboratory from fruits cultivated in Tuscany. They found that its composition was within the variability range determined for Calabrian oils. The composition of this oil is reported in Table 8.12. Drescher et al. (1984) reported the composition of two Argentinian oils obtained in laboratory by steam distillation from fruits cultivated in Entre Rios, and an industrial oil produced in Misiones, previously analyzed by Ricciardi et al. (1982). The composition of the three oils was similar, with the exception of linalool present in one of the two samples from Entre Rios at about 7%, while in the other two samples it was about 15%. The composition of the industrial oil from Misiones is reported in Table 8.5, while that of the two oils from Entre Rios is in Table 8.11.

8.4.3 INFLUENCE OF THE CULTIVAR AND OF THE ROOTSTOCK ON THE COMPOSITION OF THE OIL

In Calabria are cultivated three cultivars of bergamot: Femminello, Castagnaro, and Fantastico, with Fantastico the most common of the three. Under the name Fantastico

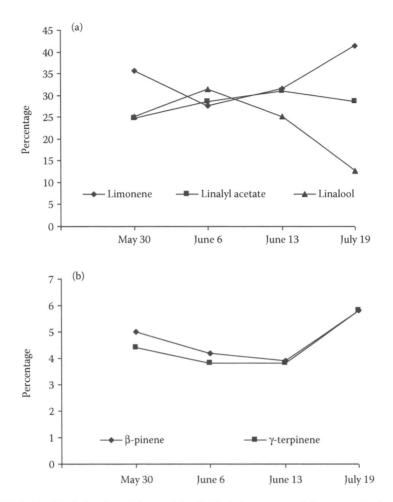

FIGURE 8.18 Variation from May to July 1965 of the content of limonene, linalool, and linalyl acetate (a), β-pinene and γ-terpinene (b) in bergamot oil from Uruguay. (From the results of Dellacassa, E., et al. *J. Essent. Oil Res.* 9, 419–426, 1997.)

are included a certain number of spontaneous hybrids, not yet sufficiently characterized. As shown in Figure 8.24, relative to oils produced industrially during the same productive seasons, the oils obtained from the cv. Fantastico have higher amounts of monoterpene and sesquiterpene hydrocarbons and a lower amount of alcohols and esters than the oils extracted from the other cvs. The cv. with highest amount of alcohols and esters is Femminello (Verzera et al. 1996). However, the differences due to the different cvs. of the fruits are exceeded by the influence of the location where the fruits were cultivated, as shown in Figure 8.25. This graph compares the average values of the class of components determined in oils obtained from fruits of Fantastico from two different sectors (Verzera et al. 1996).

Verzera et al. (2000) analyzed oils cold-pressed in laboratory from fruits of the cvs. Castagnaro, Femminello, Fantastico, and PCF (a new clone of the cv.

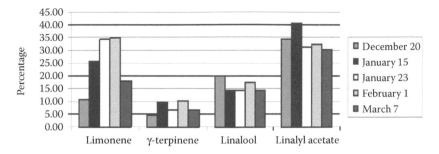

FIGURE 8.19 Variation from December to March (average values of two productive seasons) of limonene, γ-terpinene, linalool, and linalyl acetate in bergamot oil from Greece. (From Melliou, E., et al. *Molecules* 14, 839–849, 2009.)

Femminello) harvested in February 1999 from trees cultivated in the experimental field of the University of Catania, Sicily (two samples for each cv.). These authors found the highest amount of limonene, β-pinene + sabinene in oils obtained from the cv. Castagnaro, and the lowest amount of these components in the cvs. Femminello and PCF; the lowest content of linalool was determined in cvs. Castagnaro and PCF, the highest one in cv. Femminello; the lowest content of linalyl acetate was detected in cv. Castagnaro, the highest in cvs. Femminello and PCF. These results are reported in Table 8.11. The main differences in the composition of the volatile fraction of the four cvs. are summarized in Table 8.12. The same article compares oils obtained from the cvs. Castagnaro, Fantastico, and Femminello cultivated in Sicily and in Calabria. The Sicilian oils of Castagnaro and Fantastico, compared to those from Calabria, show higher amounts of limonene and a noticeably lower amount of linalool while linalyl acetate is at a comparable amount; the Sicilian oil from the cv. Femminello

TABLE 8.12

Comparison between Sicilian Oils Laboratory Cold-Pressed from the Cultivars Castagnaro, Fantastico, Femminello and PCF

	Castagnaro	Fantastico	Femminello	PCF
Limonene	51.32	43.80	40.48	40.89
β-Pinene + Sabinene	7.09	6.63	3.70	5.90
γ-Terpinene	7.63	7.07	4.68	6.03
Linalool	1.32	4.93	9.27	4.23
Linalyl acetate	26.01	30.91	36.43	36.67
Monoterpene Hydrocarbons	69.94	60.74	51.27	55.54
Sesquiterpene Hydrocarbons	1.15	1.03	0.94	1.09
Carbonyl compounds	0.60	0.85	0.89	0.75
Alcohols	1.47	5.11	9.44	4.39
Esters	27.14	31.85	37.19	37.77
Linalool/Linalyl acetate	0.05	0.16	0.25	0.12

Source: Verzera, A., La Rosa, G., Zappalà, M., and Cotroneo, A., *Ital. J. Food Sci.* 4(12), 493–502, 2000.

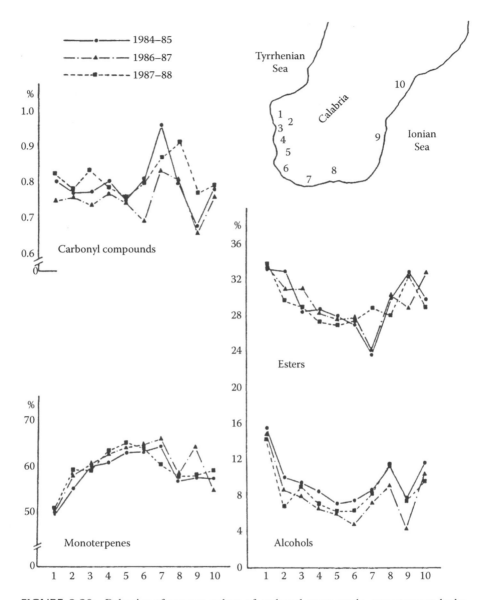

FIGURE 8.20 Behavior of average values of carbonyl compounds, monoterpene hydro-carbons, esters, and alcohols in bergamot oil produced in Calabria during three productive seasons, related to origin of the fruits. (From Dugo, G, et al. *Flavour Fragr. J.* 6, 39–56, 1991.)

shows a comparable amount of limonene and linalool to the Calabrian ones, while linalyl acetate is higher. It must be underlined that Sicilian samples were extracted in February (when linalool is at its minimum and linalyl acetate at its maximum value) while the Calabrian oils were representative of an entire industrial productive season.

Verzera et al. (2003) studied the influence of the rootstock (sour orange, Carrizzo citrange, Alemow, Trifoliata orange, Volkammerian lemon, Troyer cintrange) on the

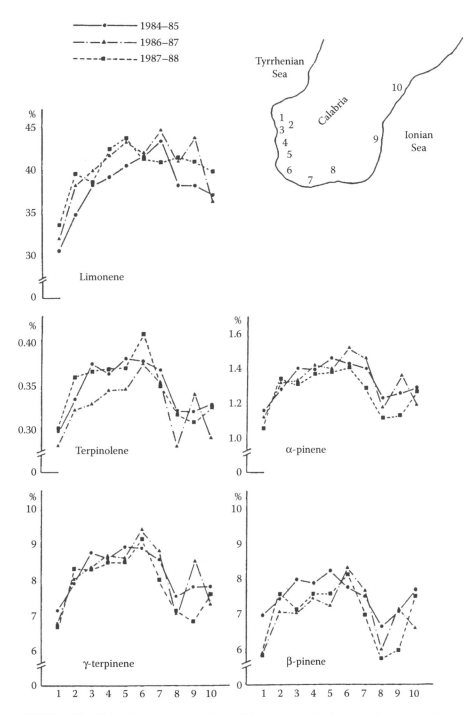

FIGURE 8.21 Behavior of average values of limonene, terpinolene, α-pinene, γ-terpinene, and β-pinene in bergamot oil produced in Calabria during three productive seasons, related to origin of the fruits. (From Dugo, G., et al. *Flavour Fragr. J.* 6, 39–56, 1991.)

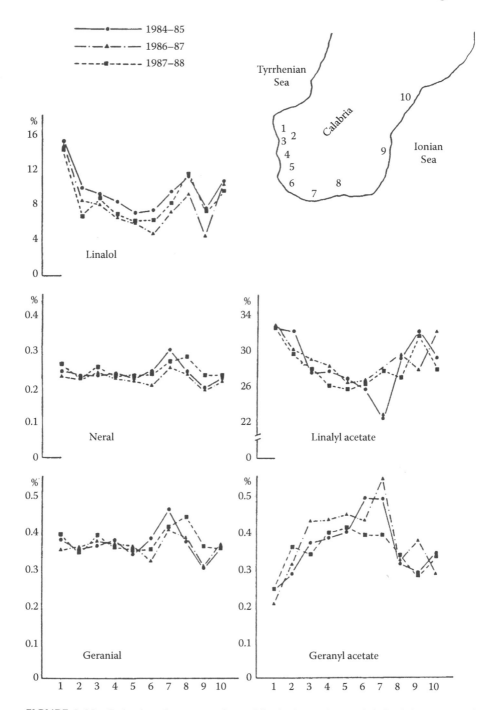

FIGURE 8.22 Behavior of average values of linalool, neral, geranial, linalyl acetate, and geranyl acetate in bergamot oil produced in Calabria during three productive seasons, related to origin of the fruits. (From Dugo, G., et al. *Flavour Fragr. J.* 6, 39–56, 1991.)

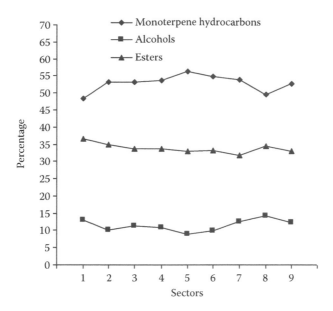

FIGURE 8.23 Behavior of monoterpene hydrocarbons, linalyl acetate, and linalool in bergamot oil produced in Calabria during the productive season 1996–97, related to origin of the fruits. (From the results of Verzera, A., et al. *EPPOS* 25, 17–38, 1998.)

composition of bergamot oil. From this investigation, carried out on 203 samples of oils extracted in laboratory from fruits harvested from 1997 to 2000 from plants cultivated in the experimental station of Acireale (Sicily), it was evident that the sour orange rootstock gave the highest content of linalyl acetate (36.6%) in the essential oil and the amount of linalool was reasonable (13.7%); the oils more similar to these, in terms of linalool and linalyl acetate content, were those from fruits grown on bergamot plants grafted on Alemow and Volkammerian lemon which, due to their production and quality of the oil, could be considered possible substitutes of the sour orange rootstock.

Huet (1981) compared the oils extracted from the cvs. Coci and Divo cultivated in the Ivory Coast. The composition of the two oils was very similar; the oil obtained from the cv. Divo had a slightly higher amount of limonene and less linalool. The composition of these two oils is reported in Table 8.5.

8.4.4 COMPOSITION AND GEOGRAPHIC ORIGIN OF THE OILS

The information given in this section is reported only for completeness; the differences observed by the authors for oils of different geographic origins could be due to different parameters. Often, in fact, the period of production of the oils and the cv. of the fruits are not indicated. The former parameter, as said in Section 8.4.1, strongly influences the composition of the essential oil.

Huet (1981) compared the results of two oils produced in Ivory Coast, already discussed in this chapter, with oils from Corsica and Italy. This author observed that α- and β-pinene, sabinene, and γ-terpinene had higher values in the oil from Italy

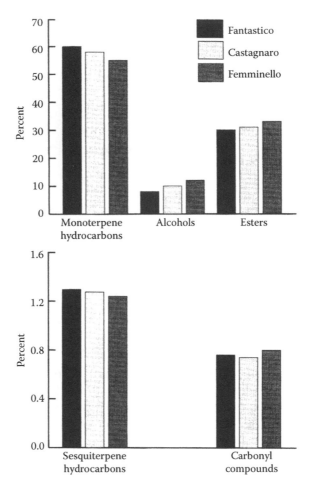

FIGURE 8.24 Average content of classes of compounds for Fantastico, Castagnaro, and Femminello bergamot oils produced in Calabria in three productive seasons. (From Verzera, A., et al. *Perfum Flav.* 21(6), 19–34, 1996.)

and Corsica than those from the Ivory Coast; the amounts of linalool and nerol were higher in the oils from the Ivory Coast and limonene was higher in the Italian ones; linalyl acetate was present in all the oils at comparable amounts. The composition of the Italian oil is reported in Table 8.3; those of the oils from Corsica and the Ivory Coast are in Table 8.4. The composition of the Argentinean oils (Dresher et al. 1984) discussed in Section 8.4.2 and that of the Brazilian ones analyzed by Koketsu et al. (1983) and by Franceschi et al. (2004) was within the range of variability determined for Italian oils for almost all the components; all of these oils, however, had values of β-pinene and γ-terpinene close to the minima and in some cases below the values determined for genuine oils from Calabria. The composition of the oils analyzed by Koketsu et al. (1983) and Franceschi et al. (2004) is reported in Table 8.5.

Sawamura et al. (1999b) and Sawamura (2000) compared the composition of two oils cold-extracted in laboratory, one from fruits of the cv. Fantastico cultivated

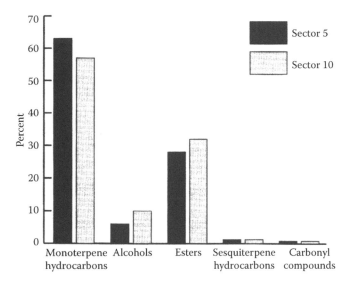

FIGURE 8.25 Average content of classes of compounds for Calabrian Fantastico bergamot oil from sectors 5 and 10 (see Figure 8.7). (From Verzera, A., et al. *Perfum Flav.* 21(6), 19–34, 1996.)

in Calabria, the other from fruits of the cv. Balotin cultivated in Japan. The fruits from Calabria were harvested in March, while the Japanese one was collected in December of the same year. In the Japanese oil were values of limonene, β-pinene, and γ-terpinene lower than the Italian oil; linalyl acetate was higher and linalool much higher in the oils from Japanese fruits. These results are reported in Table 8.11.

Baser et al. (1995) and Kirbaslar et al. (2000, 2001) analyzed oils cold-pressed in laboratory from fruits cultivated in Turkey. In all these oils β-pinene and γ-terpinene had values lower than the minima of Calabrian oils, while linalool and linalyl acetate were within the range of variability of the Italian oils. In one of these samples, Kirbaslar et al. (2000) reported a very high amount of δ-3-carene (0.9%), probably due to a wrong identification and/or a peak overlap of this component. This compound is not reported in the other samples analyzed by these authors (Kirbaslar et al. 2001). The composition of Turkish oils is reported in Table 8.11.

In the Chinese samples of oils extracted in laboratory by distillation or cold-pressing (Huang et al. 1987), β-pinene and γ-terpinene are close to the minima of the Calabrian oils, as was linalool linalyl acetate and limonene were within the range of variability of Calabrian oils. The high values of (*E*)-β-farnesene (0.45%–0.57%) determined in these oils must be highlighted, since this component is absent or present at trace levels in Calabrian oils. In the same article is reported the composition of two oils extracted in laboratory by distillation where the amount of monoterpene alcohols differing from linalool, and of esters differing from linalyl acetate are much greater than in Calabrian oils. The results obtained by Huang et al. (1987) are reported in Table 8.11.

The oils analyzed by Melliou et al. (2009), reported in Table 8.11, when compared to Calabrian oils show lower values of limonene and esters different from linalyl acetate and much lower for β-pinene. The minima of linalyl acetate and of linalool are higher than Calabrian oils, while the maxima of linalool is lower. In some samples are also indicated an unusual amount of valencene (0.63%).

As already observed in Section 8.3.2 of this chapter, most of the oils produced in regions different from Calabria show values of β-pinene and γ-terpinene lower or close to the minima determined for Calabrian oils.

8.5 ADDITIONAL ANALYTICAL INVESTIGATION ON THE VOLATILE FRACTION OF BERGAMOT OIL

Goretti and Liberti (1972) evaluated the performance of glass capillary columns coated with FFAP with i.d. 0.28 μm and length variable between 40 and 80 m, for the analyses of different citrus oils, including bergamot. The chromatograms obtained for bergamot allowed the separation of numerous peaks (from 326 to 345). Goretti et al. (1977) used graphite-glass columns, coated with PEG 20 M, with i.d. 0.24 μm and length varying between 15 and 25 m, with different load of stationary phase, to separate mixtures of monoterpene hydrocarbons and oxygenated derivatives in bergamot and lemon oils. In the article were indicated the most appropriate columns and optimal condition to separate these mixtures.

Cartoni et al. (1987) described the advantages of the use of two columns in series, one apolar (SE-54) and one polar (Carbowax 20 M), and a particular multi-step temperature program to resolve critical pairs in different essential oils; this approach allowed simultaneous separation in bergamot oil of the chromatographic pairs sabinene/β-pinene and linalyl acetate/geraniol (difficult to separate on SE-54) and α-pinene/α-thujene, linalool/geranyl acetate and geraniol/β-bisabolene (difficult to separate on Carbowax 20 M). The chromatogram obtained by the two columns in series is reported in Figure 8.26.

Lakszner and Szepesy (1988) illustrated the advantage in gas chromatography of a selective detector for oxygenated compounds (O-FID) for quality and genuineness assessment of the essential oils, including bergamot. They compared the chromatograms of a synthetic and a natural bergamot oil. The chromatograms of these two oils do not indicate the identified peaks. GC-O-FID, a technique of potential interest if operated under chromatographic conditions with improved separation over that obtained by Lakszner and Szepesy (1988), has been used very little since then.

Lamonica et al. (1990) with neutral alumina (activity II) columns, using as eluents penthane and ethyl ether, separated the volatile fraction of bergamot oil into four fractions of class of components based on their increasing polarity: hydrocarbons, esters, aldehydes, and alcohols. The GC chromatograms of these fraction were simpler than that of the whole oil; thus the identifications and quantitative analysis was simpler. In Figure 8.27 is reported the chromatogram of whole bergamot oil compared to the four fractions separated on alumina.

Micali et al. (1990) and Lanuzza et al. (1991), using a multidimensional LC-GC system, separated by LC and transferred to GC fractions of aliphatic hydrocarbons and sesquiterpene hydrocarbons (Figure 8.28). In the two fractions were identified respectively the linear saturated hydrocarbons from C_{20} to C_{33} and isoalkanes C_{21}–C_{31} (fraction I, Figure 8.28a) and the sesquiterpene hydrocarbons bergamotene*,

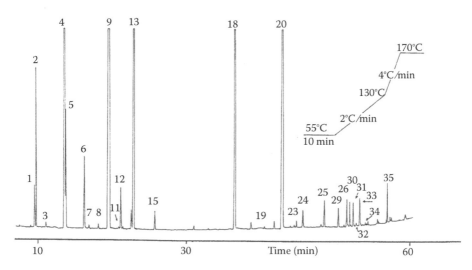

FIGURE 8.26 GC-FID chromatogram of bergamot oil on columns of SE 54 and Carbowax 20M in series. (1) α-thujene; (2) α-pinene; (3) camphene; (4) β-pinene; (5) sabinene; (6) myrcene; (7) α-phellandrene; (8) β-phellandrene; (9) limonene; (11) octanal; (12) *p*-cymene; (13) γ-terpinene; (15) terpinolene; (18) linalool; (19) terpinen-4-ol; (20) linalyl acetate; (23) α-terpineol; (24) neral; (25) geranial; (26) β-caryophyllene; (29) unknown; (30) neryl acetate; (31) bergamotene*; (32) geraniol; (33) geranyl acetate; (34) α-humulene; (35) β-bisabolene. (From Cartoni, G. P., et al. *Chromatographia* 23, 790–795, 1987.)

bisabolene*, β-caryophyllene, and α-humulene (fraction II, Figure 8.28b). Using a multidimensional LC-GC system couplet to an Ion Trap mass spectrometer (ITD), Mondello et al. (1994) separated by normal phase HPLC bergamot oil into four fractions (Figure 8.29), each transferred to the GC/MS (Figures 8.30 and 8.31). These authors emphasize that the MS analysis of the fractions allowed a more reliable identification of the oil than direct GC/MS of the whole oil. Researchers in the same group (Mondello et al. 1995a) used the same system for a systematic study of terpene hydrocarbons in citrus oils. In bergamot oil they identified the following monoterpene hydrocarbons, listed in decreasing amount order: limonene, γ-terpinene, β-pinene, α-pinene, myrcene, sabinene, *p*-cymene, α-thujene, terpinolene, (*E*)-β-ocimene, α-terpinene, camphene, α-phellandrene, (*Z*)-β-ocimene, tryciclene, and δ-3-carene, and the following sesquiterpene hydrocarbons, listed by decreasing amount: β-bisabolene, β-caryophyllene, *trans*-α-bergamotene, germacrene D, (*Z*)-β-farnesene, *cis*-α-bergamotene, α-humulene, β-santalene, germacrene B, (*Z*)-γ-bisabolene, and, tentatively, (*E*)-β-farnesene and β-sesquifellandrene. The GC chromatogram relative to the sesquiterpene hydrocarbons is reported in Figure 8.32.

In the same year Mondello et al. (1995b) proved for bergamot oil the importance of the interactive use of linear retention indices calculated on apolar and polar columns to identify by GC/MS volatile components in a real complex matrix. The results reported by Mondello et al. (1994) and Mondello et al. (1995b) are summarized in a review by Mondello et al. (1998). Tranchida et al. (2006), using headspace solid-phase microextraction-gaschromatography (HS-SPME-GC) with capillary columns

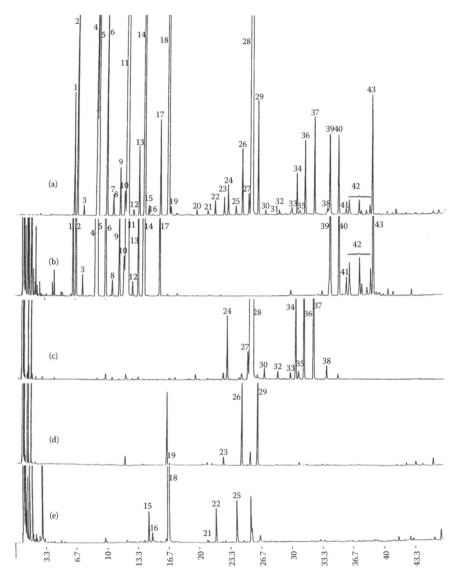

FIGURE 8.27 Chromatogram on SE-52 column of a whole bergamot oil and of the fractions obtained by separation on neutral alumina column. (a) Whole oil; (b) hydrocarbon fraction; (c) esters; (d) carbonyl compounds; (e) alcohols. (1) α-thujene; (2) α-pinene; (3) camphene; (4) sabinene; (5) β-pinene; (6) myrcene; (7) octanal; (8) α-phellandrene; (9) α-terpinene; (10) *p*-cymene; (11) limonene; (12) (*Z*)-β-ocimene; (13) (*E*)-β-ocimene; (14) γ-terpinene; (15) *cis*-sabinene hydrate; (16) octanol; (17) terpinolene; (18) linalool; (19) nonanal; (20) citronellal; (21) terpinen-4-ol; (22) α-terpineol; (23) decanal; (24) octyl acetate; (25) nerol+citronellol; (26) neral; (27) unknown; (28) linalyl acetate; (29) geranial; (30) bornyl acetate; (31) undecanal; (32) nonyl acetate; (33) linalyl propanate; (34) α-terpinyl acetate; (35) citronellyl acetate; (36) neryl acetate; (37) geranyl acetate; (38) decyl acetate; (39) β-caryophyllene; (40) *trans*-α-bergamotene; (41) α-humulene; (42) sesquiterpene hydrocarbons; (43) β-bisabolene. (From Lamonica, G., et al. *Chimica Oggi*, May, 59, 63, 1990.)

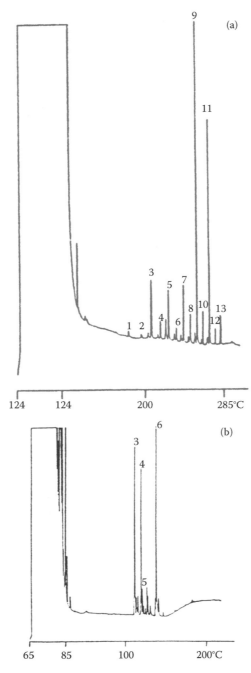

FIGURE 8.28 Online coupling HPLC-HRGC chromatogram of the (a) paraffin fraction and (b) sesquiterpene hydrocarbons fraction. Peak identification in (a): (1–13) n-alkanes C_{21}-C_{23}; the peaks that precede the n-alkanes are the respective isoalkanes. Peak identification in (b): (3) β-caryophyllene; (4) bergamotene*; (5) α-humulene; (6) bisabolene*. (From Lanuzza, F., et al. *Flavour Fragr. J.* 6, 29-37, 1991.)

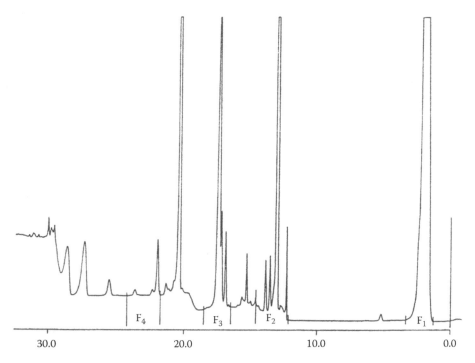

FIGURE 8.29   Normal phase HPLC chromatogram of a bergamot oil. (F_1) hydrocarbons;
(F_2) aliphatic aldehydes and esters; (F_3) monoterpene aldehydes, sesquiterpenes alcohols,
monoterpene alcohols (linalool, terpinen-4-ol); (F_4) other monoterpene alcohols. (From
Mondello, L., et al. *J. Microcol. Sep.* 6, 237–244, 1994.)

conventional and narrow bore coated with SE-52, separated in 32 and 9 minutes,
respectively, the volatile components present in bergamot oil, obtaining by the two
methods (conventional and fast) identical quantitative results.

Salvatore and Tateo (1992) compared the amount of main components in natural
bergamot oil and in a reconstituted one and found they had very similar compositions. They surprisingly asserted that reconstituted oils also had very similar odor
properties to natural bergamot oils. Tateo et al. (2000) analyzed by conventional GC
and by GC-O a genuine sample and a reconstituted oil obtained from natural bergamot oil treated with methanol to eliminate linalool, linalyl acetate, and other oxygenated compounds, and then adding synthetic racemic linalool and linalyl acetate.
These authors did not find a significant difference in the composition from the sensorial profiles, and asserted that the oils were similar to each other. These conclusions
appear surprising, due to the evident decrease of all the oxygenated compounds in
the reconstituted oil, which surely contribute significantly to the flavor, and also in
view of the well-known difference in the olphactive properties of the two enantiomers of linalool.

Benincasa et al. (1990) analyzed lemon and bergamot oils by normal phase and
reversed phase HPLC. In normal phase they identified not only coumarins and psoralens, but in bergamot oil they also determined limonene, myrcene α- and β-pinene,

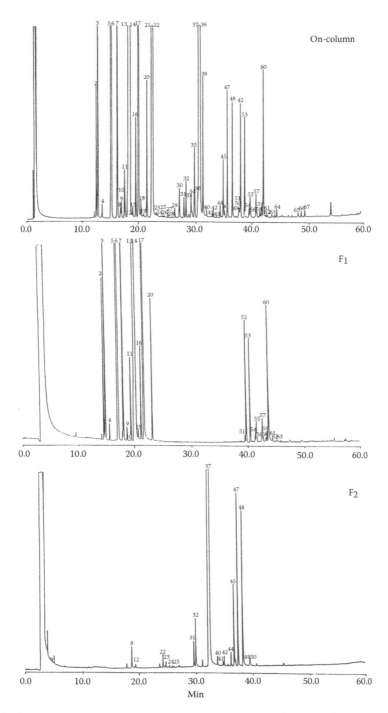

FIGURE 8.30   GC chromatogram of whole bergamot oil and of its F_1 and F_2 fractions from the HPLC separation of Figure 8.29. (From Mondello, L., et al. *J. Microcol. Sep.* 6, 237–244, 1994.)

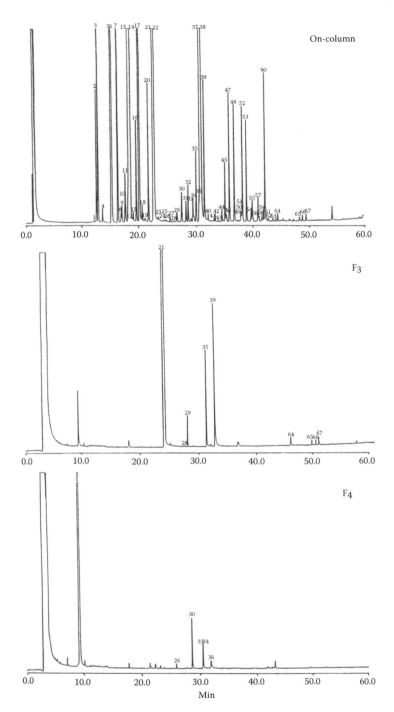

FIGURE 8.31 GC chromatogram of whole bergamot oil and of its F$_3$ and F$_4$ fractions from the HPLC chromatogram of Figure 8.29. (From Mondello, L., et al. *J. Microcol. Sep.* 6, 237–244, 1994.)

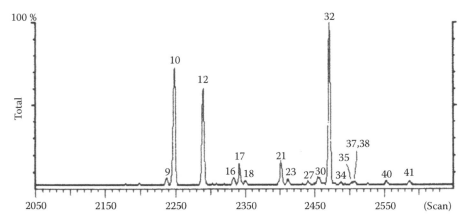

FIGURE 8.32 Sesquiterpene region of the GC chromatogram obtained from LC transfer of the hydrocarbons fraction for bergamot oil. (9) *cis*-α-bergamotene; (10) β-caryophyllene; (12) *trans*-α-bergamotene; (16) α-humulene; (17) (Z)-β-farnesene; (18) β-santalene; (21) germacrene D; (32) β-bisabolene; (34) (Z)-α-bisabolene; (37) (E)-β-farnesene; (38) β-sesquiphellandrene; (41) germacrene B; (23, 27, 30, 35, 40) unknown sesquiterpene hydrocarbons. (From Mondello, L., et al. *Flavour Fragr. J.* 10, 33–42, 1995a.)

TABLE 8.13
Yield and Composition of Bergamot Oil

	a	b
Yield of essential oil (fruit)	0.74%	0.77%
Yield of essential oil (peels)	2.80%	2.33%
Limonene	30.25%	31.38%
Myrcene	0.98%	–
β-Pinene	6.87%	6.82%
Sabinene	1.53%	–
γ-Terpinene	6.90%	7.78%
Terpinolene	0.30%	–
α-Thujene	0.19%	–
trans-α-Bergamotene	2.51%	0.39%
β-Bisabolene	0.13%	–
β-Caryophyllene	0.39%	0.31%
Geranial	0.43%	–
Neral	0.14%	–
Linalool	8.86%	10.00%
Linalyl acetate	33.74%	30.22%
Neryl acetate	0.39%	0.38%
α-Terpinyl acetate	0.28%	0.20%

Sources: (a) Intrigliolo, F., Caruso, A., Russo, G., et al., *Commu. Soil Sci. Plant Anal.* 30(13/14), 2035–2044, 1999; (b) Intrigliolo, F., Caruso, A., Giuffrida, A., et al., *Essenz. Deriv. Agrum.* 67, 42–61, 1997.

sabinene, terpinolene, γ-terpinene, caryophyllene*, neral, geranial, and neryl and geranyl acetates.

Intrigliolo et al. (1999) correlated the yield and quality of bergamot oil with some parameters (number of fruits for each tree, average weight of fruits, thickness of the peel, total solids and acidity) and the leaves' nutritional levels (macro and micro elements) with the content of main components in the oil. A study carried out for three years at 25 different bergamot producers' farms led to the results summarized in Table 8.13. These results are slightly different from those previously obtained by the same authors for a similar study (Intrigliolo et al. 1997) and reported in the same table.

Mazza (1987b) carried out a study on the oxidation products of selected mono-terpenes in bergamot oil. In this article are reported the oxidation schemes of limonene, linalool, and linalyl acetate suggested by the author, and the main products of primary and secondary oxidation (given following in parentheses) for limonene (*cis*- and *trans*-carveol, carvone, *cis*- and *trans-p*-mentha-2,8-dien-1-ol, *cis*- and *trans*-iso-piperitenol, *cis*- and *trans*-limonene-1,2-epoxy, *cis*- and *trans* limonene-8,9-epoxy); γ-terpinene (*p*-cymene, *p*-menth-4-ene-1,2-epoxy, *p*-menth-1-ene-4,5-epoxy); α- and β-pinene (myrtenal, myrtenol, *trans*-pinocarveol); (*E*)- and (*Z*)-β-ocimene (*cis*- and *trans*-ocimene-5,6-epoxy; 2,6-dimethyl-1,5,7-octatrien-3-ol); linalool (*cis*- and *trans*-linalool oxide), pyranoid and furanoid forms, 2-vinyl-2-methyl-tetrahydrofuran-5-one, *trans*-3,7-dimethyl-octa-3,5-dien-3,7-diol, 3,7-dimethylocta-1,7-dien-3,6-diol); linalyl acetate (3-acetoxy-3,7-dimethylocta-2,7-dien-6-ol); neryl and geranyl acetate (epoxyneryl and epoxygeranyl acetate, *cis*- and *trans*-1-acetoxy-3,7-dimethylocta-2,7-dien-6-ol).

La Scala and La Scala (1991) identified carbonyl compounds present in cit-rus oil by chromatography of their oxime derivatives obtained by treatment with

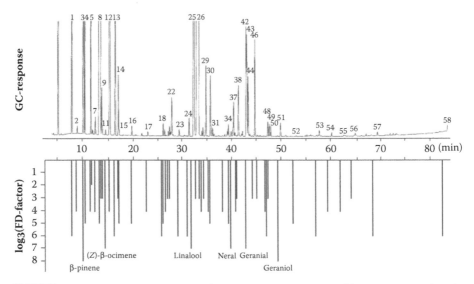

FIGURE 8.33 Gas chromatogram and aromagram (FD-factor) of bergamot essential oil. (From Sawamura, M., et al. *Flavour Fragr. J.* 21, 609–615, 2006.)

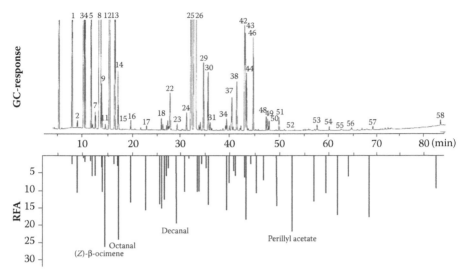

FIGURE 8.34 Gas chromatogram and aromagram (RFA) of bergamot essential oil. (From Sawamura, M., et al. *Flavour Fragr. J.* 21, 609–615, 2006.)

TABLE 8.14
Percentage Composition of a Commercial Bergamot Oil and of Bergamot Oils Obtained by Naviglio Extractor

	Commercial Oil	Oils Obtained with Naviglio Extractor	
		1	2
Limonene	40.95	18.92	29.85
β-Pinene	6.27	1.65	5.26
Sabinene	0.94	0.42	0.65
γ-Terpinene	6.99	4.38	5.94
α-Bergamotene*	0.41	1.05	0.33
β-Caryophyllene	0.43	1.10	0.27
Neral	0.44	1.11	0.58
Geranial	0.11	0.09	0.23
Solanone	0.07	0.11	0.16
Linalool	11.22	3.58	17.45
α-Terpineol	0.30	0.90	0.22
Geranyl acetate	0.19	0.18	0.41
Linalyl acetate	29.64	40.53	36.11

Source: Naviglio, D., Raia, C., Russo, M., et al., *Ingredienti Alimentari* 2 (October), 13–18, 2003.
Notes: * Correct isomer not characterized. (1) Recovered by diethyl ether. (2) Centrifuged.

hydroxylamine. The authors asserted that this technique allowed detection not only of aldehydes, but also methyl N-methyl anthranylate, which in their opinion is present with other high-boiling nitrogen-containing compounds in all citrus oils (bergamot, sweet orange, bitter orange, mandarin, and grapefruit). This statement is not clearly confirmed by the chromatograms published in the article.

Sawamura et al. (2006) determined by GC-O the flavor dilution factor (FD) and calculated the relative flavor activity (RFA) for the active odor components. The chromatograms obtained are shown in Figures 8.33 and 8.34.

Naviglio et al. (2003) proposed a new method of extraction of bergamot oil based on the effect of water pressure on the peels to cause breakage of the utricles and consequent release of the oil. The procedure avoids the release of undesired substances. The oil was separated from the aqueous emulsion either by solvent extraction or centrifugation. In Table 8.14, the percentages of the main components obtained by the method proposed by Naviglio et al. are compared with those of a commercial sample cold-pressed by the traditional method. Based on the composition of the oils extracted by diethyl ether or by centrifugation, the authors concluded that the proposed extraction method could represent a valid substitute for the traditional extraction procedures. However, from the results reported it is difficult to explain some differences of the composition, mainly due to linalool (3.58% in the oil extracted with diethyl ether and 17.45% in the oil separated by centrifuge).

Theile et al. (1960) highlighted that in bergamot oil the ratio of absorbance at 835 cm^{-1} and 801 cm^{-1} is between 0.9 and 1.14. If the absorbance at 801 cm^{-1} increases, with a consequent value of this ratio below 0.9, it is indicative of adulteration by addition of monoterpene hydrocarbons. Bellanato and Hidalgo (1971) reported in table the identification of IR bands of citrus oils. Carmona et al. (1974) proposed an IR method to quantitatively determine linalool, linalyl acetate, and limonene in bergamot oil. Bovalo et al. (1985) determined the Theile's IR ratio (Theile et al. 1960) in oils of bergamot considered genuine, with an optical rotation at 15°C lower than +28, in bergamot oils of less appreciated odor character, still considered genuine with an optical rotation between +26 and +28, and in anomalous bergamot oils which presented optical rotation above +30. Theile's ratio in genuine high quality oils was between 0.75 and 1.09; for genuine low quality oils it was between 0.69 and 0.92; and for anomalous oils it was between 0.63 and 0.79.

Additional information on these topics is given in Chapter 7, this volume.

8.6   COMPONENTS IDENTIFIED IN BERGAMOT OIL

Table 8.15 lists the components to our knowledge identified to date in bergamot oil. The components are listed with the same criteria used in the previous tables. Multifunctional compounds are listed based on their main chemical function; the only exceptions are oxides of alcohols, which are listed within oxides in accordance with the previous tables. With a numerical note are indicated the components identified in a single article or in more articles by the same research group. Letters (a, b, and c) indicate components identified respectively in oils extracted in laboratory, recovered oils, and oils that underwent oxidation as stated by the authors of the

TABLE 8.15
Compounds Identified in Bergamot Oils

Hydrocarbons

Aliphatic

4,8-dimethyl-1,3(*E*),7-nonatriene[8], 4,8-dimethyl-1,3(*Z*),7-nonatriene[8], dodecane[8], pentadecane[a,16], tridecane[12], *n*-C_{20}-C_{33} and *iso*-C_{21}-C_{31}[9]

Monoterpene

camphene, δ-2-carene[12], δ-3-carene, *m*-cymene[a,18], *p*-cymene, *p*-cymenene[7], limonene, 1,3,8-*p*-menthatriene[12], myrcene, allo-ocimene, (*E*)-β-ocimene, (*Z*)-β-ocimene, α-phellandrene, β-phellandrene, α-pinene, β-pinene, sabinene, α-terpinene, β-terpinene[11], γ-terpinene, terpinolene, α-thujene, tryciclene

Sesquiterpene

γ-acoradiene[7], aromadendrene[12], *cis*-α-bergamotene, *trans*-α-bergamotene, *trans*-β-bergamotene, bicyclogermacrene, β-bisabolene, (*E*)-α-bisabolene, (*Z*)-α-bisabolene, (*E*)-γ-bisabolene, (*Z*)-γ-bisabolene, δ-cadinene, β-caryophyllene, ar-curcumene[5], γ-curcumene[8], β-elemene, γ-elemene[8], δ-elemene, (*E,E*)-α-farnesene, (*E*)-β-farnesene, (*Z*)-β-farnesene, germacrene B[8], germacrene D, α-humulene, α-murolene[7], γ-murolene, δ-murolene[13], α-santalene[5], β-santalene, epi-β-santalene[7], α-selinene[7], β-selinene, β-sesquiphellandrene, sesquithujene[b,15], valencene[a,14].

Aldehydes

Aliphatic

decanal, (*E*)-2-decenal, dodecanal, heptanal[3], hexanal[a,17], (*E*)-2-hexenal[17], nonanal, octanal, tetradecanal[8], tridecanal, undecanal

Monoterpene

citronellal, cuminaldehyde[c,7], isogeranial[8], geranial, *p*-menth-1-en-9-al[c,7], neral, perilla aldehyde

Sesquiterpene

bergamotenal[11], lanceal*[,11], α-sinensal[a], β-sinensal[11]

Ketones

Aliphatic

cis- and-*trans*-octa-5-en-2-one[1], 6-methyl-5-hepten-2-one

Monoterpene

camphor, carvone, dihydrocarvone[c,7], isomenthone[12], 2,6-dimethyl-6-acetoxy-octa-1-en-7-one[5], 2,6-dimethyl-6-acetoxy-octa-7-en-3-one[5], 2,6-dimethyl-6-acetoxy-1,7-dien-3-one[5], menthone[12], 3-methyl-3-acetoxyocta-1,5-dien-7-one[2], piperitone, pulegone[a,16]

Sesquiterpene

camphorenone[11], 8,9-didehydronootkatone, nootkatone

Alcohols

Aliphatic

decanol, dodecanol, hexanol, (*E*)-2-hexen-1-ol[7], (*Z*)-3- hexen-1-ol, 6-methyl-3-heptanol[a,17], nonanol, octadecanol[12], octanol, octen-3-ol[12], 1-pentanol[7], 2-pentanol[7], tetradecanol

Monoterpene

borneol[4], *cis*- and *trans*-carvacrol[7], *cis*- and *trans*-carveol, citronellol, *p*-cymen-8-ol, cumin alcohol, dihydrocitronellol[12], dihydrocumin alcohol*[,3], 3,7-dimethyl-3-acetoxy-octa-1,5-dien-7-ol[7], 3,7-dimethyl-3-acetoxy-octa-1,7-dien-6-ol[7], 3,7-dimethyl-3-acetoxy-octa-2,7-dien-6-ol[7], dihydroxylinalool[c,16], 2,6-dimethyl-octa-2,6-dien-1-ol[a,13], *cis*- and *trans* 2,6-dimethylocta-1,5,7-trien-3-ol[5], 2,6-dimethyl-1,7-dien-3,6-diol[2], 3,7-dimethylocta-1,5-dien-3,7-diol[7], *trans*-3,7-dimethylocta-1,6-dien-3,8-diol[7], 3,7-dimethylocta-1,7-dien-3,6-diol[7], *trans*-3,7-dimethylocta-3,5-dien-3,7-diol[7],

continued

TABLE 8.15   (CONTINUED)
Compounds Identified in Bergamot Oils

cis- and *trans*-acetoxylinalool[7], *cis*- and *trans*-3,7-dimethyl-1-acetoxy-octa-2,7-dien-6-ol[7], 2,6-dimethyl-4-acetoxy-octa-2,7-dien-6-ol[2], 3,7-dimethyl-3-acetoxy-octa-2,7-dien-6-ol[c,7], endophencol[a16], hotrienol[7], geraniol, hydroxydihydrocarveol[2], hydroxylinalool[13], *cis*- and *trans*-isoperitenol[7], isopulegol, lavandulol[a,16], limonene-4-ol[7], limonene-10-ol[c,7], linalool, *cis*- and *trans*-*p*-menth-2-en-1-ol, *p*-menth-1-en-9-ol[5], *p*-menth-1,8-dien-4-ol[7], *p*-menth-1,8-dien-9-ol[5], *cis*- and *trans*-*p*-menth-2,8-dien-1-ol[7], menthol[12], myrtenol[c,7], neomenthol[12], nerol, perylla acohol, *trans*-pinocarveol[7], α-terpineol, terpinen-4-ol, thymol, *cis*- and *trans*-sabinene hydrate, verbenol[13]

Sesquiterpene
α-bisabolol, (45,8R)-8-epi-α-bisabolol[10], β-bisabolol, epi-β-bisabolol[8], (4S,8R)-4-epi-β-bisabolol[10], α-cadinol[5], T-cadinol[6], campherenol, cedrol[13], caryophyllene alcohol[5], caryophyllenol I and II[5], α-copaneol[d,12], β-eudesmol[6], (*E,E*)- farnesol[12], (*Z,Z*)-farnesol[d,12], guaienol[11], globulol [d,12], (*E*)-nerolidol, (*Z*)-nerolidol[13], 2,3-dimethyl-3-(4-methyl-3-pentenyl-2-norbornanol, β-photosantalol*, β-photosantalol A[5], two santalene hydrates[11], (*Z*)-β-santalol[12], *cis*- and *trans*-sesquisabinene hydrate[8], spathulenol

Esters
Aliphatic
butyl acetate[a,16], decyl acetate, heptyl acetate, *cis*-hex-3-en-1-yl acetate[7], hexyl acetate, nonyl acetate, octyl acetate, 3-(3,4,5-trimethoxyphenyl)-propenyl acetate[7], undecyl acetate[8]

Monoterpene
bornyl acetate, carvyl acetate[12], citronellyl acetate, 3-hydroxy-citronellyl acetate[2], citronellyl formate[a,16], 2,6-dimethylocta-7-en-1-yl-acetae[2], *cis*- and *trans*-2,6-dimethyl-octa-2,7-dien-1-yl acetate[2], 3,7-dimethyl-3-hydroxy-1,6-octadienylformate[a,18] geranyl acetate, 2,3-epoxygeranyl acetate[5], dihydro-2,3-epoxygeranyl acetate[c,7], geranyl formate[a,16], geranyl propanate, hotrienyl acetate[5], isobornyl acetate, isomenthyl acetate[12], linalyl acetate, 6,7-epoxylinalyl acetate[2], 7-hydroxylinalyl acetate[2], linalyl propanate, *p*-menth-1-en-8-yl acetate[a,18], *p*-menth-1-en-9-yl acetate[5], *p*-mentha-1,7 (10)-dien-2-yl acetate[5], *p*-mentha-1,7-dien-4-yl acetate[5], *p*-mentha-1,3-dien-7-yl acetate[5], *p*-menth-1,8-dien-9-yl acetate[5], methyl geranate, neomenthyl acetate[12], neryl acetate, 2,3-epoxynerylacetate[6], dihydro-2,3-epoxyneryl acetate[c,7], neryl propanate[7], *trans*-pinocarvyl acetate[7], perillyl acetate[13], *cis*-sabinenehydrate acetate[8], *trans*-sabinene hydrate acetate, sabinyl acetate[7], α-terpinyl acetate, α-terpinyl isobutyrate[a,18], terpinen-4-yl acetate[5]

Ethers and Oxides
Monoterpene
1,4-cineole[a,16], 1,8-cineole, dehydrocineole[2], *cis*- and *trans*-limonene oxide, *cis*- and *trans*-limonene 8,9-oxide[c,7], *cis*- and *trans*-linalool oxide (furanoid), *cis*- and *trans*-linalool oxide (piranoid)[7], *p*-mentha-1-en-4,5-oxide[7], *p*-mentha-4-en-1,2-oxide[7], 8-hydroxy-*p*-menth-2-en-1-yl-methyl ether, *cis*- and *trans*-ocimene-2,3-oxide[5], *cis*- and *trans*-ocimene-5,6-oxide[c,7]

Sesquiterpene
caryophyllene oxide*, humulene oxide I and II[5], isocaryophyllene oxide[5]

Acids
Aliphatic
acetic acid[7], octanoic acid, n-C_9-C_{11}, C_{16}, C_{22}-C_{26} carboxilic acids[1]

Monoterpene
dihydrogeranic acid[1], *cis*- and *trans*-geranic acids[1]

TABLE 8.15 (CONTINUED)
Compounds Identified in Bergamot Oils
Others

ascaridole*[,8], divinylbenzene[a,18], cresol*[c,7], γ-heptalactone[a16], indole, *cis-* and *trans*-jasmone[1], lilial[5], methylacetophenone[c,7], methyl N-methylanthranilate, perillene[c,7], phenylethyl alcohol[a,18], pirrole[3], (*E*)-solanone[13], 2-vinyl-2-methyltetrahydrofuran-5-one[7]

Notes: [1] Sundt et al. (1964); [2] Mookherjee (1969); [3] Di Giacomo and Calvarano (1970); [4] Herrisset et al. (1973); [5] Schenk and Lamparsky (1981); [6] Ehret and Maupetit (1982); [7] Mazza (1986, 1987a, 1987b); [8] Dugo et al. (1987, 1991, 1999, 2012); Dugo (1994); Lamonica et al. (1990); Verzera et al. (1996, 1998, 2003); Mondello et al. (1994, 1995a,b, 1998); Sciarrone (2009); Costa et al. (2010); [9] Micali et al. (1990); Lanuzza et al. (1991); [10] Ohloff et al. (1986); [11] Ohloff (1994); [12] Chouchi et al. (1995); [13] Sawamura et al. (1999a,b, 2006); [14] Melliou et al. (2009); [15] Mondello (2006); [16] Huang et al. (1986); [17] Baser et al. (1995); [18] Kirbaslar et al. (2000, 2001).

[t] tentative; [*] correct isomer not characterized.
[a] laboratory oils.
[b] recovered oils.
[c] oils surely oxidised as indicated by the authors.
[d] identified only in fractions obtained by supercritical CO_2 pre-fractionation.

original article. However, it should be highlighted that among the components listed in Table 8.15 are mentioned, even if not clearly stated, compounds which are the products of transformation processes on naturally occurring components.

REFERENCES

Adams, R. P. 2007. *Identification of essential oil components by gas chromatography/mass spectrometry,* 4th ed. Carol Stream, IL: Allured Publishing Corporation.

Baser, K. H. C., Özek, T., and Tutas, M. 1995. Composition of cold-pressed bergamot oil from Turkey. *J. Essent. Oil Res.* 7:341–342.

Bellanato, J., and Hidalgo A. 1971. *Infrared analysis of essential oils.* London: Heyden and Son Ltd./Stadler Research Lab. Inc.

Belsito, E. L., Carbone, C., and Di Gioia, M. L., et al. 2007. Comparison of the volatile constituents in cold pressed bergamot oil and a volatile oil isolated by vacuum distillation. *J. Agric. Food Chem.* 55:7847–7851.

Benincasa, M., Buiarelli, F., Cartoni, G. P., and Coccioli F. 1990. Analysis of lemon and bergamot essential oils by HPLC with microbore columns. *Chromatographia* 30(5/6):271–276.

Binder, V. G., König, W. A., and Czygan, F. C. 2001. Ätherische Öle. *Deutsche Apotheker Zeitung.* 141(37):49–58.

Bovalo, F., Cappello, C., Sarlo, C., and Di Giacomo, A. 1985. A proposito dello spettro I.R. dell'essenza di bergamotto. *Essenz. Deriv. Agrum.* 55(1):36–47.

Calabrò, G., and Currò, P. 1972. Costituenti degli olii essenziali. Nota III. Costituenti sesquiterpenici e paraffinici dell'essenza di bergamotto. *Ann. Fac. Econ. Commer. Univ. Studi Messina* 10:68–79.

Calvarano, I. 1961. Ricerche sull'essenze dei frutti immaturi di bergamotto "bergamottella." *Essenz. Deriv. Agrum.* 31:109–124.

Calvarano, M. 1963. La composizione delle essenze di bergamotto. Nota I. Gli idrocarburi monoterpenici. *Essenz. Deriv. Agrum.* 33:67–101.

Calvarano, M. 1965. La composizione delle essenze di bergamotto. Nota III. *Essenz. Deriv. Agrum.* 35:197–211.

Calvarano, M. 1968. Variazioni nella composizione dell'essenza di bergamotto durante la maturazione del frutto. *Essenz. Deriv. Agrum.* 38:3–20.

Calvarano, M., and Calvarano, I. 1964. La composizione delle essenze di bergamotto. Nota II. Contributo all'indagine analitica mediante spettrofotometria nell'UV e gascromatografia. *Essenz. Deriv. Agrum.* 34:71–92.

Calvarano, M., Rafanelli, G., and Di Giacomo, A. 1984. Sulle caratteristiche di un'essenza di bergamotto ottenuta da frutti prodotti in Toscana. *Essenz. Deriv. Agrum.* 54:220–227.

Carmona, P., Bellanato, J., and Hidalgo, A. 1974. Infrared spectroscopic study of bergamot essential oils. *Rev. Agroquim. Tecnol. Aliment.* 14(4):565–576.

Cartoni, G. P., Goretti, G., and Russo, M. V. 1987. Capillary columns in series for the gas chromatographic analysis of essential oils. *Chromatographia* 23(11):790–795.

Chouchi, D., Barth, D., Reverchon, E., and Della Porta, G. 1995. Supercritical CO_2 desorption of bergamot peel oil. *Ind. Eng. Chem. Res.* 34:4508–4513.

Cosimi, S., Rossi, E., Cioni, P. L., and Canale, A. 2009. Bioactivity and qualitative analysis of some essential oils from Mediterranean plants against stored-product pests: Evaluation of repellency against *Sitophilus zeamais* Motschulsky, *Cryptolestes ferrugineus* (Stephens) and *Tenebrio molitor* (L.). *Journal of Stored Products Research* 45:125–132.

Costa, R., Dugo, P., Navarra, M., Dugo, G., and Mondello, L. 2010. Study on the chemical composition variability of some processed bergamot (*Citrus bergamia*) essential oils. *Flavour Fragr. J.* 25:4–12.

Dellacassa, E., Lorenzo, D., Moyna, P., Verzera, A., and Cavazza, A. 1997. Uruguayan essential oils. Part V. Composition of bergamot oil. *J. Essent. Oil Res.* 9:419–426.

Di Corcia, A., Liberti, A., and Samperi, R. 1975. Cromatografia gas-liquido-solido: sua applicazione allo studio degli olii essenziali. *Essenz. Deriv. Agrum.* 45:89–99.

Di Giacomo, A. 1990. Valutazione della qualità delle essenze agrumarie cold-pressed in relazione al contenuto in composti cumarinici e psoralenici. *Essenz. Deriv. Agrum.* 60:313–334.

Di Giacomo, A., Rispoli, G., and Tracuzzi, M. L. 1963. Contributo dell'analisi strumentale alla conoscenza dei costituenti terpenici delle essenze agrumarie. *EPPOS* 45:269–281.

Di Giacomo, A., and Calvarano, M. 1970. I componenti degli agrumi. *Essenz. Deriv. Agrum.*, 40:344–378.

Di Giacomo, A., and Mincione, B.1994. Gli oli essenziali agrumari in Italia. Laruffa Editore: Reggio Calabria.

Dixon, C. W., Malone, C. T., and Umbreit, G. R. 1968. A scheme for the preparative chromatographic isolation and concentration of trace components in natural products using 4 inch diameter columns. *J. Chromatogr.* 35:475–488.

Drescher, R. W., Beñatena, H. N., De Rossin, G. G. T., Ubiergo, G. O., and Retamar, J. A. 1984. Evaluaciòn de le calidad perfumistica del aceite esencial de *Citrus bergamia* Risso (Bergamota) cultivada en la E.E.A. Inta Concordia (Entre Rios). *Essenz. Deriv. Agrum.* 54:192–199.

Dugo, G. 1994. The composition of the volatile fraction of the Italian citrus essential oils. *Perfum. Flav.* 19(6):29–51.

Dugo, G., Lamonica, G., Cotroneo, A., et al. 1987. Sulla genuinità delle essenze agrumarie. Nota XVII. La composizione della frazione volatile dell'essenza di bergamotto Calabrese. *Essenz. Deriv. Agrum.* 57:456–534.

Dugo, G., Cotroneo, A., Verzera, A., et al. 1991. Genuineness characters of Calabrian bergamot essential oil. *Flavour Fragr. J.* 6:39–56.

Dugo, G., Bartle, K. D., Bonaccorsi, I., et al. 1999. Advanced analytical techniques for the analysis of citrus essential oils. Part I. Volatile fraction: HRGC/MS Analysis. *Essenz. Deriv. Agrum,* 69:79–111.

Dugo, G., Cotroneo, A., Verzera, A., and Bonaccorsi, I. 2002. Composition of the volatile frac-
    tion cold pressed citrus peel oils. In *Citrus*, eds. G. Dugo and A. Di Giacomo, 201–317.
    London and New York: Taylor & Francis.
Dugo, G., Cotroneo, A., Bonaccorsi, I., and Trozzi, A. 2011. Composition of the volatile
    fraction of cold pressed citrus peel oils. In *Citrus oils. Composition, advanced analyti-
    cal techniques, contaminants and biological activity*, eds. G. Dugo and L. Mondello.
    London and New York: Taylor & Francis.
Dugo, G., Bonaccorsi, I., Sciarrone, D., et al. 2012. Characterization of cold-pressed and
    processed bergamot oil by GC/FID, GC/MS, enantio-GC, MDGC, HPLC, HPLC/MS,
    GC-C-IRMS. *J. Essent. Oil Res.* 24:93–117.
Ehret, C., and Maupetit, P. 1982. Two sinapyl alcohol derivatives from bergamot essential oil.
    Phytochemistry 21:2984–2985.
Fantin, G., Fogagnolo, M., Maietti, S., and Rossetti, S. 2010. Selective removal of monoter-
    penes from bergamot oil by inclusion in deoxycholic acid. 2010. *J. Agric. Food Chem.*
    58:5438–5443.
Feger, W., Brandauer, H., and Ziegler, H. 2001. Germacrenes in citrus peel oils. *J. Essent. Oil
    Res.* 13:274–277.
Ferrini, A. M., Mannoni, V., Hodzic, S., Salvatore, G., and Aureli, P. 1998. Activitè antimi-
    crobienne de l'huile de bergamote par rapport à la composition chimique et l'origine.
    EPPOS (Numero speciale):139–153.
Figoli, A., Donato, L., Carnevale, R., et al. 2006. Bergamot essential oil extraction by pervapo-
    ration. *Desalination* 193:160–165.
Formàcek, V., and Kubeczka, K. H. 1982. *Essential oils analysis by capillary gas chromatog-
    raphy and carbon-13 NMR spectroscopy.* Chichester, UK: Wiley.
Franceschi, E., Grings, M. B., Frizzo, C. D., Oliveira, J. V., and Dariva, C. 2004. Phase behavior
    of lemon and bergamot peel oils in supercritical CO_2. *Fluid Phase Equilibria* 226:1–8.
Gionfriddo, F., Mangiola, C., Siano, F., and Castaldo, D. 2000. Le caratteristiche dell'essenza
    di bergamotto prodotta nella campagna 1998/99. *Essenz. Deriv. Agrum.* 70:133–145.
Gionfriddo, F., Catalfamo, M., Siano, F., et al. 2003. Determinazione delle caratteris-
    tiche analitiche e della composizione enantiomerica di oli essenziali agrumari ai fini
    dell'accertamento della purezza e della qualità. Nota I – Essenze di arancia amara, aran-
    cia dolce e bergamotto. *Essenz. Deriv. Agrum.* 73:29–39.
Goretti, G., and Liberti, A. 1972. Sui costituenti degli olii essenziali di agrumi. *Essenz. Deriv.
    Agrum.* 42:223–231.
Goretti, G., Liberti, A., and Ciardi, M. 1977. Colonne capillari di vetro grafitate per l'analisi di
    olii essenziali. *Essenz. Deriv. Agrum.* 47:269–285.
Guerrini, A., Lampronti, I., Bianchi, N., et al. 2009. Bergamot (*Citrus bergamia* Risso) fruit
    extracts as γ-globin gene expression inducers: Phytochemical and functional perspec-
    tives. *J. Agric. Food Chem.* 57:4103–4111.
Hérisset, A., Jolivet, J., Rey, P., and Lavault, M. 1973. Differénciation de quelques huiles
    essentielles présentant une constitution voisine. Essences de divers *Citrus*. *Plant. Med.
    Phytoter.* 7:306–318.
Hochmuth, D. 2006. Mass spectral library. In *Terpenoids and related constituents of essential
    oils*. Hamburg, Germany: Library of MassFinder 3.0.
Huang, Y., Wen, M., Xiao, S., Zhao, H. et al. 1986. Studies on the chemical constituents of
    the steam-distilled leaf and peel essential oil from *Citrus bergamia*. *Acta Botanica
    Yunnanica* 8:471–476.
Huang, Y., Wen, M., Xiao, S., et al. 1987. Chemical constituents of the essential oil from the
    peel of *Citrus Bergamia* Risso. *Acta Botanica Sinica* 29:77–83.
Huet, R. 1970. Essai d'introduction d'une industrie de la bergamote au Mali. *Fruits* 25:709–715.
Huet, R. 1981. Etude comparative de l'huile essentielle de bergamote provenant d'Italie, de
    Corse et de Cote d'Ivoire. *EPPOS* 63(6):310–313.

Huet, R. 1991a. Les huiles essentielles d'agrumes. *Fruits.* 46(4):501–513.

Huet, R. 1991b. Les huiles essentielles d'agrumes. *Fruits.* 46(5):551–576.

Huet, R. 1991c. Les huiles essentielles d'agrumes. *Fruits.* 46(6):671–683.

Huet, R., and DuPuis, C. 1968. L'huile essentielle de bergamote en Afrique et en Corse. *Fruits* 23:301–311.

Huet, R., and DuPuis, C. 1969. Evolution de la composition chimique de l'huile essentielle de clementine (hybride de *Citrus reticulata* Blanco) et de l'huile essentielle de bergamote (*Citrus aurantium* LIN. Subsp. *Bergamia* Risso et Poiteau Engler) au cours de la croissance du fruit. *La France et ses Parfums* 12(63):123–130.

Ikeda, R. M., Stanley, W. L., Vannier, S. H., and Spitler, S. H. 1962. The monoterpene hydrocarbons of some essential oils. *J. Food Sci.* 27:455–458.

Inoma, S., Miyagi, Y., and Akieda, T. 1989. Characterization of citrus oils (mainly orange oil, mandarin oil, tangerine oil). *Kanzei Chuo Bunsekishoho* 29:87–97.

Intrigliolo, F., Caruso, A., Giuffrida, A., et al. 1997. Effetto dei parametri produttivi e nutrizionali sulla resa e sulla qualità dell' essenza di bergamotto. *Essenz. Deriv. Agrum.* 67:42–61.

Intrigliolo, F., Caruso, A., Russo, G., et al. 1999. Pedologic parameters related to yield and quality of bergamot oil. *Commu. Soil Sci. Plant Anal.* 30(13/14):2035–2044.

Kirbaslar, S. I., Kirbaslar, F. G., and Dramur, U. 2000. Volatile constituents of Turkish bergamot oil. *J. Essent. Oil Res.* 12:216–220.

Kirbaslar, F. G., Kirbaslar, S. I., and Dramur, U. 2001. The composition of Turkish bergamot oils produced by cold pressing and steam distillation. *J. Essent. Oil Res.* 13:411–415.

Kiwanuka, P., Mottram, D. S., and Baigrie, B. D. 2000. The effects of processing on the constituents and enantiomeric composition of bergamot essential oil. In *Flavour and fragrance chemistry*, eds. V. Lanzotti and O. Taglialatela-Scafati, 67–75. Netherlands: Kluwer Academic Publishers.

Koketsu, M., Magalhaes, M. T., Wilberg, V. C., and Donaliso, M. G. R. 1983. Oleos essenciais de frutos citricos cultivados no Brazil. *Bol. Pesqui EMBRAPA Cent. Technol. Agric. Aliment.* 7:3–21.

Kubeczka, K. H., and Formàček, V. 2002. *Essential oils analysis by capillary gas chromatography and carbon – 13NMR spectroscopy.* 2nd ed. New York: John Wiley & Sons.

La Face, F. 1972. Il bergamotto e la sua industria. *Dragoco report* 2:23–53.

Lakszner, K., and Szepesy, L. 1988. Application of the O-FID oxygenes analyzer in the cosmetic industry. *Chromatographia* 26:91–96.

Lamonica, G., Stagno d'Alcontres, I., Donato, M. G., and Merenda, I. 1990. On the Calabrian bergamot essential oil. *Chimica Oggi* May:59–63.

Lanuzza, F., Micali, G., Currò, P., and Calabrò, G. 1991. On-line HPLC-HRGC coupling in the study of citrus oils: Sesquiterpene and paraffin hydrocarbons. *Flavour Fragr. J.* 6:29–37.

La Scala, L., and La Scala, G. 1991. Preparazione dei derivati ossimici dei composti carbonilici presenti negli oli essenziali di agrumi e loro registrazione gas-cromatografica. *Essenz. Deriv. Agrum.* 2:113–129.

Lawrence, B. M. 1979, 1982, 1983, 1987, 1988, 1991, 1994, 1999, 2002. Progress in essential oils. *Perfum. Flav.* 4(3):50–52; 7(5):43–48; 8(3):65–74; 12(2):67–72; 13(2):67–78; 16(5):75–82; 19(6):57–62; 24(5):45–63; 27(6):46–64.

Liberti, A., and Conte, G. 1956. Possibilità di applicazione della cromatografia in fase gassosa allo studio delle essenze. *Atti I Congresso Internazionale di Studi e Ricerche sulle Essenze*, Reggio Calabria, Italy, March 1956.

Liberti, A., and Cartoni, G. P. 1958. Analysis of essential oils by gas chromatography. In *Gas chromatography*, ed. D. H. Desty, 321–329. London: Butterworth's Scientific Publications.

Liberti, A., and Goretti, G. 1974. Moderni criteri sulla valutazione degli olii essenziali: l'olio di bergamotto. *Essenz. Deriv. Agrum.* 44:197–208.

Lis-Balchin, M., Deans, S. G., and Eaglesham, E. 1998. Relationship between bioactivity and chemical composition of commercial essential oils. *Flavour Fragr. J.* 13:98–104.

MacLeod, W. D., and Buigues, N. M. 1964. Sesquiterpenes. I. Nootkatone, a new grapefruit flavor costituent. *J. Food Sci.* 29:565–568.

Mazza, G. 1986. Etude sur la composition aromatique de l'huile essentielle de bergamote (*Citrus aurantium* subsp. *Bergamia* Risso et Poiteau Engler) par chromatographie gazeuse et spectrometrie de masse. *J. Chromatogr.* 362:87–99.

Mazza, G. 1987a. Identificazioni di nuovi composti negli oli agrumari mediante GC/MS. *Essenz. Deriv. Agrum.* 57:19–33.

Mazza, G. 1987b. Studio sulle ossidazioni dei monoterpeni negli oli essenziali. *Essenz. Deriv. Agrum.* 57:5–18.

Melliou, E., Michaelakis, A., Koliopoulos, G., Skaltsounis A. L., and Magiatis, P. 2009. High quality bergamot oil from Greece. Chemical analysis using chiral gas chromatography and larvicidal activity against the West Nile virus vector. *Molecules.* 14:839–849.

Menichini, F., Tundis, R., Loizzo, M. R. 2010. In vitro photo-induced cytotoxic activity of *Citrus bergamia* and *C. medica* cv. Diamante peel essential oils and identified active coumarins. *Pharmaceutical Biology* 48:1059–1065.

Micali, G., Lanuzza, F., Currò, P., and Calabrò, G. 1990. Separation of alkanes in citrus essential oils by on-line coupled high performance liquid chromatography–high-resolution gas chromatography. *J. Chromatogr.* 514:317–324.

Mondello, L., Bartle, K. D., Dugo, P., Gans, P., and Dugo, G. 1994. Automated LC-GC: A powerful method for essential oils analysis. Part IV. Coupled LC-GC-MS (ITD) for bergamot oil analysis. *J. Microcol. Sep.* 6:237–244.

Mondello, L., Dugo, P., Bartle, K. D., Dugo, G., and Cotroneo, A. 1995a. Automated HPLC-HRGC: A powerful method for essential oils analysis. Part V. Identification of terpene hydrocarbons of bergamot, lemon, mandarin, sweet orange, bitter orange, grapefruit, clementine and Mexican lime oils by coupled HPLC-HRGC-MS (ITD). *Flavour Fragr. J.* 10:33–42.

Mondello, L., Dugo, P., Basile, A., Dugo, G., and Bartle, K. D. 1995b. Interactive use of linear retention indices, on polar and apolar columns, with a MS-library for reliable identification of complex mixtures. *J. Microcol. Sep.* 7:581–591.

Mondello, L., Dugo, P., Cotroneo, A., Proteggente, A. R., and Dugo, G. 1998. Multidimensional advanced techniques for the analysis of bergamot oil. *EPPOS* 20:3–27.

Mondello, L., Casilli, A., Tranchida, P. Q., et al. 2003. Comparison of fast and conventional GC analysis for citrus essential oils. *J. Agr. Food Chem.* 51:5602–5606.

Mondello, L., Tranchida, P. Q., Cicero, L., Sakkà, A., and Dugo, G. 2004a. Identificazione di componenti di miscele complesse mediante GC/MS e indici di ritenzione lineari. In *Qualità e sicurezza degli alimenti,* 471–475. Milano: Morgan Edizioni Scientifiche.

Mondello, L., Casilli, A., Tranchida, P. Q., et al. 2004b. Fast GC for the analysis of citrus oils. *J. Chromatogr. Sci.* 42:410–416.

Mookherjee, B. D. 1969. Occurrence of bifunctional monoterpene compounds in bergamot oil. 158th American Chemical Society Meeting, Paper N. 37. In Lawrence, B. M. 1979. Progress in essential oils. *Perfum. Flav.* 4(3):50–52.

Naviglio, D., Raia, C., Russo, M., et al. 2003. Estrazione dell'olio essenziale di bergamotto. *Ingredienti Alimentari* 2(October):13–18.

Nichols, M. A. 2008. Personal communication.

Oberhofer, B., Nikiforov, A., Buchbauer, G., Jirovetz, L., and Bicchi, C. 1999. Investigation of the alteration of the composition of various essential oils used in aroma lamp applications. *Flavour Fragr. J.* 14:293–299.

Ohloff, G. 1994. Scent and fragrances. *The fascination of odors and their chemical perspectives.* Translated by W. Piekenhagen and B. M. Lawrence, 129–139. Berlin: Springer-Verlag.

Ohloff, G., Giersch, W., Näf, R., and Delay, F. 1986. The absolute configuration of β-bisabolol. *Helvetica Chimica Acta* 69:698–703.

Onishi, Y., Minh Tu, N. T., Ikemoto, J., Ukeda, H., and Sawamura, M. 2003. Studies on characteristic odor components of bergamot essential oil. *47th TEAC*, Tokyo.

Ortiz, J. M., Kumamoto, J., and Scora, R. W. 1978. Possible relationship among sour oranges by analysis of their essential oils. *Internat. Flav. Food Addit.* 9:224–226.

Poiana, M. 2010. Compositional analysis of citrus oils. Europe. In *Citrus essential oils. Flavor and Fragrance*, ed. M. Sawamura. Hoboken, NJ: John Wiley and Sons.

Poiana, M., Mincione, A., Gionfriddo, F., and Castaldo, D. 2003. Supercritical carbon dioxide separation of bergamot essential oil by a countercurrent process. *Flavour Fragr. J.* 18:429–435.

Ricciardi, A. I. A., Agrelo de Nassif, A. E., Olivetti de Bravi, M. G., Peruchena de Godoy, N. M., and Moll, E. 1982. *SAIPA*, 8–13. In Dresher, R. W., Beñatena, H. N., De Rossin, G. G. T., Ubiergo, G. O., and Retamar, J. A. 1984. Evaluaciòn de le calidad perfumistica del aceite esencial de *Citrus bergamia* Risso (Bergamota) cultivada en la E. E. A. Inta Concordia (Entre Rios). *Essenz. Deriv. Agrum.* 54:192–199.

Russo, M. T., Antonelli, A., and Carnacini, A. 2001. Experiences with solid CO_2 concentration of bergamot cold pressed essential oil (*Citrus bergamia* Risso). *J. Essent. Oil Res.* 13:247–249.

Salvatore, G., and Tateo, F. 1992. Oli essenziali e corrispondenti ricostituiti: correlazioni, realtà strutturali ed applicative. *Industrie delle bevande.* 21(February):5–12.

Sawamura, M., Sun, S. H., Oraki, K., Ishikawa, J., and Ukeda, H. 1999a. Inhibitory effects of citrus essential oils and their components on the formation of N-nitrosodimethylamina. *J. Agric. Food Chem.* 47:4868–4872.

Sawamura, M., Poiana, M., Kawamura, A., et al. 1999b. Volatile components of peel oils of Italian and Japanese lemon and bergamot. *Ital. J. Food Sci.* 11:121–130.

Sawamura, M. 2000. Volatile components of essential oils of the Citrus genus. *Recent Res. Dev. Agric. Food Chem.* 4:131–164.

Sawamura, M., Song, M-S., Choi, H-S., Sagawa, K., and Ukeda, H. 2001. Characteristic aroma components of Tosa-butan (*Citrus grandis* Osbeck forma Tosa) fruit. *Food Sci. Technol. Res.* 7(1):45–59.

Sawamura, M., Onishi, Y., Ikemoto, J., Minh Tu, N. T., and Lan Phi, N. T. 2006. Characteristic odour components of bergamot (*Citrus bergamia* Risso) essential oil. *Flavour Fragr. J.* 21:609–615.

Schenk, H. P., and Lamparsky, D. 1981. Analysis of citrus oils, especially bergamot oil. *Seifen, Öle, Fette, Wachse* 107:363–369.

Schwob, R. 1965. Etude sus l'essence de bergamote d'Afriqua du Nord. *Fruits* 10:263–270.

Sciarrone, D. 2009. Personal communication.

Statti, G. A., Conforti, F., Sacchetti, G., et al. 2004. Chemical and diversity of bergamot (*Citrus bergamia*) in relation to environmental biological factors. *Fitoterapia.* 75:212–216.

Sundt, E., Willhalm, B., and Stoll, M. 1964. Analyse des parties de l'essence de bergamote saponifiée. *Helvetica Chimica Acta* 47:408–413.

Tateo, F., Bononi, M., and Lubian, E. 2000. Enantiomeric distribution and sensory evaluation of linalool and linalyl acetate in flavourings. *Ital. J. Food Sci.* 12:371–375.

Theile, F. C., Dean, D. E., and Suffis R. 1960. The evaluation of bergamot oil. *Dry and cosmetic industries* 86:758–759.

Tranchida, P. Q., Lo Presti, M., Costa, R., et al. 2006. High-throughput analysis of bergamot essential oil by fast solid-phase microextraction-capillary gas chromatography-flame ionization detection. *J. Chromatogr. A.* 1103:162–165.

Veriotti, T., and Sacks, R. 2002. High-speed characterization and analysis of orange oils with tandem-column stop-flow GC and time-of-flight MS. *Anal. Chem.* 74:5635–5640.

Vernin, G., and Vernin, G. 1966. Détection et évaluation de l'anthranilate de méthyle et de ses dérivés méthyles dans differénts échantillons naturels et de syntése par C.C.M. et G.L.C. *France et ses Parfumes* 9:429–448.

Verzera, A., Lamonica, G., Mondello, L., Trozzi, A., and Dugo, G. 1996. The composition of bergamot oil. *Perfum. Flav.* 21:19–34.

Verzera, A., Trozzi, A., Stagno d'Alcontres, I., et al. 1998. The composition of the volatile fraction of Calabrian bergamot essential oil. *EPPOS* 25:17–38.

Verzera, A., La Rosa, G., Zappalà, M., and Cotroneo, A. 2000. Essential oil composition of different cultivar of bergamot grown in Sicily. *Ital. J. Food Sci.* 4(12):493–502.

Verzera, A., Trozzi, A., Gazea, F., Cicciarello, G., and Cotroneo, A. 2003. Effect of rootstock on the composition of bergamot (*Citrus bergamia* Risso et Poiteau) essential oil. *J. Agric. Food Chem.* 51:206–210.

Wei, A., and Shibamoto, T. 2010. Antioxidant/lipoxygenase inhibitory activities and chemical composition of selected essential oils. *J. Agric. Food Chem.* 58:7218–7225.

Wenninger, J. A., Yates, R. L., and Dolinsky, M. 1967. High resolution infrared spectra of some naturally occurring sesquiterpene hydrocarbons. *J. Ass. Off. Agric. Chem.* 50:1313–1335.

Yoshida, J., Ikawa, S., and Yasunaga, K. 1971. On the production of bergamot oil in the district of the Seto Inland Sea. *Japan. J. Trop. Agric.* 14:199–203.

Zani, F., Massimo, G., Benvenuti, S., et al. 1991. Studies on the genotoxic properties of essential oils with *Bacillus subtilis rec.* assay and *Salmonella* microsome reversion assay. *Planta Medica* 57:237–241.

Ziegler, E. 1971. The examination of citrus oils. *Flavour Ind.* (November):647–653.

9 Composition of Leaf Oils

Giovanni Dugo and Ivana Bonaccorsi

CONTENTS

9.1 INTRODUCTION

Petitgrain oils are obtained by steam distillation of the leaves, small branches, and immature fruits of different citrus species. The production of these oils is today quite limited and their physico-chemical characteristics are not sufficiently standardized. The yield and composition of these oils is surely dependent on their botanical origin. It is also affected by the age of the plants, period of harvest, and conditions used for distillation. The lack of attention paid to the selection of raw material used for distillation, often not homogeneous, should also be considered. Due to its organoleptic properties, the most appreciated petitgrain is the one obtained from bitter orange. This is produced in Mediterranean countries, and in larger amounts in Paraguay. Petitgrain oils obtained from lemon, mandarin, and bergamot are produced at small amounts exclusively in Italy, but only occasionally.

Most of the components present in petitgrains are the same found in the peel oils obtained by cold extraction from the fruit. The two oils do have noticeable differences. These are due to the matrix used for the extraction and to the distillation process, which can cause transformation of the naturally occurring components in the leaves and also determines the absence of the nonvolatile components present in the cold-extracted peel oils. Compared to cold-pressed peel oils, petitgrains are characterized by higher amounts of oxygenated compounds, and a generally more complex composition. As mentioned above, the most-appreciated citrus petitgrain is bitter orange; Di Giacomo (1974) asserted, however, that the freshness and fragrance of bergamot petitgrain is probably superior to those of bitter orange. In addition, bergamot petitgrain has the advantage to mitigate the sourness of lemon petitgrain, resulting in an excellent association. The same author (Di Giacomo 1974) states that bergamot petitgrain is produced in Calabria between February and April, with an average yield of 0.35% in February and of 0.23% in April.

Citrus petitgrain are less studied than the corresponding cold-pressed peel oils. Their composition is therefore not well standardized, due to the complexity and the nature of the samples—different for geographic origin, cultivar, extraction technology, age and freshness of the raw material—hitherto studied. Bergamot petitgrain, in particular, is the less studied citrus petitgrain (Peyron 1965; Calvarano 1968; Karawya et al. 1970; Ortiz et al. 1978; Cheng and Lee 1981; Huang et al. 1986, 2000; de Rocca Serra et al. 1998; Adami et al. 2000; Kirbaslar and Kirbaslar 2006; Melliou et al. 2009; Bonaccorsi et al. 2013).

Di Giacomo (1974) also reports values of physico-chemical parameters and the components identified at that time in bergamot petitgrain, obtained by classical methods and by chromatography:

- Monoterpene hydrocarbons: Camphene, δ-3-carene, *p*-cymene, limonene, myrcene, β-ocimene*, α-phellandrene, α- and β-pinene, sabinene, α- and γ-terpinene, terpinolene, α-thujene
- Aldehydes: Decanal, heptanal, octanal, nonanal, citronellal, geranial, neral
- Alcohols: citronellol, geraniol, linalool, nerol, terpinen-4-ol, α-terpineol
- Esters: ethyl isovalerate, octyl acetate, citronellyl acetate, geranyl acetate, linalyl acetate, neryl acetate, terpenyl acetate*
- Acids: acetic acid
- Others: methyl N-methylanthranilate

Among all these components, ethyl isovalerate and acetic acid were never reported in successive studies.

9.2 INDUSTRIAL OILS

Information available in literature on industrially produced bergamot petitgrain is very scant. They refer to an article by Peyron (1965), relative to semiquantitative results, to those relative to oils produced in Calabria, Italy, reported by Calvarano (1968), and those recently obtained by Bonaccorsi et al. (2013).

Peyron (1965) reports in bergamot petitgrain the presence of the following components listed in order of decreasing amount: linalyl acetate, linalool, limonene, α-terpineol, β-pinene, geranyl acetate + neryl acetate, γ-terpinene.

The results obtained by Calvarano (1968), and Bonaccorsi et al. (2013) are reported in Table 9.1. In Figure 9.1 is the chromatogram of a bergamot petitgrain industrially produced in Calabria. It should be noticed the very high content of methyl N-methyl anthranilate (6.80%–7.95%) reported by Calvarano (1968), which was determined by classical methods.

The more recent results (Bonaccorsi et al. 2013), are, relative to the main components, in good agreement with Calvarano (1968), although these were obtained many years before. The only disagreement is the extremely lower amount (0.02%–0.03%) of methyl N-methylanthranilate. The composition of the oils analyzed by Bonaccorsi et al. (2013) and reported in Table 9.1 can therefore be considered representative of the Calabrian industrial bergamot petitgrain. The composition of these oils shows it is possible to exclude contaminations or addition of different citrus petitgrain, such as sweet orange, for the low amount of sabinene; lemon, for the low amounts of limonene and β-pinene; mandarin, for the low amounts of γ-terpinene, and of methyl

---

* Correct isomer not identified

TABLE 9.1
Percentage Composition of Industrial and Laboratory Extracted Bergamot Petitgrain Oil

	Industrial					Laboratory					
	1	2	3	4	5	6	7	8	9	10	11
Hydrocarbons											
Monoterpene											
Camphene	0.01	–	0.01	0.25	–	–	tr	–	tr	tr	–
δ-3-Carene	2.04[a]	–	0.05	–	–	–	0.05	–	tr	0.5	–
p-Cymene	1.02	0.33	0.36	–	0.31	–	tr	–	0.05	–	–
Limonene	2.80	2.46	3.39	10.91	0.63	1.21	1.16	1.0	1.79	1.8	0.65
Myrcene	2.04[a]	0.48	1.15	0.83	1.04	1.63	1.17	2.6	2.10	2.2	1.83
β-Ocimene*	0.18	–	–	3.25	1.33	1.54	0.32	–	–	–	–
(E)-β-Ocimene	–	0.03	0.77	–	–	–	–	2.4	0.92	1.4	2.44
(Z)-β-Ocimene	–	0.03	0.34	–	–	–	–	1.1	0.52	0.5	1.13
α-Phellandrene	0.04	–	tr	–	1.99	–	tr	–	tr	0.3	–
β-Phellandrene	–	–	–	–	–	0.04	–	–	tr	0.1	–
α-Pinene	0.16	0.07	0.19	3.16	–	0.09	0.06	–	0.07	0.1	–
β-Pinene	1.11	0.84	1.71	8.24	0.37	1.00	0.81	0.9	0.54	0.5	0.05
Sabinene	0.12	0.15	0.47	–	0.05	0.15	0.17	–	0.17	0.3	–
α-Terpinene	0.09	–	0.02	–	tr	–	–	–	tr	0.1	–
γ-Terpinene	1.42	0.07	0.22	–	tr	–	0.07	–	tr	0.1	0.88
Terpinolene	0.42	–	0.13	–	0.21	0.31	0.06	0.6	0.17	0.2	0.68
α-Thujene	tr	–	0.03	–	–	–	tr	–	tr	tr	–
Sesquiterpene											
cis-α-Bergamotene	–	0.01	0.01	–	–	–	–	–	0.07	tr	–

continued

TABLE 9.1 (CONTINUED)
Percentage Composition of Industrial and Laboratory Extracted Bergamot Petitgrain Oil

	Industrial				Laboratory						
	1	2	3	4	5	6	7	8	9	10	11
trans-α-Bergamotene	–	0.10	0.08	–	–	–	–	–	–	0.1	–
β-Bisabolene	–	0.17	0.16	–	–	–	0.15	–	–	0.1	–
δ-Cadinene	–	–	0.04	–	–	–	–	–	0.12	tr	–
β-Caryophyllene	–	0.48	0.82	–	–	–	0.44	–	1.35	0.5	0.48
β-Elemene	–	0.02	0.02	–	–	–	–	–	tr	–	–
δ-Elemene	–	0.01	0.01	–	–	–	–	–	–	–	–
(E,E)-α-Farnesene	–	–	0.01	–	–	–	–	–	tr	tr	–
(E)-β-Farnesene	–	0.02	0.01	–	–	–	0.04	–	–	–	–
Germacrene B	–	–	–	–	–	–	–	–	0.18	–	0.05
α-Humulene	–	0.07	0.09	–	–	0.90	0.06	–	0.17	0.2	0.05
Aldehydes											
Aliphatic											
Decanal	0.07	–	–	–	tr	–	0.01	–	tr	–	0.1
Nonanal	0.02	–	tr	–	–	–	–	–	0.03	–	–
Monoterpene											
Citronellal	0.02	–	–	1.33	–	0.02	0.03	–	0.05	tr	–
Geranial	0.14	0.36	0.12	5.33	2.49c	3.15	1.24	–	0.44	2.6	0.73
Neral	0.03	0.14	0.05	1.50	tr	0.07	0.84	–	0.30	0.1	0.15
Ketones											
Monoterpene											
6-Methyl-5-hepten-2-one	–	0.01	0.01	–	–	0.03	0.06	–	tr	tr	–

Alcohols											
Aliphatic											
(Z)-3-Hexenol	–	–	–	–	1.02	–	tr	–	–	–	–
Octanol	–	–	0.07[d]	–	–	–	–	–	tr	–	–
Monoterpene											
Citronellol	0.50	–	0.03	2.17	2.49[c]	3.27	–	–	0.07	–	–
Geraniol	0.09	–	–	0.92	1.76	2.77	22.51	6.8	5.58	2.1	–
Isopulegol	–	–	–	–	–	–	0.03	–	tr	–	–
Linalool	18.82	14.42	13.99	1.33	55.16	22.39	41.24	39.7	29.19	22.4	34.62
Nerol	1.74	0.70	0.51	10.16[b]	4.05	1.21	1.92	2.4	2.23	1.3	1.85
cis-Sabinene hydrate	–	–	0.07[d]	–	–	–	–	–	0.02	tr	–
Terpinen-4-ol	0.21	0.06	0.08	–	1.60	2.53	0.11	–	0.13	1.8	0.07
α-Terpineol	5.69	3.02	2.38	0.67	5.17	4.21	7.89	12.1	9.80	6.2	6.95
Thymol	–	–	–	–	tr	–	–	–	tr	–	–
Sesquiterpene											
(E)-Nerolidol	–	0.17	0.12	–	–	–	0.20	–	0.21	0.1	–
Spathulenol	–	0.24	0.13	–	–	–	–	–	0.25	0.1	–
Esters											
Aliphatic											
Octyl acetate	+	–	0.03	–	–	–	0.02	–	–	–	–
Monoterpene											
Citronellyl acetate	+	0.04	0.04	–	–	0.42	0.14	–	0.08	0.1	–
Geranyl acetate	+	3.14	2.48	1.17	–	0.08	4.32	5.7	5.83	2.5	9.44

continued

TABLE 9.1 (CONTINUED)
Percentage Composition of Industrial and Laboratory Extracted Bergamot Petitgrain Oil

	Industrial				Laboratory						
	1	2	3	4	5	6	7	8	9	10	11
Linalyl acetate	+	65.13	66.27	0.83	22.30	51.64	11.31	19.9	28.21	49.6	29.8
Linalyl propanate	–	–	0.10	–	–	–	–	–	–	–	0.38
Neryl acetate	+	2.19	2.11	–	–	0.27	3.00	3.3	6.87	1.3	4.85
Terpinyl acetate*	+	–	–	–	–	–	tr	–	–	–	–
α-Terpinyl acetate	–	–	0.14	–	–	–	–	–	0.14	0.1	–
Ethers and oxides											
Monoterpene											
1,8-Cineole	–	0.02	–	–	–	0.64	–	–	tr	tr	–
cis-Linalool oxide	–	0.13	–	–	–	–	tr	–	0.05	0.1	–
Sesquiterpene											
Caryophyllene oxide*	–	0.62	0.14	–	–	–	–	–	0.42	0.1	0.06
Others											
methyl N-Methyl anthranilate	6.80–7.95	0.02	0.03	–	–	–	–	–	–	–	–

Note: tr, traces; *, correct isomer not characterized; +, present, not quantified.

a δ-3-Carene + myrcene
b Nerol + unknown
c Geranial + citronellol
d Octanol + cis-sabinene hydrate

Appendix to Table 9.1:

1. Calvarano (1968). Calabria, Italy; six samples; steam distillation; chemical and TLC fractionation; GC/FID on stainless steel capillary column (45 m × 0.5 mm) coated with UCON LB 550X; the original paper reports the composition of the fractions into which the oil (a mixture of the six samples) was separated and the amount of each fraction; the relative percentage has been calculated on the basis of these values. Calvarano also found heptanal (0.01%), octanal (0.01%), and a total content of esters of 48.70%–55.40%. The latter results as well as the percentage of methyl N-methyl anthranilate were obtained by conventional laboratory procedures.

2, 3. Bonaccorsi et al. (2013). Calabria, Italy; two industrial samples; GC/FID and GC/MS on capillary column (25 m × 0.25 mm × 0.25 μm) coated with SE-52; relative percentage of peak areas. Bonaccorsi et al. also found in sample of column 3 4,8-dimethyl-1,3(*E*),7-nonatriene (0.01%), bicyclogermacrene (0.02%). (*Z*)-α-bisabolene (0.02%). (*Z*)-β-farnesene (0.01%), germacrene D (0.02%), β-santalene (0.01%), α-bisabolol (0.03%), campherenol (0.02%), 2,3-dimethyl-3-(4-methyl-3-pentenyl)-2-norbornanol (0.02%), nonyl acetate (0.02%), bornyl acetate (0.01%), geranyl formate (0.05%), neryl formate (0.02%), and *trans*-limonene oxide (0.02%), and trace amounts of tricyclene, trans-β-bergamotene, carvone, *trans*-sesquisabinene hydrate, trans-β-bergamotene, heptyl acetate, and hexyl acetate.

4. Karawya et al. (1970). Giza, Egypt; one laboratory sample; steam distillation; column chromatography; TLC; GC on stainless steel capillary column (90 m × 0.25 mm) coated with Nujol; wt%.

5. Ortiz et al. (1978). USA; one laboratory sample; steam distillation; GC on packed column of LAC 446; relative percentage of peak areas.

6. Cheng and Lee (1981), from Lawrence (1993). Taiwan; one laboratory sample; prefractionation techniques; GC. Cheng and Lee also found borneol (0.02%).

7. Huang et al. (1986). China; one laboratory sample; steam distillation; GC/FID on capillary column (50 m × 0.25 mm) coated with OV-101; GC/MS on capillary column (50 m × 0.25 mm) coated with SE-54; LRI on OV-101 are reported; relative percentage of peak areas. Huang et al. also found pentadecane (0.06%), α-bergamotene* (0.06%), pulegone (0.06%), endo-fenchol (0.01%), sabinene hydrate* (0.07%), farnesol* (0.05%). (*Z*)-nerolidol (0.03%), decyl acetate (0.03%), and trace amounts of hexanol, n-butyl acetate, 1,4-cineole.

8. de Rocca Serra et al. (1998). Corsica, France; one laboratory sample hydrodistilled from the cv. Castagnaro; GC on polar and apolar capillary columns; ^{13}C-NMR; relative percentage of peak areas.

9. Huang et al. (2000). China; one laboratory sample; steam distillation; GC/FID and GC/MS on capillary columns (50 m × 0.25 mm) coated with OV-101 and OV-17; relative percentage of peak areas. Huang et al. also found undecanal (0.02%) and trace amounts of methyl geranate.

10. Kirbaslar and Kirbaslar (2006). Antalya, Turkey; one laboratory sample; steam distillation; GC/FID and GC/MS on capillary columns (60 m × 0.25 mm × 0.25 μm) coated with DB-5 and Carbowax; Wiley and NIST MS libraries; relative percentage of peak areas. Kirbaslar et al. also found *cis-p*-menth-2-en-1-ol (0.1%) and trace amounts of isoterpinolene, *trans-p*-menth-2-en-1-ol, β-terpineol*, α-bisabolol, δ-cadinol, and linalyl butyrate.

11. Melliou et al. (2009). Greece; one laboratory sample hydrodistilled from leaves collected in January 2008; GC/FID and GC/MS on capillary column (30 m × 0.32 mm × 0.25 μm) coated with HP 5MS; Wiley MS library; relative percentage of peak areas. Melliou et al. also found nonadecane (0.09%).

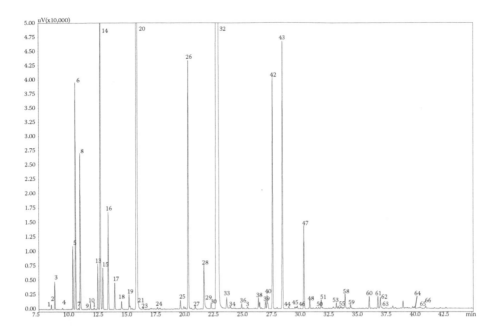

FIGURE 9.1 GC-FID chromatogram of bergamot petitgrain oil. (1) tricyclene; (2) α-thujene; (3) α-pinene; (4) camphene; (5) sabinene; (6) β-pinene; (7) 6-methyl-5-hepten-2-one; (8) myrcene; (9) α-phellandrene; (10) δ-3-carene; (11) hexyl acetate; (12) α-terpinene; (13) *p*-cymene; (14) limonene; (15) (*Z*)-β-ocimene; (16) (*E*)-β-ocimene; (17) γ-terpinene; (18) *cis*-sabinene hydrate + octanol; (19) terpinolene; (20) linalool; (21) nonanal; (22) heptyl acetate; (23) 4,8-dimethyl-1,3(E),7-nonatriene; (24) *trans*-limonene oxide; (25) terpinen-4-ol; (26) α-terpineol; (27) octyl acetate; (28) nerol; (29) citronellol; (30) neral; (31) carvone; (32) linalyl acetate; (33) geranial; (34) neryl formate; (35) bornyl acetate; (36) geranyl formate; (37) nonyl acetate; (38) linalyl propionate; (39) δ-elemene; (40) α-terpinyl acetate; (41) citronellyl acetate; (42) neryl acetate; (43) geranyl acetate; (44) β-elemene; (45) methyl N-methylanthranilate; (46) *cis*-α-bergamotene; (47) β-caryophyllene; (48) *trans*-α-bergamotene; (49) (*Z*)-β-farnesene; (50) (*E*)-β-farnesene; (51) α-humulene; (52) β-santalene; (53) germacrene D; (54) *trans*-β-bergamotene; (55) bicyclogermacrene; (56) (*Z*)-α-bisabolene; (57) (*E*,*E*)-α-farnesene; (58) β-bisabolene; (59) δ-cadinene; (60) (*E*)-nerolidol; (61) spathulenol; (62) caryophyllene-oxide; (63) *trans*-sesquisabinene hydrate; (64) 2,3-dimethyl-3-(4-methyl-3-pentenyl)-2-norbornanol; (65) campherenol; (66) α-bisabolol. (From Bonaccorsi et al., *J. Essent. Oil Res.* in press, 2013.)

N-methylanthranilate (Dugo et al. 1996). However, the composition of the industrial bergamot petitgrain reported in Table 9.1 is highly similar to that of bitter orange petitgrain for the low amount of monoterpene hydrocarbons and the high amount of linalyl acetate and linalool (Mondello et al. 1996).

9.3 LABORATORY OILS

Information on the composition of oils extracted in laboratory from bergamot leaves by distillation is also scant and limited to articles by Karawya et al. (1970); Ortiz et al. (1978); Cheng and Lee (1981); Huang et al. (1986, 2000); de Rocca Serra et al. (1998); Kirbaslar and Kirbaslar (2006); and Melliou et al. (2009).

All the oils obtained by distillation in laboratory show differences if compared to the industrial samples, with a wide range of variability; only two samples show a content of linalyl acetate higher than linalool (Cheng and Lee 1981; Kirbaslar and Kirbaslar 2006). The Egyptian oil analyzed by Karawya presents high amounts of monoterpene hydrocarbons (ca. 27%), of monoterpene aldehydes (ca. 8%), and of monoterpene alcohols different from linalool (ca. 13%), while linalool and linalyl acetate are ca. 1% each. The two Chinese oils analyzed by Huang et al. (1986, 2000) differ from each other mainly in the content of linalool, geraniol, and linalyl acetate. Ortiz et al. (1978) found a high amount α-phellandrene (1.99%), never confirmed in any other bergamot leaf oil, probably due to an incorrect identification. The oil analyzed by Cheng and Lee (1981) is the most similar to the industrial oils analyzed by Bonaccorsi et al. (2013), at least relative to the main compounds; however, this oil presents high amounts of terpinen-4-ol (2.53%) and of 1,8-cineole (0.64%).

Bergamot petitgrain was also studied by infrared (IR) spectroscopy by Di Giacomo and Romeo (1974). These authors, in addition to the signals of alcohols and esters, found in the IR spectra the N-H stretching of methyl N-methylanthranilate and signals relative to the presence of α-terpineol, neral and geranial, limonene, γ-terpinene, ocimene*, mircene, α- and β-pinene, and p-cymene.

REFERENCES

Adami, M., Arceri, G., Di Giacomo, G. et al. 2000. Estrazione di petitgrain con CO_2 supercritica. *Essenz. Deriv. Agrum.* 70:193–200.

Bonaccorsi, I., Trozzi, A., Cotroneo, A., and Dugo, G. 2013. Composition of industrial petitgrain produced in Calabria. *J. Essent. Oil Res.* in press.

Calvarano, I. 1968. Le essenze italiane di petitgrain. Nota II. I petitgrain bigarade e bergamotto. *Essenz. Deriv. Agrum.* 38:31–48.

Cheng, Y. S., and Lee, C. S. 1981. Composition of leaf essential oils from ten *Citrus* species. *Proc. Natl. Sci. Counc.* B. ROC, 5:278–283. Reproduced in Lawrence, B. M. 1993. *Perfum. Flav.* 18(5):43–48.

Di Giacomo, A. 1974. *Gli oli essenziali degli Agrumi.* Milan: EPPOS.

Di Giacomo, A., and Romeo, G. 1974. Esame per spettroscopia IR dei petitgrains prodotti in Italia. *Essenz. Deriv. Agrum.* 44:217–235.

Dugo, G., Mondello, L., Cotroneo, A., et al. 1996. Characterization of Italian citrus petitgrain oils. *Perfum. Flav.* 21(3):17–28.

Huang, Y., Wen, M., Xiao, S., et al. 1986. Studies on the chemical constituents of the steam-distilled leaf and peel essential oil from *Citrus bergamia. Acta Botanica Yunnanica* 8:471–476.

Huang, Y., Pu, Z., and Chen, Q. 2000. The chemical composition of the leaf essential oils from 110 Citrus species, cultivars, hybrids and varieties of Chinese origin. *Perfum. Flav.* 25:53–66.

Karawya, M. S., Balbaa, S. I., and Hifnawy, M. S. 1970. Study of the leaf essential oils of bitter orange and bergamot growing in Egypt. *American Perfumer and Cosmetics* 85(11): 29–32.

Kirbaslar, S. I., and Kirbaslar, F. G. 2006. Composition of Turkish mandarin and bergamot leaf oils. *J. Essent. Oil Res.* 18:318–327.

Melliou, E., Michaelis, A., Koliopoulos, G., Staltsommis, A.-L., and Magiatis, P. 2009. High quality oil from Greece: Chemical analysis using chiral gas chromatography and larvicidal activity against the West Nile virus vector. *Molecules* 14: 839–849.

Mondello, L., Dugo, P., Dugo, G., and Bartle, K. D. 1996. Italian citrus petitgrain oils. Part I. Composition of bitter orange petitgrain oil. *J. Essent. Oil Res.* 8:597–609.

Ortiz, J. M., Kumamoto, J., and Scora, R. W. 1978. Possible relationships among sour oranges by analysis of their essential oils. *Int. Flav. Food Addit.* 5:224–226.

Peyron, L. 1965. Petitgrain oils in perfumery. *Soap Perfum. Cosmet.* 38:769–780.

de Rocca Serra, D., Lota, M-L., Tomi, F., and Casanova, J. 1998. Essential oils and taxonomy among *Citrus*. Example of bergamot. *Riv. Ital. EPPOS* (Numero speciale):38–43.

10 Bergamot Peel and Leaf Extracts by Supercritical Fluids
Technology and Composition

Mariateresa Russo and Rosa Di Sanzo

CONTENTS

10.1 SUPERCRITICAL FLUID TECHNOLOGY

The term "supercritical fluid" (SCF) describes a gas or liquid at conditions above its critical temperature and pressure—in other words, above the critical point. Two researchers, Hannay and Hogarth, at a meeting of the Royal Society (London) in 1879, reported for the first time the supercritical fluids' property of pressure-dependent dissolving power: the higher the pressure, the higher their dissolving power. They described and summarized their findings as follows: "We have the phenomenon of a solid dissolving in a gas, and when the solid is precipitated by reducing the pressure, it is brought down as a 'snow' in the gas" (Hannay and Hogarth 1879). Those assertions gave rise to a series of disputes in subsequent society meetings.

In 1906 Eduard Buchner (Nobel Prize in Chemistry 1907) became the first in a long line of researchers to measure the solubility in supercritical carbon dioxide of a model compound, naphthalene (Buchner 1906). Starting in the 1960s, different research groups over the world, primarily in Europe, in particular in Germany, and later in the United States examined SCFs for developing extraction processes. They and Indian researchers emphasized SC-CO$_2$ extraction for decaffeinating coffee and tea and for flavors and essential oils extraction from spices and herbs. Meanwhile, in Asia, researchers developed SC-CO_2 extraction mainly on phytopharmaceuticals (Bott 1980; Stahl and Wilke 1980; Lack and Seidlitz 1996; Subramaniam et al. 1997; Tomasula 2003).

The special characteristics of supercritical fluids represented, and still represent, a very attractive alternative to the conventional extraction and refining techniques. In fact, techniques such as extraction with organic solvents and/or vacuum and steam distillations can damage some valuable substances, mainly for reasons related to the high temperatures involved, or lead to solvent toxic residues in the extracts.

The intense interest in SCF technology resulted in a number of reviews on SCFs' properties, theories on the SCF state, and current applications of SCFs in pharmaceutical, chemical, and food industries (Stahl and Wilke 1980; Brunner 1988, 1994; Phelps et al. 1996; Mukhopadhyay 2000; Marr and Gamse 2000; Perrut 2000; Ashihara and Crozier 2001; Chang et al. 2001; Berna et al. 2001; Bertucco and Vetter 2001; Lang and Wai 2001; Arai et al. 2002; Raventós et al. 2002; King 2002; Chafer et al. 2002; Da Cruz Francisco and Szwajcer 2003; Gaspar et al. 2003; Zougagh et al. 2004; Brunner 2005; Gamse 2005; Velasco et al. 2007; Li et al. 2007; Martínez 2008; Angela and Meireles 2008; Xu et al. 2011; Fornari et al. 2012; Schaber et al. 2012).

In the phase diagram of CO_2 can be distinguished three different curves that separate three different areas and which intersect at a common point called the triple point or critical point (Figure 10.1). This is where all the three states of aggregation (solid, liquid, and gaseous) are in equilibrium (Mukhopadhyay 2000; Martínez 2008). The temperature at this point is called the "critical temperature" (T_C) and is the maximum temperature of liquefaction of the vapor at a specific value of pressure,

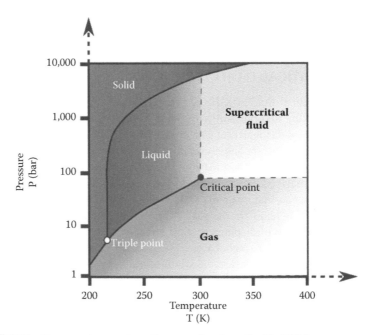

FIGURE 10.1   Pressure-temperature diagram of carbon dioxide (CO_2).

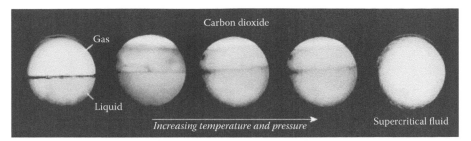

FIGURE 10.2 Frames which illustrate the sequence of the transition to carbon dioxide superfluid (NASA; http://science.nasa.gov/science-news/science-at-nasa/2003/20aug_supercriticalco2).

critical pressure (P_C). At critical temperature and pressure, liquid and vapor form a homogeneous phase in which they have the same density and are indistinguishable (Figure 10.2). The substance at this stage is defined as a fluid at supercritical state (SCF). The critical temperature (T_C) and pressure (P_C) at which this happens are unique to each pure substance. In the supercritical state, the "fluid" takes on many of the properties of both gas and liquid.

The most important properties of a SCF are density, viscosity, diffusivity, heat capacity, and thermal conductivity. Manipulating the temperature and pressure above the critical points affects these properties and enhances the ability of the SCF to penetrate and extract targeted molecules from raw materials. As a consequence, the rates of extraction and phase separation can be significantly faster than conventional extraction processes (Arai et al. 2002; McKenzie et al. 2004). A comparison of typical values for density, viscosity and diffusivity of gases, liquids, and SCFs is presented in Table 10.1.

The range of densities of the most common supercritical fluids is from 0.1–0.9 g/mL under normal working pressures (75–450 bar). Small changes in pressure or temperature near the critical point can greatly modify the density and, hence, the solubilizing power of the supercritical fluid. Close to the critical point, slight changes in the operational conditions may cause drastic variations in its density, affecting consequently the solubility of the solute in the supercritical phase (Figure 10.3). Thus the physicochemical properties of supercritical fluids can be varied significantly without changing the molecular structure of the substance. This ability to change the

TABLE 10.1
Comparison of Physical and Transport Properties of Gases, Liquids, and SCFs

	Density (kg/m^3)	Viscosity (cP)	Diffusivity (mm^2/s)
Gas	1	0.01	1–10
SCF	100–800	0.05–0.1	0.01–0.1
Liquid	1000	0.5–1.0	0.001

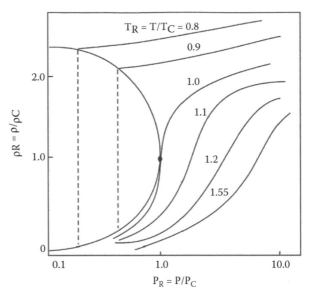

FIGURE 10.3 Variation of the reduced density of the pure compounds in the vicinity of their critical points. ρR = relative critical density; T_R = relative critical temperature; P_C = critical pressure; T_C = critical temperature; ρC = critical density.

properties of the supercritical fluids by changing the temperature and pressure in the vicinity of the critical point provides the equivalent of a series of different solvents, thus providing selective extraction properties (Brenneck and Eckert 1989; Henry and Yonker 2006).

Many solvents may be used for supercritical extraction; particularly interesting are the substances with low molecular weight having a critical temperature close to ambient temperature ($T_C \sim 10°C{-}40°C$) and a critical pressure that is not too high ($P_C \sim 40{-}60$ bar) (Table 10.2). For this reason and others outlined below, carbon dioxide (CO_2) is the main SCF solvent used in supercritical fluid extraction techniques (SC-CO_2), especially when the target molecules are apolar.

The basic properties and advantages of CO_2 in supercritical extraction include that it is inert, relatively nontoxic, easily available, odorless, tasteless, and environment friendly; thus it does not contaminate products and environment and does not damage the ozone layer. It is a "generally recognized as safe" (GRAS) solvent and widely available at a relatively low cost and high purity (Da Cruz and Szwajcer Dey 2003; Diaz-Reinoso et al. 2006; Herrero et al. 2006), with convenient critical parameters ($T_C = 31°C$, $P_C = 7.38$ MPa/72.1 bar). The working temperature is close to the ambient temperature, particularly suitable for thermolabile material. Above all, CO_2 is easy to remove from the extract because at atmospheric pressure and ambient temperature, it is gaseous. CO_2 is not flammable or explosive, avoiding the danger caused by common extraction with organic solvents. Its recovery is simple and can save energy resources, combining the extraction and removal in a unique technique, greatly shortening the processing.

TABLE 10.2
Critical Properties of Some Pure Solvents

Solvent	Molecular Weight (g/mol)	Critical Temperature (T_C) (°C)	Critical Pressure (P_C) MPa (atm)	Density (g/cm³)
Methane (CH₄)	16.04	−82.8	4.60 (45.4)	0.162
Ethylene (C₂H₄)	28.05	9.4	5.04 (49.7)	0.215
Carbon dioxide (CO₂)	44.01	31.1	7.38 (72.8)	0.469
Ethane (C₂H₆)	30.07	32.3	4.87 (48.1)	0.203
Propylene (C₃H₆)	42.08	91.75	4.60 (45.4)	0.232
Propane (C₃H₈)	44.09	96.8	4.25 (41.9)	0.217
n-Hexane (C₆H₁₄)	86.18	234.5	3.01 (29.7)	0.655
Acetone (C₃H₆O)	58.08	235.1	4.70 (46.4)	0.278
Methanol (CH₃OH)	32.04	239.6	8.09 (79.8)	0.272
Ethanol (C₂H₅OH)	46.07	240.9	6.14 (60.6)	0.276
Ethyl acetate (C₄H₈O₂)	88.11	250.2	3.83 (37.8)	0.901
Water (H₂O)	18.02	374.1	22.12 (218.3)	0.348
Trifluoromethane (Fluoroform) (CHF₃)	70.01	26.14	48.6 (479.64)	0.457
Chlorotrifluoromethane (CClF₃)	104.47	28.84	38.7 (381.94)	0.694
Trichlorofluoromethane (CCl₃F)	137.7	198.04	44.1 (435.92)	0.504
Toluene (C₇H₈)	92.14	318.64	41.0 (404.64)	0.87

Source: Reid, R. C., Prausnitz, J. M., and Poling, B. E., *The properties of liquids and gases*, 4th ed. New York: McGraw-Hill, 1998.

Despite these advantages, SC-CO₂ has an important limitation with respect to organic solvents—a lower capacity for solubilizing polar and high molecular weight compounds. Thus, in order to increase the efficiency of the process through the enhancement of the solvent's polarity and the solubility of such compounds in supercritical carbon dioxide, small percentages of polar or nonpolar co-solvents known as modifiers or entrainers may be added (Diaz-Reinoso et al. 2006). Consider that the polarity of supercritical fluids can be significantly increased without affecting the supercritical condition by adding 1%–10% volume of a polar substance (Hubert and Vitzhum 1978; Hawthorne 1990; Clifford et al. 1999; Menaker et al. 2004; Wang and Weller 2006). Page et al. (1992) drew up a comprehensive review of commonly used modifiers. Ethanol is the modifier of choice in the food industry because of its nontoxic and GRAS status. Additional variables to be taken into account in SC-CO₂ are solvent to feed ratio, particle size, modifier concentration, extraction temperature, pressure, and time and flow rate (Reverchon and De Marco 2006). In general, the SC-CO₂ extraction process consists of two steps: the extraction itself and the separation between extracted components and the solvent. Simplified flow sheets can be observed in Figures 10.4 and 10.5.

The subsequent *escursus* of the scientific literature data will focus on supercritical fluid technology for extraction of citrus peels and leaves and for fraction action of citrus oils, in particular bergamot.

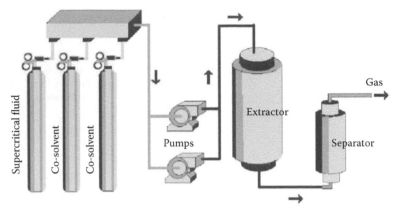

FIGURE 10.4   A schematic diagram of a supercritical fluid batch extraction.

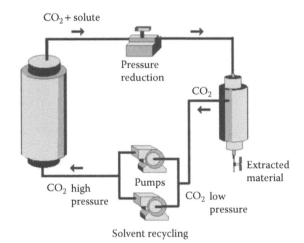

FIGURE 10.5   A schematic diagram of a supercritical fluid continuous extraction.

10.2   SUPERCRITICAL CO_2 EXTRACTS FROM CITRUS FRUIT PEELS

Scientific interest in the application of supercritical CO_2 extraction of citrus peel essential oils developed in the 1980s, and most studied were the citrus species traditionally used to extract the oils more largely used (lemon and sweet orange) as flavoring agents in beverages, pharmaceuticals, and cosmetics (Sargenti and Lancas 1998; Dos Santos and Atti-Serafini 2000; Roy Bhupesh et al. 2007; Nautiyal and Tiwari 2011). Supercritical technology, compared to other extraction technologies, represents a possible alternative to the traditional cold-pressing process commonly applied to extract citrus peel oils. Therefore, the possibility of directly extracting the citrus fruit peels was explored, testing all the operating parameters that could influence and/or make the process efficient. Several studies were conducted over the years to take advantage of the solvent power of supercritical CO_2 (Atti dos-Santos and Atti-Serafini 2000; Roy Bhupesh et al. 2005, 2007). The easy management of these processes allowed knowledge to be acquired at a relatively low cost (Schultz et al. 1974; Ely and Baker 1983; Paulaitis et al. 1983; Stahl et al. 1984).

A clear synthesis of the parameters that influence the SC-CO$_2$ of solid matrices was reported by Martínez (2008). In summary, solubility of compounds increases when the extraction pressure at constant temperature is increased (Bartle et al. 1991; Martínez 2008) and, at a pressure close to the critical pressure, by decreasing the temperature; the separation conditions depend on the solubility of the compounds at different pressures and temperatures.

The solvent to feed ratio is a very important parameter that depends on many factors such as concentration of the solute in the matrix, the solute solubility in the supercritical solvent, distribution of solute in the material, and type of material. Low solvent to feed ratios imply lower operating costs and higher production capacity. Generally, the industrial processes target solvent to feed ratios lower than 30. However, higher solvent to feed ratios are justified for high added-value products. The size and morphology of the solid material have a direct effect on the mass transfer rate. As a rule, increasing the surface area increases the extraction rate. Therefore, smaller particle size or geometry favors higher mass transfer, decreasing the batch time. Particle size needs to be evaluated case by case based on the type of material to be processed. Moisture content must be also evaluated case by case; high content of moisture is usually not desirable because it acts as a mass transfer barrier. On the other hand, moisture expands the cell structure, facilitating the mass transfer of the solvent and the solute through the solid matrix (Martínez 2008; Huang et al. 2012; Medina 2012; Sovová 2012). The first studies on citrus peels involved tests on the different solubilities in supercritical CO$_2$ of the components of the essential oils, and the direct extraction from the different citrus peels containing essential oils with different chemical compositions.

10.3 SUPERCRITICAL CO$_2$ EXTRACTS FROM BERGAMOT FRUIT PEELS

In light of the first scientific evidence that supercritical CO$_2$ might be used as a highly selective solvent without the disadvantages of commonly used processes and the perfumery industry's need for high-quality products, the Consortium of the Bergamot of Reggio Calabria (Italy) foresaw in this innovative technology a possible way to enhance bergamot essential oil. In the early 1990s the Consortium bought a supercritical CO$_2$ extraction semipilot plant (Muller Company GMBM Extract, Coburg, Germany). The plant was installed on site at the Faculty of Agriculture of University of Reggio Calabria (Italy). This insight gave rise to studies, some of which developed in partnership between the Experimental Station for Essential Oil and Citrus Products, a special Agency of the Chamber of Commerce in Reggio Calabria, the Universities of Reggio Calabria and Messina, and other universities and research centers, both Italian and outside of Italy.

The first study on the extraction of essential oil from bergamot fruit peels dates back to 1992 and was developed with the Consortium plant (Mincione et al. 1992). This study described the process conditions, and the yields and composition of the extracts from bergamot peels of cultivar Fantastico (see Table 10.3). The process conditions were based on the results of previous studies on the extraction of

TABLE 10.3
Percentage Composition (GC analysis) of SC-CO_2 Extracts from Fresh Bergamot Peels

	Mincione et al. (1992) 300 bar, 25°C		Poiana et al. (1994)					Poiana et al. (1999)					
Pressure / Temperature	6 h	4 h	8 MPa 45°C–50°C	10 MPa 45°C–50°C	15 MPa 45°C–50°C	25 MPa 45°C–50°C	25 MPa 37°C	80 bar 40°C	80 bar 40°C	90 bar 50°C	90 bar 50°C	100 bar 60°C	100 bar 60°C
	Area (%)	Area (%)	Area (%)	Area (%)	Area (%)	Area (%)	Area (%)	Area (%)	Area (%)	Area (%)	Area (%)	Area (%)	Area (%)
CO_2 consumed (kg)								0.37	4.26	0.25	4.37	0.29	3.82
α-Thujene	0.227	0.167	0.30	0.30	0.40	0.50	0.20	0.20	0.20	0.40	0.30	0.40	0.30
α-Pinene	0.882	0.636	1.00	1.20	1.50	1.70	1.40	0.90	0.90	1.30	1.00	1.60	1.20
Camphene	0.029	0.02						tr	tr	0.10	tr	0.10	0.10
Sabinene	1.171	0.849						1.20	0.90	1.10	0.90	1.30	1.00
Pinene	6.243	4.539						7.20	6.00	7.20	5.90	7.50	6.90
Sabinene + β-Pinene			8.80	9.00	10.30	10.80	8.80						
β-Myrcene	0.841	0.784	0.70	0.70	0.90	0.90	0.70	1.00	0.80	1.20	0.90	1.20	1.00
Octanal + α-Phellandrene	0.053	0.047						0.10	0.10	0.10	0.10	0.10	0.10
α-Terpinene	0.094	0.071						0.20	0.20	0.20	0.20	0.30	0.20
p-Cymene	0.464	0.327						1.20	1.10	1.10	0.30	0.50	0.40
Limonene	33.665	32.163	40.30	41.00	41.10	42.00	42.20	45.40	40.00	47.90	40.90	45.80	41.00
Z-(β)-Ocimene	0.028	0.03						tr	tr	tr	tr	tr	0.10
E-(β)-Ocimene	0.307	0.28						0.30	0.20	0.30	0.30	0.30	0.30
γ-Terpinene	7.423	6.271	8.30	8.40	8.50	8.10	9.20	9.60	8.90	8.80	8.30	9.20	9.30
trans-Sabinene hydrate	0.061	0.053						0.10	0.10	tr	0.10	tr	0.10
Octanol	0.005	0.189											
Terpinolene	0.293	0.259						0.40	0.40	0.40	0.40	0.50	0.50

Extraction Time: Mincione et al. (1992) columns are 6 h and 4 h.

	6.291	9.418	11.10	10.10	6.40	6.10	6.40	5.20	7.80	4.60	7.80	6.60	8.10
Linalool	6.291	9.418						tr	tr	tr	tr	tr	0.10
Nonanal	0.027	0.028						0.10	0.10	tr	0.10	0.20	0.30
Citronellal	0.017	0.019						0.10	0.10	0.10	0.10	0.10	0.20
Terpinen-4-ol	0.043	0.046						0.10	0.10	tr	0.10	0.10	0.10
α-Terpineol	0.048	0.051						0.10	0.10	0.10	0.10	0.10	0.10
Decanal	0.072	0.072											
Octyl acetate	0.19	0.161											
Nerol	0.049	0.123											
Neral	0.27	0.283						0.10	0.20	0.10	0.30	0.20	0.30
Linalyl acetate	36.027	30.286	26.40	26.60	26.80	26.50	27.40	24.10	28.80	22.90	29.00	21.60	25.60
Geranial	0.436	0.46						0.20	0.30	0.20	0.40	0.20	0.40
Bornyl acetate	0.033	0.024						0.10	0.10	tr	tr	tr	0.10
Undecanal	0.01	0											
Nonyl acetate	0.024	0.019						tr	tr	tr	tr	tr	tr
Linalyl propionate	0.03	—						0.10	0.10	0.10	0.10	0.10	0.10
α-Terpinyl acetate	0.344	0.345						0.20	0.30	0.30	0.30	0.30	0.30
Citronellyl acetate	0.053	0.133						0.20	0.30	0.20	0.30	0.20	0.30
Neryl acetate	0.678	0.859	0.30	0.40	0.60	0.40	0.20	0.30	0.30	0.30	0.30	tr	0.40
Geranyl acetate	0.739	0.678	0.80	0.50	0.80	0.80	0.70	tr	0.50	0.30	0.40	0.20	0.10
Dodecanal + Decyl acetate								tr	0.10	0.10	0.10	tr	tr
Unknown sesquiterpene													
β-Caryophyllene	0.633	0.598	0.20	0.20	0.30	0.30	0.40	0.30	0.30	0.30	0.30	0.30	0.30
Trans-α-Bergamotene	0.416	0.387	0.20	0.20	0.30	0.40	0.50	0.40	0.40	0.20	0.30	0.30	0.30
α-Humulene	0.046	0.041						0.10	0.10	0.10	0.10	0.10	0.10
Germacrene D								0.10	0.10	0.10	0.10	0.10	0.10
Unknown sesquiterpene	0.079	0.078						tr	tr	0.10	0.10	0.10	0.10

continued

TABLE 10.3 (CONTINUED)
Percentage Composition (GC analysis) of SC-CO_2 Extracts from Fresh Bergamot Peels

	Mincione et al. (1992)		Poiana et al. (1994)				Poiana et al. (1999)					
	Area (%)	Area (%)	Area (%)	Area (%)	Area (%)	Area (%)	Area (%)	Area (%)	Area (%)	Area (%)	Area (%)	Area (%)
Unknown sesquiterpene	0.086	0.02										
β-Bisabolene	0.60		0.30	0.70	0.30	0.60	0.50	0.60	tr	0.30	0.40	0.40
α-Bisabolol	0.592	0.548										
(E)-Nerolidol							tr	tr	tr	tr	tr	tr
Tetradecanal							tr	tr	tr	0.10	tr	tr
Nootkatone							0.10	0.20	tr	0.10	0.10	0.10
Linalool/linalyl acetate	0.592	0.548					0.214	0.269	0.201	0.268	0.304	0.317

Notes: tr, traces.

the essential oil from citrus peels tracing back to Sugiyama and Saito (1982), who improved the operating conditions for lemon peel extraction to obtain a product with a high content of oxygenated substances and low content of limonene; and the results obtained by Calame and Steiner (1982), who studied the extraction from lemon peels with SC-CO_2 at 300 bar and 40°C. Mincione et al. (1992) obtained the yield of 0.9%, similar to that obtained by mechanical cold extraction. The composition of SC-CO_2 extract was also similar to cold-pressed oil.

Additional points of reference were the studies of Robey and Sunder (1984), which deepened the study about the use of supercritical CO_2 at low temperature to increase the yield of extracts with a high content of oxygenated compounds. In 1994 Poiana et al. carried out trials with the aim to develop specific process conditions to extract, with SC-CO_2, deterpenated and psoralens-free essential oils directly from bergamot peels (Table 10.3). The tests were performed on fresh crushed peels extracted at different pressure and temperature conditions. Under the conditions of 8 MPa and 45°C–50°C (density 0.22–0.25g/cm^3) was obtained an extract with lower content of bergapten, about 1/6 of the content in the cold-pressed oil. The change in the volatile fraction composition was not significant compared to a cold-pressed essential oil, while the percentage yield varied according to the density of the supercritical CO_2.

In 1999 Mira and co-workers tested the extraction by SC-CO_2 from cut orange dried peels in different conditions of temperature and pressure. Particular attention was paid to the best conditions for the extraction of two compounds: limonene and linalool. The work evaluated the effect of SC-CO_2 flow and the influence of the size of the peel cuts on the speed of extraction.

In the same period Poiana et al. (1997 and 1999) conducted SC-CO_2 extraction from minced bergamot peels, both fresh and dried, in three different extraction conditions of pressure and temperature (see Tables 10.3 and 10.4). In addition, to characterize the volatile fraction of the extracts, the authors focused their attention on the nonvolatile fraction, reaching two conclusions: the dried peels draw less bergapten than fresh ones, and the extraction reaches 90% when the weight ratio peels/CO_2 is about 1:10.

During the same years a large number of studies were carried out on the behavior of individual components of citrus essential oils and, for its peculiarity, of some major components of bergamot essential oil. These were focused on the determination of the effectiveness of the process parameters and the cost/benefit ratio attained using SC-CO_2 extraction. Berna et al. (2000) studied the influence of the height of the bed of particles on the kinetics of the SC-CO_2 by using dried orange peel. In this regard, the authors concluded that the flow was homogeneous and that the height of the bed was not significant. Franceschi et al. (2004) studied the behavior of binary systems of CO_2/lemon and CO_2/bergamot essential oils, highlighting the relative balances transition (vapor-liquid, liquid-liquid, vapor-liquid-liquid). Based on their observations, a thermodynamic model was produced using the equation of state of Peng-Robinson (1976) that was in good agreement with the experimental data. However, the studies demonstrated that SF-CO_2 extraction from the citrus fruit peels was not a competitive process compared to the traditional cold-pressing process. Conversely, SF-CO_2 extraction will represent an interesting alternative for the production of fractionated/terpeneless/psolarene-free extracts to be used for obtaining semifinished high added-value products as an alternative to distillation and solvent extraction that is traditionally employed.

TABLE 10.4

Percentage Composition (GC Analysis) of SC-CO_2 Extracts from Bergamot Dried Peels

Pressure	80 bar		90 bar		100 bar	
Temperature	40°C		50°C		60°C	
CO_2 consumed (kg)	0.43	4.31	0.36	5.01	0.35	5.43
	Area (%)	Area (%)	Area (%)	Area (%)	Area (%)	Area (%)
α-Thujene	0.7	0.3	0.7	0.3	0.6	0.3
α-Pinene	2.7	1.2	2.8	1.3	2.2	1.2
Camphene	0.1	tr	0.1	tr	0.1	tr
Sabinene	1.7	1	1.1	0.9	1.6	1.1
β-Pinene	12.3	7.6	12.2	7.4	10.5	7.1
Myrcene	1.5	1	1.4	1	1.4	1
Octanal + α-Phellandrene	0.1	0.1	0.1	0.1	0.1	0.1
α-Terpinene	0.3	0.2	0.3	0.2	0.2	0.2
p-Cymene	0.2	0.1	0.5	0.2	0.2	0.1
Limonene	52.7	40.2	50.8	39.7	51.1	41.2
Z-(β)-Ocimene	tr	tr	tr	0.1	tr	tr
E-(β)-Ocimene	0.3	0.3	0.3	0.3	0.3	0.3
γ-Terpinene	10.4	8.6	9.3	8.5	9.7	8.5
trans-Sabinene hydrate	tr	tr	tr	tr	tr	tr
Terpinolene	0.4	0.4	0.4	0.4	0.4	0.4
Linalool	2.9	4	3.6	5	3.6	4.6
Citronellal	tr	tr	tr	tr	tr	tr
Terpinen-4-ol	tr	tr	tr	tr	tr	tr
α-Terpineol	tr	0.1	tr	tr	tr	tr
Decanal	tr	0.1	tr	0.1	tr	0.1
Octyl acetate	0.1	0.1	tr	0.1	0.1	0.1
Neral	0.1	0.2	0.1	0.1	0.1	0.2
Linalyl acetate	13	31.2	15.7	31.9	17	30.4
Geranial	0.1	0.2	0.1	0.2	0.1	0.2
Bornyl acetate	tr	tr	tr	tr	tr	tr
Nonyl acetate	tr	tr	tr	tr	tr	0.1
Linalyl propionate	tr	tr	tr	0.1	tr	0.1
α-Terpinyl acetate	0.1	0.3	0.1	0.2	0.1	0.2
Neryl acetate	0.1	0.3	0.1	0.3	0.1	0.3
Geranyl acetate	0.1	0.5	0.1	0.3	0.1	0.4
Dodecanal + decyl acetate	tr	0.1	tr	tr	tr	tr
Unknown sesquiterpene	tr	tr	tr	tr	tr	tr
β-Caryophyllene	0.1	0.4	0.1	0.3	0.1	0.3
trans-α-Bergamotene	0.1	0.4	0.1	0.3	0.1	0.3
α-Humulene	tr	0.1	0.1	0.1	tr	0.1
Germacrene D	tr	0.1	tr	tr	tr	0.1
Unknown sesquiterpene	0.1	0.1	0.1	0.1	0.1	0.1

TABLE 10.4 (CONTINUED)

Percentage Composition (GC Analysis) of SC-CO_2 Extracts from Bergamot Dried Peels

	Area (%)	Area (%)	Area (%)	Area (%)	Area (%)	Area (%)
β-Bisabolene	0.1	0.6	0.1	0.3	0.1	0.4
(E)-Nerolidol	tr	0.1	tr	tr	tr	tr
Tetradecanal	tr	tr	tr	0.1	tr	tr
Nootkatone	tr	0.1	tr	0.1	tr	0.1
Linalool/Linalyl acetate	0.22	0.13	0.229	0.155	0.208	0.153

Source: Poiana et al., *Flavour Fragr. J.* 14, 358–366, 1999.
Notes: tr, traces.

10.4 SUPERCRITICAL CO_2 EXTRACTS FROM BERGAMOT LEAVES

Petitgrain oils are typically obtained by steam distillation of leaves, twigs, and little unripe fruits of some citrus species. The average percentage of oil in different citrus species materials is about 0.3%. Petitgrain oils are composed of mono and sesquiterpene hydrocarbons and their oxygenated derivatives (alcohols, aldehydes, ketones, esters, and oxides). Generally, the amount of oxygenated compounds in petitgrain is higher than that in the corresponding citrus peel oil. It should be noted that there is a lack of scientific literature about petitgrain oils, both in relation to their chemical composition and to the extraction technologies, even though they are important in the perfume and cosmetic industries. Bergamot petitgrain oil and supercritical fluid extraction from petitgrain have not been widely researched. The extraction technologies applied for petitgrain oils were revised by Peyron and Bonaccorsi (2002) and their composition by Dugo et al. (1996, 2002). Bergamot petitgrain composition is discussed in Chapter 9, this volume.

Extraction with CO_2 in the supercritical state could be an alternative to the traditional steam distillation process used to obtain petitgrain from fresh material. The chemical composition of petitgrain oils (Dugo et al. 2002, 2012; Table 9.1 of Chapter 9, this volume) is mainly represented by oxygenated compounds of low and medium molecular weight that are extremely soluble in supercritical CO_2. The solubility of these oxygenated compounds, as part of the same family, generally decreases with increasing molecular weight; chlorophyll, carotenoids, sugars, amino acids, and most of the inorganic salts are insoluble in supercritical CO_2. It is thus possible to attain high extraction selectivity for terpene hydrocarbons, which don't give any contribution to organoleptic characteristics and whose presence in the extract may be detrimental.

The composition of leaf extracts obtained by SF-CO_2 is quite different from that obtained by distillation. The conditions applied during distillation, in fact, can cause transformations of the components naturally present in the vegetable material, while the conditions applied during SC-CO_2 extraction warrant the preservation of the natural characteristics of the extracts.

The first and the only application of SC-CO_2 extraction to bergamot leaves was carried out by Adami and co-workers (2000). They described the SC-CO_2 process conditions to extract also leaves of bitter orange and mandarin. The authors high-lighted the great difference in quality between petitgrain oils extracted by distillation and the product obtained by SC-CO$_2$ extraction and pointed out the great similarity of the latter to its natural composition even if the yield of extraction with CO$_2$ was about 75% of that obtained by hydrodistillation (Table 10.5). Six years later Ferri

TABLE 10.5
Percentage Composition (GC Analysis) of SC-CO$_2$- Bergamot Leaves Extract

Extraction Parameters

Extractor:

Pressure	105 bar
Temperature	40°C

Steps:

1st separator	60 bar/−10°C
2nd separator	50 bar/+28°C
3rd separator	50 bar/+28°C

	Area (%)
α-Pinene	0.00
Sabinene	0.006
β-Pinene	0.23
Myrcene	0.09
Limonene	0.18
Z-(β)-Ocimene	0.04
E-(β)-Ocimene	0.08
Linalool oxide	0.00
Terpinolene	0.00
Linalool	1.09
α-Terpineol	0.00
Nerol	0.04
Linalyl acetate	85.0
Hydroxy linalool	0.19
Neryl acetate	0.94
Geranyl acetate	0.20
β-Elemene	0.06
β-Caryophyllene	4.93
cis-Bergamotene	0.99
Sesquiphellandrene	0.29
α-Humulene	0.39
Nerolidol	0.20
Germacrene B	1.85
δ-Elemene	1.64
Isocaryophyllene	0.08
β-Bisabolene	0.52
Spatulenol	0.39
Caryophyllene oxide	0.08
Unknown	0.44

Source: Adami et al., *Essenz. Deriv. Agrum.* 70, 193–200, 2000.

and Franceschini (2006) determined experimentally the best operating conditions of SC-CO$_2$ extraction to obtain an extract reach of oxygenated compounds. In particular, they studied the optimal condition of temperature and pressure able to enhance the extraction of oxygenated compounds. They also evaluated the quantitative yield of the extract as a function of the total quantity of CO$_2$ used relative to the amount of matrix processed. The trials were carried out on mandarin and bitter orange leaves. Extraction process parameters such as pressure, temperature, and CO$_2$ inlet flow rate were optimized in order to promote enrichment in oxygen compounds, which are largely responsible for its fragrance. The experimental trials were carried out in two subsequent phases. In the first phase a desktop instrument was used to set up optimal conditions of extraction, necessary for the scale-up on a pilot plant by which it was also possible to purify the extract through removing the cuticular waxes. The extraction obtained at 120 bar and 40°C (density of 0.72 g CO$_2$/mL) gave the best results in terms of both yield and olfactory quality. To our knowledge no further scientific work about citrus leaves extraction by supercritical fluid has been published to date.

10.5 CONVENTIONAL FRACTIONATION PROCEDURES FOR CITRUS OILS

There is a huge interest in the chemical composition and properties of citrus essential oils because of the role they play in food and nonfood industries. Citrus peel oils are added as flavoring agents to pharmaceuticals and drugs as well as to herbal medicines in order to mask their unpleasant tastes (Lota et al. 2002). As a result of their freshness, lightness, and aroma, there are increasingly widespread applications of citrus peel oils in the food and beverage industries, in particular soft drinks, confectionery, bakery, and household extract, as well as in some nonfood applications (Swisher and Swisher 1977; Mira et al. 1999). They played an important role in the perfumery and cosmetic industries and in the manufacturing of soap and paper soap (Swisher and Swisher 1977). The traditional process for recovering essential oils from citrus fruit peels is the cold-pressing mechanical process. Vacuum, steam, or hydrodistillation and solvent extraction of dried or preprocessed fruit peels are not commonly used to extract some species of citrus or citrus by-products. The extraction technologies for citrus essential oil production were revised by Di Giacomo (2002), Di Giacomo and Di Giacomo (2002), and Haro Guzman (2002), and those specifically applied to obtain bergamot oil are treated in detail in Chapter 6, this volume.

The characteristic aroma of citrus fruits can be ascribed to high-boiling sparingly water-soluble oils. Chemically, cold-pressed citrus peel essential oils consist of mixtures of 100 to 300 compounds. The composition of the volatile fraction is described in Chapter 8, this volume, and the nonvolatile one in Chapter 12, this volume. The cold-pressing process promotes the release of quantities, sometimes not negligible, of heavy compounds, such as waxes and terpenoids, which can polymerize easily to resinous materials, promoting alteration of the aroma and darkening the product. The individual odor of the different citrus species' fruits and cultivars is closely related to the various chemical components' proportions, with the most important contribution given by the oxygenated compounds (mainly monoterpene), usually present at low levels. This fraction includes mainly aldehydes, alcohols, esters, ketones, oxides, and acids. Among aldehydes, geranial and neral are the major ones. However, citronellal and numerous aliphatic aldehydes such as octanal, decanal, and dodecanal, present in small amounts, strongly contribute to the characteristic aroma of citrus peel oils. Among monoterpene alcohols, linalool, octanol and α-terpeneol, nerol, and geraniol are the main compounds while ketones, esters, oxides, and acids, even if less represented, make appreciable contributions to flavor.

Another important flavor compound of certain citrus essential oils is the esters: linalyl acetate, neryl acetate, geranyl acetate, and bornyl acetate. Terpene hydrocarbons, especially monoterpenes, contribute poorly to the whole flavor on their own. Moreover, unsaturated hydrocarbons are unstable. They may easily degrade via heat, light, and oxygen to form undesirable off-flavor compounds and aromas. Therefore, amounts more or less abundant of hydrocarbons are usually removed by deterpenation, a common industrial practice that allows an increase in the concentration of the aroma compounds while increasing the stability of the oil (Raeissi and Peters 2005; Raeissi et al. 2008).

Sesquiterpene hydrocarbons, including trans-β-farnesene, trans-α-bergamotene, β-caryophyllene, germacrene A to D, and β-bisobolene, present in very small amounts, make appreciable contributions to flavor and odor. Farnesenes exert the important apple flavor; sinensals, aldehydes of the corresponding farnesenes, are also found in many citrus oils and because of their strength of odor, even trace amounts contribute to the entire citrus flavor.

The sesquiterpene ketone (+)-nootkatone possesses a citrus-like aroma and a bitter taste, and occurs in many citrus peel oils contributing to the overall aroma character. The nonvolatile fraction also includes coumarins and psoralens. Among these, substances such as bergapten, characteristic of bergamot essential oil, are fitted with a well-established phototoxic activity (Zaynoun et al. 1977; Naganuma et al. 1985). Bergamot essential oil finds its main destination in the perfume industry. It is applied in lotions, colognes, soaps, creams, makeup, and tanning lotions, giving a delicate and refreshing aroma to these products. The biggest problem for its application in the perfumery industry is just the presence of bergapten. Its use in creams, lotions, and perfumes is restricted. Biological activity of bergapten and restriction of its use are discussed in Chapters 21, 22, and 24, this volume. Thus, above all, bergamot essential oil is usually subjected to further extraction processes, both physical and chemical, with the goal to eliminate bergapten from the essential oil. Industries are strongly interested in bergaptene-free essential oils, all the better if purified from the fraction of monoterpene and sesquiterpene hydrocarbon. The processes commonly used for this purpose are vacuum distillation, extraction with alcohol, the partition between two solvents with different polarities, and preparative adsorption chromatography. All these processes have a disadvantage: the organic solvents must be removed; the distillation involves the thermal degradation of some thermolabile substances and, in general, deterioration in the quality of the product. For this reason the alternative use of SF-CO_2 has been investigated.

10.6 DETERPENATION OF CITRUS OILS BY SC-CO_2

The need to develop alternate refining processes has given rise to an extensive research in the field of supercritical CO_2 extraction, covering all aspects from fundamentals to practical process design and employing different approaches from experimental to modeling and optimization (Goto et al. 1997; Raeissi and Peters 2005). The separation approaches cover many types of techniques, from semibatch to continuous, and from simple countercurrent extraction to more elaborate extraction with reflux and temperature gradients. This interest is due to the possibility of obtaining solvent-free extracts at low temperatures, which is an attractive way to preserve the quality of thermosensitive products (Paviani et al. 2006, Pereira et al. 2010).

In SFE processes, knowledge of the solubility and phase behavior of the components are the most important factors for design and analysis. Literature data demonstrate that there isn't any possibility to predict solubility or phase equilibria behavior without experimental investigations (Marr and Gamse 2000). As evidence of this, concurrently with the growing interest towards the application of SC-CO_2 fractionation of citrus essential oils are many studies on the behavior of the mixture of the main compounds of citrus oils in supercritical CO_2, and in particular on the high-pressure phase behavior both of the binary system CO_2 with limonene (Stahl and Gerard 1985; Matos et al. 1989; Di Giacomo et al. 1989; Abdoul et al. 1991; Iwai et al. 1994, 1996; Marteau et al. 1995; Gironi and Lamberti 1995; Vieira de Melo et al. 1996, 1999; Akgün et al. 1999; Chang and Chen 1999; Kim and Hong 1999; Berna et al. 2000a, 2000b; Gamse and Marr 2000; Sovová et al. 2001; Corazza et al. 2003) and ternary system CO_2 with limonene and the other main representative, terpene compounds of citrus essential oils (Da Cruz Francisco and Sivik 2002; Fonseca et al. 2003; Raeissi and Peters 2005).

Essential oils from citrus fruits, such as lemon, orange, grapefruit, and tangerine, may contain terpene hydrocarbons, unsaturated compounds as high as 95%–98%, from which 95% is limonene. Bergamot essential oil is quite different. The cold-pressed bergamot essential oil presents in its composition a larger amount of oxygenates than normally found in other citrus oils. Bergamot essential oil is the only citrus oil in which limonene is not the dominant component, as it does not exceed 55% and does not contribute to the aroma of the oil but acts as a carrier for the oxygenates. It facilitates oxidation of the oil when in contact with air. The main constituents of bergamot essential oil volatile fraction are linalool and linalyl acetate, which are present in the oil in concentrations ranging from 15% to 40% and 1.5% to 20%, respectively.

Oxygenated derivatives of the hydrocarbons of caryophyllene, germacrene D, farnesene, and bisabolene also contribute to the typical odor of bergamot. Supercritical CO_2 deterpenation of bergamot essential oil is of great commercial interest, since it can result in an almost terpeneless aroma fraction (Sato et al. 1994). As previously stated, the fraction of oxygenated compounds is strongly odoriferous, while monoterpene hydrocarbons do not significantly contribute to the overall flavor profile. The latter readily decompose by heat, light, and oxygen, resulting in an undesirable taste and odor profile; the hydrocarbons make the product very slightly soluble in aqueous or alcohol systems, like fruit drinks and beverages, so it is mandatory to separate them from the citrus oils to obtain a more stable product. The deterpenation of citrus essential oils lowers the terpene content to 25%–50% by weight. For citrus essential oils, deterpenation usually employs molecular or flash distillation and solvent extraction processes. Thermal degradation results in development of some off-flavors, and the presence of residual solvent often makes these processes unsuitable for good quality

of beverages, perfumery, and cosmetic preparations. Goto (2003) highlights fractionation by supercritical carbon dioxide as the prominent technology for the deterpenation process. It is suitable because it requires lower temperatures than distillation, thereby preserving the quality of cold-pressed citrus oils.

While there is limited scientific literature concerning the SC-CO_2 extraction of the essential oil from the citrus fruit peels, a great deal of research covers theoretical aspects related to the phase equilibria among the main compounds of essential oils and CO_2 in supercritical phase and many works concern studies on operating conditions for essential oils fractionation and/or deterpenation (Goto et al. 1997; Budich and Brunner 1999; Stuart et al. 2000; Espinosa et al. 2000, 2005; Goto 2003). Many of these involve theoretical studies to develop models (Diaz et al. 2003, 2005; Espinosa et al. 2005) and others concern the actual behavior in different operating conditions, carried out on specific essential oils (Goto 2003; Gironi and Maschietti 2005). The studies about the use of SC-CO_2 to fractionate and concentrate essential oils in some compounds or classes of compounds involved two different approaches: the supercritical column fractionation and ad/desorption. A supercritical fluid fractionation flow sheet is reported in Figure 10.6. It is well known that supercritical fluid properties can be tuned by manipulating temperature and pressure to attain essential oils with different chemical compositions. Fractionation of the liquid feed is typically carried out in a continuous countercurrent column high-pressure tower (Figure 10.7)

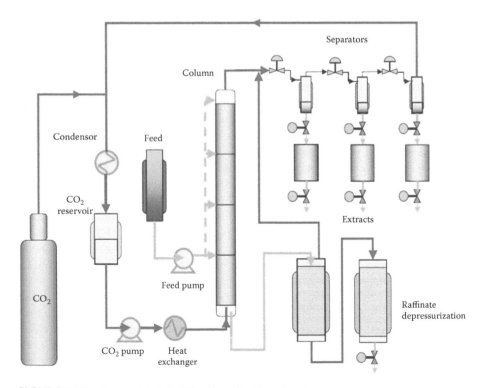

FIGURE 10.6 Supercritical fluid fractionation flow sheet.

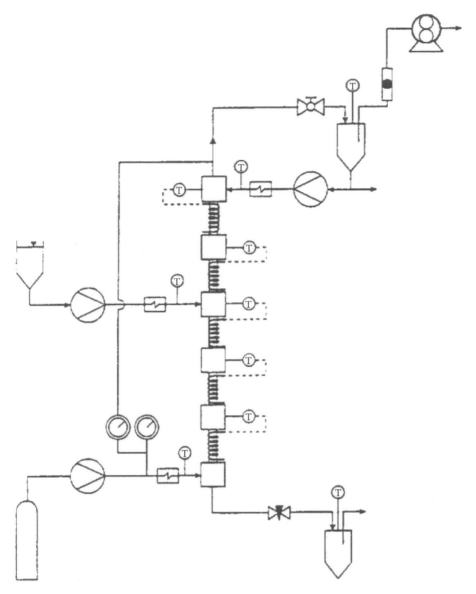

FIGURE 10.7   Continuous countercurrent high-pressure tower for the fractionation of liquid feeds by supercritical CO_2. (From Reverchon, E., J. *Supercrit. Fluids*, 10, 1–37, 1997b.)

but can be carried out also in the semibatch mode by adsorbing the liquid to be treated on a suitable material that selectively retains the different compounds' families.

In continuous mode, the citrus essential oil to be processed is introduced continuously from the middle or the top of the column while the supercritical fluid is introduced from the bottom of the column. Extract was partly refluxed to the top of the extraction column. The extract, light compounds, are collected at the top from where solvent leaves the column. These terpenes are separated by decreasing the pressure and the CO_2, which is recycled to the column from the bottom. The heavier material is collected from the bottom. Normally, continuous supercritical CO_2 deterpenation may be carried out at a relatively lower temperature of 40°C–60°C using a counter-current column. For citrus oils, when conducting deterpenation using supercritical CO_2, it is necessary to operate the fractionating column below 110 bar at 60°C due to the formation of a homogenous phase with SC-CO_2 at higher pressures. The temperature at the top of the fractionating column is usually held at a higher temperature than at the bottom of the extractor. This allows the less-volatile components to condense as a result of a decrease in their solubility with an increase in temperature, thus providing the reflux required for selective separation in the rectification column.

As with conventional countercurrent column processes, the contact between phases can be favored by adding random or structured packing material. In addition, reflux of extract improves selectivity in the extraction process. Process design is based on phase equilibrium data, which determine the number of theoretical stages necessary to perform a specific separation; height of the column, which is related to mass transfer or height equivalent to a theoretical plate; and diameter of the column, which determines the capacity and is related to hydrodynamic behavior of the mixture in contact with the packing. Pressure, temperature, tower packing, length-to-diameter ratio, recycle ratio, and feed-to-solvent ratio must be selected with regards to matrix and in function of the required fraction (Reverchon 1997a, 1997b).

As in continuous countercurrent mode, both liquid and supercritical CO_2 can be used for citrus essential oil deterpenation in semibatch mode with and without an adsorption/desorption column. Studies carried out in 1984 and 1985 (Robey and Sunder 1984; Gerard 1984; Stahl et al. 1984; Stahl and Gerard 1985) deepened knowledge on citrus essential oil supercritical fractionation with continuous processes, both countercurrent and high pressure column. The essential oils of bitter and sweet orange were treated at a pressure of 80 bar, with the highest temperature in the central part of the column (85°C), while the temperatures of the head and the bottom of the column were respectively set at 75°C and 60°C. Starting from an essential oil containing 90% hydrocarbons, the efficiency of supercritical fluid deterpenation was demonstrated by obtaining at the bottom of the column an essential oil with only 42% hydrocarbons.

Temelli et al. (1988) used SC-CO_2 to concentrate oxygenated terpenes from orange oil by deterpenation in a semibatch mode. They suggested 83 bar and 70°C as the SC-CO_2 condition for minimum loss of odoriferous constituents in spite of low extraction yields. Even a 20-fold increase in aroma concentration of orange peel oil retains a significant quantity of terpenes. Temelli highlighted that to evaluate the feasibility of using SC-CO_2 to separate limonene from linalool, it was necessary to obtain vapor-liquid equilibrium (VLE) data at different conditions of pressure and temperature. Later, Temelli et al. (1990) and Sato et al. (1996, 1998a, 1998b) considered orange peel oil a binary mixture of its two key compounds, limonene and linalool, representing the terpene and the oxygenated fractions, respectively.

Sato et al. (1994) used a model of essential oil consisting of limonene, linalool, and citral in an 1800-mm length column with an internal diameter (ID) of 20 mm filled with 3-mm Dixon rings, operating both in semibatch and in continuous modes. The semibatch procedure entailed the charge supply from the bottom of the column while the supercritical CO_2 column fed from the top. The operations of fractionation were made with and without an axial temperature gradient along the column. It was found that at pressures above 110 bar and temperatures of 60°C between the super-critical CO_2 and the essential oil was formed a single phase, but no separation was possible under these operating conditions. In continuous mode, the authors operated at 88 and 98 bar and confirmed that at fixed pressure and temperature the selectivity increases with the solvent to feed ratio. In a subsequent work, the same authors (Sato et al. 1995) employed a different semibatch procedure using a fractionating column (ID 9 mm, length 1000 mm) connected at the bottom to an extractor of 70 cm^3 charged with a mixture in equal parts of limonene, linalool, and citral (neral and gerianal). At the same pressure and temperature conditions used in the previous work (Sato et al. 1994) they flowed SC-CO_2 into the extractor, dissolving part of the load, and then passed it through the fractionating column. The best operating conditions were found to be 88 bar and a temperature range from 40°C to 60°C. At these operating conditions, limonene, linalool, and citral were extracted in sequence with a good selectivity. In the same work, the authors tested the fractionation of orange essential oil. In the same year, Perre et al. (1995) described a pilot plant for the fractionation of the essential oil in continuous countercurrent. The plant was able to treat 250 dm^3 of oil with pressures between 70 to 90 bar and declared a maximum reduction of 95% of monoterpene hydrocarbons.

In 1997, following the reviews on SFE of flavors and fragrances by Stahl et al. (1980), Moyler (1993), and Kerrola (1995), Reverchon (1997b) analyzed solubility data of pure compounds of citrus essential oils. The author proposed mathematical models for SC-CO_2 extraction of fragrance compounds (Reverchon 1997b) demonstrating that SF-CO_2 extraction allows higher-quality products to be obtained, characterized by the absence of artifacts and by a better reproduction of the original fragrance. However, Reverchon pointed out the need for deeper knowledge about solubility and partition factor between the phases. This issue is still the subject of theoretical and applied studies.

In 2012, Gironi and Maschietti carried out a new experimental gas–liquid phase equilibrium of the system CO_2–lemon oil. This study was performed in a constant-volume apparatus operated at 50°C and 70°C, and pressure ranges of interest for the deterpenation process (in the ranges of 8.6–10.1 MPa and 9.7–13.5 MPa, respectively). The essential oil used for the experiments is composed of 93.7% monoterpene hydrocarbons, 4.3% oxygenated compounds, and 2.0% sesquiterpene hydrocarbons (wt.%). In the range of pressures under investigation, no evidence of equilibria other than the (two-phase) gas–liquid equilibrium was found. Both the composition of the gas and the liquid phase were determined, treating the essential oil as a three-component system (monoterpene hydrocarbons, oxygenated compounds, sesquiterpene hydrocarbons). A thermodynamic model based on the Peng–Robinson equation of state was developed in order to represent the behavior of the system supercritical CO_2–lemon oil. The model uses the regression parameters calculated only on data of selected binary subsystems, but the proposed thermodynamic model is capable of providing a good representation of the experimental data. Reverchon (1997b) reported unpublished results of testing on a continuous countercurrent tower consisting of five sections with IDs of 1.75 mm and heights of 30.5 mm, operating with recycle. Reverchon (1997b) reported unpublished results of testing on a continuous countercurrent tower consisting of five sections with IDs of 1.75 mm and heights of 30.5 mm. It operates with a recycle of the flow exiting at the top of the column. Results demonstrate that the fractionation of hydrocarbon and oxygenated terpenes was comparable with supercritical desorption. This plant was also employed to carry out preliminary tests of deterpenation of the bergamot essential oil.

In 1998, Sato et al. (1998b) replayed the fractionation of a cold-pressed orange essential oil with an extraction column under reflux. The essential oil was fractionated into terpenes, oxygenated compounds, and waxes. The three fractions were periodically collected from the top, middle, and bottom of the column. Parameters such as temperature and flow of SC-CO_2 were varied to evaluate the process. The column was filled with steel Dixon rings.

Budich and Brunner (1999) and Budich et al. (1999) fractionated orange essential oil into two fractions, "terpenes" and "aroma," through the use of supercritical CO_2 in a column filled with inert structures such as Sulzer EX. Phase equilibrium and density measurements were carried out at 323, 333, and 343 K in a pressure range from 7 to 13 MPa. For countercurrent column experiments, continuous apparatus with a 4 m column of 25 mm ID was used. The results of this study proved that 333 K and 10.7 MPa are required to produce a 20-fold concentrate containing 68.8 wt.% of terpenes from a feed material of 98.25 wt.%. Under these condition it was possible to

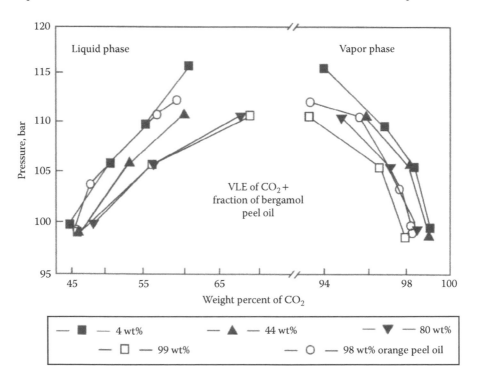

FIGURE 10.8 Vapor-liquid equilibrium (VLE) data for mixture of bergamot oil and orange peel at 60°C. (From Heilig et al., Counter-current supercritical fluid extraction of bergamot peel oil. In *Proceedings of the 5th Meeting on Supercritical Fluids*, 445. Nice, France, 1998.)

obtain a terpene fraction with a purity of 99.8 wt.%. The large amounts of linalool and linalyl acetate influence the selectivity of deterpenation, as can be seen from Figure 10.8, where vapor-liquid equilibrium (VLE) data of bergamot oil and orange peel oil are compared at 60°C (Heilig et al. 1998). On the basis of previous studies, Kondo et al. (2000) employed supercritical CO_2 for the fractionation of bergamot essential oil in a rectifying column by semibatch and continuous extraction processes. The semibatch processes were used to study the equilibrium between CO_2-oxygenated terpenes while the continuous process in countercurrent at 333 K was used to study the influence of pressure on selectivity. In conclusion, in the continuous process with a solvent to feed ratio equal to 63.2, the concentration of monoterpene hydrocarbons was reduced to 1%.

Further studies were carried out on citrus essential oils. Jaubert et al. (2000) used a system consisting of a column, 400-mm long with inner diameter of 23 mm, filled with 261 g of glass pearls with an average diameter of 5 mm. The authors chose the ternary system SC-CO_2-limonene-citral as a basis to extract terpenes from lemon essential oil with supercritical CO_2. The extractions were carried out at different temperatures and pressures to evaluate the influence of these parameters on the efficiency of separation. A theoretical model based on the amended Peng-Robinson's equation of state was used to simulate the thermodynamic aspects of the extraction mass transfer. Benvenuti et al. (2001) developed a study on the thermodynamic properties of the system CO_2/lemon essential oil and a model for the simulation of a continuous and semicontinuous extraction process for deterpenation of essential oil blends. They determined the composition of vapor and liquid phases at different extraction times, using an extractor capacity of 200 cm^3, in a semicontinuous extraction process with a single stage device, at constant temperature (316 K) and using two different pressures values: 8.0 and 8.5 MPa. They also developed a mathematical model for the thermodynamic properties, calculated by the Peng-Robinson equation of state of the system CO_2/lemon essential oil using as a model a mixture of the most important five components: limonene, γ-terpinene, citral, linalool, and β-caryophyllene. The experimental semicontinuous process was modeled assuming that the vapor phase flowing out of the apparatus was always in equilibrium with the liquid stationary phase inside the vessel. The mathematical model was validated by comparing the simulation results with the experimental data. Using the same thermodynamic approach, they also simulated the behavior of a continuous multistage extraction, testing the response of the process for different operative conditions. They demonstrated that both semicontinuous and continuous processes are able to produce enrichment of the refined essential oil phase with oxygenated compounds. The continuous countercurrent extraction offered higher levels of performance in terms of separation and yield. The results showed recoveries higher than 70% in weight and, by tuning the expansion pressure, the additional possibility to optimize the percentage of oxygenated compounds in the final extract.

Kondo et al. (2002a, 2002b) studied the phase behavior of various mixtures of limonene-linalyl acetate, with a process optimized using the simulator SimSci PRO/II for the fractionation column at 333 K and 8.8 MPa. Goto (2003) tested two processes by using SC-CO_2, the countercurrent extraction process and the pressure swing adsorption process. The first process was tested with cold-pressed orange essential

TABLE 10.6
Percentage Composition (GC Analysis) of Bergamot Essential Oil SC-CO_2 Fractions at Different Operating Process Conditions

	Essential Oil	Extraction Condition 1	Extraction Condition 2	Extraction Condition 3	Extraction Condition 4
Pressure		8 MPa	8 MPa	8 MPa	8 MPa
Temperature gradient		46-50-54°C	46-50-54°C	52-56-60°C	42-46-50°C
CO_2 density (g/cm^3)		205.6	205.6	213.6	219.6
CO_2 flow (kg/h)		5.01	4.9	3.9	3.52
Feed (mL/h)		48	46	24	155
Yield (%)		74.11	76.7	91	27.5
	Area (%)	Area (%)	Area (%)	Area (%)	Area (%)
α-Thujene	0.40	0.40	0.50	0.50	0.01
α-Pinene	1.50	1.30	1.80	1.80	3.00
Camphene	tr	tr	0.10	0.10	0.10
Sabinene	0.90	1.10	1.20	1.20	1.90
β-Pinene	7.00	6.70	7.10	7.10	10.60
Myrcene	1.00	1.40	1.20	1.20	1.50
Octanal	tr	tr	tr	tr	0.10
α-Phellandrene	tr	0.10	0.10	0.10	0.10
α-Terpinene	0.20	0.20	0.30	0.30	0.40
p-Cymene	0.70	0.70	0.70	0.60	0.70
Limonene	32.10	35.30	31.70	31.60	41.10
Z-(β)-Ocimene	tr	0.30	0.10	0.10	0.10
E-(β)-Ocimene	0.30	0.60	0.40	0.40	0.40
γ-Terpinene	8.50	8.70	9.40	9.00	10.40
trans-Sabinene hydrate	0.10	0.10	0.10	0.10	0.10
Octanol	tr	tr	0.10	0.10	tr
Terpinolene	0.40	0.50	0.60	0.60	0.50
Linalool	12.10	11.40	13.90	13.10	9.60
Terpinen-4-ol	tr	tr	0.10	0.10	tr
α-Terpineol	0.10	0.10	0.20	0.20	0.10
Decanal	0.10	0.10	0.10	0.10	tr
Octyl acetate	0.10	0.10	0.20	0.20	0.10
Nerol	0.10	tr	0.10	0.10	tr
Neral	0.20	tr	0.20	0.20	0.10
cis-Sabinene hydrate acetate	0.10	0.10	0.10	0.10	0.10
Linalyl acetate	29.70	28.40	27.90	27.40	17.30
Geranial	0.30	0.20	0.30	0.30	0.10
Linalyl propionate	0.10	tr	0.10	0.10	tr
α-Terpinyl acetate	0.30	0.20	0.20	0.30	0.10
Neryl acetate	0.50	0.50	0.30	0.50	0.20
Geranyl acetate	0.40	0.60	0.20	0.50	0.10

TABLE 10.6 (CONTINUED)
Percentage Composition (GC Analysis) of Bergamot Essential Oil SC-CO$_2$ Fractions at Different Operating Process Conditions

	Area (%)	Area (%)	Area (%)	Area (%)	Area (%)
β-Caryophyllene	0.60	0.30	0.40	0.60	0.10
trans-α-Bergamotene	0.60	0.20	0.30	0.60	0.10
β-Bisabolene	0.80	0.20	0.20	0.80	0.10
Nootkatone	0.10	tr	tr	0.10	tr
Bergaptene	0.50	tr	tr	tr	tr

Note: tr, traces.
Source: Poiana et al., *Flavour Fragr. J.* 18, 429–435, 2003.

oil and bergamot essential oil as a feed. Bergamot essential oil contained 40% monoterpene hydrocarbons (25% limonene) and 60% oxygenated compound (25% linalyl acetate). He used a 2400-mm-length column with 20 mm ID. The column was packed with stainless steel 3-mm Dixon packing in a length of 1800 mm. Pressure was set at 9.8 MPa and temperature in a gradient 303–333 K, in semibatch mode. The second was countercurrent without temperature gradient. The work reported the extraction curves and the change in composition of each fraction. The results showed that in the semibatch operation oxygenated compounds were separated more selectively than under countercurrent mode with uniform temperature column.

In the same year, Poiana et al. (2003) focused on SC-CO$_2$ fractionation of bergamot essential oil and carried out tests in countercurrent mode to obtain high value extracts with the lowest bergaptene content. The column was filled with Rashig for fractionation. The process in countercurrent with supercritical CO$_2$ was carried out in conditions of low density CO$_2$ of 206 g/dm^3 at 8 MPa pressure and temperature gradient along the column of 46°C–50°C–54°C according to Sato et al. (1994). The operating process conditions led to essential oil extracts of high quality (see Table 10.6) with yields of 74%–77% and bergapten content less than 0.01%.

A year later Fang et al. (2004) suggested a combination of supercritical CO$_2$ and vacuum distillation for bergamot essential oil fractionation. The authors pointed out that SC-CO$_2$ did not replace vacuum distillation for the industrial process mainly because of the low yield of the process. They proposed the employment of the two techniques, used in series to allow deterpenation (the first) and to gain the purest oxygenate compounds (the second), recovering the waxes and pigments (Table 10.7). The combined process allowed high recovery of oxygenated compound. The total recovery was higher than 85% and the extracts did not contain macromolecular impurities or phototoxic components. In conclusion, Fang et al. (2004) highlighted that the process of deterpenation can be performed successfully by vacuum distillation. The impossibility to recover the residual oxygenated compounds by vacuum distillation can be implemented through the treatment with SC-CO$_2$, obtaining high quality, terpeneless, pigmentless, waxless, and phototoxiless extracts. The combination of supercritical CO$_2$ extraction and desorption makes the deterpenation process

TABLE 10.7
Percentage Composition (GC Analysis) of Bergamot Oil Analyzed with GC-MS Results of Combining of Vacuum Distillation (350–400 MPa) with Supercritical CO₂ Fractionation

Vacuum Distillation Condition					308 K/350–400 MPa			
Supercritical CO₂ Fractionation Condition					313–348 K/12.5 MPa			
		Content of main compounds (%)			OCs recovery (%)	Content of phototoxic compound (%)		
Sample	Mass (g)	Limonene	Linalool	Linalyl acetate		Bergamottin	Bergapten	Citropten
Bergamot oil	100	24.58	14.2	30.69		1.501	0.0717	0.349
MTs	25–27	78–82	0.78–2.1	0.05–1.52		ND	ND	ND
Deterpenated oil	63–66	<0.79	19–22	40–48	95–98	2.03–2.35	0.10–0.12	0.48–0.60
OCs	59–62	<0.45	20–26	42–50	90–93	<0.03	<0.03	<0.03
Residual of SC-CO₂ fractionation	4.0–7.0	<0.05	9.0–15.0	23.01–35.6		15–26	0.85–1.8	3.8–6.0
Total recovery					85–89			

Source: Fang et al., *J. Agric. Food Chem.* 52, 5162–5167, 2004.

more efficient. Since oxygenated compounds can be selectively adsorbed on silica gel polar sites, it is possible to perform a fractional desorption. Hydrocarbon terpenes are desorbed at low CO_2 densities; oxygenated terpenes are desorbed by increasing the operating pressure.

Yamauchi and Saito (1990) used a semipreparative supercritical chromatographic system with silica gel column fractionated cold-pressed lemon essential oil. The essential oil was injected directly into a column 50 mm × 7.3 mm ID containing silica gel. The column temperature was set at 40°C while the outlet pressure was increased from 100 to 200 bar. Ethanol was added as co-solvent to obtain a fraction containing 95% of the oxygenated aroma compounds.

Cully et al. (1990) carried out the desorption with supercritical CO_2 of citrus essential oils for testing different adsorbents: silica gel, aluminum oxide, diatomaceous earth, cellulose, bentonite, and magnesium silicate. Temperatures varied from 50°C to 70°C and the pressure from 70 to 90 bar; the maximum abatement of terpenes reported was of 95%. Using lemon essential oils, Knez et al. (1991) tested the deterpenation at increasing pressures. Barth et al. (1994) used supercritical desorption to deterpenate and to remove coumarins and psoralens. The desorbed fractions were precipitated by means of two high-performance cyclonic separators, thermostated and operating in series. The enrichment of oxygenated compounds was approximately 20-fold.

Barth et al. (1994) also investigated deterpenation of cold-pressed citrus oils with different amounts of oxygenated compounds on a pilot plant scale using a silica column at 40°C and a pressure gradient of 78–100 bar, getting excellent enrichment of aroma compounds with good yields. In 1995 the first application on bergamot essential oil was reported by Chouchi et al. (1995), who developed the first attempt at modeling the supercritical desorption with silica gel (Table 10.8). They proposed a simple kinetic scheme to model the desorption yield curves during the deterpenation of bergamot peel oil. They considered the oil as composed by two families of key compounds and measured the extraction yield of these families against time of desorption. This model was based on the one previously proposed by Tan and Liou (1988, 1989) for supercritical desorption of benzene and toluene from activated carbon. The same research group reported the application on mandarin, lemon (Della Porta et al. 1995), and bitter orange (Chouchi et al. 1996). The latter, carrying out desorption of bitter orange peel oil from a polar adsorbent, was performed by supercritical CO_2 to improve the oil quality by selectively eliminating hydrocarbon terpenes and coumarins.

In 1995, Dugo et al. proposed the supercritical deterpenation of sweet orange and lemon essential oils with different adsorbents. The authors tested different adsorbents and showed that sand and magnesium sulfate produced no fractionation; celite was very selective for aldehydes; and linalool was almost completely extracted with hydrocarbon terpenes. The calcium sulphate gave a good deterpenation, but the results were highly correlated to the water content in the adsorbent, confirming that silica gel is the best adsorbent for deterpenation. They used SC-CO_2 to elute hydrocarbon terpenes from a silica gel column coated with orange peel oil.

Later, oxygenated compounds could be eluted by increasing the temperature and pressure of the SC-CO_2 stream. However, liquid CO_2 is not selective, so extraction of orange oil from a precoated silica-gel column with liquid CO_2 at 67.5 bar and 15°C resulted in reduction of limonene from 95% to 40% (Ferrer and Mathews 1987). The work was conducted in two steps: at 122 bar and 40°C for 20 minutes, and at 405 bar and 60°C for 100 minutes. Most of the hydrocarbons were desorbed during the first phase. The fraction desorbed at the higher density of CO_2 still contained 20%–30% of terpene hydrocarbons. The extract obtained using supercritical desorption was compared with terpeneless oils produced by vacuum distillation. This produced the loss of the more volatile aliphatic aldehydes.

An interesting case study was that of Vega-Bancel and Subra (1995). The authors conducted experiments of supercritical desorption on a model of citrus essential oil consisting of six terpene hydrocarbons and six oxygenated. They absorbed and then desorbed the model mixture with SC-CO_2 with a density between 0.50 and 0.75 g/cm3 obtaining the solubility curves for each compound. Nonselectivity was observed at the highest density of SC-CO_2. At 0.50 g/cm^3 it was possible to distinguish two distinct families of curves, one for terpene hydrocarbons and one for oxygenated terpenes, and at constant density of CO_2, the authors found that the selectivity decreased slightly with increasing temperature. These results confirmed the operating conditions used by Barth et al. (1994) due to evidence that when lower-density CO_2 is used even better separation can be obtained.

TABLE 10.8
Percentage Composition (GC Analysis) of Bergamot Peel Oil (Crude Oil) and SC-CO$_2$ Desorption Fractions

Source		Cold-Pressed Oil	
Pressure		75 bar	75 bar
Temperature		40°C	40°C
Desorption time		37 min	140 min
	Crude Oil	Fraction 2	Fraction 10
	Area (%)	Area (%)	Area (%)
α-Thujene	0.25	0.4	—
α-Pinene	0.99	1.57	—
Camphene	0.11	0.27	—
β-Pinene	6.49	11.48	—
Octen-3-ol	0.08	—	tr
Myrcene	2.33	2.08	4.26
δ-Carene	0.11	—	0.1
α-Phellandrene	0.18	0.26	—
p-Cymene	1.29	3.89	0.6
Limonene	32.14	55.33	2.16
Z-(β)-Ocimene	0.43	0.21	1.05
E-(β)-Ocimene	1.06	0.81	2.03
γ-Terpinene	7.54	11.58	0.24
Terpinolene	0.72	1.08	0.47
Linalool	11.63	3.36	24.39
1,3,8-*p*-Menthatriene	0.17	0.07	0.39
Menthone	0.06	0.27	0.3
Isomenthone	0.01	0.04	0.04
Neomenthol	0.02	0.06	0.07
Menthol	0.16	0.6	0.72
Terpinen-4-ol	tr	tr	tr
α-Terpineol	0.09	—	0.1
Decanal	0.16	—	0.32
Neral	0.28	0.11	0.45
cis-Sabinene hydrate acetate	0.07	—	0.11
Linalyl acetate	30.06	6.13	56.77
Geranial	0.43	—	0.31
Tridecane	0.05	—	0.15
Isomenthyl acetate	0.1	0.31	0.41
α-Terpinyl acetate	0.18	—	0.32
Neryl acetate	0.47	—	0.78
Geranyl acetate	0.5	—	0.87
β-Caryophyllene	45.04	0.08	0.78
Aromadendrene	0.36	—	0.67

TABLE 10.8   (CONTINUED)

Percentage Composition (GC Analysis) of Bergamot Peel Oil (Crude Oil) and SC-CO2 Desorption Fractions

Source		Cold-Pressed Oil	
Pressure		75 bar	75 bar
Temperature		40°C	40°C
Desorption time		37 min	140 min
	Crude Oil	Fraction 2	Fraction 10
	Area (%)	Area (%)	Area (%)
α-Humulene	0.03	—	0.05
cis-β-Farnesene	0.08	—	0.14
γ-Muurolene	0.07	—	0.07
β-Bisabolene	0.52	—	0.92

Source: Chouchi et al., *Ind. Eng. Chem. Res.* 34, 4508–4513, 1995.
Note: tr, traces.

Reverchon (1997a) carried out a selective desorption of binary mixture limonene-linalool, chosen to represent hydrocarbon and oxygenated terpene families, respectively, on silica gel using supercritical CO_2 as a basis for creation of a mathematical model for the deterpenation of essential oil of citrus fruits. Desorption isotherms were obtained using a Langmuir-like empirical equation; the process was modeled by integrating the differential mass balance written for the fluid and solid phases for both compounds. An equilibrium-based model fit the experimental data fairly well.

In the same year, Reverchon and Iacuzio (1997) and Reverchon et al. (1998) proposed the same approach to model supercritical desorption-deterpenation on bergamot peel oil and other citrus peel oils. Selective adsorption and cyclic adsorption/desorption processes in supercritical CO_2 have been attempted also by Subra and Vega (1997) and Subra et al. (1998), who performed a removal of psoralens and coumarins from bergamot essential oil by means of chromatography with SC-CO_2, with particular reference to bergaptene (Table 10.9).

Subra et al. (1998) tested the adsorption of a mixture of 13 terpenes from supercritical carbon dioxide on silanized silica with the aim of simulating essential-oil fractionation. Adsorption isotherms were obtained at temperatures of 310 and 320 K and a carbon dioxide density of 750 kg/m³. The measurements of CO_2 adsorptive capacities for these mixtures showed that the solvent indeed competes for adsorption.

In 2002, Goto and co-workers suggested the use of pressure swing adsorption (PSA) in SC-CO_2. The PSA process is based on the regeneration of adsorber by a difference in adsorbent amounts as a function of pressure. The same research group had already developed (Sato et al. 1998) a PSA process in supercritical carbon dioxide for deterpenation of orange oil using silica gel as adsorbent of election for oxygenated compounds. The extraction column of 600 mm in length and 9 mm ID was

TABLE 10.9

Percentage Composition (GC Analysis) of Psoralens- and Coumarins-Free Bergamot SC-CO$_2$ Extracts

	Cold-Pressed Oil	Total Effluent	Adsorbed Content
Pressure: 105 bar			
Temperature: 47°C			
	Area (%)	Area (%)	Area (%)
α-Pinene	1.06	1.00	0.39
β-Pinene	4.64	4.90	1.83
Myrcene	0.77	0.96	0.31
Limonene	26.4	37.80	28.25
γ-Terpinene	5	5.96	2.43
β-Caryophyllene	0.18	0.23	0.33
Linalool	15.8	13.3	22.84
Citral	1.04	0.72	1.61
Linalyl acetate + Geraniol	43.3	34.70	35.91
Citronellyl acetate	0.16	0.11	0.21
Geranyl acetate	0.28	0.22	0.17
Bergamottin	1.04	0	4.16
Bergaptene	0.14	0	0.34
Citropten	0.17	0	0.62

Source: Subra, P., and Vega, A., *J. Chromatogr. A* 771, 241–250, 1997.

equipped ahead of the adsorption column in order to completely dissolve the bergamot essential oil in SC-CO$_2$ and to remove waxes and pigment. Two adsorption columns of 500 mm in length and ID 9 mm, packed with silica gel, were used as adsorbents. The authors studied the effects of operating parameters (half cycle time, pressure ratio, flow rate ratio, feed concentration on the purity, yield, and recovery) on the bergamot essential oil deterpenation. They checked the adsorption behavior of bergamot oil at 8.8 MPa and 313 K. The breakthrough curve of main compounds indicated that terpenes were slightly adsorbed on the silica gel and immediately eluted from the column. Instead most of the oxygenated aroma compounds were more strongly adsorbed on the silica gel. The behavior demonstrated that the oxygenated component and terpenes can be separated with the proposed process. The same author (Goto 2003) deepened the behavior of bergamot essential oil fractionated with the simple two-bed PSA process in combination with SC-CO$_2$ to make the adsorption/desorption process continuous. The extraction column was equipped before the adsorption column in order to dissolve the raw oil completely in supercritical carbon dioxide and to remove waxes and pigments. A pair of adsorption columns (500 mm long and 9 mm ID) packed with silica gel were used as adsorbent. The effect of desorption-to-adsorption pressure ratio (PD/PA) on the purity of product oil, the recovery of oxygenated compounds, and the yield of product oil at a constant PD for the desorption step was investigated. The author highlights that an increase in the

PD/PA pressure ratio, purity, recovery, and yield were increased in the desorption step. The purity of 82.5% was obtained in the desorption step at a PD/PA ratio of 2.5.

In the same year Araújo and Farias (2003) tested the selectivity of supercritical CO_2 for the extraction of the limonene in bergamot essential oil to concentrate the oxygenated fraction (linalool and linalyl acetate) and reduce the amount of bergapten. They used silica gel as adsorbent to improve the fractionation between the components of interest. The authors studied the effects of temperature, time of extraction, and flow of CO_2. By varying the extraction conditions, substantial amounts of limonene could be removed and oxygenated compounds with low amounts of bergapten could be collected sequentially. They found that with the increase of temperature the efficiency of fractionation decreases, regardless of the applied pressure. Therefore, the best conditions were 50°C/0.70 g/mL (151 bar) at a flow extraction of 3 mL/min and dynamic time of 15 minutes. In these conditions, the authors obtained 51.97% of linalool and 69.38% of linalyl acetate, with an abatement of 72% in limonene and a reduction of 92.7% of bergapten. The results were very interesting and show the potential for effective use of this technology for the production of high added value of bergamot extracts.

Despite the promising results obtained with this technology, in Calabria bergamot essential oil is still extracted with the cold-pressing mechanical process and terpeneless oils by distillation.

ACKNOWLEDGMENTS

Apart from the authors, this work has involved contributions from Drs. Vittoria Cefaly and Demetrio Serra. We thank them.

REFERENCES

Abdoul, W., Rauzy, E., and Péneloux, A. 1991. Group-contribution equation of state for correlating and predicting thermodynamic properties of weakly polar and non-associating mixtures: Binary and multicomponent systems. *Fluid Phase Equilibr.* 68:47–102.

Adami, M., Arcuri, L., Di Giacomo, G., et al. 2000. Estrazione di petit-grain con CO_2 supercritica. *Essenz. Deriv. Agrum.* 70:193–200.

Akgün, M., Akgün, N. A., and Dinçer, S. 1999. Phase behaviour of essential oil components in supercritical carbon dioxide. *J. Supercrit. Fluids* 15:117–125.

Angela, M., and Meireles, A. 2008. Supercritical fluid extraction of medicinal plants. *Electron. J. Envir. Agr. and Food Chem.* 7:3254–3258.

Arai, Y., Sako, T., and Takebayashij, Y. 2002. *Supercritical fluids. Molecular interactions, physical properties, and new applications.* New York: Springer.

Araújo, J. M. de A., and Farias, A. P. S. F. 2003. Limonene and bergapten reduction from bergamot essencial oil adsorbed in silica-gel by supercritical carbon dioxide. *Ciênc. Tecnol. Aliment.* 23:112–115.

Ashihara, H., and Crozier, A. 2001. Caffeine: A well-known but little mentioned compound in plant science. *Trends Plant Sci* 6:407–413.

Atti-Dos Santos, A., and Atti-Serafini, L. 2000. Supercritical carbon dioxide extraction of Mandarin (Citrus deliciosa tenore) from South Brazil. *Perf. Flav.* 25(3):26–36.

Barth, D., Chouchi, D., Della Porta, G., Reverchon, E., and Perrut, M. 1994. Desorption of lemon oil by supercritical carbon dioxide: Deterpenation and psoralens elimination. *J. Supercrit. Fluids* 7:177–183.

Bartle, K. D., Clifford, A. A., Jafar, S. A., Shilstone, G. F. 1991. Solubilities of solids and liquids of low volatility in supercritical carbon dioxide. *J. Phys. Chem. Ref. Data* 20:713–756.

Benvenuti, F., Gironi, F., and Lamberti, L. 2001. Supercritical deterpenation of lemon essential oil, experimental data and simulation of the semicontinuous extraction process. *J. Supercrit. Fluids* 20:29–44.

Berna, A., Cháfe, A. R., and Montón, J. B. 2000a. Solubilities of essential oil components of orange in supercritical carbon dioxide. *J. Chem. Eng. Data* 45:724–727.

Berna, A., Tarrega, A., Blasco, M., and Subirats S. 2000b. Supercritical CO_2 extraction of essential oil from orange peel; effect of the height of the bed. *J. Supercrit. Fluids* 18:227–237.

Berna, A., Chafer, A., Monton, J. B., and Subirats, S. 2001. High-pressure solubility data of system ethanol (1) plus catechin (2) plus CO_2 (3). *J. Supercrit. Fluids* 20:157–162.

Bertucco, A., and Vetter, G. 2001. *High pressure process technology: Fundamentals and applications*. Amsterdam: Elsevier.

Bott, T. R. 1980. Supercritical gas extraction. *Chem. Ind.* March 15:228–232.

Brenneck, J. F., and Eckert, C. A. 1989. Phase equilibria for supercritical fluid process design. *AIChE J.* 35:1409–1427.

Brunner, G. 1988. In *Extraction of caffeine from coffee with supercritical solvents*. Proceedings of the First International Symposium on Supercritical Fluids, 691–698. Nice, France.

Brunner, G. 1994. *Gas extraction. An introduction to fundamentals of supercritical fluids and the application to separation processes*. New York: Springer.

Buchner, E. G. 1906. Die beschrankte Mischbarkeit von Flussigkeiten das System Diphenyamin und Kohlensaure. *Z. Phys. Chem.* 56:257–260.

Budich, M., and Brunner, G. 1999. Vapor-liquid equilibrium data and flooding point measurements of the mixture carbon dioxide + orange peel oil. *Fluid Phase Equilibr* 158:759–773.

Budich, M., Heilig, S., Wesse, T., Leibküchler, V., and Brunner G. 1999. Countercurrent deterpenation of citrus oils with supercritical CO_2. *J. Supercrit. Fluids* 14:105–114.

Calame, J. P., and Steiner, R. 1982. CO_2 extraction in the flavour and perfumery industries. *Chem. Ind.* June 19:399–402.

Chafer, A., Berna, A., Monton, J. B., and Munoz, R. 2002. High-pressure solubility data of system ethanol (1) plus epicatechin (2) plus CO_2 (3). *J. Supercrit. Fluids* 24:103–109.

Chang, C. M. J., and Chen, C. C. 1999. High-pressure densities and P.T.x.y diagrams for carbon dioxide + linalool and carbon dioxide + limonene. *Fluid Phase Equilibr* 163:119–126.

Chang, C. J., Chiu, K. L., Chen, Y. L., Chang, C. Y., and Yang, P. W. 2001. Effect of ethanol content on carbon dioxide extraction of polyphenols from tea. *J. Food Comp. Anal.* 14:75–82.

Chouchi, D., Barth, D., Reverchon E., and Della Porta, G. 1995. Supercritical CO_2 desorption of bergamot peel oil. *Ind. Eng. Chem. Res.* 34:4508–4513.

Chouchi, D., Barth, D., Reverchon E., and Della Porta, G. 1996. Bigarade peel oil fractionation by supercritical carbon dioxide desorption. *J. Agric. Food. Chem.* 44:1100–1104.

Clifford, A. A., Basile, A., and Al-Saidi, S. H. R. 1999. A comparison of the extraction of clove buds with supercritical carbon dioxide and superheated water. *Anal. Chem.* 364:635–637.

Corazza, M. L., Filho, L. C., Antunes, O. A. C., and Dariva, C. 2003. High pressure phase equilibria of the related substances in the limonene oxidation in supercritical CO_2. *J. Chem. Eng. Data* 48:354–358.

Cully, J., Schutz, E., and Volbrecht, H.-R. 1990. Verfahren zur Entfernung von Terpenen aus Etherisehen Olen. *European Patent no.* A0 363-971.

Da Cruz Francisco, J., and Sivik, B. 2002. Solubility of three monoterpenes, their mixtures and eucalyptus leaf oils in dense carbon dioxide. *J. Supercrit. Fluids* 23:11–19.

Da Cruz Francisco, J., and Szwajcer Dey, E. 2003. Supercritical fluids as alternative, safe, food-processing media: An overview. *Acta Microbiol Pol* 52:35–43.

Della Porta, G., Reverchon, E., Chouchi, D., and Barth, D. 1995. Citrus peel oils processing by SC-CO2 desorption: Deterpenation and high molecular weight compounds elimination. *Atti del III Congresso sui fluidi supercritici e le loro applicazioni,* 139. Grignano (Trieste), September 3–6.

Diaz, M. S., Espinosa, S., and Brignole, E. A. 2003. Optimal solvent cycle design in supercritical fluid processes. *Latin Am Appl Res* 33:161–165.

Diaz, S., Espinosa, S., and Brignole, E. A. 2005. Citrus peel oil deterpenation with supercritical fluids: Optimal process and solvent cycle design. *J. Supercrit. Fluids* 35:49–61.

Diaz-Reinoso, B., Moure, A., Dominguez, H., and Parajó, J. C. 2006. Supercritical CO_2 extraction and purification of compounds with antioxidant activity. *J. Agric. Food Chem.* 54:2441–2469.

Di Giacomo, A. 2002. Flowsheet showing steps in the processing of Citrus fruits. In *Citrus*, eds. G. Dugo and A. Di Giacomo. London and New York: Taylor & Francis.

Di Giacomo, A., and Di Giacomo, G. 2002. Essential oil production. In *Citrus*, eds. G. Dugo and A. Di Giacomo. London and New York: Taylor & Francis.

Di Giacomo, G., Brandani, V., Del Re, G., and Mucciante, V. 1989. Solubility of essential oil components in compressed supercritical carbon dioxide. *Fluid Phase Equilibr.* 52:405–411.

Dugo, P., Mondello, L., Bartle, K. D., et al. 1995. Deterpenation of sweet orange and lemon essential oils with supercritical carbon dioxide using silica gel as an adsorbent. *Flavour Fragr. J.* 10:51–58.

Dugo, G., Mondello, L., Cotroneo, A., et al. 1996. Characterization of Italian citrus petitgrain oils. *Perfum. Flav.* 21:17–28.

Dugo, G., Cotroneo, A., Verzera, A., and Bonaccorsi, I. 2002. Composition of the volatile fraction of cold-pressed citrus peel oils. In *Citrus*, eds. G. Dugo and A. Di Giacomo. London and New York: Taylor & Francis.

Dugo, G., Cotroneo, A., Bonaccorsi, I., and Trozzi, A. 2011. Composition of the volatile fraction of cold pressed citrus essential oils. In *Citrus Oils. Composition, Advanced Analytical Techniques, Contaminants and Biological Activity,* eds. G. Dugo and L. Mondello. London and New York: Taylor & Francis.

Ely, J. F., and Baker, J. K. 1983. A review of supercritical fluid extraction. NBS Technical Note 1070, U.S. Dept. of Commerce, National Bureau of Standards.

Espinosa, S., Diaz, S., and Brignole, E. A. 2000. Optimal design of supercritical fluid processes. *Comput Chem Eng* 24:1301–1307.

Espinosa, S., Diaz, M. S., and Brignole, E. A. 2005. Process optimization for supercritical concentration of orange peel oil. *Latin Am Appl Res* 35:321–326.

Fang, T., Goto, M., Sasaki, M., and Hirose, T. 2004. Combination of supercritical CO_2 and vacuum distillation for the fractionation of bergamot oil. *J. Agric. Food Chem.* 52:5162–5167.

Ferrer, O. J., and Matthews, R. F. 1987. Terpene reduction in cold-pressed orange oil by frontal analysis-displacement adsorption chromatography. *J. Food Sci.* 52:801–805.

Ferri, D., and Franceschini, N. 2006. Supercritical CO_2 extraction of "petit grain." *Italian Society of Chemistry and Cosmetic Sciences (SICC) Proceedings.* November 16 (www.siccproceedings.org/documents/9Ferri_MS.pdf).

Fonseca, J., Simoes, P. C., and Nunes da Ponte, M. 2003. An apparatus for high-pressure VLE measurements using a static mixer. Results for (CO_2 + limonene + citral) and (CO_2 + limonene + linalool). *J. Supercrit. Fluids* 25:7–17.

Fornari, T., Vicente, G., Vázquez, E., García-Risco, M. R., and Reglero, G. 2012. Isolation of essential oil from different plants and herbs by supercritical fluid extraction. *J. Chromatogr. A* 1250:34–48.

Franceschi, E., Grings, M. B., Frizzo, C. D., Vladimir Oliveira, J., and Dariva, C. 2004. Phase behavior of lemon and bergamot peel oils in supercritical CO_2. *Fluid Phase Equilibr.* 226:1–8.

Gamse, T. 2005. Industrial applications and current trends in supercritical fluid technologies. *Hem. Ind.* 59:207–212.

Gamse, Th., and Marr, R. 2000. High-pressure phase equilibria of the binary systems carvone-carbon dioxide and limonene-carbon dioxide at 30, 40 and 50°C. *Fluid Phase Equilibr.* 171:165–174.

Gaspar, F., Lu, T., Santos R., and Al-Duri, B. 2003. Modelling the extraction of essential oils with compressed carbon dioxide. *J. Supercrit. Fluids* 25:247–260.

Gerard, D. 1984. Continuous removal of terpenes from essential oils by countercurrent extraction with compressed carbon dioxide. *Chem. Ing. Tech.* 56:794–795.

Gironi, F., and Lamberti, L. 1995. Solubility of lemon oil components in supercritical carbon dioxide. *Proc. 3rd Italian Conference on I Fluidi Supercritici e le loro Applicazioni*, 165, Trieste.

Gironi, F., and Maschietti, M. 2005. Supercritical carbon dioxide fractionation of lemon oil by means of a batch process with an external reflux. *J. Supercrit. Fluids* 35:227–234.

Gironi, F., and Maschietti, M. 2012. Phase equilibrium of the system supercritical carbon dioxide–lemon essential oil: New experimental data and thermodynamic modeling. *J. Supercrit. Fluids* 70:8–16.

Goto, M., Fukui, G., Wang, H., Kodama, A. and Hirose, T. 2002. Deterpenation of bergamot oil by pressure swing adsorption in supercritical carbon dioxide. *J. Chem. Eng. Jpn.* 35:372–376.

Goto, M. 2003. Supercritical fluid fractionation for citrus oil processing. *Theories and Applications of Chem. Eng.* 9:1852–1855.

Goto, M., Sato, M., Kodama, A., and Hirose, T. 1997. Application of supercritical fluid technology to citrus oil processing. *Physica B* 239:167–170.

Hannay, J. B., and Hogarth, J. 1879. On the solubility of solids in gases. *Proc. Roy. Soc.* (London) 29:324.

Haro-Guzman, L. 2002. Production of distilled peel oils. In *Citrus,* eds. G. Dugo and A. Di Giacomo. London and New York: Taylor & Francis.

Hawthorne, S. B. 1990. Analytical–scale supercritical fluid extraction. *Anal. Chem.* 62:633–642.

Heilig, S., Budich, M., and Brunner, G. 1998. Counter-current supercritical fluid extraction of bergamot peel oil. In *Proceedings of the 5th Meeting on Supercritical Fluids*, 445. Nice, France.

Henry, M. C., and Yonker, C. R. 2006. Supercritical fluid chromatography, pressurized liquid extraction, and supercritical fluid extraction. *Anal. Chem.* 78(12):3909–3915.

Herrero, M., Cifuentes, A., and Ibañez, E. 2006. Sub- and supercritical fluid extraction of functional ingredients from different natural sources: Plants, food by-products, algae and microalgae. *Food Chem.* 98:136–148.

Huang, Z., Shi, X.-H., and Jiang, W.-J. 2012. Theoretical models for supercritical fluid extraction. *J. Chromatogr. A* 1250:2–26.

Hubert, P., and Vitzthum, O. G. 1978. Fluid extraction of hops and tobacco with supercritical gases. *Agnew. Chem. Int. Ed. Engl.* 17:710–715.

Iwai, Y., Hosotami, N., Morotomi, T., Koga, Y. and Arai, Y. 1994. High-pressure vapor-liquid equilibria for carbon dioxide + linalool. *J. Chem. Eng. Data* 39:900–902.

Iwai, Y., Morotomi, T., Sakamoto, K., Koga, Y., and Arai, Y. 1996. High-pressure vapor liquid-equilibria for carbon dioxide + limonene. *J. Chem. Eng. Data* 41:951–952.

Jaubert, J.-N., Margarida, G. M., and Barth, D. 2000. A theoretical model to simulate supercritical fluid extraction: Application to the extraction of terpenes by supercritical carbon dioxide. *Ind. Eng. Chem. Res.* 39:4991–5002.

Kerrola, K. 1995. Literature review: Isolation of essential oils and flavor compounds by dense carbon dioxide. *Food Rev. Int.* 11:547–573.

Kim, K. H., and Hong, J. 1999. Equilibrium solubilities of spearmint oil components in supercritical carbon dioxide. *Fluid Phase Equilibr.* 164:107–115.

King, J. W. 2002. Supercritical fluid extraction: Present status and prospects. *Grasas Aceites* 5:8–21.

Knez, Z., Posel, F., and Golob, J. 1991. Extraction of plant materials with supercritical CO_2. In *Proceedings of the Second International Symposium on Supercritical Fluids*, ed. M. A. McHugh, 101–104. John Hopkins University, Baltimore, MD.

Kondo, M., Goto, M., Kodama, A., and Hirose, T. 2000. Fractional extraction by supercritical carbon dioxide for the deterpenation of bergamot oil. *Ind. Eng. Chem. Res.* 39:4745–4748.

Kondo, M., Akgun, N., Goto, M., Kodama, A., and Hirose, T. 2002a. Semi-batch operation and countercurrent extraction by supercritical CO_2 for the fractionation of lemon oil. *J. Supercrit. Fluids* 23:21–27.

Kondo, M., Goto, M., Kodama, A., Hirose, T. 2002b. Separation performance of supercritical carbon dioxide extraction column for the citrus oil processing: Observation using simulator. *Separ. Sci. Technol.* 37:3391–3406.

Lack, E., and Seidlitz, H. 1996. Standardized industrial scale high pressure extraction plant. *National Scientific Conference on Analytical and Technological Use of Supercritical Fluids*, May 16. Budapest, Hungary.

Lang, Q., and Wai, C. M. 2001. Supercritical fluid extraction in herbal and natural product studies: A practical review. *Talanta* 53:771–782.

Li, S., Lambros, T., Wang, Z., Goodnow, R., and Ho, C.-T. 2007. Efficient and scalable method in isolation of polymethoxyflavones from orange peel extract by supercritical fluid chromatography. *J. Chromatogr. B* 846:291–297.

Lota, M-L., de Rocca Serra, D., Tomi, F., Jacquemond, C., and Casanova, J. 2002. Volatile components of peel and leaf oils of lemon and lime species. *J. Agric. Food Chem.* 50:796–805.

Marr, R., and Gamse, T. 2000. Use of supercritical fluids for different processes including new developments. *Chem. Eng. Process.* 39:19–28.

Marteau, Ph., Obriot, J., and Tufeu, R. 1995. Experimental determination of vapor-liquid equilibria of CO2 + limonene and CO2 + citral mixtures. *J. Supercrit. Fluids* 8:20–24.

Martínez, J. L. 2008. *Supercritical fluid extraction of nutraceutical and bioactive compounds.* London and New York: Taylor & Francis.

Matos, H. A., De Azevedo, E. G., Simones, P. C., Carrondo, M. T., and Da Ponte, M. N. 1989. Phase equilibria of natural flavours and supercritical solvents. *Fluid Phase Equilibr* 52:357–364.

McKenzie, L. C., Thompson, J. E., Sullivan, R., and Hutchison, J. E. 2004. Green chemical processing in the teaching laboratory: A convenient liquid CO_2 extraction of natural products. *Green Chem.* 6:355–358.

Medina, I. 2012. Determination of diffusion coefficients for supercritical fluids. *J. Chromatogr. A* 1250:124–140.

Menaker, A., Kravets, M., Koel, M., and Orav, A. 2004. Identification and characterization of supercritical fluid extracts from herbs. *C. R. Chim.* 7:629–633.

Mincione, B., Di Giacomo, A., Leuzzi, U., et al. 1992. Estrazione dell'essenza di bergamotto con CO_2 supercritica. *Essenz. Deriv. Agrum.* 62:28–38.

Mira, B., Blasco, M., Berna, A., and Subirats, S. 1999. Supercritical CO_2 extraction of essential oil from orange peel. Effect of operation conditions on the extract composition. *J. Supercrit. Fluids* 14:95–104.

Moyler, D. A. 1993. Extraction of essential oils with carbon dioxide. *Flavour. Fragr. J.* 8:235–247.

Mukhopadhyay, M. 2000. *Natural extracts using supercritical carbon dioxide.* Boca Raton: CRC Press.

Naganuma, M., Hirose, S., Nakayama, K., Katajirna, K., and Somega, T. A. 1985. A study of the phototoxicity of lemon oil. *Arch. Dermatol. Res.* 278:31–36.

Nautiyal, O. H., and Tiwari, K. K. 2011. Supercritical carbon dioxide extraction of Indian orange peel oil and hydro-distillation comparison on their compositions. *Sci Technol* 1:29–33.

Page, S. H., Sumpter, S. R., and Lee, M. L.1992. Fluid phase-equilibria in supercritical fluid chromatography with CO_2-based mixed mobile phase—a review. *J. Microcol. Sep.* 4:91–122.

Paulaitis, M. E., Krukonis, V. J., Kurnik, R. T., et al. 1983. Supercritical fluid extraction. *Rev. Chem. Eng.* 1(2):179–250.

Paviani, L., Pergher, S. B. C., and Dariva, C. 2006. Application of molecular sieves in the fractionation of lemongrass oil from high-pressure carbon dioxide extraction. *Braz. J. Chem. Eng.* 23:219–225.

Peyron, L, and Bonaccorsi, I. 2002. Extracts from the bitter orange flowers (Citrus aurantium L): Composition and adulteration. In *Citrus*, eds. G. Dugo and A. Di Giacomo, 413–424. London and New York: Taylor & Francis.

Pereira, C. G., Angela, M., and Meireles, A. 2010. Supercritical fluid extraction of bioactive compounds: Fundamentals, applications and economic perspectives. *Food Bioprocess Technol.* 3:340–372.

Perre, C., Delestre, G., Schrive, L., and Carles, M. 1995. Deterpenation process for citrus oils by supercritical CO_2 extraction in a packed column. *Proceedings of the Third International Symposium on Supercritical Fluids* 2:465–470. Strasbourg, October 17–19.

Perrut, M. 2000. Proceedings of the 5th International Symposium on Supercritical Fluids, April 8–12, Atlanta, GA. Proceedings distributed on CD-ROM.

Phelps, C. L., Smart, N. G., and Wai, C. M. 1996. Past, present, and possible future applications of supercritical fluid extraction technology. *J. Chem. Educ.* 73:1163.

Poiana, M., Reverchon, E., Sicari, V., Mincione, B., and Crispo, F. 1994. Supercritical carbon dioxide extraction of bergamot oil: Bergaptene content in the extracts. *Ital. J. Food Sci.* 4:459–466.

Poiana, M., Sicari, V., Mincione, B., and Crispo, F. 1997. Prospettive di utilizzazione della tecnica estrattiva con fluidi supercritici applicata ai derivati della coltura del bergamotto. *Essenz. Deriv. Agrum.* 67:27–41.

Poiana, M., Fresa, R., and Mincione, B. 1999. Supercritical carbon dioxide extraction of bergamot peels. Extraction kinetics of oil and its components. *Flavour Fragr. J.* 14:358–366.

Poiana, M., Mincione, A., Gionfriddo, F., and Castaldo, D. 2003. Supercritical carbon dioxide separation of bergamot essential oil by a countercurrent process. *Flavour Fragr. J.* 18:429–435.

Raeissi, S., and Peters, C. J. 2005. Experimental determination of high-pressure phase equilibria of the ternary system carbon dioxide + limonene + linalool. *J. Supercrit. Fluids* 35:10–17.

Raeissi, S., Diaz, S., Espinosa, S., Peters, C. J., and Brignole, E. A. 2008. Ethane as an alternative solvent for supercritical extraction of orange peel oils. *J. Supercrit. Fluids* 45:306–313.

Raventós, M., Duarte, S., and Alarcón, R. 2002. Application and possibilities of supercritical CO_2 extraction in food processing industry: An overview. *Food Sci. Technol. Int.* 8:269–284.

Reid, R. C., Prausnitz, J. M., Poling, B. E. 1998. *The Properties of Liquids and Gases,* 4th ed. New York: McGraw-Hill.

Reverchon, E. 1997a. Supercritical desorption of limonene and linalool from silica gel: Experiments and modelling. *Chem. Eng.* Sci. 52:1019–1027.

Reverchon, E. 1997b. Supercritical fluid extraction and fractionation of essential oils and related products. *J. Supercrit. Fluids* 10:1–37.

Reverchon, E., and Iacuzio, G. 1997. Supercritical desorption of bergamot peel oil from silica gel: Experiments and mathematical modelling. *Chem. Eng. Sci.* 52:3553–3559.

Reverchon, E., Lamberti, G., and Subra, P. 1998. Modelling and simulation of the supercritical adsorption of complex terpene mixtures. *Chem. Engin. Sci.* 53:3537–3544.

Reverchon, E., and De Marco, I. 2006. Supercritical fluid extraction and fractionation of natural matter. *J. Supercrit. Fluids* 38:146–166.

Robey, R. J., and Sunder, S. 1984. Application of supercritical processing to the concentration of citrus oil fractions. Paper presented at *Air Products and Chemicals* (AIChE) Meeting, San Francisco, CA.

Roy Bhupesh, C., Sasaki, M., and Goto, M. 2005. Extraction of citrus oil from peel slurry of Japanese citrus fruits with supercritical carbon dioxide. *J. Applied Sci.* 5:1350–1355.

Roy Bhupesh, C., Munehiro, H., Hiro, U., et al. 2007. Supercritical carbon dioxide extraction of the volatiles from the peel of Japanese citrus fruits. *J. Essent. Oil Res.* 19:78–84.

Sargenti, S. R., and Lancas, F. M. 1998. Influence of extraction mode and temperature in the supercritical fluid extraction of *Citrus sinensis* (Osbeck). *J. Chromatogr. Sci.* 36:169–174.

Sato, M., Goto, M., and Hirose, T. 1994. Fractionation of citrus oil by supercritical fluid extraction tower. *Proceedings of the Third International Symposium on Supercritical Fluids* 2:83. Strasbourg, October 17–19.

Sato, M., Goto, M., and Hirose, T. 1995. Fractional extraction with supercritical carbon dioxide for the removal of terpenes from citrus oil. *Ind. Eng. Chem. Res.* 34:3941–3946.

Sato, M., Goto, M., and Hirose, T. 1996. Supercritical fluid extraction on semibatch mode for the removal of terpene in citrus oil. Ind. *Eng. Chem. Res.* 35:1906–1911.

Sato, M., Goto, M., Kodama, A., and Hirose, T. 1998a. New fractionation process for citrus oil by pressure swing adsorption in supercritical carbon dioxide. *Chem. Eng. Sci.* 53:4095–4104.

Sato, M., Kondo, M., Goto, M., Kodama, A., and Hirose, T. 1998b. Fractionation of citrus oil by supercritical countercurrent extractor with side-stream withdrawal. *J. Supercrit. Fluids* 13:311–317.

Schaber, P. M., Larkin, J. E., Pines, H. A. K., et al. 2012. Supercritical fluid extraction versus traditional solvent extraction of caffeine from tea leaves: A laboratory-based case study for an organic chemistry course. *J Chem Educ* 89:1327–1330.

Schultz, W. G., Schultz, T. G., Carlson, R. A., et al. 1974. Pilot-plant extraction with liquid CO_2. *Food Technol.* 6:28–32.

Sovová, H., Stateva, R. P., and Galushko, A. A. 2001. Essential oils from seeds: Solubility of limonene in supercritical CO2 and how it is affected by fatty oil. *J. Supercrit. Fluids* 20:113–129.

Sovová, H. 2012. Modeling the supercritical fluid extraction of essential oils from plant materials. *J. Chromatogr. A* 1250:27–33.

Stahl, E., Schutz, E., and Mangold, H. M. 1980. Extraction of seed oils with liquid and supercritical carbon dioxide. *J. Agric. Food Chem.* 28:1153–1157.

Stahl, E., and Willing, E. 1980. Extraction of natural materials with supercritical and condensed gases—quantitative determination of the solubility of opium alkaloids. *Mikrochim. Acta* 2 (5–6):465–474.

Stahl, E., Quirin, K. W., Glatz, A, Gerard, D., and Rau, G. 1984. New developments in the field of high-pressure extraction of natural products with dense gases. *Congrès Meeting of the Deutsche Bunsen-Gesellschaft für Physikalische Chemie Supercritical Fluid Solvents* 88:900–907.

Stahl, E., and Gerard, D. 1985. Solubility behaviour and fractionation of essential oils in dense carbon dioxide. *Perf. Flav.* 10(2):29–37.

Stuart, G. R., Dariva, C., and Vladimir Oliveira, J. 2000. High-pressure vapor-liquid equilibrium data for CO_2 orange peel oil. *Braz. J. Chem. Eng.* 17:181–189.

Subra, P., and Vega, A. 1997. Retention of some components in supercritical fluid chromatography and application to bergamot peel oil fractionation. *J. Chromatogr. A* 771:241–250.

Subra, P., Vega-Bancel, A., Reverchon, E. 1998. Breakthrough curves and adsorption isotherms of terpene mixtures in supercritical carbon dioxide. *J. Supercrit. Fluids* 12:43–57.

Subramaniam, B., Rajewski, R. A., and Snavely, K. 1997. Pharmaceutical processing with supercritical carbon dioxide. *J. Pharmaceutical Sci.* 86:885–890.

Sugiyama, K., and Saito, M. 1982. Simple microscale supercritical fluid extraction and its application to gas chromatography-mass spectrometry of lemon peel oil. *J. Chromatogr. A* 442:121–131.

Swisher, H. E., and Swisher, L. H. 1977. Specialty citrus products. In *Citrus Science and Technology*, ed. S. Nagy, P. E. Shaw, and M. K. Velduis, 290–345. Westport, CO: AVI Publ. Comp., Inc.

Tan, C. S., and Liou, D. C. 1988. Desorption of ethyl acetate from activated carbon by supercritical carbon dioxide. *Ind. Eng. Chem.* Res. 28:988–992.

Tan, C. S., and Liou, D. C. 1989. Modeling of desorption at supercritical conditions. *AIChE Journal* 35:1029–1031.

Temelli, F., Braddock, R. J., Chen, C. S., and Nagy, S. 1988. Supercritical carbon dioxide extraction of terpenes from orange essential oil. *Supercritical Fluid Extraction and Chromatography*, eds. B. A. Charpentier and M. R. Sevenants, ACS Symposium Series 366:109–126.

Temelli, F., O'Connel, J. P., Che, C. S., and Braddock, R. J. 1990. Thermodynamic analysis of supercritical carbon dioxide extraction of terpenes from cold-pressed orange oil. *Ind. Eng. Chem. Res.* 29:618–624.

Tomasula, P. M. 2003. Supercritical fluid extraction of foods. *Encyclopedia Agr. Food Biol Eng* 964–967.

Vega-Bancel, A., and Subra, P. 1995. Adsorption and desorption of some hydrocarbon and oxygenated terpenes under supercritical conditions. *Atti del III Congresso sui fluidi supercritici e le loro applicazioni* 149. Grignano (Trieste), September 3–6.

Velasco, R. J., Villada, H. S., and Carrera, J. E. 2007. Aplicaciones de los fluidos supercríticos en la agroindustria. *Información Tecnológica* 18:53–65.

Vieira de Melo, S. A. B., Pallado, P., Bertucco, A., and Guarise, G. B. 1996. High-pressure phase equilibria data of systems containing limonene, linalool and supercritical carbon dioxide. *Process Technology Proceedings* (12): *High Pressure Chemical Engineering*, 411–412. Elsevier Science.

Vieira de Melo, S. A. B., Pallado, P., Guarise, G. B., and Bertucco, A. 1996. High pressure vapor-liquid data for binary and ternary systems formed by supercritical CO_2, limonene and linalool. *Braz. J. Chem. Eng.* 16:7.

Wang, L., and Weller, C. L. 2006. Recent advances in extraction of nutraceuticals from plants. *Food Sci Technol* 17:300–312.

Xu, L., Zhan, X., Zeng, Z., et al. 2011. Recent advances on supercritical fluid extraction of essential oils. *Afr J Pharm Pharmacol* 5:1196–1211.

Yamauchi, Y., and Saito, M. 1990. Fractionation of lemon-peel oil by semi-preparative supercritical fluid chromatography. *J. Chromatogr. A* 505:237–246.

Zaynoun, S. T., Johnson, B. E., and Frain-Bell, W. 1977. A study of oil of bergamot and its importance as a phototoxic agent. II. Factors which affect the phototoxic reaction induced by bergamot oil and psoralen derivatives. *Contact Dermatitis* 5:225–239.

Zougagh, M., Valcarcel, M., and Rios, A. 2004. Supercritical fluid extraction: A critical review of its analytical usefulness. *Trends Anal. Chem.* 23:399–405.

11 Enantiomeric Distribution of Volatile Components in Bergamot Peel Oils and Petitgrain

Luigi Mondello, Danilo Sciarrone,
Carla Ragonese, and Giovanni Dugo

CONTENTS

11.1 INTRODUCTION

The enantiomeric distribution of the components of an essential oil can be an important indication in the evaluation of its authenticity and quality, and can be sometimes related to the extraction technology, geographic origin, and period of production. High resolution gas chromatography (HRGC), using modified cyclodextrins (CD) stationary phases, pure or diluted in polysiloxanes, can be considered the most suitable approach for the determination of the enantiomeric ratios of the volatile fraction of essential oils. This fraction represents a quite complex mixture, difficult to elucidate in a single gas chromatographic analysis; thus the determination of the enantiomeric ratio of its components is a hard task. The separation of the enantiomers in an essential oil by gas chromatography, exploiting a chiral column, is sometimes possible, mainly for the most representative components. However, this is a difficult process and may lead to the wrong results due to the co-elution of the enantiomers of interest with other components; therefore, the choice of the experimental conditions is critical.

Sometimes it can be convenient to use precolumns, packed with conventional stationary phases, in series with the column that allows the chiral discrimination (Dugo et al. 1992, 1993). However, especially for minor components, a pre-separation step allows a way to obtain chiral compounds with an elevated purity grade, and thus more reliable results. Pre-separation can be carried out by means of high performance thin layer chromatography (HPTLC) or high performance liquid cromatography (HPLC) coupled "off-line" or "on-line" with gas chromatography (Dugo et al. 1994a, 1994b; Mondello et al. 1996). As an example, the chromatograms relative to the on-line HPLC-HRGC analyses of linalool and terpinen-4-ol in mandarin essential oils are reported in Figure 11.1. Nowadays, multidimensional gas chromatography (MDGC) performed using a capillary column equipped with a conventional stationary phase (polar or nonpolar) in the first dimension, followed by a cyclodextrine-based stationary phase in the second dimension, can be considered the best approach for the determination of the enantiomeric ratios of the volatile fraction components in such complex samples. The aforementioned technique has been widely applied to the essential oil analysis, particularly to bergamot essential oil, by several research groups, especially in Italy (Mondello et al. 1997, 1998a; Dugo et al. 2001; Mangiola et al. 2009), in Germany (Mosandl et al. 1990; Mosandl 1995; Juchelka and Mosandl

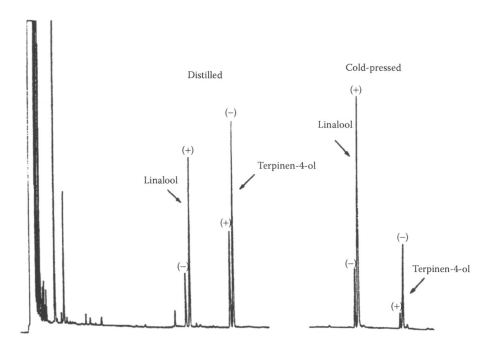

FIGURE 11.1   On-line HPLC-HRGC chromatograms of linalool and terpinen-4-ol in distilled and cold pressed mandarin oils. (From Mondello, L., Dugo, G., Dugo, P., and Bartle, K. D. *Perfum. Flav.* 21:25–49, 1996.)

1996; Mosandl and Juchelka 1997a) and in France (Casabianca and Graff 1994; Casabianca et al. 1995, 1998).

Figures 11.2 and 11.3 report the chromatograms relative to the separation operated in the first achiral (a) and to the second chiral column (b) of linalool, linalyl acetate, and three monoterpene hydrocarbon enantiomers in bergamot oils, obtained by means of a multidimensional system (Juchelka and Mosandl, 1996). Figures 11.4 and 11.5 show two examples of identification of several volatile compounds in bergamot essential oil in a single analysis, obtained by means of totally automated multidimensional systems. The chromatograms reported in Figure 11.4 are relative to the analysis carried out on a system with a mechanical valve interface (Mondello et al. 1998a), whereas Figure 11.5 (Dugo et al. 2012) is relative to a multidimensional system equipped with a pressure balanced interface developed by Mondello et al. (2006, 2008).

It is well known that the enantiomeric ratios of linalool and linalyl acetate in bergamot oil can be determined with good results by means of direct enantioselective analysis of the entire fraction (Cotroneo et al. 1992; Dellacassa et al. 1997) (Figure 11.6); moreover, the recent improvements of chiral column performances allow the determination of the enantiomeric ratio even of some minor components (Sciarrone 2009, personal communication) (Figure 11.7). Sciarrone et al. (2010) proposed to overcome the problem relative to the co-elutions occurring between enantiomers and nonchiral components in direct enantioselective analysis, correcting the results on the basis of the percentage of the co-eluting components calculated from the nonchiral GC-FID analysis.

In this chapter, industrial cold-pressed, commercial, recovered from residues of cold-pressed extraction, bergapten-free, deterpenated, laboratory-extracted, and petitgrain essential oils will be separately treated. The results relative to the different types of oils will be discussed and reported in a table and in an appendix. The table will report the data extracted from the original papers relative to the results of a single sample, to average values or, when possible, to ranges of variability, expressed with one decimal or without decimal figures if this is the approximation used in the original paper. Information concerning botanical and geographical origin, the number of samples related to the reported results, and the analytical technique, when available, will be reported in the appendix. If given, the column size will be reported in brackets as follows: the first number relative to the length (meters), the second to the internal diameter (millimeters), and the third to the stationary phase film thickness (micrometers). With regards to chiral columns, if given in the original papers, the chiral stationary phase modifier will be reported.

11.2 COLD-PRESSED OILS

The determination of the enantiomeric ratio of volatile fraction target components is a key point in the authenticity assessment of bergamot essential oil. Since the volatile fraction composition may undergo considerable quantitative variations depending on the collection period and on the origin of the fruits, as described in Chapter 8, this volume, the composition of volatile fraction and its seasonal variations cannot allow

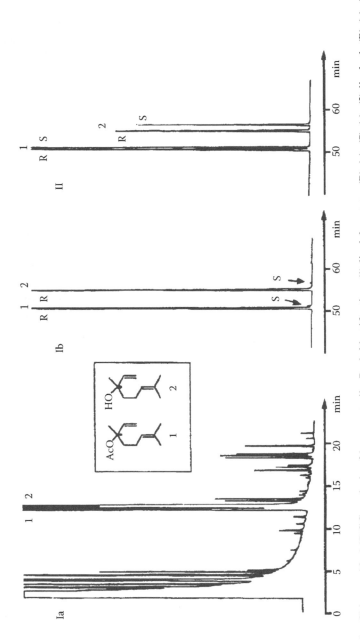

FIGURE 11.2 Chirospecific MDGC analysis of bergamot oils. Peak identification: (1) linalyl acetate: (R)-(−), (S)-(+); (2) linalool: (R)-(−), (S)-(+); (I) Authentic bergamot oil: precolumn (Ia), main column (Ib); Commercial bergamot oil: main column (II). (From Juchelka, D., and Mosandl, A. *Pharmazie* 51:417–422, 1996.)

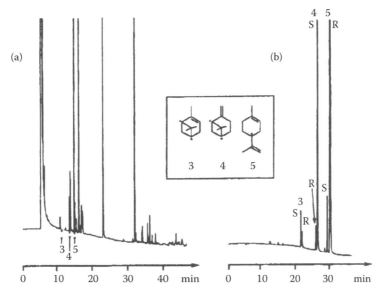

FIGURE 11.3 Chirospecific MDGC analyses of a genuine bergamot oil; (a) precolumn, (b) main column. Peak identification: (3) α-pinene: (S)-(−), (R)-(+); (4) β-pinene: (R)-(+), (S)-(−); (5) limonene: (S)-(−), (R)-(+). (From Juchelka, D., and Mosandl, A. In *Pharmazie* 51:417–422, 1996.)

a reliable authenticity assessment of bergamot essential oil, as opposed to the case of lemon and mandarin oils.

On the contrary, the enantiomeric distribution of some components, such as linalool and linalyl acetate, varies within narrow limits, not depending on the production period and geographical origin of the oil; thus, it can be considered a key reference for the genuineness assessment. The enantiomeric ratio of linalool in industrial genuine bergamot oils was determined for the first time in Messina by Cotroneo et al. (1992). Previously, the enantiomeric distribution of linalool was determined on commercial samples from Mosandl and Schubert (1991) and Bernreuther and Schreier (1991). Mosandl and Schubert (1991) also analyzed a solvent-extracted oil, while Bernreuther and Schreier (1991) analyzed an oil obtained from dried peel by steam distillation. Based on the results obtained from the GC analysis of genuine and reconstituted bergamot oil samples carried out on a β-cyclodextrin column, Cotroneo et al. (1992) established that the maximum percentage of dextrorotatory linalool in genuine oils did not exceed the 0.5% of total linalool. Therefore, oils with higher values were considered mixed with products containing linalool from different origins or not obtained by means of mechanical extraction of bergamot oil. Figures 11.8 and 11.9 show the chromatograms relative to the chiral separation obtained by Cotroneo et al. (1992) for a genuine oil and for mixtures of a genuine oil with reconstituted oils using synthetic linalool. These results were confirmed by other researchers (Verzera et al. 1996; Casabianca and Graff 1996; König et al. 1997; Mosandl and Juchelka 1997a; Mondello et al. 1997, 1998a; Dellacassa et al. 1997; Dugo et al. 2001, 2012; Sciarrone 2009; Mangiola et al. 2009), although for some authors values of (+)-linalool up to

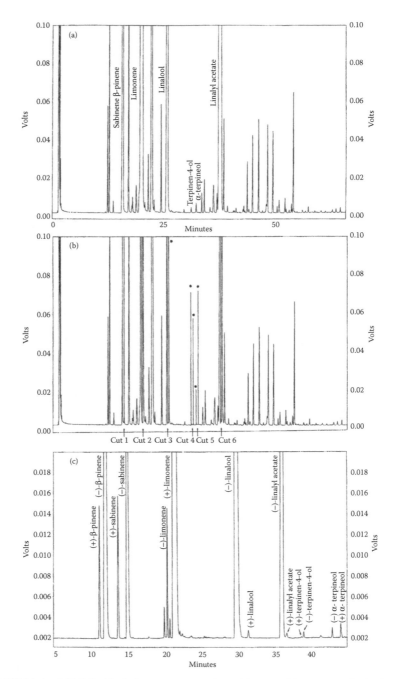

FIGURE 11.4 MDGC chiral chromatogram of a cold-pressed bergamot oil obtained on a system that uses mechanical valves as the interface. (a) Chromatogram on the SE-52 column; (b) chromatogram on the SE-52 column with the six heart-cuts; (c) GC-GC chiral chromatogram of the transferred components. (From Mondello, L., Verzera, A., Previti, P., Crispo, F., and Dugo, G. *J. Agric. Food Chem.* 46:4275–4282, 1998.)

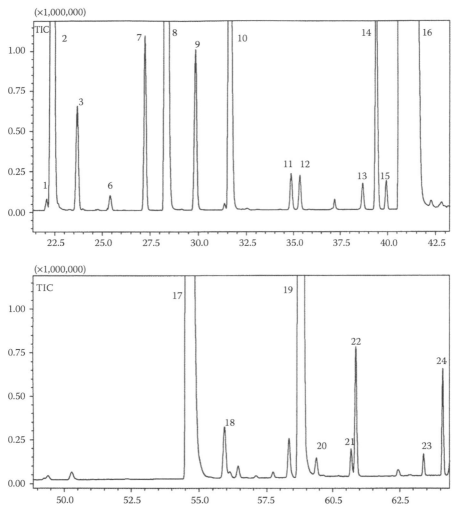

FIGURE 11.5 MDGC chiral chromatogram of a cold-pressed bergamot oil relative to an instrument with an interface made by a pressure balanced system developed by Mondello et al. Peak identification: (1) (+)-α-thujene; (2) (−)-α-thujene; (3) (−)-camphene; 6 (+)-camphene; (7) (+)-β-pinene; (8) (−)-β-pinene; (9) (+)-sabinene; (10) (−)-sabinene; (11) (−)-α-phellandrene; (12) (+)-α-phellandrene; (13) (−)-β-phellandrene; (14) (−)-limonene; (15) (+)-β-phellandrene; (16) (+)-limonene; (17) (−)-linalool, (18) (+)-linalool; (19) (−)-linalyl acetate; (20) (+)-linalyl acetate; (21) (+)-terpinen-4-ol; (22) (−)-terpinen-4-ol; (23) (−)-α-terpineol; (24) (+)-α-terpineol. (From Dugo, G., Bonaccorsi, I., Sciarrone, D., et al. *J. Essent. Oil Res.* 24:93–117, 2012.)

1% of total content of linalool are still acceptable for genuine bergamot oils (Juchelka and Mosandl 1996; Mosandl and Juchelka 1997a, 1997b). A similar behavior was observed for linalyl acetate, with a dextrorotatory isomer percentage generally lower than 0.3% of its total content (König et al. 1997; Mondello et al. 1997, 1998a; Dugo et al. 2001, 2012; Sciarrone 2009).

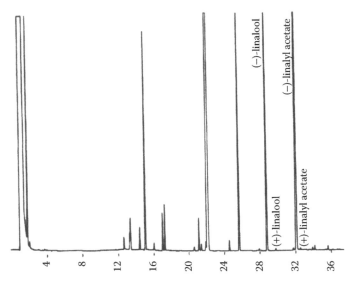

FIGURE 11.6 Chiral chromatogram of an industrial bergamot oil: Enantiomeric ratio of linalool and linalyl acetate. (From Dellacassa, E., Lorenzo, D., Moyna, P., Verzera, A., and Cavazza, A. *J. Essent. Oil Res.* 9:419–426, 1997.)

Table 11.1 shows the results available in literature on cold-pressed bergamot oils of industrial production. The table appendix contains information on the origin of the samples and on the techniques utilized. The results in Table 11.1 refer to the Calabrian oils produced by peeling machine (Pelatrice) with the exception of Dellacassa et al. (1997), whose results are related to an experiment conducted in Uruguay with an industrial Sfumatrice machine, and with the further exception of a few oils produced on the Ivory Coast (Casabianca and Graff 1996; Casabianca and Chau 1997; Casabianca et al. 1998; Sciarrone 2009). Most values are related to a rather small number of samples, while those reported by Mondello et al. (1998a) and Dugo et al. (2001, 2012) are representative of entire growing seasons and of the production areas of Calabria. The data reported by different authors are generally in good agreement, with the exception of the enantiomeric ratio of terpinen-4-ol reported by Mangiola et al. (2009) (column 15 of Table 11.1), probably due to typing or calculation errors; in fact, the chromatogram reported in the paper shows an S-(+)/R-(−)-terpinen-4-ol ratio of about 25:75, thus in agreement with other literature data. The values of R-(+)-α-terpineol reported by Mangiola et al. (2009) ranging between 31.5% and 43.7% fall within the range previously established for cold-pressed oils from Calabria. It's worth noting, however, that in this case there were not significant variations of R-(+)-α-terpineol during the season, whereas Mondello et al. (1998a) reported an increase of R-(+)-α-terpineol from 30.6% to 82.5%.

As can be noticed in Table 11.1, most papers report only the enantiomeric ratio of linalool and linalyl acetate, while the other components were determined by a limited number of authors. For example, the enantiomeric distributions of citronellol and

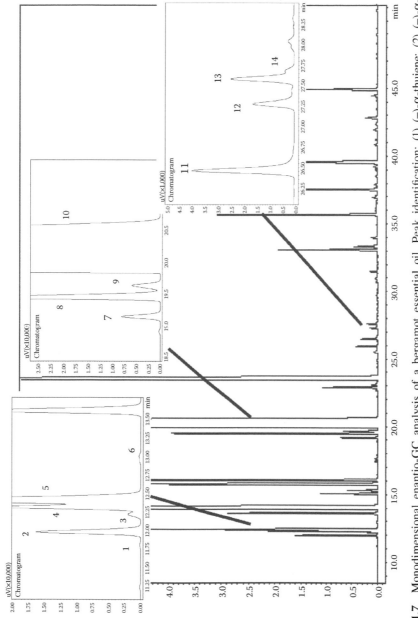

FIGURE 11.7 Monodimensional enantio-GC analysis of a bergamot essential oil. Peak identification: (1) (−)-α-thujene; (2) (−)-α-thujene; (3) (−)-camphene; (4) (+)-α-pineve; (5) (−)-α-pinene; (6) (+)-camphene; (7) (−)-β-phellandrene; (8) (−)-limonene; (9) (+)-β-phellandrene; (10) (+)-limonene; (11) (−)-linalool; (12) (+)-linalool; (13) (−)-citronellal; (14) (+)-citronellal. (From Sciarrone, D. Personal communication, 2009.)

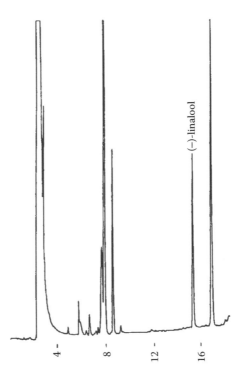

FIGURE 11.8 Chiral chromatogram of a genuine cold-pressed bergamot oil. (From Cotroneo, A., Stagno d'Alcontres, I., and Trozzi, A. *Flavour Fragr. J.* 7:15–17, 1992.)

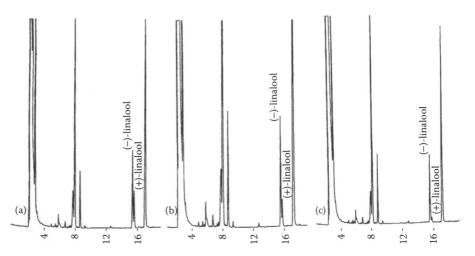

FIGURE 11.9 Chiral chromatogram of mixtures of reconstituted and natural bergamot oils. (a) Reconstituted oil 60%, natural oil 40%; (b) reconstituted oil 30%, natural oil 70%; (c) reconstituted oil 5%, natural oil 95%. (From Cotroneo, A., Stagno d'Alcontres, I., and Trozzi, A. *Flavour Fragr. J.* 7:15–17, 1992.)

TABLE 11.1

Enantiomeric Distribution of Some Volatile Components of Industrial Cold-Pressed Bergamot Oils

						Italy				
		1, 2	3a, 4a	5	6, 7	8, 9	10, 11	12a	13	14a
α-Thujene	S-(+)[a]							0–1.0		0.5–1.0
	R-(–)[a]							100–99.0		99.5–99.0
α-Pinene	R-(+)			28.9–32.7	26–34			29.3–35.0		31.0–33.6
	S-(–)			71.1–67.3	74–66			70.7–65.0		69.0–66.4
Camphene	1S,4R-(–)									85.7–90.1[d]
	1R,4S-(+)									14.3–9.9[d]
β-Pinene	R-(+)			7.4–8.9	<9	6.8–8.9	6.8–9.5	7.3–8.6		8.2–10.3
	S-(–)			92.6–91.1	>91	93.2–91.1	93.2–90.5	92.7–91.4		91.8–89.7
Sabinene	R-(+)					14.1–16.0	14.1–18.8	15.0–18.6	15	15.8–17.4
	S-(–)					85.9–84.0	85.9–81.2	85.0–81.4	85	84.2–82.6
α-Phellandrene	R-(–)									52.0–54.7
	S-(+)									48.0–45.3
β-Phellandrene	R-(–)									26.3–36.0
	S-(+)									73.7–64.0
Limonene	S-(–)	>99.5		2.0–2.6	>97	2.0–2.7		1.7–2.2	3	1.5–1.7
	R-(+)	<0.5		98.0–97.4	<3	98.0–97.3		98.3–97.8	97	98.5–98.3
Linalool	R-(–)		99.9–100	99.0–99.7	>99	99.5–99.7	99.4–99.7	99.2–99.5	99.2	99.5–99.7
	S-(+)		0.1–0	1.0–0.3	<1	0.5–0.3	0.6–0.3	0.8–0.5	0.8	0.5–0.3
Terpinen-4-ol	S-(+)					13.4–25.4	9.7–26.3			22.4–31.5
	R-(–)					86.6–74.6	90.3–73.7			77.6–68.5
α-Terpineol	S-(–)					31.9–50.7	17.5–69.4	36.4	58	14.2–55.4
	R-(+)					68.1–49.3	82.5–30.6	63.6	42	85.8–44.6

continued

TABLE 11.1 (CONTINUED)
Enantiomeric Distribution of Some Volatile Components of Industrial Cold-Pressed Bergamot Oils

		Italy									Uruguay
		1, 2	3a, 4a	5	6, 7	8, 9	10, 11	12a	13	14a	18
Citronellol	S-(−)							12			
	R-(+)							88			
Linalyl acetate	R-(−)		99.5–100	>99	>99	99.7–99.8	99.7–99.9	99.7–99.8	>99	99.8–99.9	
	S-(+)		0.5–0	<1	<1	0.3–0.2	0.3–0.1	0.3–0.2	>1	0.2–0.1	
α-Terpinyl acetate	S-(−)							44			
	R-(+)							56			
(E) Nerolidol	S-(+)				>80						
	R-(−)				<20						

		Italy					Ivory Coast		
		15	16	17a	17b	17c	3b, 4b	12b	14b
α-Thujene	S-(+)[a]			0.4–0.7[d]	0.5–0.7[d]	0.5–0.7[d]		1.0	0.5–1.3
	R-(−)[a]			99.6–99.3[d]	99.5–99.3[d]	99.5–99.3[d]		99.0	99.5–98.7
α-Pinene	R-(+)			32.5–33.0[b]	30.2–32.6[b]	24.2–30.0[b]		32.5–38.4	33.9–36.1
	S-(−)			67.5–67.0[b]	69.8–67.4[b]	75.8–70.0[b]		67.5–61.6	66.1–63.9
Camphene	1S,4R-(−)			83.5–91.0[d]	84.7–88.7[d]	82.0–89.6[d]			88.0–89.4[d]
	1R,4S-(+)			16.5–9.0[d]	15.3–11.3[d]	18.0–10.4[d]			12.0–10.6[d]
β-Pinene	R-(+)	8.7–9.5	9.5	9.7–10.4[d]	8.4–9.5[d]	8.2–9.6[d]		8.6–8.3	7.6–7.9
	S-(−)	91.3–90.5	99.5	90.3–89.6[d]	91.6–90.5[d]	91.8–90.4[d]		91.4–91.7	92.4–92.1

Compound	Config									
Sabinene	R-(+)	13.7–16.3	14.6	16.9–17.5[d]	16.0–17.1[d]		15.9–17.1[d]	15.0–15.0	16.4–19.8	
	S-(−)	86.3–83.7	85.4	83.1–82.5[d]	84.0–82.9[d]		84.1–82.9[d]	85.0–85.0	83.6–80.2	
α-Phellandrene	R-(−)			50.4–55.3[d]	50.7–54.7[d]		48.8–66.4[d]		43.1–46.6	
	S-(+)			49.6–44.7[d]	49.3–45.3[d]		51.2–33.6[d]		56.9–53.4	
β-Phellandrene	R-(−)			42.4–55.9[d]	43.7–49.1[d]		48.2–66.7[d]		33.8–36.9[d]	
	S-(+)			57.6–44.1[d]	56.3–50.9[d]		51.8–33.3[d]		66.2–63.1[d]	
Limonene	S-(−)	1.9–2.1	1.8	1.8–1.9[d]	1.7–1.9[d]		1.8–2.3[d]	2.1–1.7	1.2–1.4	
	R-(+)	98.1–97.9	98.2	98.2–98.1[d]	98.3–98.1[d]		98.2–97.7[d]	97.9–98.3	98.8–98.6	
Linalool	R-(−)	99.3–99.5	99.4	99.4–99.5[d]	99.4–99.6[d]	100	99.4–99.7[d]	99.4–99.5	98.4–99.6	99.4–99.6
	S-(+)	0.7–0.5	0.6	0.6–0.5[d]	0.6–0.4[d]	0	0.6–0.3[d]	0.6–0.5	1.6–0.4	0.6–0.4
Terpinen-4-ol	S-(+)	44.7–67.7[c]		12.4–24.8[d]	16.0–19.2[d]		12.3–44.3[d]		23.7–25.0	
	R-(−)	55.3–32.3[c]		87.6–75.2[d]	84.0–80.8[d]		87.7–55.7[d]		76.3–75.0	
α-Terpineol	S-(−)	56.3–68.5		22.7–44.6[d]	20.5–54.7[d]		24.1–64.0[d]	23.0–25.8	14.0–27.2	
	R-(+)	43.7–31.5		77.3–55.4[d]	79.5–45.3[d]		75.9–36.0[d]	77.0–74.2	86.0–72.8	
Citronellol	S-(−)							16.0–20.0		
	R-(+)							84.0–80.0		
Linalyl acetate	R-(−)	99.6–99.7	99.7	99.6–99.7[d]	99.7–99.8[d]	100	99.5–99.8[d]	99.7–100	99.0–99.8	99.5–99.8
	S-(+)	0.4–0.3	0.3	0.4–0.3[d]	0.3–0.2[d]	0	0.5–0.2[d]	0.3–0	1.0–0.2	0.5–0.2
α-Terpinyl acetate	S-(−)							36.0–42.6		
	R-(+)							64.0–57.4		
(E)-nerolidol	S-(−)									
	R-(+)									

a Correct enantiomer not confirmed assigned according to Casabianca and Chau (1997)
b Range of the values of E-GC analyses; in MDGC analyses the enantiomers of α-pinene are co-eluted
c See text
d Range of the values of MDGC analyses

continued

TABLE 11.1 (CONTINUED)
Enantiomeric Distribution of Some Volatile Components of Industrial Cold-Pressed Bergamot Oils

Appendix to Table 11.1

1, 2.   Cotroneo et al. (1992), Verzera et al. (1996). Range of the values of 150 samples of genuine industrial cold-pressed bergamot oils produced from 1988 to 1991; direct enantio-GC on capillary column (25 m × 0.25 mm) coated with 2,3,6-tri-*O*-methyl-β-CD (30%) and OV-17 (70%).

3, 4.   Casabianca and Graff (1996), Casabianca et al. (1998). Three samples from (a) Italy, (b) one sample from Ivory Coast; MDGC/FID; precolumn: capillary column (25 m × 0.25 mm × 0.25 μm) coated with BP-20, main columns: capillary columns (25 m × 0.25 mm × 0.25 μm) coated with 2,6-diethyl-6-*tert*-butyldimethylsilyl-β-CD or with 2,6-dimethyl-3-pentyl-β-CD; enantio-GC/MS on capillary column (25 m × 0.25 mm × 0.25 μm) coated with 2,6-dimethyl-3-pentyl-β-CD.

5.   Juchelka and Mosandl (1996). Range of the values of eight genuine industrial cold-pressed bergamot oils produced in different periods of a productive season; MDGC; precolumn: capillary column (30 m × 0.32 mm × 0.23 μm) coated with SUPELCOWAX, main column: capillary column (26 m × 0.23 m × 0.25 μm) coated with 2,3-di-*O*-acetyl-6-*O*-*tert*-butylmethylsilyl-β-CD (50%) and OV-1701 (50%) (for analysis of linalool and linalyl acetate); precolumn: capillary column (30 m × 0.23 mm × 0.23 μm) coated with OV-225, main column: capillary column (25 m × 0.23 mm × 0.25 μm) coated with 2,3-di-*O*-methyl-6-*O*-*tert*-butylmethylsilyl-β-CD (50%) and OV-1701 (50%) (for analysis of α-pinene, β-pinene, and limonene).

6, 7.   Mosandl and Juckelka (1997a, 1997b). MDGC; for experimental conditions see point 5 of this appendix.

8, 9.   Mondello et al. (1997, 1998d). Range of the values of 52 samples of genuine industrial cold-pressed bergamot oils; MDGC; precolumn: capillary column (30 m × 0.32 mm × 0.40–0.45 μm) coated with SE-52, main column: capillary column (25 m × 0.25 mm × 0.2 μm) coated with 2,3-di-*O*-ethyl-6-*O*-*tert*-butyldimethylsilyl-β-CD (30%) and PS-086 (70%).

10, 11.   Mondello et al. (1998a), Dugo et al. (2001). Range of the values of 101 samples of genuine industrial cold-pressed bergamot oils produced during the 1996–97 productive season; for experimental conditions see point 8, 9 of this appendix.

12.   Casabianca and Chau (1997). Range of the values of 4 samples of genuine industrial cold-pressed bergamot oils from Italy (a); values of 2 genuine industrial cold-pressed bergamot oils from Ivory Coast (b); direct enantio-GC on capillary column (25 m × 0.22 mm × 0.25 μm) coated with 2,3-di-*O*-methyl-6-*O*-*tert*-butylmethylsilyl-β-CD (50%) and OV-1701 (50%) (for the analysis of monoterpene hydrocarbons), coated with 2,3-di-*O*-ethyl-6-*O*-*tert*-butylmethylsilyl-β-CD (50%) (for the analysis of linalool and linalyl acetate), coated with 2,6-di-*O*-pentyl-3-*O*-butyryl-γ-CD (for the analysis of citronellol, α-terpineol and α-terpinyl acetate).

13. Shellie and Marriott (2002). Italy; one sample. Details on experimental conditions are reported in the Chapter 7, this volume.

14. Sciarrone (2009, personal communication). (a) Range of the values of 17 genuine industrial cold-pressed bergamot oils from Italy; (b) values of 2 genuine industrial cold-pressed bergamot oils from Ivory Coast; direct enantio-GC on capillary column (25 m × 0.25 mm × 0.25 µm) coated with 2,3-di-O-ethyl-6-O-*tert*-butyldimethylsilyl-β-CD (50%) and PS-086 (50%); MDGC using the same chiral column of direct enantio-GC and a capillary precolumn (30 m × 0.25 mm × 0.25 µm) coated with SLB-5MS.

15. Mangiola et al. (2009). Range of the values of 6 genuine industrial cold-pressed bergamot oils produced from November 2007 to March 2008; MDGC; precolumn: capillary column (30 m × 0.25 mm × 0.25 µm) coated with RTX-5; main column: capillary column (25 m × 0.20 mm × 0.18 µm) coated with 2,3-di-O-ethyl-6-O-*tert*-butyldimethylsilyl-β-CD.

16. Costa et al. (2010). One genuine industrial cold-pressed bergamot oil; for experimental conditions see point 14 of this appendix.

17. Dugo et al. (2012). Range of the values of (a) 8 genuine industrial cold-pressed bergamot oils produced in 2008–09 season; (b) 11 oils produced in 2009–10 season; (c) 23 oils produced in 2010–11 season; for experimental conditions see point 14 of this appendix.

18. Dellacassa et al. (1997). Range of composition of 3 genuine industrial cold-pressed bergamot oils from Uruguay; direct enantio-GC on capillary column (25 m × 0.25 µm) coated with 2,3-di-O-ethyl-6-O-*tert*-butyldimethylsilyl-β-CD.

1–18. The oils relative to references 1, 2, 5 through 11, and 13 through 17 were extracted by Pelatrice; the oils relative to reference 18 were extracted by Sfumatrice; in references 3,4, and 12, the extraction technique is not indicated.

α-terpinyl acetate are reported only by Casabianca and Chau (1997); the enantiomeric ratios of camphene and α- and β-phellandrene are provided only by researchers in Messina (Sciarrone 2009; Dugo et al. 2012) while (E)-nerolidol is reported by
Mosandl and Juchelka (1997b).

Tateo et al. (2000) (results not included in Table 11.1) report the chromatogram
of a genuine bergamot oil, where the enantiomeric distributions of limonene, linalool, and linalyl acetate appear to be in agreement with those of cold-pressed
bergamot oils, and the chromatogram of a reconstituted oil with racemic ratios of
linalool and linalyl acetate. Surprisingly, the authors argue that the enantiomeric
distribution of these components and their origins do not affect the sensory profile
of bergamot oil.

As previously mentioned, in the case of cold-pressed bergamot oil, the enantiomeric ratios of linalool and linalyl acetate vary within very narrow limits and can
be used as a valuable reference for its authenticity assessment. The enantiomeric
distribution of limonene, which is also fairly constant, is similar to that of many
other essential oils, such as lemon (Mondello et al. 1999), mandarin (Mondello
et al. 1998b), and lime (Mondello et al. 1998c) and is slightly different from the one
of sweet orange (Mondello et al. 1997), where the percentage of levorotatory isomer
does not exceed 0.6% of its total content. The enantiomeric ratios of β-pinene and
sabinene vary within rather narrow limits and present values similar to those of
lemon oil (Mondello et al. 1999) but different from those of mandarin oil (Mondello
et al. 1998b). The same behavior is described for α-thujene, while the enantiomeric
ratios of camphene, α-pinene, and α- and β-phellandrene vary within relatively
wide limits; unfortunately, the literature data for these compounds are limited. The
enantiomeric ratios of terpinen-4-ol and α-terpineol vary within wide intervals
(Mondello et al. 1997, 1998a; Dugo et al. 2001, 2012; Sciarrone 2009). For the first
of these two compounds variations are quite irregular, while for the second one
they are related to the period of fruit harvesting. Mondello et al. (1998a) correlates
the changes in the enantiomeric distribution of β-pinene, sabinene, limonene, linalool, terpinen-4-ol, α-terpineol, and linalyl acetate with the production period of
the essential oil. The results of the latter research are summarized in Figure 11.10,
where the average values of the enantiomeric excesses of the components are investigated as a function of a fortnightly production period. The enantiomeric distributions of linalool, linalyl acetate, limonene, sabinene, and β-pinene do not undergo
significant changes during the production season; terpinen-4-ol undergoes irregular
variations, while α-terpineol varies consistently with a regular trend (see Figure
11.10). The fortnightly average value of the ratio between (−) and (+)-α-terpineol is
in fact 63.0:37.0 at the beginning of the season, changing to 21.7:78.3 at the end of
it. The value of this ratio can therefore provide information on the period of production of the oil.

Changes in the enantiomeric excess of the volatile components of bergamot oil
during the season have been recently confirmed by Dugo et al. (2012), who analyzed
a series of batches (each of 1000 kg) of essential oil produced from October 2008
to February 2009, October 2009 to March 2010, and October 2010 to March 2011
(Figure 11.11a–c). The enantiomeric ratios for linalool, linalyl acetate, limonene,
β-pinene, and sabinene remain constant while irregular variations for terpinen-4-ol,

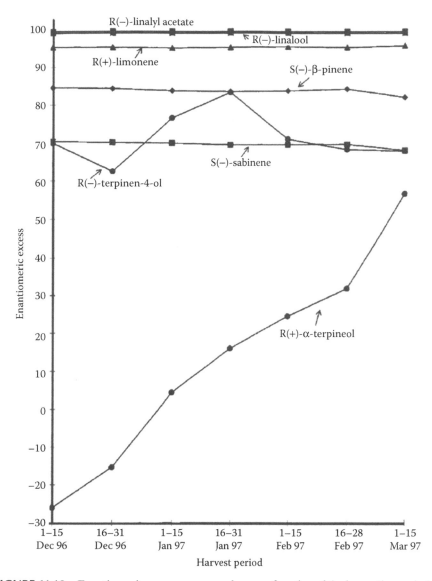

FIGURE 11.10   Enantiomeric excess average values as a function of the harvesting period of β-pinene, sabinene, limonene, linalool, terpinen-4-ol, α-terpineol, and linalyl acetate. (From Mondello, L., Verzera, A., Previti, P., Crispo, F., and Dugo, G. *J. Agric. Food Chem.* 46: 4275–4282, 1998.)

probably influenced more by the conditions of extraction than from the harvesting period of the fruits, are reported; furthermore, increased enantiomeric excesses of (+)-α-terpineol are reported, even if the values in some periods of production are different from each other and different from those reported by Mondello et al. (1998a). In the latter paper the enantiomeric ratios of α- and β-phellandrene and α-thujene show a constant behavior throughout the whole growing season.

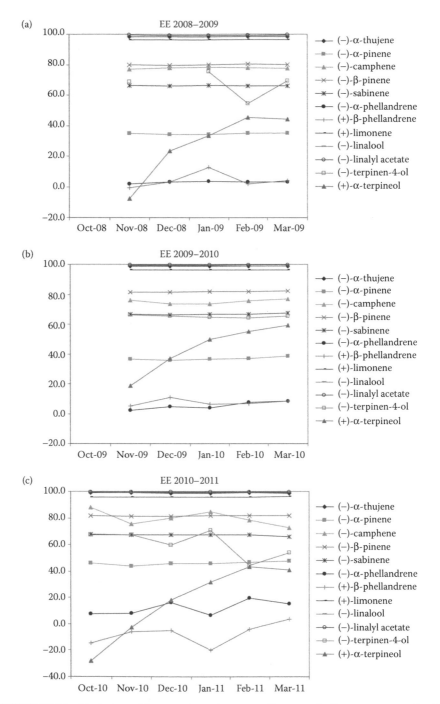

FIGURE 11.11 Variation of the enantiomeric excess of some volatile components of cold-pressed bergamot oils during (a) 2008–09, (b) 2009–10, and (c) 2010–11 productive seasons. (From Dugo, G., Bonaccorsi, I., Sciarrone, D., et al. *J. Essent. Oil Res.* 24: 93–117, 2012.)

The few results available in literature for oils from the Ivory Coast and Uruguay are in agreement with those of oils from Calabria. An exception is represented by one of the two oils from the Ivory Coast analyzed by Sciarrone et al. (2009), which shows 1.6% of (+)-linalool and 1.0% of (+)-linalyl acetate, values never found in genuine cold-pressed oils; these results are difficult to be attributed to the source of the oil and are probably related to a contamination or an adulteration.

Globally, based on the behavior of the ranges, the enantiomeric distribution can be considered an indication of the authenticity of cold-pressed oils only for some of the components as reported below:

For some other components, the enantiomeric distribution cannot be considered significant; some of them, such as terpinen-4-ol and α-terpineol, present very wide ranges of variability, while others, such as camphene, α- and β-phellandrene, citronellal, and α-terpinyl-acetate, have not been sufficiently investigated at this time. It is the authors' opinion that, even if reported in literature, the S-(+) enantiomers of linalool and linalyl acetate never reach 1% of the total contents of the compounds but, on the other hand, are never absent and are present at least in traces. Failure in the determination of traces of S-(+) enantiomers is probably due to the analytical technique.

11.3 COMMERCIAL OILS

In Table 11.2 are data from literature relative to bergamot commercial oils. As can be noticed, most samples can be considered adulterated because of the enantiomeric ratio values of linalool and/or linalyl acetate or limonene. Only one of the samples analyzed by Bernreuther and Schreier (1991), a sample analyzed by Casabianca et al. (1995), two samples analyzed by Casabianca and Chau (1997), the samples analyzed by Ravid et al. (1994, 2009), two of the eight samples analyzed by Neukom et al. (1993), and two of the seven samples analyzed by Juchelka and Mosandl (1996) show values in agreement with those of a genuine oil. The oils analyzed by Mosandl et al. (1990) and Weinrich and Nitz (1992) present a very high amount of the levorotatory isomer of limonene. This data indicates that these oils might have been reconstituted with sweet orange terpenes and added to the commercial product, commonly called (−)-limonene, in order to correct the optical rotation value. Considering the value of the enantiomeric ratio of limonene, the oil analyzed by Casabianca and Graff (1994) can be also considered an oil reconstituted using orange terpenes, without the correction of the optical rotation value with (−)-limonene. The amount of (−)-limonene in sweet orange oil accounts, in fact, only for 0.6% of the total content of limonene (Mondello et al. 1997). Due to the separation difficulty, such a small amount could not be detected during the analysis.

In addition to the data reported in Table 11.2, further information on commercial oil of bergamot is found in literature. König et al. (1997) report that 37 out of 52 samples of commercial bergamot oil were clearly adulterated due to the high content of the S-(+)-enantiomers of linalool and linalyl acetate and only 18 of them presented a content of the two S-(+)-enantiomers lower than 1%. The presence of large amounts of the S-(+)-enantiomers of linalool and linalyl acetate in

TABLE 11.2
Enantiomeric Distribution of Some Volatile Components of Commercial Bergamot Oils

		Commercial Oils																
		Adulterated											Genuine						
		1,2,3a	3b	4	5a	7	8a	9	10a	11,12	13a	14a	5b	6	8b	10b	13b	14b	15
α-Pinene	R-(+)	28									25.4–34.0						29.6–32.5		
	S-(−)	72									74.6–66.0						70.4–67.5		
β-Pinene	R-(+)	6									4.1–7.4						8.0–9.0		
	S-(−)	94									95.9–92.6						92.0–91.0		
Limonene	S-(−)	14				10		0			1.8–15.0						2.4–2.5		0–1
	R-(+)	86				90		100			98.2–85.0						97.6–97.5		100–99
Linalool	R-(−)			68.5	55.8–87.8	66	1.6–16.7	63		63.0–75.6	54.9–72.2	63	100				99.0–99.3	100	100
	S-(+)			31.5	44.2–12.2	34	86.4–83.3	37		37.0–24.4	45.1–27.8	37	0				1.0–0.7	0	0
Linalyl	R-(−)		53–69			60		75	75–98	73.5–75.0	51.9–83.0	75		100		99.5	>99		100
acetate	S-(+)		47–31			40		25	25–2	26.5–25.0	48.1–17.0	25		0		0.5	<1		0

Appendix to Table 11.2:

1,2, 3a. Mosandl et al. (1990), Mosandl (1995), Hener et al. (1990). One sample; MDGC; precolumn: capillary column (60 m × 0.32 mm × 0.25 μm) coated with SUPELCOWAX TM-10, main column: capillary column (47 m × 0.23 mm) coated with 2,3,6-tri-O-methyl-β-CD (10%) and OV-1701 (90%).

3b. Hener et al. (1990). Range of the values of tree samples; MDGC; precolumn: capillary column coated with DB-1701, main column: coated with $Ni(HFC)_2$.

4. Schubert and Mosandl (1991). One sample; MDGC; precolumn: capillary column (27 m × 0.32 mm) coated with Superox-06; main column: capillary column (41 m × 0.23 mm) coated with 2,3,6-tri-O-ethyl-β-CD (30%) and OV-1701 (70%).

5. Bernreuther and Schreier (1991). (a) Two adulterated oils; (b) one genuine oil; MDGC; precolumn: capillary column (30 m × 0.25 mm × 0.25 μm) coated with DB-5, main column: capillary column (25 m × 0.25 μm) coated with 2,6-di-O-pentyl-3-O-acetyl-β-CD.

6. Ravid et al. (1994). Several samples: separation of linalyl acetate by preparative GC on packed column of Carbowax 20M; chiral separation of linalyl acetate on capillary column (25 m × 0.25 mm) coated with 3-O-butyryl-2,6-di-O-pentyl-γ-CD.

7. Weinreich and Nitz (1992). One sample; MDGC; precolumn: capillary column (30 m × 0.25 mm × 1 µm) coated with SPB-TMS, main column: capillary column (50 m × 0.32 mm) coated with 2,3,6-tri-O-methyl-β-CD (30%) (for the analyses of limonene and linalool); precolumn: capillary column (30 m × 0.25 mm × 0.25 µm) coated with SE-54, main column: capillary column (50 m × 0.32 mm) coated with 2,6-di-O-pentyl-3-O-butyryl-γ-CD (for the analyses of linalyl acetate).

8. Neukom et al. (1983). Range of the values of (a) 6 adulterated commercial oils; (b) 2 genuine commercial oils; direct enantio-GC on capillary column (40 m × 0.25 mm × 0.25 µm) coated with 2,3,6-tri-O-methyl-β-CD (10%) and OV-1701 (90%).

9. Casabianca and Graff (1994). One sample of commercial oil; flash chromatography on Silica column for isolation of limonene, linalool, linalyl acetate; hydrogenation of linalyl acetate; MDGC; precolumn: capillary column (25 m × 0.32 mm × 0.5 µm) coated with BP-20, main column: capillary column (25 m × 0.22 mm × 0.25 µm) coated with OV-1701 (90%) and 2,3,6-tri-O-methyl-β-CD (10%).

10. Casabianca et al. (1995). Range of the values of (a) 3 commercial adulterated oils; (b) one commercial genuine oil; for experimental conditions see point 9 of this appendix.

11,12. Casabianca and Graff (1996). Casabianca et al. (1998). Two samples. For experimental methods see point 3, 4 of the appendix to Table 11.1.

13. Juchelka and Mosandl (1966). Range of the values of (a) 5 adulterated commercial oils; (b) 2 genuine commercial oils; for experimental conditions see point 5 of appendix to Table 11.1.

14. Casabianca and Chau (1997). (a) One adulterated commercial oil; (b) 2 genuine commercial oils; for experimental conditions see point 12 of appendix to Table 11.1.

15. Ravid et al. (2009). Six samples; auto-HS-SPME; direct enantio-GC/MS on capillary column (30 m × 0.25 mm × 0.25 µm) coated with 2,3,6-tri-O-methyl-β-CD.

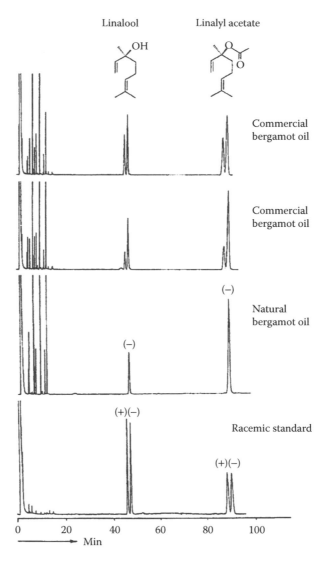

FIGURE 11.12 Enantiomeric ratio of linalool and linalyl acetate in commercial and natural bergamot oils. (From König, W. A. *Chirality* 10:499–504, 1998.)

some bergamot commercial oils was confirmed by König (1998) (Figure 11.12). Becher (1995) analyzed 53 commercial oils concluding, from the evaluation of the enantiomeric ratios of linalool and linalyl acetate, that 37 of them were adulterated. Binder et al. (2001) analyzed 24 commercial oils: six samples showed enantiomeric ratios of linalool and linalyl acetate not compatible with those of genuine oils. Begnaud and Chaintreau (2005) found a 15:85 ratio between S-(+) and R-(−)-linalool in a commercial bergamot oil.

α-Thujene	S-(+)-	<1.5%
	R-(−)-	>98.5%
α-Pinene	R-(+)-	<35%
	S-(−)-	>65%
β-Pinene	R-(+)-	<10%
	S-(−)-	>90%
Sabinene	R-(+)-	<20%
	S-(−)-	>80%
Limonene	S-(−)-	<3%
	R-(+)-	>97%
Linalool	S-(+)-	<1%
	R-(−)-	>99%
Linalyl acetate	S-(+)-	<1%
	R-(−)-	>99%

11.4   BERGAPTEN-FREE OILS

The presence on the market and the technologies of preparation of bergapten-free bergamot oils were discussed in Section 8.3.3 of Chapter 8, this volume.

König et al. (1997) claim to have analyzed some samples of bergamot oil, processed in order to remove furocoumarins, that presented distribution values of linalool and linalyl acetate matching those of the cold-pressed oils, but numerical results are not reported in that paper. The enantiomeric ratio of some components of a volatile fraction of bergapten-free bergamot oils, obtained by alkaline treatment or by distillation, has been evaluated by Mondello et al. (1998a), Dugo et al. (2001, 2012), and Costa et al. (2010). The results of these works are shown in Table 11.3. From this data, it is clear that the treatments do not affect the enantiomeric ratio values, which remain in the ranges of variability previously presented for genuine cold pressed oils (see Table 11.1).

11.5   RECOVERED OILS

The oils obtained from the residues of cold extraction (Torchiati, Ricicli, Pulizia dischi, Fecce, and Peratoner oils), and the recovery methods have been described in Section 8.3.5 of Chapter 8, this volume.

11.5.1   OILS RECOVERED BY COLD-PROCEDURES

Information on the enantiomeric distribution of oils recovered by cold procedures is quite scant. König et al. (1997) report that in some samples of Torchiati oils the enantiomeric distribution of linalool and linalyl acetate was the same as that of the cold-pressed oils. Mondello et al. (1998a) and Dugo et al. (2001) analyzed 18 recovered oils and compared the results with those obtained for genuine cold pressed oils, produced during the same period. In fact, due to the changes in enantiomeric ratios of certain components during the season

TABLE 11.3
Enantiomeric Distribution of Some Volatile Components of Bergapten-Free Bergamot Oils

		Alkali Treated			Distilled		
		1a, 2a	3a	4a	1b, 2b	3b	4b
α-Thujene	S-(+)[a]			0.6–0.7[c]			0.6[c]
	R-(−)[a]			99.4–99.3[c]			99.4[c]
α-Pinene	R-(+)			32.2–33.0[b]			31.6–32.4[b]
	S-(−)			67.8–67.0[b]			68.4–67.6[b]
Camphene	1S,4R-(−)			86.1–88.0[c]			87.7–88.9[c]
	1R,4S-(+)			13.9–12.0[c]			12.3–11.1[c]
β-Pinene	R-(+)	8.2–8.7	9.6	9.2–9.4[c]	8.1–9.2	9.5	9.0–10.1[c]
	S-(−)	91.8–91.3	90.4	90.8–90.6[c]	91.9–90.8	90.5	91.0–89.9[c]
Sabinene	R-(+)	15.3–15.7	14.3	16.8–17.3[c]	15.1–16.0	14.6	16.4–17.0[c]
	S-(−)	84.7–84.3	85.7	83.2–82.7[c]	84.9–84.0	85.4	83.6–83.0[c]
α-Phellandrene	R-(−)			49.7–52.6[c]			51.5–52.8[c]
	S-(+)			50.3–47.4[c]			48.5–47.2[c]
β-Phellandrene	R-(−)			47.3–52.2[c]			43.4–48.0[c]
	S-(+)			52.7–47.8[c]			56.6–52.0[c]
Limonene	S(−)	2.1	1.8	1.7–1.8[c]	2.1–2.2	1.8	1.7–1.9[c]
	R(+)	97.9	98.2	98.3–98.2[c]	97.9–97.8	98.2	98.3–98.1[c]
Linalool	R(−)	99.6	99.5	99.4–99.5[c]	99.5–99.6	99.5	99.3–99.4[c]
	S(+)	0.4	0.5	0.6–0.5[c]	0.5–0.4	0.5	0.7–0.6[c]
Terpinen-4-ol	S-(+)	18.4–20.1		16.1–20.0[c]	18.8–21.5		21.8–22.0[c]
	R-(−)	81.6–79.9		83.9–80.0[c]	80.2–78.5		78.2–78.0[c]
α-Terpineol	S-(−)	21.4–52.1		28.9–38.7[c]	41.4–50.5		31.0–34.2[c]
	R-(+)	68.6–47.9		71.1–61.3[c]	58.6–49.5		69.0–65.8[c]
Linalyl acetate	R-(−)	99.7–99.8	98.8	99.7–99.8[c]	99.7–99.9	99.7	99.7
	S-(+)	0.3–0.2	0.2	0.3–0.2[c]	0.3–0.1	0.3	0.3

[a] Correct enantiomer not confirmed; tentatively assigned according to Casabianca and Chau (1997).
[b] Range of the values of E-GC analysis; in MDGC analyses the enantiomers of α-pinene are co-eluted.
[c] Range of the values of MDGC analyses.

Appendix to Table 11.3:

1, 2. Mondello et al. (1998a), Dugo et al. (2001). Range of the values of 7 bergapten-free oils obtained by (a) NaOH treatment; (b) distillation; for experimental conditions see point 8, 9 of Table 11.1.

3. Costa et al. (2010). One sample of bergapten-free oil obtained by (a) NaOH treatment; (b) distillation; for experimental conditions see point 14 of appendix to Table 11.1.

4. Dugo et al. (2012). Range of the values of 3 bergapten-free oils obtained by (a) NaOH treatment; (b) distillation; for experimental conditions see point 14 of appendix to Table 11.1.

(Figures 11.10 and 11.11), the comparison must be performed between oils produced in the same period in order to evaluate the influence of the technological process.

The data reported by Mondello et al. (1998a) and Dugo et al. (2001) are shown in Table 11.4; note that in all three types of recovered oils no racemization phenomena

TABLE 11.4

Enantiomeric Distribution of Some Volatile Components of Cold-Recovered (Torchiati, Ricicli, Pulizia Dischi) Bergamot Oils and of Cold-Pressed Oils Produced in the Same Period

		February 1–15			February 16–28			
		Torchiati	Ricicli	Cold-pressed	Torchiati	Ricicli	Pulizia dischi	Cold-pressed
		(2)[a]	(2)[a]	(7)[a]	(5)[a]	(5)[a]	(3)[a]	(22)[a]
β-Pinene	R-(+)	7.7–8.5	7.8–7.9	7.5–8.9	7.6–8.0	7.4–7.8	7.8–8.0	7.1–9.0
	S-(−)	92.3–91.5	92.2–92.1	92.5–91.1	92.4–92.0	92.6–92.2	92.2–92.0	92.9–91.0
Sabinene	R-(+)	14.8–15.0	14.9–15.0	14.5–15.9	14.7–15.0	14.5–15.0	14.7–15.0	14.1–18.8
	S-(−)	85.2–85.0	85.1–85.0	85.5–84.1	85.3–85.0	85.5–85.0	85.3–85.0	84.9–81.2
Limonene	S-(−)	2.1–2.3	2.2–2.3	2.0–2.4	2.2–2.3	2.2–2.3	2.2–2.3	2.1–2.6
	R-(+)	97.9–97.7	97.8–97.7	98.0–97.6	97.8–97.7	97.8–97.7	97.8–97.7	97.9–97.4
Linalool	R-(−)	99.6	99.6	99.4–99.7	99.6	99.5–99.6	99.5–99.6	99.4–99.6
	S-(+)	0.4	0.4	0.6–0.3	0.4	0.5–0.4	0.5–0.4	0.6–0.4
Terpinen-4-ol	S-(+)	18.7–18.8	19.3–20.0	12.5–16.6	18.1–19.1	18.7–21.4	12.3–13.1	9.7–20.0
	R-(−)	81.3–81.2	80.7–80.0	87.5–83.4	81.9–80.9	81.3–78.6	87.7–86.9	90.3–80.0
α-Terpineol	S-(−)	5.1–5.2	11.3–18.0	30.6–41.9	4.9–10.5	6.4–18.7	27.5–41.8	20.7–44.1
	R-(+)	94.9–94.8	88.7–82.0	69.4–58.1	95.1–89.5	93.6–81.3	72.5–58.2	79.3–55.9
Linalyl acetate	R-(−)	99.8	99.8	99.8–99.9	99.7–99.9	99.7–99.8	99.7–99.8	99.8–99.9
	S-(+)	0.2	0.2	0.2–0.1	0.3–0.1	0.3–0.2	0.3–0.2	0.2–0.1

[a] Number of samples analyzed.

Source: The results of this table are taken from Mondello et al., *J. Agric. Food Chem.* 46, 4275–4282, 1998a; and Dugo et al., *Perfum. Flav.* 26(1), 20–35, 2001. For experimental conditions see point 8, 9 of appendix to Table 11.1.

occurred, and the values of the enantiomeric distribution fall within the ranges of variability presented by cold-pressed oils. An exception is represented by α-terpineol, for which is evident, in Torchiati and Ricicli, a levorotatory isomer average value lower than the minimum found in genuine cold-pressed oils during the same period. Obviously, this cannot be attributed to a phenomenon of interconversion of the two enantiomers by chemical and physical processes. This phenomenon should, in fact, result in the racemization of the compound and not in the increase in the enantiomeric excess of one of the two enantiomers. For a mixture of S-(−)- and R-(+)-α-terpineol (70:30), treated in hot aqueous acid environment by citric acid, racemization occurs after four hours (Mondello et al. 1998a). The authors provided two explanation of the phenomenon. The first is that R-(+)-α-terpineol could be produced by microorganisms, considering that recycled water and solid residues of cold extraction represent an excellent culture medium (Murdock et al. 1969). The second is that the phenomenon could be attributed to acid or enzymatic reactions which occur before the oil is separated from solid residues, leading to the release of α-terpineol from a glycosidic form. Both hypotheses find support in the fact that the content of α-terpineol in the Torchiati and the Recycle is generally 2 to 5 times higher than that of cold-pressed oils of the same period (see Section 8.3.5.1 in Chapter 8, this volume).

11.5.2 OILS RECOVERED BY DISTILLATION

Table 11.5 summarizes research on bergamot oil recovered by distillation. In so-called "fecce" oils, recovered by distillation at atmospheric pressure (Juchelka and Mosandl 1996; König et al. 1997; Mondello et al. 1998a; Dugo et al. 2001) linalool presents a pronounced tendency to racemization, whereas linalyl acetate and terpinen-4-ol tendency to racemization is low (Figure 11.13). In the oils recovered from "fecce" by distillation at reduced pressure (Mondello et al. 1998a; Dugo et al. 2012), the tendency of linalool and linalyl acetate and terpinen-4-ol to racemization becomes very low, and in the oils obtained with modern techniques of distillation of all the cold extraction residues under vacuum (Sciarrone 2009; Dugo et al. 2012), it is still observed for linalool and terpinen-4-ol, while linalyl acetate shows a similar distribution to that of the cold-pressed oils.

11.6 CONCENTRATED OILS

The procedures employed to obtain concentrated bergamot oils are described in Section 8.3.6 of Chapter 8, this volume. Information on the enantiomeric distribution of the volatile components of concentrated oils produced with traditional methods (distillation and fractionation) are limited to a few samples of terpeneless oils (colored oils) and terpeneless dewaxed oils (colorless oils) analyzed by Costa et al. (2010) and Dugo et al. (2012). The results relative to such oils are shown in Table 11.6. The values of the enantiomeric distribution fall within the ranges of variability of the cold-pressed oils, with the exception of a sample that presents an S-(+)-linalool enantiomer content slightly higher than 1%.

TABLE 11.5
Enantiomeric Distribution of Some Volatile Components of Distilled Bergamot Oils

		1	2	3a, 4a	3b, 4b	5	6a	6b1	6b2
α-Thujene	S-(+)[a]					0.7–1.0	0.5–0.7[b]	0.7[b]	0.7[b]
	R-(−)[a]					99.3–99.0	99.5–99.3[b]	99.3[b]	99.3[b]
α-Pinene	R-(+)	30.2				32.8–32.9	27.4–32.4[c]	32.4[c]	30.2[c]
	S-(−)	69.8				67.2–67.1	72.6–67.6[c]	67.6[c]	69.8[c]
Camphene	1S,4R-(−)					86.7–92.7[b]	83.3–91.0[b]	85.3[b]	88.5[b]
	1R,4S-(+)					13.3–7.3[b]	16.7–9.0[b]	14.7[b]	11.5[b]
β-Pinene	R-(+)	8.2		8.9	8.2	9.3–9.5	9.2–10.2[b]	9.2[b]	7.7[b]
	S-(−)	91.8		91.1	91.8	90.7–90.5	90.8–89.8[b]	90.8[b]	92.3[b]
Sabinene	R-(+)			15.9	15.2	16.8–17.0	14.6–17.0[b]	16.6[b]	15.9[b]
	S-(−)			84.1	84.8	83.2–83.0	85.4–83.0[b]	83.4[b]	84.1[b]
α-Phellandrene	R-(−)					52.8–53.1	49.5–55.8[b]	52.2[b]	53.8[b]
	S-(+)					47.2–46.9	50.4–44.2[b]	47.8[b]	46.2[b]
β-Phellandrene	R-(−)					24.1–25.5	41.4–55.1[b]	46.9[b]	46.6[b]
	S-(+)					75.9–74.5	58.6–44.9[b]	53.1[b]	53.4[b]
Limonene	S(−)	1.5		2.0	2.3	1.6	1.8–2.3[b]	1.9[b]	1.6[b]
	R(+)	98.5		98.0	97.7	98.4	98.2–97.7[b]	98.1[b]	98.4[b]
Linalool	R(−)	91.5	91.3	81.6	98.7	97.8–98.9	94.8–98.1[b]	99.3[b]	97.8[b]
	S(+)	8.5	8.7	18.4	1.3	2.2–1.1	5.2–1.9[b]	0.7[b]	2.2[b]
Terpinen-4-ol	S-(+)			31.8	27.1	29.3–30.4	28.1–30.7[b]	27.7[b]	28.1[b]
	R-(−)			68.2	72.9	70.7–69.6	71.9–69.3[b]	72.3[b]	71.9[b]
α-Terpineol	S-(−)			26.6	11.2	20.5–31.9	21.0–32.7[b]	3.4[b]	8.2[b]
	R-(+)			73.4	88.8	79.5–68.1	79.0–67.3[b]	96.6[b]	97.8[b]
Linalyl acetate	R-(−)	>99		98.9	99.1	99.8	98.8–99.8[b]	99.6[b]	99.5[b]
	S-(+)	<1		1.1	0.9	0.2	1.2–0.2[b]	0.4[b]	0.5[b]

[a] Correct enantiomer not confirmed; tentatively assigned according to Casabianca and Chau (1997).
[b] Range of the values of MDGC analyses.
[c] Range of the values of E-GC analyses; in MDGC analyses the enantiomers of α-pinene are co-eluted.

Appendix to Table 11.5:
1. Juchelka and Mosandl (1996). One sample probably obtained by distillation at atmospheric pressure.
2. König et al. (1997). One sample of fecce oil distilled at atmospheric pressure. Direct enantio-GC on capillary column (25 m) coated with 3-*O*-butyryl-2,6-di-*O*-pentyl-γ-CD.
3,4. Mondello et al. (1998a), Dugo et al. (2001). One sample of fecce oil distilled at atmospheric pressure (a), one sample of fecce oil distilled at reduced pressure (b); for experimental conditions see point 8, 9 of appendix to Table 11.1.
5. Sciarrone (2009, personal communication). Range of the values of 3 Peratoner oils obtained by distillation at reduced pressure; for experimental conditions see point 14 of appendix to Table 11.1.
6. Dugo et al. (2012). (a) Range of the values of 3 Peratoner oils obtained by distillation at reduced pressure; (b) two samples of fecce oils; for experimental conditions see point 14 of appendix to Table 11.1.

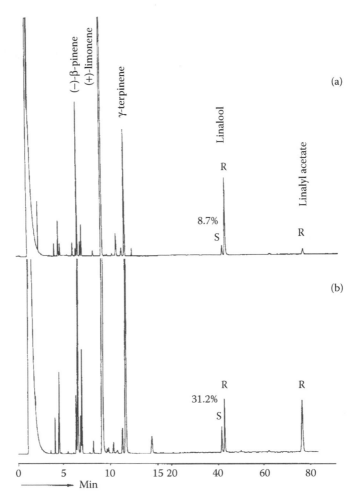

FIGURE 11.13 Chiral chromatograms of an oil obtained by hydrodistillation of (a) berga-mot fruit peel and (b) one fecce oil. (From König, W. A., Fricke, C., Saritas, Y., Momeni, B., and Hohenfeld, G. *J. High Resolut. Chromatogr.* 20:55–61, 1997.)

In addition to those reported in Table 11.6, in literature the results by Ravid et al. (2009) for concentrated oil are presented. Ravid et al. (2009) found the following enantiomeric ratios for limonene, linalool, and linalyl acetate:

Limonene	(−) 20.1	Linalool	(−) 51.5	Linalyl acetate	(−) 49.9
	(+) 79.9		(+) 48.5		(+) 50.1

These data indicate either adulteration or the employment of a technology leading to partial or total racemization of the analyzed compounds.

To the best of the authors' knowledge, there is no information in literature on the enantiomeric distribution of the volatile components of bergamot oils concentrated by means of supercritical fluids.

TABLE 11.6
Enantiomeric Distribution of Some Volatile Components of Concentrated Bergamot Oils

		1a	2a	1b	2b
α-Thujene	S-(+)[a]				10.6[c]
	R-(−)[a]				99.4[c]
α-Pinene	R-(+)				30.8–32.2[b]
	S-(−)				69.2–67.8[b]
β-Pinene	R-(+)		6.5–8.8[c]		8.9–9.8[c]
	S-(−)		93.5–91.2[c]		91.1–90.2[c]
Sabinene	R-(+)				15.7–17.7[c]
	S-(−)				84.3–82.3[c]
α-Phellandrene	R-(−)		51.5–55.3[c]		
	S-(+)		48.5–44.7[c]		
β-Phellandrene	R-(−)				46.7–50.3[c]
	S-(+)				53.3–49.7[c]
Limonene	S-(−)		1.5–2.1[c]		1.7–1.9[c]
	R-(+)		98.5–97.9[c]		98.3–98.1[c]
Linalool	R-(−)	99.4	99.3–99.5[c]	99.5	98.7–99.3[c]
	S-(+)	0.6	0.7–0.5[c]	0.5	1.3–0.7[c]
Terpinen-4-ol	S-(+)		14.9–20.8[c]		21.2–24.8[c]
	R-(−)		85.1–79.2[c]		78.8–75.2[c]
α-Terpineol	S-(−)		23.3–34.9[c]		19.2–34.0[c]
	R-(+)		76.7–65.1[c]		80.8–66.0[c]
Linalyl acetate	R-(−)	99.7	99.8[c]	99.8	99.6–99.8[c]
	S-(+)	0.3	0.2[c]	0.2	0.4–0.2[c]

[a] Correct isomer not confirmed; tentatively assigned according to Casabianca and Chau (1997).
[b] Range of the values of E-GC analyses; in MDGC analyses the enantiomers of α-pinene are co-eluted.
[c] Range of the values of MDGC analyses.

Appendix to Table 11.6:
1. Costa et al. (2009). One sample of (a) terpeneless coloured oil; (b) terpeneless colourless oil; for experimental conditions see point 14 of appendix to Table 11.1.
2. Dugo et al. (2012). Range of the values of (a) 3 samples of terpeneless coloured oils; (b) 2 samples of terpeneless colourless oils; for experimental conditions see point 14 of appendix to Table 11.1.

11.7 LABORATORY-EXTRACTED OILS

In Table 11.7 are research findings on the enantiomeric distribution of some volatile components of bergamot oils extracted in laboratory using different techniques. The oils are divided on the basis of the extraction technique. As shown, the extraction system may affect the value of the ratio between the two enantiomers of the compounds considered. The cold extraction, carried out by operating a manual pressure

TABLE 11.7

Enantiomeric Distribution of Some Volatile Components of Laboratory Extracted Bergamot Oils

		Cold-Pressed							
		1	2a	2b	3	4	5	6a	6b
α-Thujene	S-(+)[a]		0.5	1					
	R-(-)[a]		99.5	89					
α-Pinene	R-(+)		27	34		30.5		46	
	S-(−)		73	66		69.5		54	
β-Pinene	R-(+)		6.3	8.6		7.4	7.6–12.9	8	
	S-(−)		93.7	91.4		92.6	92.4–87.1	92	
Sabinene	R-(+)		14	15		17.8	15.4–21.1		
	S-(-)		86	85		82.1	86.6–78.9		
Limonene	S-(−)		2.7	1.7		98.3	1.4–2.7	2	
	R-(+)		97.3	98.3		1.7	98.6–97.3	98	
Linalool	R-(−)	99.5–99.8	99.4	99.5	99.4–99.5		99.3–99.7	99.7	90
	S-(+)	0.5–0.2	0.6	0.5	0.6–0.5		0.7–0.3	0.3	10
α-Terpineol	S-(−)							50	
	R-(+)							50	
Linalyl acetate	R-(−)	99.6–99.8	99.8	99.8	99.7–99.8		99.1–99.6	99.9	
	S-(+)	0.4–0.2	0.2	0.2	0.3–0.2		0.9–0.4	0.1	

		Solvent					SFE	
		7	8a	9	2c	10a	2d	2e
α-Thujene	S-(+)[a]				1			
	R-(−)[a]				99			
α-Pinene	R-(+)				27–35	32.2	26.3–27.5	26.8
	S-(−)				73–65	68.7	73.7–72.5	73.2
β-Pinene	R-(+)				6.2–8.5	7.1	6.3–6.6	6.2
	S-(−)				93.8–91.5	92.9	93.7–93.4	93.8
Sabinene	R-(+)				14–15		13.8–14.3	14
	S-(−)				86–85		86.2–85.7	86
Limonene	S-(−)				1.7–2.7	1.9	1	1
	R-(+)				98.3–97.3	98.1	99	99
Citronellal	S-(−)			99.5				
	R-(+)			0.5				
Linalool	R-(−)	99.5	99		99.3–99.5	>99	99.2–99.3	91.3
	S-(+)	0.5	1		0.7–0.5	<1	0.8-0.7	8.7
α-Terpineol	S-(−)							
	R-(+)							
Linalyl acetate	R-(−)		99		99.8	>99	99.8	99.8
	S-(+)		1		0.2	<1	0.2	0.2

TABLE 11.7   (CONTINUED)
Enantiomeric Distribution of Some Volatile Components of Laboratory Extracted Bergamot Oils

		Distilled						SDE		
		6b	10b	10c	10d	11	12	8b	13	14
α-Thujene	S-(+)[a]									
	R-(−)[a]									
α-Pinene	R-(+)		34.2	34.3	32.5					32.8
	S-(−)		65.8	65.7	67.5					67.2
β-Pinene	R-(+)		8.0	7.3	7.7					8.5
	S-(−)		92.0	92.7	92.3					91.5
Limonene	S-(−)		1.5	1.8	2.0				1.3–1.7	1.7
	R-(+)		98.5	98.2	98.0				98.7–98.3	98.3
Citronellal	S-(−)									>98.0
	R-(+)									<2.0
Linalool	R-(−)	90	61.9	78.1	43.7	74.3	99.9-100	78	0	99.8
	S-(+)	10	38.1	21.9	56.3	25.7	0.1–0	22	100	0.2
α-Terpineol	S-(−)						74.3			29.5
	R-(+)						25.7			70.5
Linalyl acetate	R-(−)		>99.0		>99.0		99.5–100	99	100	
	S-(+)		<1.0		<1.0		0.5–0	1.0	0	

[a]   Correct enantiomer not confirmed; tentatively assigned according to Casabianca and Chau (1997).

Appendix to Table 11.7:

1. Dellacassa et al. (1997). Range of composition of 4 Uruguayan samples laboratory cold-pressed; for experimental conditions see point 18 of appendix to Table 11.1.
2. Casabianca and Chau (1997). (a) One sample laboratory cold-pressed from fruits grown in Italy; (b) one sample laboratory cold-pressed from fruits grown in the Ivory Coast; (c) range of composition of 2 Italian samples and 1 Ivory Coast sample solvent extracted from fresh and dry fruits; (d) range of composition of 2 samples laboratory SFE extracted at 40°C and 80°C, respectively; (e) one sample laboratory SFE extracted at 120°C; for experimental conditions see point 12 of appendix to Table 11.1.
3. Verzera et al. (2000). Range of the values of 8 samples laboratory cold-pressed from fruits of the cvs. Castagnaro, Fantastico, and Femminello, PCF grown in Sicily, Italy; direct enantio-GC on capillary column (25 m × 0.25 mm × 0.25 μm) coated with 2,3-di-*O*-ethyl-*tert*-butyldimethylsilyl-β-CD (30%) and PS-086 (70%).
4. Mitiku et al. (2001). One sample cold-pressed from fruits grown in Calabria, Italy; enantiomeric ratios of α-pinene, β-pinene, and limonene were achieved using an achiral precolumn (60 m × 0.25 mm × 0.25 μm) coated with Thermon-600 T and a chiral main column (30 m × 0.25 mm) coated with 2,3,6-tri-*O*-methyl-β-CD (30%) coupled in series; enantiomeric ratio of sabinene was achieved by direct enantio-GC on the same chiral column.
5. Gionfriddo et al. (2003). Range of the values of 25 samples laboratory cold-pressed from fruits of the cvs. Castagnaro, Femminello, and Fantastico grown in Calabria, Italy; MDGC; precolumn: capillary column coated with DB-5, main column: capillary column (30 m × 0.25 mm × 0.25 μm) coated with 2,3-di-*O*-ethyl-6-*O*-*tert*-butyldimethylsilyl-β-CD.

continued

TABLE 11.7 (CONTINUED)
Enantiomeric Distribution of Some Volatile Components of Laboratory Extracted Bergamot Oils

6. Melliou et al. (2009). (a) One sample cold-pressed; (b) one sample hydrodistilled; direct enantio GC/MS on capillary column (30 m × 0.25 mm × 0.25 µm) coated with 2,3-di-*O*-methyl-6-*O*-*tert*-butyl dimethylsilyl-β-CD.
7. Schubert and Mosandl (1991). One sample solvent extracted; for experimental conditions see point 4 of appendix to Table 11.2.
8. Weinreich and Nitz (1992). (a) One sample solvent extracted from the peel; (b) one sample SDE extracted from the peel at pH 7; for experimental conditions see point 7 of appendix to Table 11.2. These authors also analyzed solvent extracted oils and SDE extracted oils from the flesh.
9. Casabianca et al. (1995). One sample; for experimental conditions see point 10 of appendix to Table 11.2
10. Kiwanuka et al. (2000). Calabria. Italy; (a) one sample solvent extracted from the peel; (b) one sample steam distilled at pH 2.5 from the peel; (c) one sample steam distilled at pH 5.3 from the peel; (d) one sample steam distilled from the whole fruits; MDGC; precolumn: capillary column (30 m × 0.53 mm × 0.25 µm) coated with DBX-5, main column: capillary column (25 m × 0.25 mm) coated with 2,3,6-tri-*O*-methyl-β-CD; temperature program of main column was dependent on the compound being resolved.
11. Bernreuther and Schreier (1991). One sample; for experimental conditions see point 5 of appendix to Table 11.2.
12. Casabianca and Graff (1996). Range of the values of 3 samples steam distilled (2 from fruits grown in Italy and one from fruits grown in Ivory Coast).
13. Ravid et al. (2009). Israel; 2 samples hydrodistilled. For experimental condition see point 15 of Table 11.2.
14. Hara et al. (1999). One sample laboratory extracted (extraction procedure not indicated in the original work).

on the skin or by means of solvents, leads to oils possessing an enantiomeric distribution equal to that of genuine cold-pressed industrial oils (see Table 11.1). The extraction carried out with various distillation techniques from the fruit peel can cause partial racemization of linalool (Bernreuther and Schreier 1991; Weinrich and Nitz 1992; Kiwanuka et al. 2000; Melliou et al. 2009), while distillation from a homogenate of the whole fruit causes a partial racemization of limonene, in addition to the racemization of linalool (Kiwanuka et al. 2000).

In addition to the results reported in Table 11.7, it's worth mentioning that König et al. (1997) did not observe the presence of the S-(+)enantiomers of linalool and linalyl acetate in a solvent-extracted oil from the peel of bergamot fruit, while they observed the total racemization of linalool and the complete loss of linalyl acetate in an oil produced by steam distillation of an homogenate obtained from the entire fruit; they also found 31.3% of S-(−)-linalool in the oil obtained by hydrodistillation, carried out for two hours, from fruit peel of bergamot. The chromatogram of this sample and the one of fecce oil analyzed by the same authors are shown in Figure 11.13. Moreover, König (1998), as well as Casabianca and Graff (1996), did not observe a racemization trend of linalool and linalyl acetate in oil extracted by distillation from the peel of the fruits.

11.8 OILS EXTRACTED FROM PHARMACEUTICAL PRODUCTS AND FLAVORED TEAS

Neukom et al. (1993), Casabianca and Graff (1994), Casabianca et al. (1995), and Ravid et al. (2009) analyzed bergamot oils extracted from flavored tea. In all the samples, linalool shows a racemic distribution or a tendency to racemization, as well as dihydrolinalool and linalyl acetate, in the samples where these two compounds were analyzed; these data are indicative of flavoring achieved with the addition of synthetic products or, otherwise, of products different from bergamot oil or, moreover, the employment of a production technology that can lead to this phenomenon.

Casabianca and Graff (1994) and Casabianca et al. (1995) also analyzed bergamot oils extracted from pharmaceutical preparations, finding that the enantiomeric distributions of limonene, linalool, and linalyl acetate were in agreement with those of genuine oils. The results of these works are shown in Table 11.8.

11.9 PETITGRAIN OILS

The data on the enantiomeric distribution of bergamot petitgrain are limited to the results obtained by Bonaccorsi et al. (2013) on one industrial oil from Calabria. These results are reported in Table 11.9.

TABLE 11.8
Enantiomeric Distribution of Some Volatile Components in Perfumed Tea and Pharmaceutical Extracts

		Tea			Pharmaceutical
		1a, 2	3	4	1b
Limonene	S-(−)			1.8–3.0	2
	R-(+)			98.2–97.0	98
Linalool	R-(−)	50	50–84.3	50.9–64.3	100
	S-(+)	50	50–15.7	49.1–35.7	0
Dihydrolinalool	(−)		39.5–57.0		
	(+)		60.5–43.0		
Linalyl acetate	R-(−)	50		49.1–62.7	99.5–100
	S-(+)	50		50.9–37.3	0.5–0

Appendix to Table 11.8:

1. Casabianca and Graff (1994). (a) Two samples extracted from perfumed tea; (b) two samples extracted from pharmaceutical preparation; for experimental conditions see point 9 of appendix to Table 11.2.
2. Casabianca et al. (1995). Two samples extracted from perfumed tea; for experimental conditions see point 10 of appendix to Table 11.2.
3. Neukom et al. (1993). Range of the values of 12 samples extracted from perfumed tea. For experimental conditions see point 8 of Table 11.2.
4. Ravid et al. (2009). Range of the values of 4 samples extracted from perfumed tea. For experimental conditions see point 15 of Table 11.2.

TABLE 11.9
Enantiomeric Distribution of Some Volatile Components of Bergamot Petitgrain Oils

		1
α-Thujene	S-(+)[a]	17.6[b]
	R-(−)[a]	82.4[b]
α-Pinene	R-(+)	23.3[c]
	S-(−)	76.7[c]
β-Pinene	R-(+)	3.4[b]
	S-(−)	96.6[b]
Sabinene	R-(+)	54.8[b]
	S-(−)	45.2[b]
β-Phellandrene	R-(−)	17.4[b]
	S-(+)	82.6[b]
Limonene	S-(−)	0.3[b]
	R-(+)	99.7[b]
Linalool	R-(−)	77.1[b]
	S-(+)	22.9[b]
Terpinen-4-ol	S-(+)	52.0[b]
	R-(−)	48.0[b]
α-Terpineol	S-(−)	26.9[b]
	R-(+)	73.1[b]
Linalyl acetate	R-(−)	99.7[b]
	S-(+)	0.3[b]

[a] Correct enantiomer not confirmed; tentatively assigned according to Casabianca and Chau (1997).
[b] MDGC values.
[c] E-GC values; in MDGC analysis the enantiomers of α-pinene are co-eluted.

Appendix to Table 11.9:
  1. Bonaccorsi et al. (2013). One industrial sample. For experimental conditions see point 14 of Table 11.1.

Comparing the enantiomeric distribution with the genuine cold-pressed peel oils, petigrain sample shows almost identical values for the enantiomeric distribution of linalyl acetate and α-pinene; similar values for β-pinene; a slightly higher enantiomeric excess for (+)-limonene and β-phellandrene; and a slight tendency to racemization of linalool, terpinen-4-ol, and α-thujene.

REFERENCES

Beker, S. 1995. Ätherische öle-gepanschte seelen. *Öko-Test* 10:41–49.
Bernreuther, A., and Schreier, P. 1991. Multidimensional gas chromatography/mass spectrometry: A powerful tool for the direct chiral evaluation of aroma compounds in plant tissues. II. Linalool in essential oils and fruits. *Phytochem. Anal.* 2:167–170.
Begnaud, F., and Chaintreau, A. 2005. Multidimensional gas chromatography using a double cool-strand interface. *J. Chromatogr. A* 1071:13–20.

Binder, G., König W. A., and Czygan, F. C. 2001. Ätherische öle. *Deutsche Apotheker Zeitung* 141:4263–4270.

Bonaccorsi, I., Trozzi, A., Cotroneo, A., and Dugo, G. 2013. Composition of industrial petit-grain produced in Calabria. *J. Essent. Oil Res.* in press.

Casabianca, H., and Graff, J.-B. 1994. Separation of linalyl acetate enantiomers: Application to the authentication of bergamot food products. *J. High Resolut. Chromatogr.* 17: 184–186.

Casabianca, H., Graff, J.-B., Jame, P., Perrucchietti, C., and Chastrette, M. 1995. Application of hyphenated techniques to the chromatographic authentication of flavors in food products and perfumes. *J. High Resolut. Chromatogr.* 18:279–285.

Casabianca, H., and Graff, J.-B. 1996. Chiral analysis of linalool and linalyl acetate in various plants. *EPPOS* 7:227–243.

Casabianca, H., and Chau, C. 1997. Analyses chirale des principaux constituants d'huiles de ber-gamotte. *16 ème journées internationals huiles essentielles*, Digne les Bains, France, 45–52.

Casabianca, H., Graff, J.B., Faugier, V., Fleig, F., and Grenier, C. 1998. Enantiomeric distribution studies of linalool and linalyl acetate. *J. High Resol. Chromatogr.* 21:107–112.

Costa, R. Dugo, P., Navarra, M., et al. 2010. Study on the chemical variability of some processed bergamot (*Citrus bergamia*) essential oils. *Flavour Fragr. J.* 25:4–12.

Cotroneo, A., Stagno d'Alcontres, I., and Trozzi, A. 1992. On the genuineness of citrus essential oils. Part XXXIV. Detection of added reconstituted bergamot oil in genuine bergamot essential oil by high resolution gas chromatography with chiral capillary columns. *Flavour Fragr. J.* 7:15–17.

Dellacassa, E., Lorenzo, D., Moyna, P., Verzera, A., and Cavazza, A. 1997. Uruguayan essential oils. Part V. Composition of bergamot oil. *J. Essent. Oil Res.* 9:419–426.

Dugo, G., Stagno d'Alcontres, I., Cotroneo, A., and Dugo, P. 1992. On the genuineness of citrus essential oils. Part XXXV. Detection of added reconstituted mandarin oil in genuine cold-pressed mandarin essential oil by high resolution gas chromatography with chiral capillary columns. *J. Essent. Oil Res.* 4:589–594.

Dugo, G., Stagno d'Alcontres, I., Donato, M. G., and Dugo, P. 1993. On the genuineness of citrus essential oils. Part XXXVI. Detection of added reconstituted lemon oil in genuine cold-pressed lemon essential oil by high resolution gas chromatography with chiral capillary columns. *J. Essent. Oil Res.* 5:21–26.

Dugo, G., Verzera, A., Trozzi, A., and Cotroneo, A. 1994a. Automated HPLC-HRGC: A powerful method for essential oils analysis. Part I. Investigation on enantiomeric distribution of mono-terpene alcohols of lemon and mandarin essential oils. *Essenz. Deriv. Agrum.* 94:35–44.

Dugo, G., Verzera, A., Cotroneo, A., et al. 1994b. Automated HPLC-HRGC: A powerful method for essential oil analysis. Part II. Determination of the enantiomeric distribution of linalol in sweet orange, bitter orange and mandarin essential oils. *Flavour Fragr. J.* 9:99–104.

Dugo, G., Mondello, L., Cotroneo, A., Bonaccorsi, I., and Lamonica, G. 2001. Enantiomeric distribution of volatile components of citrus oils by MDGC. *Perfum. Flav.* 26(1):20–35.

Dugo, G., Bonaccorsi, I., Sciarrone, D., et al. 2012. Characterization of cold-pressed and processed bergamot oils by using GC/FID, GC/MS, GC-C-IRMS, enantio-GC, MDGC, HPLC and HPLC-MS-IT-TOF. *J. Essent. Oil Res.* 24:93–117.

Gionfriddo, F., Catalfamo, M., Siano, F., et al. 2003. Determinazione delle caratteris-tiche analitiche e della composizione enantiomerica di oli essenziali agrumari ai fini dell'accertamento della purezza e della qualità. Nota I – Essenze di arancia amara, dolce e bergamotto. *Essenz. Deriv. Agr.* 73:29–39.

Hara, F., Shinohara, S., Toyoda, T., and Kanisawa, T. 1999. The analysis of some chiral compo-nents in Citrus volatile compounds. *Proceedings 43rd TEAC Meeting*, Oita, Japan 360–362.

Hener, U., Hollnagel, A., Kreis, P., et al. 1990. Direct enantiomer separation of chiral volatiles from complex matrices by multidimensional gas chromatography. In *Flavour Science and Technology*, eds. Y. Bessière and A. F. Thomas, 25–28. Chichester, West Sussex, England: Wiley.

Juchelka, D., and Mosandl, A. 1996. Authenticity profiles of bergamot oil. *Pharmazie* 51: 417–422.

Kiwanuka, P., Mottram, D. S., and Baigrie, B. D. 2000. The effects of processing on the constituents and enantiomeric composition of bergamot essential oil. In *Flavour Fragrance Chemistry*, eds. V. Lanzotti and O. Tagliatela-Scafati, 67–75. The Netherlands: Kluwer Academic Publishers.

König, W. A., Fricke, C., Saritas, Y., Momeni, B., and Hohenfeld, G. 1997. Adulteration or natural variability? Enantioselective gas chromatography in purity control of essential oils. *J. High Resolut. Chromatogr.* 20:55–61.

König, W. A. 1998. Enantioselective capillary gas chromatography in the investigation of stereochemical correlation of terpenoids. *Chirality* 10:499–504.

Mangiola, C., Postorino, E., Gionfriddo, F., Catalfamo, M., and Manganaro, R. 2009. Evaluation of the genuineness of cold-pressed bergamot oil. *Perfum. Flav.* 34(5):26–32.

Melliou, E., Michaelakis, A., Koliopoulos, G., Skaltsounis A. L., and Magiatis, P. 2009. High quality bergamot oil from Greece; chemical analysis using chiral gas chromatography and larvicidal activity against the West Nile virus vector. *Molecules* 14:839–849.

Mitiku, S. B., Ukeda, H., and Sawamura, M. 2001. Enantiomeric distribution of α-pinene, β-pinene, sabinene and limonene in various citrus essential oils. In *Food Flavors and Chemistry: Advances of the New Millennium*, eds. A. M. Spanier, F. Shahidi, T. H. Parliament, C. Mussinam, C-T Ho, and E. Tratras Contis, 216–231. Cambridge: Royal Society of Chemistry.

Mondello, L., Dugo, G., Dugo, P., and Bartle, K. D. 1996. On-line HPLC-HRGC in the analytical chemistry of citrus essential oils. *Perfum. Flav.* 21(4):25–49.

Mondello, L., Catalfamo, M., Dugo, P., Proteggente, A. R., and Dugo, G. 1997. La gascromatografia multidimensionale per l'analisi di miscele complesse. Nota preliminare. Determinazione della distribuzione enantiomerica di componenti degli olii essenziali agrumari. *Essenz. Deriv. Agrum.* 67:62–85.

Mondello, L., Verzera, A., Previti, P., Crispo, F., and Dugo, G. 1998a. Multidimensional capillary GC-GC for the analysis of complex samples. 5. Enantiomeric distribution of monoterpene hydrocarbons, monoterpene alcohols, and linalyl acetate of bergamot (*Citrus bergamia* Risso et Poiteau) oils. *J. Agric. Food Chem.* 46:4275–4282.

Mondello, L., Catalfamo, M., Proteggente, A. R., Bonaccorsi, I., and Dugo, G. 1998b. Multidimensional capillary GC-GC for the analysis of real complex samples. 3. Enantiomeric distribution of monoterpene hydrocarbons and monoterpene alcohols of mandarin oils. *J. Agric. Food Chem.* 46:54–61.

Mondello, L., Catalfamo, M., Dugo, P., and Dugo, G. 1998c. Multidimensional capillary GC-GC for the analysis of real complex samples. Part II. Enantiomeric distribution of monoterpene hydrocarbons and monoterpene alcohols of cold-pressed and distilled lime oils. *J. Microcol. Sep.* 10:203–212.

Mondello, L., Dugo, P., Cotroneo, A., Proteggente, A. R., and Dugo, G. 1998d. Multidimensional advanced techniques for the analysis of bergamot oil. *EPPOS* 26:3–27.

Mondello, L., Catalfamo, M., Cotroneo, A., et al. 1999. Multidimensional capillary GC-GC for the analysis of real complex samples. Part IV. Enantiomeric distribution of monoterpene hydrocarbons and monoterpene alcohols of lemon oil. *J. High Resolut. Chromatogr.* 22:350–356.

Mondello L., Casilli A., Tranchida P. Q., et al. 2006. Fast enantiomeric analysis of a complex essential oil with innovative multidimensional gas chromatographic system. *J. Chromatogr. A* 1105:11–16.

Mondello L., Casilli A., Tranchida P. Q., et al. 2008. Analysis of allergenes in fragrances using multiple heart-cut multidimensional gas chromatography-mass spectrometry. *LC GC Europe* 21:130–137.

Mosandl, A., Hener, U., Kreis P., and Schmarr, H-G. 1990. Enantiomeric distribution of α-pinene, β-pinene and limonene in essential oils and extracts. Part 1. Rutaceae and Gramineae. *Flavour Fragr. J.* 5:193–199.

Mosandl, A. 1995. Enantioselective capillary gas chromatography and stable isotope ratio mass spectrometry in the authenticity control of flavors and essential oils. *Food. Rev. Int.* 11:597–664.

Mosandl, A., and Juchelka, D. 1997a. Advances in the authenticity assessment of citrus oils. *J. Essent. Oil Res.* 9:5–12.

Mosandl, A., and Juchelka, D. 1997b. The bitter orange tree: A source of different essential oils. In *Flavour Perception*, eds. H-P. Kruse and M. Rothe, 321–331. Eigenverlag Deutsch Inst. F. Ernahrungsforsch.

Murdock, D. I., Hunter, G. L. K., Bucek, W. A., and Brent, J. A. 1969. Relation of bacterial contamination in orange oil recovery system to quality of finished product. *Food Technol.* 23:98–100.

Neukom, H-P, Meier, D. J., and Blum D. 1993. Nachweis von natürlichem oder rekonstituiertem bergamottöl in earl grey tess anhand der enantiomerntrennung von linalool und dihydrolinalol. *Mitt. Gebiete Lebensm. Hyg.* 84:537–544.

Ravid, U., Putievsky, E., and Katzir, I. 1994. Chiral GC analysis of enantiomerically pure (R)(−)-linalyl acetate in some Lamiaceae, myrtle and petitgrain essential oils. *Flavour Fragr. J.* 9:275–276.

Ravid, U., Elkabetz, M., Zamir, C., et al. Authenticity assessment of natural fruit flavour compounds in foods and beverages by auto-HP-SPME stereoselective GC-MS. *Flavour Fragr. J.* 25:20–27.

Schubert, V., and Mosandl, A. 1991. Chiral compounds of essential oils. VIII. Stereo differentiation of linalool using multidimensional gas chromatography. *Phytochem. Anal.* 2:171–174.

Sciarrone, D., Schipilliti, L., Ragonese, C., et al. 2010. Thorough evaluation of the validity of conventional enantio-gas chromatography in the analysis of volatile chiral compounds in mandarin essential oil: A comparative investigation with multidimensional gas chromatography. *J. Chromatogr. A* 1217:1101–1105.

Sciarrone, D. 2009. Personal communication.

Shellie, R., and Marriott, P. J. 2002. Comprehensive two-dimensional gas chromatography with fast enantioseparation. *Anal. Chem.* 74:5426–5430.

Tateo, F., Bodoni, M., and Lubian, E. 2000. Enantiomeric distribution and sensory evaluation of linalol and linalyl acetate in flavourings. *Ital. J. Food Sci.* 12:371–375.

Verzera, A., Lamonica, G., Mondello, L., Trozzi, A., and Dugo, G. 1996. The composition of bergamot oil. *Perfum. Flavor.* 21(6):19–34.

Verzera, A., La Rosa, G., Zappalà, M., and Cotroneo A. 2000. Essential oil composition of different cultivars of bergamot grown in Sicily. *J. Food Sci.* 12:493–501.

Weinrich, B., and Nitz, S. 1992. Influences of processing on the enantiomeric distribution of chiral flavour compounds. Part A: linalyl acetate and terpene alcohols. *Chem. Mikrobiol. Technol. Lebensm.* 14:117–124.

12 The Oxygen Heterocyclic Compounds

Paola Dugo, Marina Russo, Germana Torre, and Giovanni Dugo

CONTENTS

12.1 INTRODUCTION

The oxygen heterocyclic fraction of bergamot oil has undoubtedly received more attention than that of any other citrus oil. Bergamot oil is a very valuable product, largely used in perfumery as the basis of eau de cologne, and in cosmetics as a constituent of suntan products. It is known to have strong photosensitizing activity, due to the presence of psoralens and, in particular, bergapten. This has prompted the development of analytical procedures for the determination of bergapten in bergamot oil and in cosmetic products containing the oil. The early studies on the composition of the oxygen heterocyclic fraction of bergamot oil have been reviewed by Mossman and Bogert (1941) and, more recently, by Di Giacomo and Calvarano (1978a, 1978b), Lawrence (1982, 1994), Dugo and McHale (2002), and Dugo and Russo (2011).

It is also worth mentioning that, due to the phototoxicity of this psoralen, bergapten-free oils obtained using different methods are available on the market.

313

The procedures commonly used to remove bergapten from oil involve either distillation or treatment with NaOH. Distillation drastically reduces the amount of all four components of the oxygen heterocyclic fraction, while NaOH treatment affects only citropten and bergapten content.

The isolation of colorless needles from bergamot oil was first reported by Mulder (1839). In the same year, independently from Mudler's results, Ohme (1839) gave the name "bergapten" to crystals obtained during his study on bergamot oil. The composition of the crystals was studied by Crismer (1891) and by Pomerantz (1891), who found their formula to be $C_{12}H_8O_4$. Subsequently, Thoms and Baetecke (1912) proved that the compound was 5-methoxypsoralen. A second crystalline compound was isolated from bergamot oil by von Soden and Rojahn (1901) and named "bergaptin."

In the years between 1934 and 1937, Späth and co-workers confirmed the presence of the previously isolated components, and identified new compounds bergaptol (Späth and Socias 1934) and citropten (Späth and Kainrath 1937). In addition, Späth and Kainrath established that "bergaptin" (von Soden and Rojahn 1901) was identical to their compound bergamottin (5-geranyloxypsoralen). Rodighero and Caporale (1954) also found bergapten (3240 ppm), citropten (2090 ppm), and bergaptol (2330 ppm) by distilling the residue of bergamot oil under high vacuum, but they believed that the bergaptol was derived from the decomposition of bergamottin. Guenther (1949) reported the tendency of bergamottin to decompose into bergaptol during some analytical operations.

P. Dugo et al. (2011) found 6 ppm of bergaptol after a treatment of two hours at 100°C of 100 ppm of bergamottin in water acidified with citric acid and 0.2 ppm of bergaptol after the same treatment of bergapten. Bergaptol was also detected in a bergamot oil subjected to the same treatment. Günther and Ziegler (1977) detected artifact formation after the analysis of furocoumarins of lemon oil. Most of the papers on the composition of the oxygen heterocyclic fraction of bergamot oil report the identification and quantification of two coumarins (citropten and 5-geranyloxy-7-methoxycoumarin) and two psoralens (bergamottin and bergapten); however, many papers report only the determination of bergapten. In fact, this component has been more closely studied due to its reported photosensitizing properties. Some papers reported the presence of polymethoxylated flavones such as sinensetin and tetra-O-methylscuttellarein (P. Dugo et al. 1999a, 2000; G. Dugo et al. 2012), nobiletin and tangeretin (Donato et al. 2013), and 5-hydroxy-7,8,3',4'-tetramethoxyflavone (Farid 1968). A study carried out by the researchers of the Stazione Sperimentale di Reggio Calabria (Calvarano et al. 1995; Gionfriddo et al. 1997, 1998) reports the presence of five psoralens (biakangelicin, biakangelicol, 5-isopentenyloxy-8-methoxypsoralen, oxypeucedanin, and oxypeucedanin hydrate) and of 5-isopentenyloxy-7-methoxycoumarin, in addition to the other components commonly identified in bergamot oil. These components were never reported again in further studies on bergamot oil, even when advanced analytical techniques were used.

In bergamot oil, 5-isopentenyloxy-7-methoxycoumarin was found by Di Giacomo in 1990. Some papers of the same research group (Di Giacomo and Calvarano 1974; Di Giacomo 1990) considered bergaptol to be a natural component of bergamot oil. Epoxybergamottin was reported by Freròt and Decorzant (2004) and Casabianca (personal communication, 2010). Herniarin was detected in many industrial samples produced between 2008 and 2011 (G. Dugo et al. 2012); as a further confirmation, herniarin was found by the same authors in bergamot oils that were laboratory cold-pressed from fruits of Femminello and Castagnaro cultivars, while it was not detected in oils obtained from Fantastico fruits. Herniarin was previously reported by Costa et al. (2010) in a concentrated bergamot oil.

This chapter will review qualitative investigations; then quantitative studies will be discussed in separate sections dedicated to Calabrian (Italy) cold-pressed oils, commercial or different geographical origin oils, bergapten-free oils, oils recovered from the residues of cold-pressing process, and reconstituted oils. An additional section will deal with the determination of oxygen heterocyclic components in cosmetic and food products containing bergamot oil, and the presence of these components in bergamot juice will be briefly discussed.

Data reported in the tables, taken from original references, represents the composition of only one sample, average values or, when possible, variability ranges. In the tables, single components are grouped by chemical class and for every class, listed in alphabetic order. In the notes to the tables, when available the following information is reported: origin of the oils, production technology, number of samples relative to reported data, and analytical techniques. In the case of HPLC analysis, if column dimensions are reported, the first number is for length (mm), the second is for internal diameters (mm), and the third is for particles size (µm).

12.2   QUALITATIVE DATA

Table 12.1 reports results relative to qualitative investigations carried out on the oxygen heterocyclic fraction of bergamot oil.

From the TLC analysis of bergamot oils extracted from unripe fruits, D'Amore and Calapaj (1965) observed the presence on the TLC plates of two red spots, with Rf 0.02 and 0.73, in addition to spots relative to identified components. Farid (1968) identified a polymethoxylated flavone in an oil of unspecified origin. However, this component was never reported again in studies on bergamot oil. Huet (1970) stated that TLC allows for an approximate quantification of oxygen heterocyclic components. Oils from Mali and the Ivory Coast presented $E^{1\%, 1cm}$ values at 312 nm lower than those of Italian oils, maybe due to a lower amount of bergamottin. Di Giacomo and Calvarano (1974) did not report the amount of the three identified components (bergaptol, citropten, 5-geranyloxy-7-methoxycoumarin) but only their relative percentage. They observed that the variability range for the three studied components varied within narrow ranges and did not present significant variations during the whole productive season. Latz and Ernes (1978) stated that varying the wavelength used to obtain HPLC chromatograms using a fluorescence detector increases the possibility of identifying single components.

Dugo et al. (1996) compared results obtained using OPLC with those previously obtained by the same research group by HPLC (Mondello et al. 1993). Their conclusion was that OPLC method could be a tool for a rapid screening to detect contaminations or adulterations.

Bonaccorsi et al. (1999, 2000) developed a fast HPLC method for the separation, under the same experimental conditions, of oxygen heterocyclic components of different

TABLE 12.1
Qualitative Composition of Oxygen Heterocyclic Fraction of Bergamot Oil

Trivial Name	1	2	3	4	5	6	7	8	9	10	11, 12	13, 14	15
Coumarins													
Citropten	+		+	+	+	+	+	+	+	+	+	+	+
5-Geranyloxy-7-methoxycoumarin	+		+	+	+	+	+		+	+	+	+	+
5-Isopentenyloxy-7-methoxycoumarin									+				
Psoralens													
Bergamottin	+		+	+		+	+	+	+	+	+	+	+
Bergapten	+		+	+		+	+	+	+	+	+	+	+
Bergaptol					+				+				
Polimethoxyflavones													
5-Hydroxy-7,8,3',4'-tetramethoxyflavone			+										
Sinensetin												+	
Tetra-*O*-methylscutellarein												+	

Appendix to Table 12.1:

1. D'Amore and Calapaj (1965). Ten genuine cold-pressed oils; preparative TLC on silica gel using different mixtures of hexane/ethyl acetate as eluents; identification of the isolated compounds by UV spectra between 220 and 380 nm.
2. Farid (1968). Isolation of the studied compound from the solid residue of bergamot oil and its identification by ^{1}H-NMR.
3. Huet (1970). Fourteen authentic oils from Mali and 7 authentic oils from the Ivory Coast; analytical and preparative TLC on silica gel using cyclohexane:ethyl acetate (75:25) as eluents; identification of the isolated compounds by UV spectra between 200 and 360 nm.
4. Stanley and Jurd (1971).
5. Di Giacomo and Calvarano (1974). Two hundred genuine cold-pressed oils extracted in 1972–73 productive season; TLC on silica gel using hexane:ethyl acetate (75:25) as eluents; detection by in situ spectrofluorimetry.
6. Latz and Ernes (1978). NP-HPLC; column: μ-Porasil (300 × 4 mm); linear programmed gradient starting with 20% chloroform in hexane and ending with 100% chloroform; UV (254 nm) detection and selective fluorescence detection at $\lambda_{ecc}/\lambda_{em}$ 315/480 nm, 308/435 nm, 335/400 nm.
7. Herpol-Borremans et al. (1985). Analytical and preparative TLC; NMR; NP-HPLC/UV (254 nm); pre-column: HC pellosil (60 × 2.2 mm); column: Zorbax Sil (250 × 4.6 mm); eluents: hexane:ethyl acetate (70:30).
8. Benincasa et al. (1990). One genuine cold-pressed oil; RP-HPLC/UV (320 and 250 nm) and RP-HPLC/MS (plasma-spray); column: Spherisorb ODS$_2$ (250 × 2.1 mm, 5 μm); eluents: acetonitrile:water; NP-HPLC/UV (320 and 220 nm); column: Spherisorb CN (250 × 1.1 mm, 5 μm); eluents: hexane:isopropanol (98:2).
9. Di Giacomo (1990).
10. Dugo et al. (1996). One genuine cold-pressed oil; OPLC on aluminum-backed silica gel 60 F_{254} plates; eluents: chloroform:n-butyl acetate (9:1) diluted 40:60 with hexane; Rf values and fluorescence were observed at 366 nm.

TABLE 12.1 (CONTINUED)
Qualitative Composition of Oxygen Heterocyclic Fraction of Bergamot Oil

11. Bonaccorsi et al. (1999). One genuine cold-pressed oil; fast RP-HPLC/PDA (315 nm); column C18 (30 × 4.6 mm, 3 µm), gradient elution program starting with 70% water in acetonitrile and ending with 20% water.
12. Bonaccorsi et al. (2000). One genuine cold-pressed oil; analyses were performed as in point 11 of this appendix and also with a conventional Luna C18 column (150 × 4.6 mm, 5 µm).
13. Dugo et al. (1999a). One sample; NP-HPLC/UV (315 nm) and NP-HPLC/MS (APCI⁺); column: µ-Porasil (150 × 3.9 mm, 10 µm); gradient elution program starting with 98% of hexane:ethyl acetate (9:1) and 2% of hexane:ethyl alcohol (9:1) and ending with 5% of hexane:ethyl acetate (9:1) and 95% of hexane:ethyl acetate (9:1).
14. Dugo et al. (2000). RP-HPLC/MS (APCI⁺) on C18 Pinnacle column (250 × 4.6 mm, 5 µm) using a gradient elution program starting with 100% of tetrahydrofuran:acetonitrile:methanol:water (15:5:22:58) and ending with 30% of this mixture and 70% of acetonitrile.
15. Cavazza et al. (2001). Capillary electrochromatography/UV (214 and 254 nm) on fused silica capillary (450 × 0.1 mm) packed with 3 µm ODS1 Spherisorb C18 using acetonitrile: hydroxymethyl (methylamine) pH 7.8 (80:20) and, for nonacqueous experiment, mixtures of acetonitrile/isopropanol/hexane in different ratios as eluents.

cold-pressed citrus oils, including bergamot oil. The method permits the detection of possible cross-contaminations between different oils and/or adulteration. Figure 12.1 shows HPLC chromatograms obtained by Bonaccorsi et al. (1999) for the different citrus oils.

Dugo et al. (1999a) identified two polymethoxylated flavones (sinensetin and tetra-O-methylscutellarein) for the first time in bergamot oil, using the HPLC-APCI-MS technique. The same authors confirmed this finding in a later study (P. Dugo et al. 2000). Figure 12.2 reports HPLC-UV and HPLC-MS acquired at 20 V chromatograms of a genuine bergamot oil and the chromatogram extracted at m/z = 343 + 373. The same figure also shows the MS spectra obtained at different cone voltages for the peaks identified as tetra-O-methylscutellarein and sinensetin (P. Dugo et al. 1999a). Recently, using the same analytic technique, Donato et al. (2013) identified nobiletin and tangeretin in bergamot oil.

Cavazza et al. (2001) used capillary electrochromatography (CEC) for the separation of coumarins and psoralens of bergamot oil. The authors concluded that, when compared to HPLC, this technique offers advantages such as speed and efficiency.

12.3 QUANTITATIVE DATA

12.3.1 INDUSTRIAL COLD-PRESSED CALABRIAN OILS

Table 12.2 reports results on the quantitative determination of the oxygen heterocyclic fraction of Calabrian bergamot oils. Two coumarins (citropten and 5-geranyloxy-7-methoxycoumarin) and two psoralens (bergamottin and bergapten) are the main components, identified and quantified in most of the papers cited in Table 12.2. As can be seen, major attention was dedicated to bergapten due to its photosensitizing properties. Variability ranges of these four components are very large.

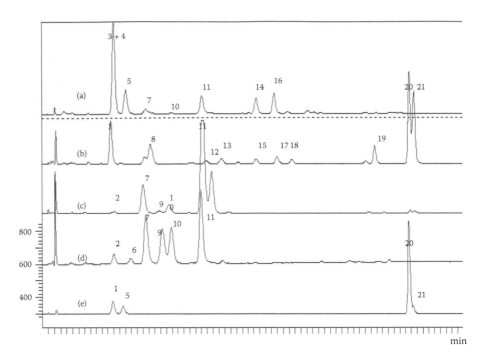

FIGURE 12.1 HPLC chromatograms obtained under identical conditions for five citrus essential oils: (a) bitter orange, (b) lemon, (c) mandarin, (d) sweet orange, and (e) bergamot. P-E C18, 3 cm × 4.6 mm, 3 μm column. (1) citropten; (2) sinensetin; (3) meranzin; (4) isomeranzin; (5) bergapten; (6) 3,3′,4′,5,6,7-hexamethoxyflavone; (7) nobiletin; (8) oxypeucedanin; (9) tetra-*O*-methylscutellarein; (10) 3,3′,4′,5,6,7,8-heptamethoxyflavone; (11) tangeretin; (12) imperatorin; (13) phellopterin; (14) osthol; (15) isoimperatortin; (16) epoxybergamottin; (17) 5-isopent-2′-enyloxy-7-methoxycoumarin; (18) 5-isopent-2′-enyloxy-8-(2′,3′-epoxyisopentyloxy)psoralen; (19) 8-geranyloxycoumarin, (20) bergamottin; (21) 5-geranyloxy-7-methoxycoumarin. (Reproduced from Bonaccorsi et al., *J. Agric. Food Chem.* 47, 4237–4239, 1999.)

Excluding the bergapten value reported by Calvarano (1961) obtained after a long and laborious procedure, the very high values of citropten and bergapten obtained by Vernin et al. (1979) by GC with packed columns, results obtained by Gionfriddo et al. (1997, 1998) that report the presence of some psoralens and one coumarin not detected even in successive studies of the same research group (Mangiola et al. 2009), and bergamottin amount always lower than 10,000 mg/L, variability ranges of the main four components, as reported in Table 12.2, can be determined as follows:

Citropten	1200–3813 mg/L
5-geranyloxy-7-methoxycoumarin	400–2700 mg/L
Bergamottin	10,000-27,500 mg/L
Bergapten	1100-4750 mg/L

More recent results obtained from analysis of oils of the productive seasons 2007–08 (Mangiola et al. 2009), 2008–09 (Sciarrone, D., personal communication, 2009;

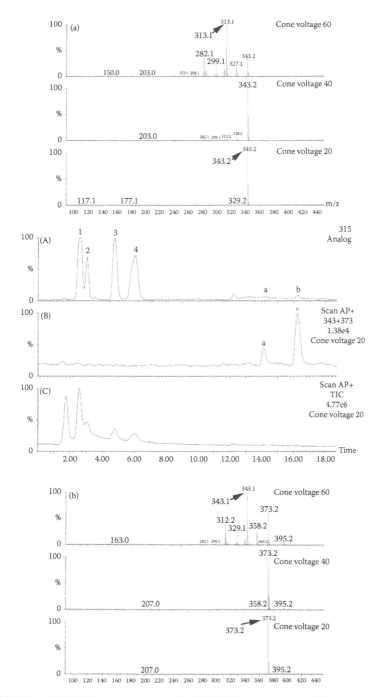

FIGURE 12.2 HPLC-UV chromatogram (A), HPLC-TIC chromatogram (C) and extracted chromatogram at m/z = 343 + 373 (B) of the oxygen heterocyclic compounds of bergamot oil, with cone voltage fragmentation of peaks (a) and (b), identified as tetra-O-methylscutellarein and sinensetin, respectively. (Reproduced from Dugo et al., *J. Liq. Chrom. & Rel. Technol.* 22, 2991–3005, 1999a.)

Citrus bergamia

TABLE 12.2

Quantitative Composition (mg/L) of Oxygen Heterocyclic Fraction of Calabrian Cold-Pressed Bergamot Oil

Trivial Name	1	2	3	4	5	6	7	8
Coumarins								
Citropten	2500–3500	2400–3200	2580		1400–2400		7000[b]	1600–3000
5-Geranyloxy-7-methoxycoumarin		400–600	1000		1000–1500			
Herniarin								
5-Isopentenyloxy-7-methoxycoumarin								
Psoralens								
Bergamottin	4400–7500	15,000–19,000			14,000–22,000			11,400–27,300
Bergapten		3000–3600		3570	1500–3300	2040–4750[a]	8500[b]	1560–4040
Bergaptol								
Byakangelicin								
Biakangelicol								
Epoxybergamottin								
Isoimperatorin								
5-Isopentenyloxy-8-methoxypsoralen								
Oxypeucedanin								
Oxypeucedanin hydrate								
Polimethoxyflavones								
Sinensetin								
Tetra-*O*-methylscutellarein								

Trivial Name	9	10	11	12	13a	13b	13c	14
Coumarins								
Citropten		2100	1400–3500	1400–3500	1176–3207	1550–2616	2366–3566	2200
5-Geranyloxy-7-methoxycoumarin		1100	800–2200	800–2200	534–1072	529–869	555–923	1300
Herniarin								
5-Isopentenyloxy-7-methoxycoumarin				20–80	40–169	62–93	118–168	
Psoralens								
Bergamottin	2300–3000[c]	18,300	10,200–27,500	10,200–27,500	6327–8263	8389–8822	7814–9123	18700
Bergapten		2700	1100–3200	1100–3200	1893–3310	1711–3371	1548–2526	2100
Bergaptol								
Byakangelicin				60–80	23–90	53–73	169–227	
Biakangelicol				150–410	165–895	788–805	514–869	
Epoxybergamottin								
Isoimperatorin								
5-Isopentenyloxy-8-methoxypsoralen[d]				tr–60	0–15	0–8	2–12	
Oxypeucedanin				440–640	180–475	343–392	172–261	
Oxypeucedanin hydrate				120–190	22–59	46–54	65–83	
Polimethoxyflavones								
Sinensetin								
Tetra-O-methylscutellarein								

continued

TABLE 12.2 (CONTINUED)
Quantitative Composition (mg/L) of Oxygen Heterocyclic Fraction of Calabrian Cold-Pressed Bergamot Oil

Trivial Name	15	16	17	18	19	20	21	22	23
Coumarins									
Citropten		1200–3500	1588–2708	1340–2124	2582	1927	2467–3813	1504–3032	2232
5-Geranyloxy-7-methoxycoumarin		900–2700	1195–1818	878–1035	1120	1423		803–1459	1065
Herniarin								8–385	67
5-Isopentenyloxy-7-methoxycoumarin									
Psoralens									
Bergamottin		10,000–19,400	18,417–23,923	10,969–14,089	21,419	21,685	16,557–25,154	15,203–23,147	19,605
Bergapten	3200	1100–2900	1321–2534	1384–2089	2374	2070	2460–3642	1319–3329	2474
Bergaptol							19–286		
Byakangelicin									
Biakangelicol									
Epoxybergamottin							73–702		
Isoimperatorin							27–542		
5-Isopentenyloxy-8-methoxypsoralen[d]									
Oxypeucedanin									
Oxypeucedanin hydrate							10–34		
Polymethoxyflavones									
Sinensetin		+						35–194	103
Tetra-O-methylscutellarein		+						20–298	195

continued

a   HPLC values; TLC values were 1780–3860.

b   GC values.

c   Ranges of the values found by five different laboratories for the same oil.

d   Reported as 5-isopentenyloxy-7-methoxypsoralen in Calvarano et al. 1995 and Gionfriddo et al. 1997.

e   mg/kg.

+, Identified but not quantitatively determined.

Appendix to Table 12.2:

1. Calvarano (1961). Eleven genuine cold-pressed oils. Isolation of bergapten and citropten by alkaline treatment, followed by separation of the acqueous phase, acidification and extraction with ethyl ether; preparative paper chromatography; quantitative determination by UV measurement at 310 nm (bergapten) and 325 nm (citropten).

2. Cieri (1969). Six genuine cold-pressed oils; isolation of coumarins and psoralens by preparative TLC on silica gel of the nonvolatile residue using hexane:ethyl acetate (75:25) as eluents. Quantitative determination by UV measurement at 310 nm (bergapten and bergamottin) and 325 nm (citropten and 5-geranyloxy-7-methoxycoumarin).

3. Madsen and Latz (1970). One sample; TLC; qualitative and quantitative detection by in situ fluorimetry.

4. Porcaro and Shubiak (1974). One sample; NP-HPLC with differential refractometry and UV detectors; column: Corasil II (500 × 2.3 mm, 37–50 μm); eluents: hexane:chloroform (75:25).

5. Calabrò and Currò (1975). Fourteen genuine cold-pressed oils; TLC on silica gel using hexane:ethyl acetate (75:25) as eluents; detection by in situ fluorimetry.

6. Di Giacomo and Calvarano (1978a). One hundred twenty-five genuine cold-pressed oils; 1976–77 productive season; NP-HPLC/UV (254 nm); column: Cyano-Sil-X-1 (250 × 2.6 mm); eluents: hexane:isopropanol (94:6). TLC on silica gel F 254 using hexane:ethyl acetate (75:25) as eluents; detection by in situ fluorimetry; NP-HPLC/UV (254 nm); column: Cyano-Sil-X-1 (250 × 2.6 mm); eluents: hexane:isopropanol (94:6). The same results are reported by Di Giacomo and Calvarano (1978b), Di Giacomo (1981, 1990), Di Giacomo and Di Giacomo (1999).

7. Vernin et al. (1979). HPTLC; NP-HPLC; RP-HPLC; GC on packed column of SE-30. Results reported in the table refer to GC analyses.

8. Calvarano et al. (1979). One hundred genuine cold-pressed oils; 1976–77 productive season; NP-HPLC/UV (254 nm); column: Cyano-Sil-1 (250 × 2.6 mm); eluents: hexane:isopropanol (98:2). The same results are reported by Di Giacomo and Di Giacomo (1999).

9. Analytical Methods Committee (1987). One sample; NP-HPLC/UV (305 nm); column: Silica Gel Si-60 (250 × 4.6 mm, 5 μm); eluents: hexane:ethyl acetate (75:25 or 80:20), ciclohexane:tetrahydrofuran (85:15), hexane:chloroform (85:15 or 75:25), hexane:isopropanol (94:6).

10. McHale and Sheridan (1989). One sample; NP-HPLC/PAD (315 nm); column: M Zorbax Sil (250 × 4.6 mm; 6 μm); gradient starting with (hexane:ethyl acetate, 9:1), (hexane:ethanol, 9:1) (98:2) and ending with a ratio of 5:95.

11. Mondello et al. (1993). One hundred twenty-eight genuine cold-pressed oils; 1991–92 productive season; NP-HPLC/UV (315 nm); column: μ Porasil (300 × 3.9 mm, 10 μm); gradient starting with hexane:ethyl acetate 9:1 and ending with 5% of this mixture and 95% of hexane:ethyl alcohol (9:1).

12. Calvarano et al. (1995). Twenty-five genuine cold-pressed oils; NP-HPLC/UV (315 nm); column: Erbasil Spherical Silica (250 × 4.6 mm, 10 μm); gradient starting with hexane:ethyl acetate (9:1) and ending with hexane:ethyl alcohol (9:1).

TABLE 12.2 (CONTINUED)
Quantitative Composition (mg/L) of Oxygen Heterocyclic Fraction of Calabrian Cold-Pressed Bergamot Oil

13. Gionfriddo et al. (1997, 1998). Thirty-seven genuine cold-pressed oils; cv. Fantastico; 1996–97 productive season; (a) 19 samples obtained by Pelatrice machine; (b) 12 samples obtained by Calabrese machine; (c) 6 samples obtained by sponge technique. For experimental conditions, see point 12 of this appendix.

14. Dugo et al. (1997). Average values of 128 genuine cold-pressed oils; 1991–92 productive season; for experimental conditions see point 11 of this appendix. The same results are reported by Cavazza et al. (2000).

15. Martin et al. (1998). One sample; RP-HPLC/UV (254 nm); column: Lichrosorb RP 18 (250 × 4.6 mm, 5 μm); elution program: 0–12 min 100% (isocratic) of water:acetonitrile: tetrahydrofuran (70:25:5), 12–25 min gradient to 10% of the initial mixture + 90% of acetonitrile:tetrahydrofuran (90:10), 25–35 isocratic, then from 35 to 73 min gradient to the 100 % of the initial mixture, followed by conditioning (37–55 min).

16. Dugo et al. (1999b). Four hundred fifty-one genuine cold-pressed oils; 1996–97 productive season. For experimental conditions see point 13 of appendix to Table 12.1.

17. Sciarrone (personal communication, 2009). Sixteen cold-pressed oils produced in 2008–09 productive season. For experimental conditions see point 19 of this appendix.

18. Mangiola et al. (2009). Six genuine cold-pressed oils; 2007–08 productive season; NP-HPLC/PDA (315 nm); column: Phenomenex Luna (250 × 4.6 mm, 5 μm); eluents: hexane:ethyl acetate (90:10).

19. Dugo et al. (2009). One genuine cold-pressed oil; RP-HPLC/PAD (315 nm); column: Ascentis Express C18 (150 × 4.6 mm, 2.7 μm); solvent A: water:acetonitrile:THF (85:10:5), solvent B: acetonitrile:methanol:THF (65:30:5); gradient: 0–5 min, 0% B; 5–25 min, 0–40% B; 25–45 min, 40-90% B; 45–55 min, 90% B; 55–60 min, 0% B.

20. Costa et al. (2010). One genuine cold-pressed oil; RP-HPLC/PAD (315 nm); column: Discovery-HS C18 (250 × 4.6 mm; 5 μm); solvent A: water, solvent B: acetonitrile; gradient from 30% B to 100% B and then to 30% B.

21. Casabianca (personal communication, 2010). Seven genuine cold-pressed oils; RP-HPLC/UV (254 nm); column: Zorbax Eclipse XDB-C18, rapid resolution HT (100 × 4.6 mm, 1.8 μm). Solvent A: water:acetonitrile:tetrahydrofuran (85:10:5), solvent B: acetonitrile:methanol:tetrahydrofuran (65:30:5); gradient starting with 100% of A and ending with 100% of B.

22. Dugo G. et al. (2012). Forty-one cold-pressed oils produced in 2008–09, 2009–10, 2010–11 productive seasons. RP-HPLC/PAD (315 nm); column: Ascentis Express C18 (150 × 4.6 mm, 2.7 μm); solvent A: water:methanol:THF (85:10:5), solvent B: methanol:THF (95:5); gradient: 0–5 min, 0% B; 5–25 min, 0%–40% B; 25–45 min, 40%–90% B; 45–55 min, 90% B; 55–60 min, 0% B.

23. Russo M. et al. (2012). One cold-pressed oil. For experimental conditions see point 22 of this appendix.

G. Dugo et al. 2012), 2009–10 and 2010–11 (G. Dugo et al. 2012) indicate that the content of the four main components in bergamot oil is as follows:

Citropten	1340–3032 mg/L
5-Granyloxy-7-methoxycoumarin	803–1818 mg/L
Bergamottin	10,969–23,923 mg/L
Bergapten	1319–3329 mg/L

From the date reported in Table 12.2, it is possible to see that the first quantitative results on the composition of the oxygen heterocyclic fraction of bergamot oil were obtained using preparative TLC and UV absorbance of separated components, and by TLC with spectrofluorimetric detection in situ; from the mid-1970s onward, HPLC was used. Porcaro and Shubiak (1974) found that HPLC was faster and more reliable than TLC.

The Analytical Method Committee (1987) reviewed HPLC methods reported in literature for the quantitative determination of bergapten, and concluded that silica columns are better than reversed phase columns because they allow fast analyses of bergamot oil under isocratic conditions; moreover, UV detection at 305 nm is better than that at 254 nm because lower wavelengths present more interference. Normal-phase HPLC has been used by numerous authors, including McHale and Sheridan (1989), who performed an excellent characterization of oxygen heterocyclic fraction of different cold-pressed citrus oils. In the last decade, however, the reversed phase has been preferred to normal phase HPLC because of its higher repeatability. Figures 12.3a and 12.3b compare HPLC chromatograms of a bergamot oil obtained under normal- and reversed-phase modes.

Among the values reported in Table 12.2, it is worth mentioning that the results reported by Calvarano et al. (1995) for the four main components are surprisingly identical to those reported two years before by Mondello et al. (1993); Calvarano et al. (1995) and Gionfriddo et al. (1997, 1998) identified 5 psoralens and a coumarin never reported again by other authors (an HPLC chromatogram obtained by these authors is reported in Figure 12.4); bergaptol was reported in some of the first studies carried out on bergamot oil, while three psoralens (epoxybergamottin, isoimperatorin, and oxypeucedanin hydrate) were reported by Casabianca (personal communication, 2010); herniarin, detected by Costa et al. (2010), in a concentrated bergamot oil was confirmed by G. Dugo et al. (2012) in many of the samples industrially produced between 2008 and 2012. Herniarin was also detected in cold-pressed laboratory extracted oils from fruits of the Femminello and Castagnaro cultivars, while it was not detected in laboratory extracted oils from Fantastico fruits (Dugo et al. 2011).

Gionfriddo et al. (1997) compared quantitative values obtained for oils extracted using different technologies. They found that the total amount of nonvolatile residue was lower (3.87%) in oils extracted using sponges than that of oils extracted using the Calabrese machine (5.86%) and Pelatrice (5.57%), while the amount of oxygen heterocyclic components was lower (1.37%) in Pelatrice oils, where recycling water is used, than that of oils extracted using sponges (1.58%) and the Calabrese machine (1.54%) that do not use water during the extraction process.

Gionfriddo et al. (1997) observed that the maturation (production) period does not influence the content of the most abundant oxygen heterocyclic components, even

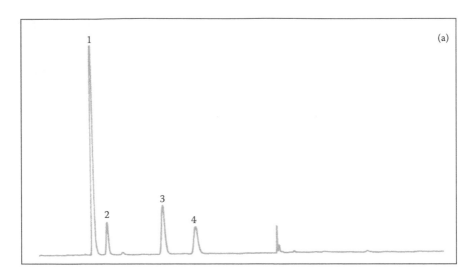

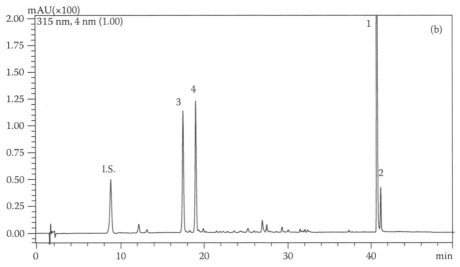

FIGURE 12.3 (a) NP-HPLC chromatogram of a genuine bergamot essential oil, solvent A: hexane:ethyl acetate, 9:1; solvent B: hexane:ethyl alcohol, 9:1; gradient program: 0–7 min: = 0% B; 7–27 min: 0%–95% B; 27–32 min: 95% B. Flow rate 1 mL/min; column 300 × 3.9 mm i.d., 10 μm μ-Porasil. (b) RP-HPLC chromatogram of a bergamot essential oil. Solvent A: water:acetonitrile:THF, 85:10:5; solvent B: acetonitrille:methanol:THF, 60:35:5; gradient program: 0–5 min: = 0% B; 5–25 min: 0%–40% B; 25–45 min: 40%–90% B; 45–55 min: 90% B. Flow rate 1 mL/min; column 150 × 4.6 mm i.d., 2.7 μm Ascentis express C18. (I.S.) internal standard; (1) bergamottin; (2) 5-geranyloxy-7-methoxycoumarin; (3) citropten; (4) bergapten. ((a) reproduced from Mondello et al., *Flavour Fragr. J.* 8, 17–24, 1993; (b) reproduced from Dugo et al., *J. Agric. Food Chem.* 57, 6543–6551, 2009.)

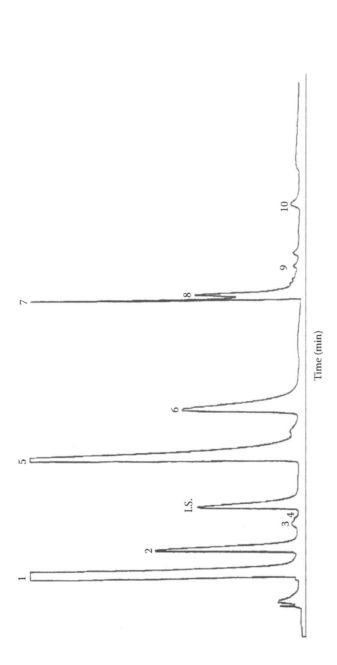

FIGURE 12.4 HPLC chromatogram of bergamot essential oil. (1) bergamottin; (2) 5-geranyloxy-7-methoxycoumarin; (3) 5-isopentenyloxy-7-methoxycoumarin; (4) 5-isopentenyloxy-7-methoxycoumarin; (5) citropten; (6) bergapten; (7) oxypeucedanin; (8) byakangelicol; (9) oxypeucedanin hydrate; (10) byakangelicin; (I.S.) coumarin (internal standard). (Reproduced from Gionfriddo et al., *Riv. Ital. EPPOS* (Spec. Num.) 96–104, 1998.)

though their total content seems to decrease slightly from December to March. In fact, all the other papers reporting quantitative data relative to different periods of the production season (Mondello et al. 1993; P. Dugo et al. 1999b; G. Dugo et al. 2012) do not indicate significant correlation between the total amount of the main coumarins and psoralens and the period of the productive season. Only herniarin content (G. Dugo et al. 2012) seems to be linked to the production period, and its amount decreases during the season (Figure 12.5). Mondello et al. (1993) did not find any correlation between different Calabrian productive areas and the composition of the oxygen heterocyclic fraction; P. Dugo et al. (1999b) observed that oils from areas 6 and 9 (see Figure 8.7 of Chapter 8, this volume) showed the highest content of coumarins and psoralens, while those from area 2 presented the lowest content.

The results reported in Table 12.2 columns 22 and 23 were attained by G. Dugo et al. (2012) and Russo et al. (2012) with the same method used previously (P. Dugo et al. 2009), replacing acetonitrile with methanol and thermostatting the column at 40°C instead of 30°C. The method, applicable to all

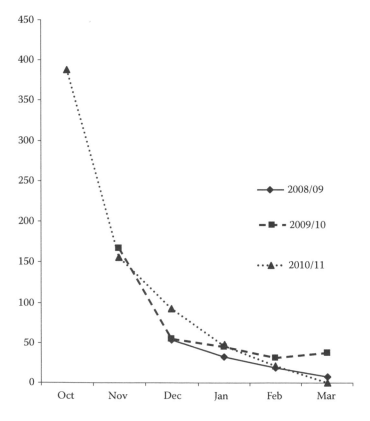

FIGURE 12.5 Seasonal variation (monthly) of herniarin determined in 50 samples of cold pressed bergamot oils produced in 2008–09, 2009–10, and 2010–11 productive seasons. (Reproduced from G. Dugo et al. 2012).

citrus essential oils using the same experimental conditions, enabled an excellent baseline separation of all the heterocyclic oxygenated compounds present in the various oils and an increase in resolution for some critical psoralen pairs. A comparison between two chromatograms, attained using the two approaches on a standard mixture, is illustrated in Figures 12.6a and 12.6b (presence of acetonitrile and methanol in the eluent mixture, respectively), while essential oil chromatograms, relative to the analysis carried out with the methanol method, are compared in Figure 12.7.

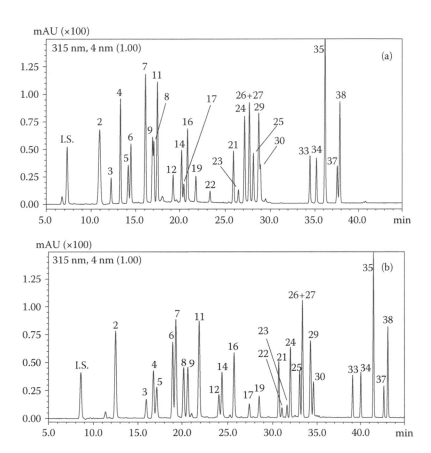

FIGURE 12.6 RP-HPLC chromatograms of a separation on a C18 Ascentis Express column of a standard mix of coumarins, psoralens, and polymethoxyflavones obtained with two different methods: with (a) acetonitrile and (b) methanol. (I.S.) internal standard; (2) herniarin; (3) byakangelicin; (4) 8-methoxypsoralen; (5) psoralen; (6) oxypeucedanin hydrate; (7) citropten; (8) isopimpinellin; (9) meranzin; (11) bergapten; (12) isobergapten; (14) byakangelicol; (16) oxypeucedanin; (17) nobiletin; (19) heptamethoxyflavone; (21) isoimperatorin; (22) tangeretin; (23) epoxyaurapten; (24) imperatorin; (25) cnidilin; (26) phellopterin; (27) osthol; (29) epoxybergamottin; (30) 5-isopentenyloxy-7-methoxycoumarin; (33) cnidicin; (34) 8-geranyloxypsoralen; (35) aurapten; (37) bergamottin; (38) 5-geranyloxy-7-methoxycoumarin. (Reproduced from Russo et al., *J. Essent. Oils Res.* 24, 119–129, 2012.)

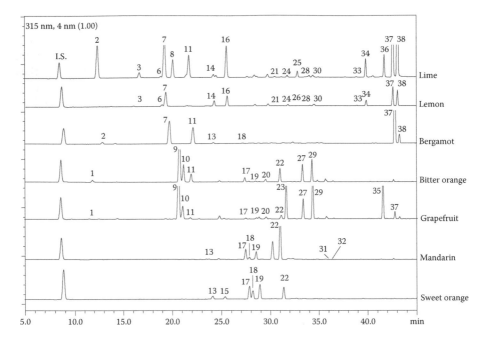

FIGURE 12.7 RP-HPLC chromatograms of seven *Citrus* essential oils obtained by means of a methanol-based method. (I.S.) internal standard; (1) meranzin hydrate; (2) herniarin; (3) byakangelicin; (6) oxypeucedanin hydrate; (7) citropten; (8) isopimpinellin; (9) meranzin; (10) isomeranzin; (11) bergapten; (13) sinensetin; (14) byakangelicol; (15) hexamethoxyflavone; (16) oxypeucedanin; (17) nobiletin; (18) tetra-*O*-methyl-scutellarein; (19) heptamethoxyflavone; (20) epoxybergamottin hydrate; (21) isoimperatorin; (22) tangeretin; (23) epoxyaurapten; (24) imperatorin; (25) cnidilin; (26) phellopterin; (27) osthol; (28) 5-(isopent-2'-eniloxy)-8-(2',3'-epoxy)-isopentenyloxypsoralen; (29) epoxybergamottin; (30) 5-isopentenyloxy-7-methoxycoumarin; (31) 5-demethyltangeretin; (32) 5-demethylnobiletin; (33) cnidicin; (34) 8-geranyloxypsoralen; (35) aurapten; (36) 5-geranyloxy-8-methoxypsoralen; (37) bergamottin; (38) 5-geranyloxy-7-methoxycoumarin. (Reproduced from Russo et al., *J. Essent. Oils Res.* 24, 119–129, 2012.)

12.3.2 LABORATORY-EXTRACTED CALABRIAN OILS

Information on the composition of the oxygen heterocyclic fraction of laboratory-extracted bergamot oils is scant except for a few samples recently analyzed by P. Dugo et al. (2012). In addition, Donato et al. (2013) published the results relative to only polymethoxyflavones.

The composition of laboratory-extracted oils obtained from fruits of the three most common cultivars present in Calabria (Castagnaro, Femminello, and Fantastico) coming from different areas are reported in Table 12.3. For every cultivar, the composition presents large intervals, and variability ranges for the different components are almost always the same. However, some differences can be detected: oils of Fantastico fruits do not contain herniarin, which is present in the oils obtained from the other two cultivars, while oils of Femminello fruits present the lowest content of

TABLE 12.3
Quantitative Composition (mg/L) of Oxygen Heterocyclic Fraction of Laboratory Extracted Calabrian Bergamot Oils

Trivial Name	1a	1b	1c	2a	2b	2c
Coumarins						
Citropten	2146–2276	4539–4902	1657–3351			
5-Geranyloxy-7-methoxycoumarin	800–1563	651–866	604–1021			
Herniarin	13–30	149–172	–			
Psoralens						
Bergamottin	23,997–35,381	13,085–13,778	15,064–21,693			
Bergapten	1306–2572	2327–2675	1490–2399			
Polymethoxyflavones						
Sinensetin	344–415	697–840	749–802	249	253	504
Nobiletin				76	234	259
Tetra-*O*-methylscutellarein	284–444	320–443	155–416	59	180	187
Tangeretin				297	406	306

Appendix to Table 12.3:
1. Dugo P. et al. (2011). Hand-pressed oils; February 2011; (a) 3 samples from the cultivar Castagnaro; (b) 3 samples from the cultivar Femminello; (c) 4 samples from the cultivar Fantastico. For experimental conditions see point 22 of appendix to Table 12.2.
2. Donato et al. (2013). Hand-pressed oils; February 2011; (a) 3 samples from the cultivar Castagnaro; (b) 3 samples from the cultivar Femminello; (c) 4 samples from the cultivar Fantastico. For experimental conditions see point 22 of appendix to Table 12.2.

bergamottin. Figures 12.8a through 12.8c compare HPLC chromatograms obtained for oils of the three cultivars by Dugo et al. (Dugo, P., Russo, M., and Dugo, G., personal communication, 2011).

12.3.3 COMMERCIAL OILS

Table 12.4 reports the quantitative composition of the nonvolatile fraction of commercial oils of unspecified origin. These results fall within the ranges reported in the previous studies for genuine bergamot oils.

In addition to data presented in Table 12.4, a paper by Full et al. (1991) reported the use of an HPLC-HRGC-FTR for the identification and quantification of bergapten and citropten in three commercial oils; two of them presented values in agreement with those of cold-pressed oils, while the other showed a content of citropten of 30 mg/kg and bergapten of 20 mg/kg, in agreement with values presented by bergapten-free oils obtained by distillation or alkaline treatment.

12.3.4 BERGAPTEN-FREE OILS

Table 12.5 reports results for bergapten-free oils. Most of the studies report only the amount of bergapten; from values presented by the four main components or

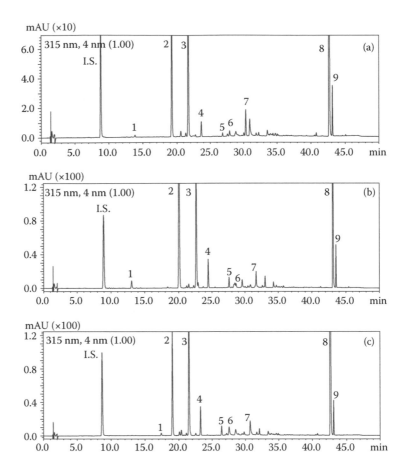

FIGURE 12.8 RP-HPLC chromatograms of laboratory hand-pressed bergamot oils from the cultivars (a) Castagnaro, (b) Femminello, and (c) Fantastico. (I.S.) internal standard; (1) herniarin; (2) citropten; (3) bergapten; (4) sinensetin; (5) nobiletin; (6) tetra-*O*-methylscutellarein; (7) nobiletin; (8) bergamottin; (9) 5-geranyloxy-7-methoxycoumarin.

by bergamottin and bergapten only, the extraction technology can be determined. Results reported in columns 7a (P. Dugo et al. 1999b), 8 (Frèrot and Decorzant 2004), and 12 (Costa et al. 2010) confirmed that oils were obtained by alkaline treatment, while those reported in columns 7b (P. Dugo et al. 1999b) and 10 (Casabianca, personal communication, 2010) confirmed that oils were obtained by distillation. In fact, only bergapten and citropten are significantly reduced after alkaline treatment, while all the four components are drastically reduced after distillation.

Poiana et al. (1994) extracted bergamot oil from peels using supercritical CO_2 under different pressure and temperature conditions in an attempt to get an oil with low bergapten content. The lowest amount of bergapten (500 mg/L) was found in the oils extracted using 8 MPa and 45°C–50°C; oils extracted using 25 MPa and 37°C presented a bergapten content similar to that of oils obtained using conventional extraction technologies of around 3000 mg/L. Poiana et al. (2003) analyzed

TABLE 12.4
Quantitative Composition (mg/L) of Oxygen Heterocyclic Fraction of Commercial Bergamot Oils

Trivial Name	1	2	3	4	5	6	7
Coumarins							
Citropten	2100–2500	1550				1700	2030[a]
5-Geranyloxy-7-methoxycoumarin	400	754					
Psoralens							
Bergamottin	11,000–17,000					10,400	
Bergapten		2400–3200	1620–2360	2800–3300	2000	1400	

[a] The value obtained by different pulse voltametry (DPV); the result obtained by LC-UV was very close.

Appendix to Table 12.4:
1. Cieri (1969). Three commercial oils. For experimental conditions see point 2 of appendix to Table 12.2.
2. Madsen and Latz (1970). One commercial oil. For experimental conditions see point 3 of appendix to Table 12.2.
3. Porcaro and Shubiak (1974). Five commercial oils. For experimental conditions see point 4 of appendix to Table 12.2.
4. Shu et al. (1975). Two commercial oils; NP-HPLC/UV (254 nm) columns packed with Carbowax 400/Corasil, μ Porasil, Zorbax Sil, Corasil II using IPA:n-heptane (3:97) or THF:n-heptane (5:95), isooctane:ethyl acetate:isopropanol (80:10:10), isooctane:ethyl acetate:isopropanol (80:10:10), chloroform:n-hexane (25:75) + 0.125% of methanol, as eluents, respectively.
5. Girard et al. (1981). One commercial oil; RP-HPLC/UV (220 nm); column: Spherisorb C6 (200 × 30 mm, 5 μm); eluents: 0–10 min, ammonium acetate 10^{-2} M (pH 4):acetonitrile (70:30), 11–20 min, gradient until 60% of acetonitrile.
6. Subra and Vega (1997). One commercial oil; NP-HPLC/UV (313 nm); column: Lichrosorb Si (250 × 4.6 mm, 5 μm); eluents: heptane: chloroform (80:20).
7. Wang and Liu (2009). One commercial oil; differential pulse voltametry (DPV); RP-HPLC/PDA; column: Phenomenex Luna C18 (250 × 4.6 mm, 5 μm); eluents: methanol:water (45:55).

by GC two bergapten-free oils, one obtained by NaOH treatment and the other by distillation. They found trace (≤ 0.05%) amounts of bergapten, which was present at 0.5% in the untreated cold-pressed oil. The authors also found trace amount of bergapten in the fractions obtained after supercritical CO_2 treatment using different experimental conditions.

12.3.5 Concentrated Oils

As already discussed in Chapter 8, this volume, dedicated to the composition of volatile fraction, the concentration of bergamot oils is not carried out in a standard way, but only as a function of a specific need of the customers that may need a variable amount of oxygenated components.

TABLE 12.5

Quantitative Composition (mg/L) of Oxygen Heterocyclic Fraction of Bergapten-Free Oils

Trivial Name	1	2	3	4	5	6	7a	7b	8	9	10	11	12	13a	13b
Coumarins															
Citropten							0–52	0–47				7.9	tr		3–13
5-Geranyloxy-7-methoxycoumarin							1539–1975	0–349					1299	691–790	8–29
Psoralens															
Bergamottin							11,726–16,250	0–3017	16312[d]		620	52.3	18,194	13,755–16,742	3–5
Bergapten	10–180	100–140[a]	tr–420[b]	ND[c]–1050	≤5–40	1.7–5.7	0–91	0–41	8[d]	tr–700	80	9.6	tr		3–23
Epoxybergamottin									70.3[d]			3.4			
Polymethoxyflavons															
Sinensetin													0.2		

Note: tr, traces.

[a] Range of the values obtained with three different columns on the same sample. In another sample of treated bergamot oil bergapten was not detected (detection limit of the method: 5 ppm).

[b] In the original cold-pressed oils bergapten ranged between 1750–3700 mg/L.

[c] ND = not detected

[d] In the table are the results obtained by UV detection; those obtained by fluorescence and MS detection were very close.

Appendix to Table 12.5:

1. Porcaro and Shubiak (1974). Two oils labeled debergaptenized and four oils labeled distilled by the authors, almost certainly defurocumarinized by distillation. For experimental conditions see point 4 of the appendix to Table 12.2.

2. Shu et al. (1975). One commercial sample. For experimental conditions see point 4 of appendix to Table 12.4.

3. Di Giacomo and Calvarano (1978a). Seven samples alkali treated. For experimental conditions see point 6 of the appendix to Table 12.2.

4. Suzuki (1979), Suzuki et al. (1979). Seven commercial oils almost certainly defurocumarinized; bergapten was isolated by using the following method: hydroyisis with KOH in MeOH, extraction with ethyl ether, treatment of the alkaline phase with H_2SO_4, extraction with chloroform, concentration of the extract and addition of internal standard (chrysene); GC on packed column of SE-30.

5. Girard (1981). Three commercial samples. For experimental conditions see point 5 of appendix to Table 12.4.

6. Analytical Methods Committee (1987). For experimental conditions see point 9 of appendix to Table 12.2.

7. Dugo et al. (1999b). (a) Eight alkali treated oils; (b) 8 distilled oils; 1996–97 productive season. For experimental conditions see point 13 of appendix to Table 12.1.

8. Frérot and Decorzant (2004). One sample alkali treated; RP-HPLC; solvent A, water:acetonitrile:THF (85:10:5); solvent B, acetonitrile:methanol:THF (65:30:5); gradient profile: 0–5 min, 0% B; 5–20 min, 0%–32% B (linear); 20–24 min, 32% B; 24–38 min, 32%–55% B (linear); 38–40 min, 55%–90% B (linear); 40–50 min, 90% B; detection UV (310 nm), fluorescence ($\lambda_{ex}/\lambda_{em}$, 310/490 nm), MS using APCI source.

9. Gionfriddo et al. (2004). Fifteen oils alkali treated. For experimental conditions see point 13 of appendix to Table 12.2.

10. Prosen and Kočar (2008). One sample. RP-HPLC-MS/MS; column: Hypersil ODS (250 × 4.6 mm, 5 μm); (A) gradient of water (A), (B) acetonitrile, starting with 30% B and ending with 80% B.

11. Casabianca (personal communication, 2010). One sample almost certainly distilled. For experimental conditions see point 21 of appendix to Table 12.2.

12. Costa et al. (2010). One sample alkali treated. For experimental conditions see point 20 of appendix to Table 12.2.

13. Dugo G. et al. (2012). (a) Three alkali treated oils; (b) 3 distilled oils. For experimental conditions see point 22 of appendix to Table 12.2.

TABLE 12.6

Quantitative Composition (mg/L) of Oxygen Heterocyclic Fraction of Concentrated Bergamot Oils

Trivial Name	1	2a	2b
Coumarins			
Citropten	6134	4295–6673	3–5
5-Geranyloxy-7-methoxycoumarin	2877	2234–3780	18–41
Herniarin	251	57–139	
Psoralens			
Bergamottin	39,203	46,033–59,501	10–18
Bergapten	4215	3506–6031	2–8
Polimethoxyflavones			
Sinensetin	372	133–209	–
Tetra-*O*-methylscutellarein	–	63–140	–

Appendix to Table 12.6:

1. Costa et al. (2010). One terpeneless colored oil with waxes. For experimental conditions see point 20 of appendix to Table 12.2.
2. G. Dugo et al. (2012). (a) Tree terpeneless colored oils with waxes; (b) Tree terpeneless colorless oils wax-free. For experimental conditions see point 22 of appendix to Table 12.2.

Information on the oxygen heterocyclic content of concentrated oils is limited to those reported by Costa et al. (2010) and G. Dugo et al. (2012) presented in Table 12.6.

It is clear that the amount of oxygen heterocyclic components in concentrated, not wax-free, oils is higher than in cold-pressed oils and varies in agreement with the concentration degree. Concentrated wax-free oils present an amount of oxygen heterocyclic components comparable to that of bergapten-free oils.

12.3.6 RECOVERED OILS

Table 12.7 reports results relative to bergamot oils cold-recovered from the residues of the cold-pressing process. The only available results in the literature for such oils were obtained by Dugo et al. (1999b). As reported by the authors, the three types of recovered oils presented an amount of oxygen heterocyclic components similar to that of cold-pressed oils produced in the same period of the season.

Cold recovery of oils from the cold-pressing process is not frequently used today, as distillation under reduced pressure using the Peratoner method is the method of choice.

G. Dugo et al. (2012) analyzed 14 samples recovered by distillation under reduced pressure produced between 2008 and 2011, and 2 samples recovered by distillation with traditional systems. Five of the 14 Peratoner oils did not contain quantifiable amounts of oxygen heterocyclic components. The remaining nine samples contained

TABLE 12.7
Quantitative Composition (mg/L) of Oxygen Heterocyclic Fraction of Bergamot Recovered Oils

Trivial Name	1a	1b	1c	2a	2b
Coumarins					
Citropten	1400–1900	900–2200	1300–1700	5–10	14
5-Geranyloxy-7-methoxycoumarin	1200–1700	1600–1800	1100–1400	–	3
Psoralens					
Bergamottin	11,700–16,600	11,800–15,700	9300–13,500	0–3	37
Bergapten	1400–2000	1300–2900	1500–2400	16–30	268

Appendix to Table 12.7:

1. Dugo et al. (1999b). (a) Seven Torchiati oils; (b) 7 Ricicli oils; (c) 3 Pulizia dischi oils. For experimental conditions see point 13 of appendix to Table 12.1.
2. G. Dugo et al. (2012). (a) Nine Peratoner oils produced from 2008 to 2011; other 5 samples analyzed by the same authors did not contain oxygen heterocyclic components. (b) One distilled oil; another sample analyzed by the same authors did not contain oxygen heterocyclic components. For experimental conditions see point 22 to appendix to Table 12.2.

oxygen heterocyclic components in amounts comparable to those of bergapten-free oils obtained by distillation. The composition of these oils and of those recovered by traditional systems is also reported in Table 12.7.

Buiarelli et al. (2002) examined a commercially distilled bergamot oil. These authors used RP-HPLC/MS (triple quadrupole mass spectrometer operated in negative mode using an atmospheric pressure chemical ionization-heated nebulizer [APCI-HN]) for the determination of the following coumarins and psoralens in a distilled commercial bergamot oil: citropten, 230 ppm; bergamottin, 13 ppm; bergapten, 0.16 ppm. In addition, the authors found 70 ppm of 5-methoxy-7-dihydroxycoumarin never before reported in a bergamot oil. The authors did not specify if the oil was bergapten-free obtained by distillation or an oil recovered by distillation from the residues of the cold-extraction. The ratios between the single components do not appear to be in agreement with those from a bergapten-free oil by distillation, or a recovered oil by distillation from the residues of the cold-extraction.

12.3.7 RECONSTITUTED OILS

The information on the composition of the oxygen heterocyclic components of these oils is rather limited (see Table 12.8), and it is both qualitatively and quantitatively different for each of the samples. This is due to the different products used to prepare the sample. Their composition, of course, is always very different from that of cold-pressed genuine bergamot oils.

TABLE 12.8
Quantitative Composition (mg/L) of Oxygen Heterocyclic Fraction of Some Bergamot Reconstituted Oils

Trivial Name	1	2	3
Coumarins			
Citropten	400	300	100–130
5-Geranyloxy-7-methoxycoumarin	100	200	180–370
Herniarin			+
5'-Isopentenyloxy-7-methoxycoumarin			+
Psoralens			
Bergamottin	1800	0	680–1160
Bergapten	300	0	40–100
5-Geranyloxy-8-methoxypsoralen			+
8-Geranyloxypsoralen			+
Isopimpinellin			+
5-Isopentenyloxy-8-methoxypsoralen			+
Oxypeucedanin			+

+ *Identified but not quantitatively determined.*

Appendix to Table 12.8:
1. Cieri (1969). One sample. For experimental conditions see point 2 of appendix to Table 12.2.
2. Calabrò and Currò (1975). One sample. For experimental conditions see point 5 of appendix to Table 12.2.
3. Mondello et al. (1993). Three samples. For experimental conditions see point 11 of appendix to Table 12.2.

12.3.8 PERFUMES AND COSMETICS

Wisneski (1975) applied a method developed by Madsen and Latz (1970) for the determination of bergapten added to bergapten-free perfumes, at 0.001, 0.005, and 0.01% levels. The average recovery was of 88%.

Zaynoun et al. (1977) determined the content of bergapten in 108 commercial perfumes using the *Candida albicans* phototoxicity test and with spectrometry. Bergapten was detected in 62 samples in a range from 0.8 to 10.8 ppm. Authors considered the method with *Candida albicans* superior respect to the spectrophotometric one.

Suzuki (1979) and Suzuki et al. (1979) determined bergapten added to cosmetic products where its content was lower than the detection limit. The method used is described in point 4 of the appendix to Table 12.5. Recovery was of 65% for an addition of 10 ppm and 90% on average for the addition of 100 ppm. Quercia et al. (1979) determined bergapten in an emulsion using the method proposed by Porcaro and Subiak (1974), considered by the authors to be of good sensitivity and repeatability.

Ruggeri et al. (1982) proposed a method for the determination of bergapten in sunburn and suntan preparation without any sample pretreatment. The method was based on the RP-HPLC/UV at 396 nm analysis using a RP8 LiChrosorb column connected to a pre-column packed with the same stationary phase and methanol to water (60:40) mobile phase.

Verger (1983) quantified bergapten in bergamot essential oils and different cosmetic products, such as perfumes, after shave, and suntan products by NP-HPLC/UV (254 nm) using a column packed by radial compression. The detection limit was 0.1 ppm.

Bettero and Benassi (1983) optimized a fast method (around 10 min), specific and of high sensitivity, based on the use of RP-HPLC and fluorescence detection and selective for the determination of bergapten and citropten in cosmetic products. Wavelengths of excitation and emission for bergapten determination were 325 and 440 nm, respectively, while 335 and 435 nm were used for citropten. Perfumes and lotions were directly analyzed while emulsions needed a preliminary extraction with specific solvent.

Herpol-Borremans et al. (1985) identified citropten, 5-geranyloxy-7-methoxycoumarin, bergamottin, and bergapten in a Calabrian bergamot oil, by preparative TLC and NMR, and determined bergapten by NP-HPLC/UV (254 nm) using a Zorbax Sil column and a HC pellosil pre-column, in numerous suntan products (milk, cream, and oil) from Belgium. Bergapten amounts ranged between 5 and 50 ppm. Other suntan products containing bergapten at lower than 0.5 ppm were considered by the authors to be prepared using bergapten-free oils.

Martin et al. (1998) determined bergapten content in 33 samples of different perfumery products (toilet water for men and women, after-shave, perfumes, perfume water, eau de cologne), using the method described at point 16 of the appendix to Table 12.2. Bergapten was present only in 7 products with maximum concentration of 2 ppm.

12.3.9 PERFUMED FOOD PRODUCTS

12.3.9.1 Bergamot Tea

Prosen and Kŏcar (2008) detected bergamottin (0.003 mg/kg) and bergapten (under the detection limit of the method) in a bergamot-flavored tea using SPE-LC-MS/MS. Dugo et al. (2009) did not find oxygen heterocyclic components in a sample of Earl Grey tea.

12.3.9.2 Bergamot Liquors

Production of this liquor is traditionally based on the alcoholic maceration of the fruit peel. Water and sugar are the other two main ingredients of the liquors, which present an alcoholic content of about 30–32% (v/v). Dugo et al. (2009) analyzed two commercial bergamot liquors and found very different amounts of oxygen heterocyclic components. However, they found the four main components of the fraction in both the samples. The samples were extracted with ethyl acetate and analyzed by HPLC-DAD. Results (mg/L) were:

	1	2
Citropten	2.6	0.9
5-geranyloxy-7-methoxycoumarin	0.9	tr
Bergamottin	12.9	1.5
Bergapten	12.9	1.9

12.3.10 BERGAMOT JUICES

Among other citrus juices, bergamot was also studied by different authors, and results are summarized in Table 12.8. Gattuso et al. (2007) determined the distribution of flavonoids and furocoumarins in bergamot juices prepared in laboratory from fruits of three different cultivars (Castagnaro, Fantastico, and Femminello) grown in Calabria. HPLC-DAD-MS/MS analyses were carried out on samples diluted in DMF, centrifuged, and filtered through 0.45 µm membrane. Bergapten and bergamottin were also isolated by preparative HPLC and characterized by spectroscopic methods. Juice obtained from Femminello fruits presented the highest values of furocoumarins, as well as the highest values of flavonoids.

Similarly to Gattuso et al. (2007), Gardana et al. (2008) studied the composition of flavonoids and furocoumarins in bergamot juice prepared in the laboratory from Calabrian fruits at the end of their maturation period. They used HPLC-DAD-MS/MS instrumentation.

TABLE 12.9
Quantitative Composition (mg/L) of Oxygen Heterocyclic Fraction of Bergamot Juices

Trivial Name	1a	1b	1c	2	3	4	5a	5b	5c	6
Coumarins										
Citropten					0.8		0.2	0.3	0.8	
5-Geranyloxy-7-methoxycoumarin					1.1		0.6	0.5	0.6	
Psoralens										
Bergamottin	26.4–29.5	22.5–24.6	38.5–43.7	18.2	61.0	5[a]	25.6	24.5	38.0	21
Bergapten	6.4–7.2	7.2–8.5	9.5–11.8	9.0	30.4	7[a]	2.6	8.8	21.8	4

[a] mg/kg.

Appendix to Table 12.9:
  1. Gattuso et al. (2007). Calabria, Italy; laboratory samples; (a) Castagnaro; (b) Fantastico; (c) Femminello. RP-LC-ESI-MS/MS; column: Discovery C18 (250 × 4.6 mm, 5 µm); eluents: ACN and water under multistep linear gradient from 5% to 100% ACN in 45 min.
  2. Gardana et al. (2008). Calabria, Italy; one laboratory sample. RP-LC-DAD-MS/MS; column: Luna C18 (2) (150 × 2.0 mm, 3 µm); eluents: A: 0.1% formic acid, B: ACN under multisteps linear gradient from 10% B to 95% B.
  3. Dugo et al. (2009). Calabria, Italy; one laboratory sample. For experimental conditions see point 19 of appendix to Table 12.2.
  4. Giannetti et al. (2010). Calabria, Italy; one laboratory sample. RP-LC-DAD; column: Symmetry C18 (75 × 4.6 mm, 3.5 µm); eluents: methanol and 5% acetic acid aqueous solution under multistep linear gradient from 5% to 100% methanol in 25 min.
  5. Dugo P. et al. (2011). Calabria, Italy; laboratory samples; (a) Castagnaro; (b) Fantastico; (c) Femminello. For experimental conditions see point 22 of appendix to Table 12.2.
  6. Kaiwaii et al. (1999). Japan; one laboratory sample; RP-HPLC/UV (330 nm); column: Hypersil RP-18 (125 × 4 mm, 5 µm); eluents: methanol:water (70:30) for bergamottin, methanol:water (20:80) for bergapten.

More recently, Dugo et al. (2009) determined coumarins and furocoumarins in laboratory-made bergamot juice by HPLC-DAD analysis after solvent extraction using ethyl acetate. The same composition of bergamot essential oil was found, with psoralens being the main components. The ratio between bergamottin and bergapten was very different from that of the essential oil, probably due to the lower solubility of bergamottin in the aqueous juice. Dugo et al. (2009) reported higher values than previous authors. A possible explanation is that the filtration step performed by Gattuso et al. (2007) and Gardana et al. (2008) may cause loss of furocoumarins due to the poor solubility of these components in the juice.

Giannetti et al. (2010) analyzed juice, peels, and "pastazzo" that were laboratory obtained from fresh bergamot fruits, along with a commercial powder of a concentrated bergamot juice, commercially available as 500-mg capsules. In these samples, Giannetti et al. (2010) found both flavonoids and furocoumarins; the quantitative data for fresh bergamot juice are reported in Table 12.9. The concentration value of bergamottin was much higher in peels with respect to the fresh juice and pastazzo, while the bergapten content was higher in the pastazzo. The commercial powder of concentrated bergamot juice was found not to contain bergamottin, whereas it showed a bergapten content of 7 mg/kg.

Dugo et al. (2011) analyzed laboratory-extracted juices obtained from fruits of the Castagnaro, Fantastico, and Femminello cultivars; the content of coumarins was similar in the three samples, while juice obtained from Femminello fruits contained a higher amount of psoralens. Values reported in Table 12.9 show differences but range in the same order of magnitude, with the exception of very low values reported for bergamottin by Giannetti et al. (2010). These differences may be due not only to the cultivar, but also to the collection period and the geographical area of origin

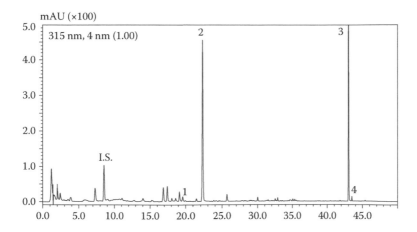

FIGURE 12.9 RP-HPLC chromatogram of laboratory hand-squeezed bergamot juice. The juice has been subjected to solvent extraction prior to HPLC analysis. The extraction procedure was carried out on 10 mL of sample extracted with three aliquots of 10 mL of ethyl acetate. (I.S.) internal standard; (1) citropten; (2) bergapten; (3) bergamottin; (4) 5-geranyloxy-7-methoxycoumarin.

TABLE 12.10
Oxygen Heterocyclic Compounds Identified in Cold-Pressed Bergamot Essential Oils

Compound (Trivial Name)	Systematic Name
Coumarins	
Herniarin[a]	7-Methoxycoumarin
Citropten	5,7-Dimethoxycoumarin
5-Geranyloxy-7-methoxycoumarin	
5-Isopentenyloxy-7-methoxycoumarin[b]	
Psoralens	
Bergapten	5-Methoxypsoralen
Bergamottin	5-Geranyloxypsoralen
Bergaptol[c]	5-Idroxypsoralen
Byakangelicol[b]	5-Methoxy-8-(2',3'-epoxyisopentyloxy)psoralen
Byakangelicin[b]	5-Methoxy-8-(2',3'-dihydroxyisopentyloxy)psoralen
Oxypeucedanin[b]	5-(2',3'-epoxyisopentyloxy)psoralen
Oxypeucedanin hydrate[b]	5-(2',3'-dihydroxyisopentyloxy)psoralen
5-Isopentenyloxy-8-methoxypsoralen[b]	
Isoimperatorin[b]	5-Isopentenyloxypsoralen
Epoxybergamottin[b]	5-(6',7'-epoxygeranyloxy)psoralen
Polymethoxyflavones	
Sinensetin	3',4',5,6,7-Pentamethoxyflavone
Nobiletin[d]	3',4',5,6,7,8-Hexamethoxyflavone
Tetra-*O*-methylscutellarein	4',5,6,7-Tetramethoxyflavone
Tangeretin[d]	4',5,6,7,8-Pentamethoxyflavone
5-Hydroxy-7,8,3',4'-tetramethoxyflavone[e]	

[a] Detected and quantified in cold-pressed oils only by G. Dugo et al. (2012), P. Dugo et al. (2011), and Russo et al. (2012).
[b] Identified and quantified only by Di Giacomo (1990), Calvarano et al. (1995), and Gionfriddo et al. (1997, 1998).
[c] Derived from the decomposition of bergamottin and bergapten.
[d] Identified and quantified for the first time by Donato et al. (2013).
[e] Identified only by Farid (1968).

rather than the analytical technique used. Figure 12.9 shows the chromatogram of a juice analyzed by Dugo et al. (2011).

Kawaii et al. (1999) separated and identified by ^{1}H-NMR and ^{13}C-NMR, citropten, bergapten, and bergamottin from juice, juice sacs and segment epidermis, and fruit peels. The quantitative determination was carried out by RP-HPLC/UV (330 nm). The amount of coumarin and psoralen found in the juice (Table 12.9) was comparable to that found by Gattuso et al. (2007) and Gardana et al. (2008). Bergamottin, bergapten, and citropten amounts in juice sacs and segment epidermis were 72.2,

10.7, and 1.7 mg/kg, respectively. The same components in dried peels resulted in 966, 1525, and 217 mg/kg, respectively.

12.4 FINAL REMARKS

The oxygen heterocyclic fraction of bergamot essential oil has been widely studied. A summary of all the components reported to be present in this fraction is listed in Table 12.10.

Many of these components were reported only in early research and not confirmed later. Some others, such as herniarin and polymethoxylated flavones, were reported only in the most recent literature. This may be due to the use of more efficient and sensitive modern analytical techniques, and they can be considered reliable.

The qualitative and quantitative composition of the oxygen heterocyclic fraction can be used as a criterion of authenticity of the oils. In addition, their presence can be correlated with the technological treatment of the oil, for example, for the case of bergapten-free oils.

Particular interest, in the case of bergamot oil, has been due to the phototoxicity of psoralens. For this reason, studies on the presence of these components (mainly bergapten) in bergamot oils as well as in other products containing citrus have been carried out to estimate benefit/risk factors in relation to the exposure of these components in foodstuffs, beverages, and cosmetic products.

REFERENCES

Analytical Methods Committee. 1987. Applications of high performance liquid chromatography to the analysis of essential oils. Part 1. Determination of bergapten (4-methoxyfuro[3,2-g]chromen-7-one) (8-methoxypsoralen) in oils of bergamot. *Analyst* 112:195–198.

Benincasa, M., Buiarelli, F., Cartoni, G. P., and Coccioli, F. 1990. Analysis of lemon and bergamot essential oils by HPLC with microbore columns. *Chromatographia* 30:271–276.

Bettero, A., and Benassi, C. A. 1983. Determination of bergapten and citropten in perfumes and suntan cosmetics by high-performance liquid chromatography and fluorescence. *J. Chromatogr.* 280:167–171.

Bonaccorsi, I., McNair, H. M., Brunner, L. A., Dugo, P., and Dugo, G. 1999. Fast HPLC for the analysis of oxygen heterocyclic compounds of citrus essential oils. *J. Agric. Food Chem.* 47:4237–4239.

Bonaccorsi, I., Dugo, G., McNair, H. M., and Dugo, P. 2000. Rapid HPLC methods for the analysis of the oxygen heterocyclic fraction in citrus essential oils. *Ital. J. Food Sci.* 4:485–491.

Buiarelli, A., Cartoni, G., Coccioli, F., Jasionowska, R., and Mazzarino, M. 2002. Analysis of limette and bergamot distilled essential oils by HPLC. *Annali di Chimica* 92:363–372.

Calabrò, G., and Currò, P. 1975. Spectrofluorimetric determination of the coumarins of bergamot essential oil. *Essenz. Deriv. Agrum.* 45:246–262.

Calvarano, M. 1961. Sulle cumarine dell'essenza di bergamotto. *Ess. Deriv. Agrum.* 31:167–174.

Calvarano, I., Ferlazzo, A., and Di Giacomo, A. 1979. The cumarinic and furocumarinic composition of bergamot oil. *Essenz. Deriv. Agrum.* 49:12–21.

Calvarano, I., Calvarano, M., Gionfriddo, F., Bovalo, F., and Postorino, E. 1995. HPLC profile of citrus essential oils from different species and geographic origin. *Essenz. Deriv. Agrum.* 65:488–502.

Cavazza, A., Bartle, K. D., Dugo, P., and Mondello, L. 2001. Analysis of oxygen heterocyclic compounds in Citrus essential oils by capillary electrochromatography and comparison with HPLC. *Chromatographia* 53:57–62.

Cieri, U. R. 1969. Characterization of the steam non-volatile residue of bergamot oil and some other essential oils. *J. Assoc. Off. Anal. Chem.* 52:719–728.

Costa, R., Dugo, P., Navarra, M., et al. 2010. Study on the chemical composition variability of some processed bergamot *(Citrus bergamia)* essential oils. *Flavour Fragr. J.* 25:4–12.

Crismer, M. L. 1891. Sur produits crystallisés des essences de citron et bergamote. *Bull. Chem.* 6:30–33.

D'Amore, G., and Calapaj, R. 1965. Le sostanze fluorescenti contenute nelle essenze di limone, bergamotto, mandarino, arancio dolce, arancio amaro. *Rassegna Chimica* 6:264–269.

Di Giacomo, A., and Calvarano, I. 1974. Estudio del aceite de bergamotta por cromatografia en lamina delgada y espectrofluorimetria. *Essenz. Deriv. Agrum.* 44:329–339.

Di Giacomo, A., and Calvarano, M. 1978a. Il contenuto di bergaptene nell'essenza di bergamotto estratta a freddo. *Essen. Deriv. Agrum.* 48:51–83.

Di Giacomo, A., and Calvarano, M. 1978b. The bergapten content in cold-extracted bergamot essential oil. *Riv. It. EPPOS* 2:169–178.

Di Giacomo, A. 1981. Situation actuelle de l'industrie de la bergamote en calabre. *Riv. Ital. EPPOS* 6:300–305.

Di Giacomo, A. 1990. Valutazione della qualità delle essenze agrumarie cold-pressed in relazione al contenuto in composti cumarinici e psoralenici. *Essenz. Deriv. Agrum.* 60:313–334.

Di Giacomo, G., and Di Giacomo, A. 1999. Cumarine e furocumarine nell'olio di bergamotto. *Essenz. Deriv. Agrum.* 60:313–334.

Donato, P., Bonaccorsi, I., Russo, M., and Dugo, P. 2013. Determination of new bioflavonoids in bergamot (*C. bergamia*) peel oil by liquid chromatography coupled to tandem ion trap-time of flight mass spectrometry. *Flav. Fragr. J.* Submitted.

Dugo, G., Bonaccorsi, I., Sciarrone, D., et al. 2012. Characterization of cold-pressed and processed bergamot oils by GC/FID, GC/MS, enantio-GC, MDGC, HPLC, HPLC/MS, GC/IRMS. *J. Essent. Oil Res.* 24:93–117.

Dugo, P., Mondello, L., Lamonica, G., and Dugo, G. 1996. OPLC analysis of heterocyclic oxygen compounds from Citrus fruit essential oils. *J. Planar Chromatogr.* 9:120–125.

Dugo, P., Mondello, L., Stagno d'Alcontres I., Cavazza A., and Dugo, G. 1997. Oxygen heterocyclic compounds of citrus essential oils. *Perf. Flav.* 22(1):25–30.

Dugo, P., Mondello, L., Sebastiani, E., et al. 1999a. Identification of minor oxygen heterocyclic compounds of Citrus essential oils by liquid chromatography-atmospheric pressure chemical ionisation mass spectrometry. *J. Liq. Chrom. & Rel. Technol.* 22:2991–3005.

Dugo, P., Mondello, L., Proteggente, A. R., and Dugo, G. 1999b. Oxygen heterocyclic compounds of bergamot essential oils. *Riv. Ital. EPPOS* 27:31–41.

Dugo, P., Mondello, L., Dugo, L., Stancanelli, R., and Dugo, G. 2000. LC-MS for the identification of oxygen heterocyclic compounds in citrus essential oils. *J. Pharm. Biomed. Anal.* 24:147–154.

Dugo, P, and McHale, D. 2002. The oxygen heterocyclic compounds of citrus essential oils. In *Citrus*, eds. G. Dugo and A. Di Giacomo, 355–390. London and New York: Taylor & Francis.

Dugo, P., Piperno, A., Romeo, R., et al. 2009. Determination of oxygen heterocyclic components in citrus products by HPLC with UV detection. *J. Agric. Food Chem.* 57:6543–6551.

Dugo, P., and Russo, M. 2011. The oxygen heterocyclic components of Citrus essential oils. In *Citrus Oils*, eds. G. Dugo and L. Mondello, 405–443. London and New York: Taylor & Francis.

Farid, S. 1968. A new flavone from bergamott oil. Weak coupling in NMR between ring and methoxyl protons. *Tetrahedron* 24:2121–2123.

Frérot, E., and Decorzant, E. 2004. Quantification of total furocoumarins in citrus oils by HPLC coupled with UV, fluorescence, and mass detection. *J. Agric. Food Chem.* 52:6879–6886.

Full, G., Krammer, G., and Schreier, P. 1991. On-line coupled HPLC-HRGC: A powerful tool for vapor phase FTIR analysis (LC-GC-FTIR). *J. High Res. Chromatogr.* 14:160–163.

Gardana, C., Nalin, F., and Simonetti, P. 2008. Evaluation of flavonoids and furanocoumarins from *Citrus bergamia* (bergamot) juice and identification of new compounds. *Molecules* 13:2220–2228.

Gattuso, G., Barreca, D., Caristi, C., Gargiulli, C., and Lezzi, U. 2007. Distribution of flavonoids and furocoumarins in juices from cultivars of *Citrus bergamia* Risso. *J. Agric. Food Chem.* 55:9921–9927.

Giannetti, V., Boccacci Mariani, M., Testani, E., and D'Aiuto, V. 2010. Evaluation of flavonoids and furocoumarins in bergamot derivates by HPLC-DAD. *J. Commodity Sci. Technol. Quality* 49(I):63–72.

Gionfriddo, F., Postorino, E., and Bovalo, F. 1997. On the authenticity of bergamot oil: HPLC profile of heterocyclic components. *Essenz. Deriv. Agrum.* 67:342–352.

Gionfriddo, F., Postorino, E., and Bovalo, F. 1998. On authenticity of bergamot oil: HPLC profiles of heterocyclic components. *Riv. Ital. EPPOS* (Spec. Num.):96–104.

Gionfriddo, F., Postorino, E., and Calabrò, G. 2004. Elimination of furocoumarins in bergamot peel oil. Composition, extraction methods and olfactory characteristics. *Perf. Flav.* 29(4):48–52.

Girard, J., Unkovic, J., Guimbard, J. P., and Bocchio, E. 1981. Intérêt en cosmétique d'une essence de bergamotte non phototoxique. *Parfums, Cosmétiques, Arômes* 38:39–43.

Guenther, E. 1949. *The essential oils*. New York: Van Nostrand.

Günther, H. O., and Ziegler, E. 1977. Formation of artefacts during thin layer chromatography of furocoumarins of citrus oils. *Essenz. Deriv. Agrum.* 46:473–484.

Herpol-Borremans, M., Masse, M. O., and Grimee, R. 1985. Furocoumarines dans les huiles essentielles identification et dosage du 5-methoxypsoralene dans le produits solaires. *J. Pharm. Belg.* 40:147–158.

Huet, R. 1970. Absorbance dans la région ultraviolette du spectre des huiles essentielles de bergamote de Côte d'Ivoire et du Mali. *Fruits* 25:5–10.

Kawaii, S., Tomono, Y., Katase, E., Ogawa, K., and Yano, M. 1999. Isolation of furocoumarins from bergamot fruits as HL-60 differentation-inducing compounds. *J. Agric. Food Chem.* 47:4073–4078.

Latz, H. W., and Ernes, D. A. 1978. Selective fluorescence detection of citrus oil components separated by high-pressure liquid chromatography. *J. Chromatogr.* 166:189–199.

Lawrence, B. M. 1982. Progress in essential oils. *Perf. Flav.* 7(3):57–65.

Lawrence, B. M. 1994. Progress in essential oils. *Perf. Flav.* 19(6):57–58.

Madsen, B. C., and Latz, H. W. 1970. Qualitative and quantitative in situ fluorimetry of citrus oil thin-layer chromatograms. *J. Chromatogr.* 50:288–303.

Mangiola, C., Postorino, E., Gionfriddo, F., Catalfamo, M., and Manganaro, R. 2009. Evaluation of the genuineness of cold-pressed bergamot oil: Traditional analyses high resolution gas chromatographic and chiral analyses. *Perf. Flav.* 34(10):26–32.

Martin, R., Bobin, M. F., Pelletier, J., and Martini, M. C. 1998. Bergaptene research in eaux de parfum, eaux de toilette et eaux de cologne. *Riv. Ital. EPPOS* (Spec. Num.):130–138.

McHale, D., and Sheridan, J. B. 1989. The oxygen heterocyclic compounds of citrus peel oils. *J. Ess. Oil Res.* 1:139–149.

Mondello, L., Stagno d'Alcontres, I., Del Duce, R., and Crispo, F., 1993. On the genuineness of citrus essential oils. Part 40. The composition of the coumarins and psoralens of Calabrian bergamot essential oil (*Citrus bergamia* Risso). *Flavour Fragr. J.* 8:17–24.

Mossman, D. D., and Bogert, M. T. 1941. *Bergamot oil.* Washington: The American Pharmaceutical Association.

Mulder, G. J. 1839. Über die Zusammensetzung einiger Stearopten und Ätherischen Öle. *Annalen* 31:67–72.

Ohme, C. 1839. Über die Zusammensetzung des Bergamottöls. *Annalen* 31:316–321.

Poiana, M., Reverchon, E., Sicari, V., Mincione, B., and Crispo, F. 1994. Supercritical carbon dioxide extraction of bergamot oil: bergapten content in the extracts. *Ital. J. Food Sci.* 6:459–466.

Poiana, M., Mincione, A., Gionfriddo, F., and Castaldo, D. 2003. Supercritical carbon dioxide separation of bergamot essential oil by a countercurrent process. *Flavour Fragr. J.* 18:429–435.

Pomerantz, C. 1891. Über das Bergapten. *Monatsh. f. Chem.*, 12:379–392.

Porcaro, P. J., and Shubiak, P. 1974. Liquid chromatographic determination of bergapten content in treated or natural bergamot oils. *J. Assoc. Off. Anal. Chem.* 57:145–147.

Prosen, H., and Kočar, D. 2008. Different sample preparation methods combined with LC-MS/MS and LC-UV for determination of some furocoumarin compounds in products containing citruses. *Flavour Fragr. J.* 23:263–271.

Quercia, V., Pierini, N., and Schreiber, L. 1979. High pressure liquid chromatography determination of 5-methoxypsoralen in cosmetic preparation. *Relat. Tech.* 11:18–21. From *Chemical Abstract 1980* 93:504.

Rodighero, G., and Caporale G. 1954. The coumarins obtained from extracts of *Citrus bergamia*. *Atti Ist. Veneto Sci. Nat.* 112:97–102.

Ruggeri, P., Ruggeri, G., and Fonseca, G. 1982. 5-Methoxypsoralene determination by RPLC in some sunburn and suntan preparations. *Rassegna Chimica* 6:195–198.

Russo, M., Torre, G., Carnovale, C., et al. 2012. A new HPLC method developed for the analysis of oxygen heterocyclic compounds in Citrus essential oils. *J. Essent. Oils Res.* 24:119–129.

Shu, C. K., Walradt, J. P., and Taylor, W. I. 1975. Improved method for bergapten determination by high performance liquid chromatography. *J. Chromatogr.* 106:271–282.

von Soden, H., and Rojahn, W. 1901. Über Bergaptin einen neuen Inhaltstoff des Bergamottöles. *Pharm. Ztg.* 46:778–779.

Späth, E., and Socias, E. L. 1934. Über Bergaptol einen neuen Inhaltstoff des Calabrischen Bergamottöles (VIII. Mitteil über naturaliche Cumarine). *Ber.* 67B:59–61.

Späth, E., and Kainrath, P. 1937. Über Bergamottin und über die Aufindung von Limettin in Bergamottöl (XXXIV. Mitteil über naturaliche Cumarine). *Ber.* 70B:2272–2276.

Stanley, W. L., and Jurd, L. 1971. Citrus coumarins. *J. Agric. Food Chem.* 19:1106–1110.

Subra, P., and Vega, A. 1997. Retention of some components in supercritical fluid chromatography and application to bergamot peel oil fractionation. *J. Chromatogr. A* 771:241–250.

Suzuki, H. 1979. Selective extraction and determination of bergapten, a photosensitive substance, from bergamot oil as well as from cosmetic containing bergamot oil. *J. SCCJ* 13:57–71. From *Chemical Abstract 1980* 92:28392.

Suzuki, H., Nakamura, K., and Iwaida, M. 1979. Detection and determination of bergapten in bergamot oil and in cosmetics. *J. Soc. Cosmet. Chem.* 30:393–400.

Thoms, H., and Baetcke, E. 1912. Die Konstitution des Bergaptens. *Ber.* 45:3705–3712.

Verger, G. 1983. Controle et dosage du bergaptène dans les cosmétiques par chromatographie liquide haute performance. *Parfums, Cosmét. Arômes* 51:63–65.

Vernin, G., Bianchini, J-P., and Siouffi, A. 1979. Methods for determining bergapten and citropten in bergamot essential oils. Comparative study. *Parfum, Cosmét. Arômes* 30:49–50, 53–55.

Wang, L.-H., and Liu, H-H. 2009. Electrochemical reduction of coumarins at a film-modified electrode and determination of their levels in essential oils and traditional Chinese herbal medicines. *Molecules* 14:3538–3550.

Wisneski, H. H. 1975. Determination of bergapten in fragrance preparations by thin layer chromatography and spectrophotofluorometry. *J. Ass. Off. Anal. Chem.* 59:547–551.

Zaynoun, S. T., Johnson, B. E., and Frain-Bell, W. 1977. A study of oil of bergamot and its importance as a photoxic agent. I. Characterization and quantification of the photoactive component. *Br. J. Dermatol.* 96:475–482.

.

13 Adulteration of Bergamot Oil

*Ivana Bonaccorsi, Luisa Schipilliti,
and Giovanni Dugo*

CONTENTS

13.1 INTRODUCTION

Bergamot peel oil is a valuable product for its olfactory properties and high commercial value. As unfortunately often happens, bergamot oil is subject to adulteration practices by industrial or commercial operators to increase their profit.

The first information on bergamot adulteration is given by De Domenico (1854) from Reggio Calabria, who wrote a book on the medical properties of bergamot oil. De Domenico reported the addition to cold-extracted bergamot oil of a mixture obtained with "turpentine" (an essential oil extracted from the pine tree also known as turpentine oil) and sweet orange or lemon oils or oils obtained by distillation or cold extraction of unripe fruits of bergamot (called bergamottella) prematurely fallen from the trees. Similar practices were probably also applied for the adulteration of other citrus oils.

Later, after Semmler and Tiemann (1892) found that linalyl acetate is the main component of bergamot oil, it became common practice to add esters obtained from sources different from bergamot, and of lower commercial value, to adulterate bergamot oil.

The first adulterants used to dilute bergamot oil and other citrus oils were turpentine, petroleum, kerosene, and mineral oils. In addition to these products are hitherto used by-products obtained from citrus transformation processes, such as recovered oils obtained from the residues from the cold extraction, and terpenes, mainly from sweet orange, obtained by fractionation of citrus oils to prepare terpene-free oils. In this last case, the "reconstitution" of the oil is obtained by addition of substances to increase the specific gravity (in the past castor oil was used); addition of mixtures of natural or synthetic substances to adjust the physicochemical properties (e.g., optical rotation, UV absorbance, etc.); addition of the adulterated oil; or addition of synthetic or natural ingredients of different botanical origin such as linalool or linalyl acetate to imitate more or less the composition of the genuine oil.

When the methods for determining the physical properties of oils were not fully developed and available, the quality assessment of the oils was limited to qualitative assays, such as treatment with fuming sulphuric acid to reveal the presence of inert paraffins contained in petroleum, mineral oils, and kerosene; to sensorial methods, which are not objective; or to simple determinations such as the nonvolatile residue by evaporation. The development and diffusion of new methods, such as the polarimetric, revealed the presence of compounds with optical rotation different from the natural oil. For example, the measure of the optical rotation of the first 10% distillate of the essential oil revealed addition of turpentine oil. In fact, the major component of turpentine is α-pinene, with a boiling point lower than most of the components present in citrus essential oils, and with an optical rotation in turpentine different from the (+)-limonene that is the main monoterpene hydrocarbon in citrus oils.

The possibility of revealing adulteration of bergamot oil, as of any other citrus cold-pressed essential oil, is mainly linked to the development and use of spectroscopic and chromatographic analytical techniques. This chapter will review applications to this field such as ultraviolet spectroscopy (UV), thin layer chromatography (TLC), high pressure liquid chromatography (HPLC), gas chromatography (GC) with conventional and chiral columns, and the coupled and multidimensional chromatographic techniques with classic detectors and mass spectrometer (conventional and isotopic mass spectrometers). Infrared spectrometry (IR), seldom applied to the analysis of bergamot oil and with unsatisfactory results, and paper chromatography will also be discussed.

13.2   POSSIBLE ADULTERATIONS OF BERGAMOT OIL

Bergamot essential oil can be adulterated in many ways. Some of these methods were detectable by analytical methods available in the 1950s and 1960s. In the 1990s it became possible to reveal most of them by the determination of enantiomeric ratios of volatile components and of the isotopic ratios.

- Addition of turpentine, petroleum, kerosene, mineral oils, castor oil, cedar wood stearin
- Addition of natural or synthetic compounds (menthyl and n-homomenthyl salicilate; methyl N-methyl anthranilate; 4-methoxychalcone)

- Addition of esters different from linalyl acetate
- Addition of citrus oils cheaper than bergamot, or their terpenes, mainly obtained from sweet orange
- Addition of oils recovered from the residues of cold extraction
- Addition of the nonvolatile residue of citrus oils, mainly lime and grapefruit
- Addition of linalool and linalyl acetate synthetic or natural (obtained from sources different from bergamot)

Also possible, although not common, is the occasional contamination with different citrus oils due to inadequate cleaning procedures of the extraction lines previously used to process different citrus.

13.3 METHODS TO REVEAL ADULTERATIONS OF BERGAMOT OIL

13.3.1 EARLIER METHODS

The oldest methods for the analysis of citrus oil, including bergamot, developed by pioneers of the first half of the past century, were revised by Guenther (1949) and by Gildermeister and Hoffman (1959). Some of these (treatment with fuming sulphuric acid, determination of specific gravity, optical rotation) are mentioned in the introduction of this chapter. Photometric and chromatographic methods will be described here. UV spectroscopy and liquid chromatography are aimed at the analysis of the oxygen heterocyclic, present in the nonvolatile residue of the oil, while gas chromatography is dedicated to the determination of the volatiles.

13.3.2 IR SPECTROSCOPY

Presnel (1953) compared the IR spectrum obtained from a genuine bergamot oil with that of an adulterated one. He assumed that this technique could be used to reveal possible frauds. Theile et al. (1960) observed that the ratio between the absorbance bands at 835 cm^{-1} and 801 cm^{-1} varied for genuine bergamot oil between 0.90 and 1.14, and could be used to reveal the addition of monoterpene hydrocarbons. Di Giacomo (1972) reported that at the Experimental Station in Reggio Calabria the Theile's ratio was considered among the genuineness parameters for bergamot oil, with a range of variability of this ratio for genuine oil between 0.830 and 1.114. Bovalo et al. (1985), following a study on numerous genuine oils, determined a range of Theile's ratio between 0.75 and 1.09. These authors also related Theile's ratio with the optical rotation.

13.3.3 UV SPECTROSCOPY

Morton (1929) found that cold-pressed lemon oils absorb UV light at 311 nm. Surprisingly, Morton's studies were independently continued only after 20 years by Cultrera et al. (1952) in Sicily and by Sale et al. (1953) in the United States. This research led to the conclusion that the oxygen heterocyclic compounds present in the nonvolatile residue were responsible for this absorbance, while distilled oils did not adsorb in the UV region. Thus, it was possible to differentiate these oils and reveal mixtures of cold-pressed oils with distilled ones even if nonvolatile substances, which did not absorb in the UV region, were added to mask the dilution. Cultrera et al. (1952) reported on the same graphic the UV transmittance curves of the maxima and minima determined for genuine oils: the resulting area between the two curves represented all the transmittance values of these oils. This is known as the "Palermo method." Instead of transmittance, Sale et al. (1953) used the values of UV absorbance and expressed the results as values of a parameter called "CD."

The studies carried out by Cultrera et al. and by Sale et al. were originally on lemon oils, but the UV characterization was soon applied for the genuineness assessment of all citrus cold-pressed oils. Van Os and Dikastra (1937) measured the UV absorbance of bergamot oil and of other citrus to explore the potential of this technique for quality assessment of the oils. La Face (1959) determined the UV absorbance of numerous genuine bergamot oils in order to establish limits of variation. In this article were reported the absorbance profiles of genuine oils, of residues from evaporation, and of selected compounds of the volatile and nonvolatile fraction of bergamot.

Theile et al. (1960) found that the absorbance at 312 nm of an alcoholic solution of genuine bergamot oil varied between 10 and 13. They believed that it could be possible to reveal the addition of mineral oils by the UV profile of bergamot oil. Calvarano and Calvarano (1964) determined for a very large number of samples of securely genuine bergamot oils the CV values and the $E^{1\%,1cm}$, proposed by Theile et al. (1960) as parameter of genuineness of the oils, and the transmittance values determined at 260 and 310 nm using the Palermo method (Cultrera et al. 1952). They found for the genuine sample a range of 9.95–15.40 for $E^{1\%,1cm}$, and of 0.83–1.04 for CD. In the same article they reported the results relative to a large number of surely adulterated or altered and reconstituted bergamot oils.

Calvarano and Calvarano (1964) concluded that the CD parameter, among those investigated, was the most reliable to reveal frauds on bergamot oil. Presently the ranges of CD values for genuine bergamot oils are fixed by the ISO 3520:1998(E) regulation between 0.760 and 1.180. In Figure 13.1 is a typical CD graph of a genuine bergamot oil and of a distilled oil.

Calabrò et al. (1977) determined the fluorimetric properties of some cold-pressed citrus oils, including bergamot, furocoumarin-free, and reconstituted oils. The values of fluorescence expressed as percentage of citropten in cold-pressed oils ranged between 0.35% and 0.45%, while they were noticeably lower in furocoumarin free oils and were below the limit of detection in the reconstituted oils. These authors concluded that the measure of emitted fluorescence, along with the CD, could represent a parameter for genuineness assessment, mainly useful to reveal reconstituted oils where the CD value was artificially brought within the limit of genuine oils by the addition of extraneous substances.

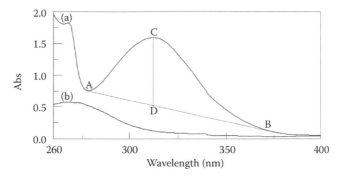

FIGURE 13.1   Typical CD line plots for cold-pressed (a) and distilled (b) bergamot oils. (From Dugo G., Bonaccorsi I., Russo M., and Dugo P., Unpublished results, 2011a.)

13.3.4   PAPER CHROMATOGRAPHY (PC) AND THIN LAYER CHROMATOGRAPHY (TLC)

Although Chakraborty and Bose (1956) demonstrated that partition chromatography on paper was a suitable technique to separate and identify coumarins and psoralens, this technique has been scantily applied to the analysis of citrus essential oils and their residues. To our knowledge applications of this technique on bergamot oil are limited to Calvarano (1961) and Chambon et al. (1969), who used paper chromatography to separate and determine bergaptene and citroptene.

Stanley and Vannier (1957a) found that limettin and other similar compounds isolated from the nonvolatile residue by open column chromatography and TLC were responsible for the UV absorbance of lemon oils. The same authors (Stanley and Vannier 1957b; Vannier and Stanley 1958) developed an analytical method for the analysis of coumarins and psoralens in different essential oils and found that by analyzing these compounds it could be possible to differentiate the pure oils from mixtures of different essential oils. Stanley (1959, 1961) using similar procedures determined some natural and synthetic compounds (methyl and n-homomethyl salicilate, methyl anthranilate, and 4-methoxcalcone) which absorb in the UV region and were used to increase the CD value of cold-pressed oils diluted with distilled ones or terpenes.

Stanley and Jurd (1971) in their review on the distribution of coumarins in citrus asserted that the open column chromatography, TLC, along with spectral absorbance, fluorescence emission, and chromatographic analysis provide useful information on the quality and genuineness of essential oils. D'Amore and Calapay (1965) used analytical TLC to determine the chromatographic characters of fluorescent compounds present in citrus essential oils, including bergamot, and preparative TLC to isolate these components and determine their UV spectroscopic properties. Rf values determined with different eluents and the wavelengths of the maxima, minima, and flex of the single fluorescent components were reported, but no quantitative results. These results are useful to detect contamination or mixtures of different oils.

Cieri (1969) determined the content of coumarins and furocoumarins in bergamot oil and in other natural and commercial essential oils and the ratio between bergapten and the other oxygen heterocyclic compounds present in the residue of the oils analyzed. These ratios, as stated by the authors, can provide information on the origin of bergapten in commercial oils and in oils of unknown origin.

Madsen and Latz (1970) developed a method to determine the content of fluorescent compounds in different essential oils, including bergamot, by direct fluorimetric scanning of the fraction separated by TLC; Di Giacomo and Calvarano (1974) used Madsen and Latz's method to determine these components in genuine bergamot oils. They observed that the variability range of these compounds was narrow and without significant variation during the productive season, and could be therefore used as a valid tool for purity assessment of the oils. In Figure 13.2 is a graph obtained by spectrofluorimetric scanning of the chromatograms where the signals of bergaptol, citroptene, and of 5-gernyloxy-7-methoxy coumarin are visible. The presence of bergaptol has been reported in literature, in addition to this article, only by Späth and Socias (1934) and Di Giacomo (1990). Rodighero and Caporale (1974) believed that bergaptol could derive from the decomposition of bergamottin, and Guenther (1949) observed the tendency of bergamottin to produce bergaptol under specific analytical conditions. Günther and Zigler (1977) reported on the possible formation of artifacts

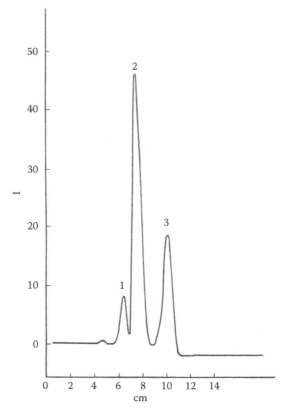

FIGURE 13.2   Graphic of a spectrofluorimetric scan of a TLC chromatogram: (1) bergaptol; (2) citropten; (3) 5-geranyloxy-7-methoxycoumarin. (From Di Giacomo, A., and Calvarano, I., *Essenz. Deriv. Agrum.* 44, 329–339, 1974.)

during the TLC separation of furocoumarins in citrus essential oils. Di Giacomo (1990) recalled that to reproduce the UV profile in bergamot oil it was common practice to add not only the above-mentioned compounds, but also p-aminobenzoic acid, commonly called anticalcone, which could be detected by TLC.

13.3.5   High Performance Liquid Chromatography (HPLC)

The development of HPLC allowed researchers to separate, identify, and quantitatively determine the oxygen heterocyclic compounds in citrus essential oil with more reliable results than the previously applied methods. The technique is simple and versatile; it is possible to operate in normal and reversed phase and under isocratic or gradient elution, thus selecting the most appropriate conditions to separate the components of interest is possible. The use of UV detectors with variable wavelengths permits optimization of the response of single compounds. The photodiode array (PDA) and the hyphenation to mass spectrometers (MS) allows identification of the components analyzed.

McHale and Sheridan (1989) in a milestone article determined the composition of the oxygen heterocyclic compounds in cold-pressed citrus oils and listed all the components present in a single oil. These authors concluded with the following sentence: "While there is not a unique oxygen heterocyclic marker for every *Citrus* species, the patterns of occurrence of the various components are sufficiently diverse to permit the detection of peel oil from one species in that of another."

Mondello et al. (1993) determined the variability range of coumarins and psoralens in 128 samples of bergamot oil produced in Calabria during an entire growing season. In the same article were analyzed some reconstituted bergamot oils which presented an anomalous composition of the oxygen heterocyclic fraction, which was explained by the authors as the adulteration with oil or nonvolatile residue of lime oil. The chromatogram of one of these oils is compared to that of a lime oil in Figure 13.3.

Researchers at the Experimental Station in Reggio Calabria (Calvarano et al. 1995; Gionfriddo et al. 1997) identified in bergamot oil oxygen heterocyclic compounds not indicated by other authors, but usually detected in lemon and lime oils. Presently, if these compounds are detected in bergamot oils, their presence is still considered a symptom of adulteration. The presence of these components in bergamot oil needs further investigation to be confirmed. Calvarano et al. (1995) confirmed, however, that the HPLC analysis of the oxygen heterocyclic compounds in bergamot oil reveals adulteration with extraneous substances with UV spectra similar to the cold-pressed oils or with different citrus species.

Bonaccorsi et al. (1999, 2000) optimized a fast HPLC method to obtain a chromatographic profile of the oxygen heterocyclic fraction of different citrus cold-pressed oils, including bergamot oil. This method permitted differentiation between citrus oils, revealing possible contaminations or cross-adulteration. An example of these chromatographic profiles of the different citrus oils is reported in Figure 13.4.

Recently Dugo et al. (2011a) obtained the HPLC chromatograms of bergamot oils to which had been added 5%, 10%, and 20% of different citrus oils (Persian and

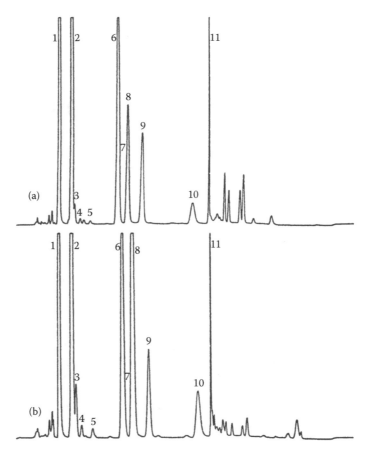

FIGURE 13.3 HPLC chromatogram of a reconstituted bergamot oil (a) and of a lime oil (b). (1) bergamottin; (2) 5-geranyloxy-7-methoxycoumarin; (3) 5-geranyloxy-8-methoxypsoralen; (4) 5-isopenthenyloxy-7-methoxycoumarin; (5) 5-isopenthenyloxy-8-methoxypsoralen; (6) citropten; (7) 8-geranyloxypsoralen; (8) herniarin; (9) bergapten; (10) isopimpinellin; (11) oxypeucedanin. (From Mondello, L., Stagno d'Alcontres, I., Del Duce, R., and Crispo, F., *Flavour Fragr. J.* 8, 17–24, 1993. Reproduced with permission.)

Key limes, lemon, grapefruit, bitter orange, sweet orange, and mandarin). In Figures 13.5a through 13.5g are chromatograms of a genuine bergamot oil and of the mixtures from which are visible the addition of oils of different citrus species.

Dugo et al. (1999), using HPLC-MS with an atmospheric pressure chemical ionization (APCI) probe, identified in bergamot oil trace amounts of two polymethoxylated flavones, tetra-*O*-methylscutellarein and sinensetin and, more recently, Donato et al. (2013) identified nobiletine and tangeretine. Therefore, trace amounts of these four compounds are not indicative of contamination or adulteration with sweet orange oil. Dugo et al. (2012) found, in many genuine samples of bergamot oil produced between 2008 and 2011, variable amounts of herniarin ranging between 8 and 385 ppm. The presence of herniarin at these concentrations must not be considered indicative of contamination or adulteration of bergamot oil.

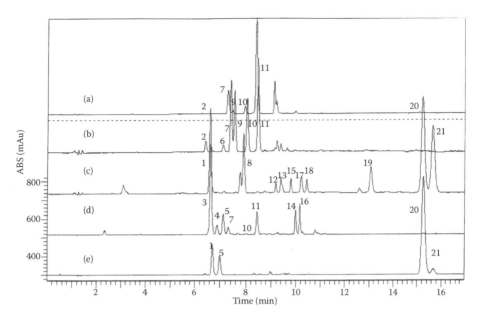

FIGURE 13.4 HPLC-PDA chromatogram of mandarin (a); sweet orange (b); lemon (c); bitter orange (d); bergamot (e) essential oils. (From Bonaccorsi, I., Dugo, G., McNair, H. M., and Dugo, P., *Ital. J. Food Sci.* 4, 485–491, 2000. Reproduced with permission.)

13.3.6 Gas Chromatography (GC) with Conventional Stationary Phases

Gas chromatography, first on packed columns and later on capillary columns, is the most appropriate analytical technique to learn the composition of the volatile fraction of essential oils. Liberti was the first to speculate on the potentiality of this technique for the analysis of essential oils in general and for citrus oils in particular. In 1956 he performed the first gas chromatogram of a bergamot oil (Liberti and Conte 1956). Since then GC, working in isothermal or in temperature programmed, with universal or selective detectors, has become the analytical technique most applied to determine the composition of the volatile fraction of all essential oils and to evaluate contaminations and adulteration of essential oils with single components or mixtures of other volatiles. Successively, the coupling of GC to mass spectrometry conjugated the high separation power of one technique (GC) with the high identification power of another (MS). The interactive use of the mass spectral data with chromatographic retention parameters as the linear retention indices (LRI) rendered more reliable the identification of components in complex mixtures. The use of stationary phases capable of discriminating chiral volatile compounds, the development of multidimensional chromatographic techniques, and the determination of the isotopic ratios by mass spectrometry presently represent the most powerful tools to reveal the adulteration of numerous natural products, including citrus essential oils.

Theile et al. (1960) determined in bergamot oil the following variability ranges: β-pinene 7%–10%; limonene 39%–46%; linalool 17%–24%; linalyl acetate 22%–30%. They assumed that the addition of terpenes (mainly from sweet orange) could

TABLE 13.1

Composition of the Oxygen Heterocyclic Fraction of a Bergamot Essential Oil and a Bergamot +5% of a Key Lime, Persian Lime, Grapefruit, and Bitter Orange Essential Oils; +10% of a Mandarin Essential Oil; and +20% of a Lemon Essential Oil (mg/L)

No.	Class	Compounds	Bergamot	+5% Key lime	+5% Persian lime	+5% Grapefruit	+5% Bitter orange	+10% Mandarin	+20% Lemon
1	CUM	Meranzin hydrate	–	–	–	262	–	–	–
2	CUM	Herniarin	301	517	670	2485	262	245	201
3	PSO	Byakangelicin	–	331	–	–	–	–	–
4	CUM	Citropten	2962	3260	3070	668	2606	2333	2224
5	CUM	Isopimpinellin	–	225	170	–	–	–	–
6	CUM	Meranzin	–	–	–	138	244	–	–
7	CUM	Isomeranzin	–	–	–	1853	65	–	–
8	PSO	Bergapten	2285	2243	2125	151	1950	1842	1584
9	PMF	Sinensetin	72	75	59	–	57	62	98
10	PSO	Byakangelicol	–	–	–	–	–	–	229
11	PSO	Oxypeucedanin	–	235	168	–	–	–	456
12	PMF	Nobiletin	–	–	–	78	192	403	–
13	PMF	Tetra-*O*-methyl-scutellarein	79	77	111	–	90	85	31
14	PMF	Heptamethoxyflavone	–	–	–	–	72	396	–
15	PMF	Tangeretin	–	–	–	–	63	683	–
16	CUM	Epoxyaurapten	–	–	–	301	–	–	–
17	PSO	Phelloperin	–	–	–	–	–	–	32
18	CUM	Osthol	–	–	–	111	75	–	–
19	PSO	Epoxybergamottin	–	–	–	877	277	–	–

20	CUM	5-Isopentenyloxy-7-methoxy-coumarin	–	–	–	–	–	–	159
21	PSO	Cnidicin	–	–	–	–	–	–	26
22	PSO	8-Geranyloxy-psoralen	–	167	68	–	–	–	224
23	CUM	Aurapten	–	–	–	298	–	–	–
24	PSO	5-Geranyloxy-8-methoxy-psoralen	–	155	64	–	–	–	–
25	PSO	Bergamottin	17,606	16,473	16,927	13,986	14,712	13,660	12,413
26	CUM	5-Geranyloxy-7-methoxy-coumarin	916	2132	2870	773	799	779	1145
		Coumarins	4179	5909	6609	5036	4052	3356	3729
		Psoralens	19,891	20,081	19,792	16,716	16,938	15,501	14,964
		Polimethoxyflavones	351	252	270	228	574	2041	129
		All	**22,070**	**24,990**	**25,401**	**21,981**	**20,564**	**23,255**	**18,822**

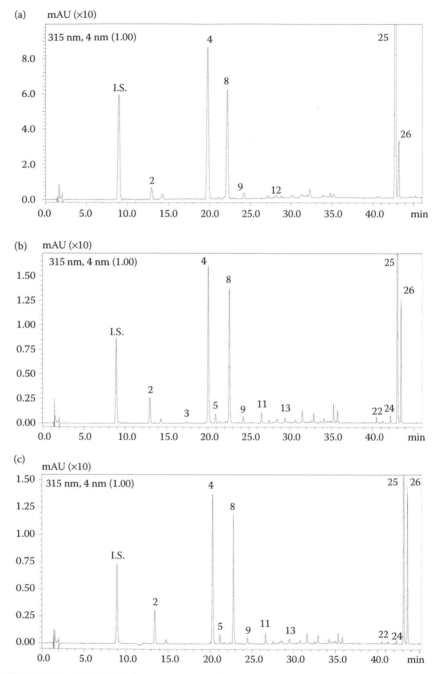

FIGURE 13.5 RP-HPLC chromatograms of a cold-pressed bergamot oil (a) and of mixtures of bergamot oil with 5% of Key lime oil (b), Persian lime oil (c), grapefruit oil (d), bitter orange oil (e), 10% of mandarin oil (f), and 20% of lemon oil (g). For peak identification see Table 13.1. (From Dugo G., Bonaccorsi I., Russo M., and Dugo P., Unpublished results, 2011a.)

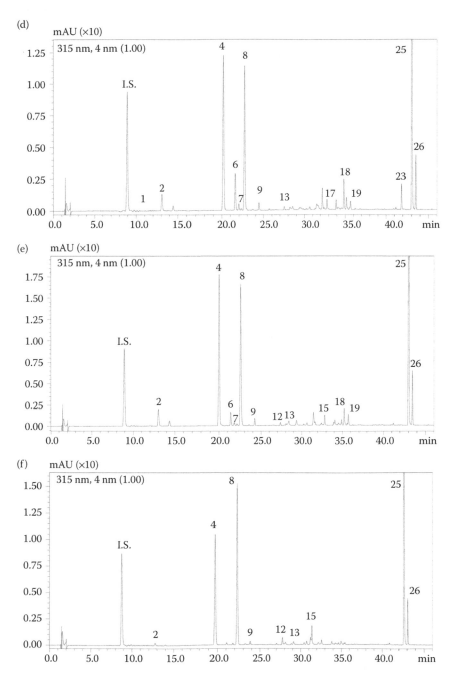

FIGURE 13.5 (CONTINUED) RP-HPLC chromatograms of a cold-pressed bergamot oil (a) and of mixtures of bergamot oil with 5% of Key lime oil (b), Persian lime oil (c), grapefruit oil (d), bitter orange oil (e), 10% of mandarin oil (f), and 20% of lemon oil (g). For peak identification see Table 13.1. (From Dugo G., Bonaccorsi I., Russo M., and Dugo P., Unpublished results, 2011a.)

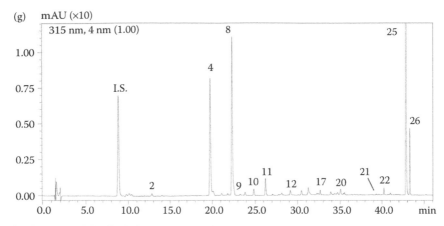

FIGURE 13.5 (CONTINUED) RP-HPLC chromatograms of a cold-pressed bergamot oil (a) and of mixtures of bergamot oil with 5% of Key lime oil (b), Persian lime oil (c), grapefruit oil (d), bitter orange oil (e), 10% of mandarin oil (f), and 20% of lemon oil (g). For peak identification see Table 13.1. (From Dugo G., Bonaccorsi I., Russo M., and Dugo P., Unpublished results, 2011a.)

be detected by evaluation of the decrease in β-pinene and increase in limonene, and by the lower percentage of linalool and linalyl acetate compared to the above-mentioned values. The presence of other adulterants could also be found by the detection of extraneous peaks in the chromatogram.

Calvarano and Calvarano (1964) determined in genuine and adulterated bergamot oils the percentages of some monoterpene hydrocarbons as well as of linalool and linalyl acetate, and the values of some physicochemical parameters. They observed that the physicochemical indices, including the total free alcohols, determined in the adulterated samples were compatible with those determined in genuine oils. On the other hand, the GC analysis showed higher values of limonene and lower values of linalool, leading to the assumption that these oils were adulterated by the addition of terpenes, probably of sweet orange, and by the addition of alcohols different from linalool. The same researchers (Calvarano 1965, 1968; Calvarano and Calvarano 1968) corroborated the importance and validity of gas chromatography to detect adulteration, perpetuated at that time on cold-pressed bergamot oil, from the addition of recovered oils to mixtures of synthetic compounds. In particular, these authors found that terpinen-4-ol is contained at higher percentages in recovered oils than in cold-pressed ones; the ratio of the peak area relative to terpinen-4-ol and that of the unidentified immediately following peak on a chromatogram obtained on UCON LB 550X column (Figure 13.6) never exceeded 0.50 in cold-pressed oils, while it increased in recovered oils.

Dugo et al. (1987, 1991), following the analysis of many hundreds of samples of genuine bergamot oils cold-extracted during different productive seasons, found that the quantitative composition varied during the season within such a wide range that genuine assessment could not be based uniquely on this analytical approach. Verzera et al. (1996) calculated for 1081 samples of genuine bergamot oil the ranges of variability of the following ratios of components: citronellal/

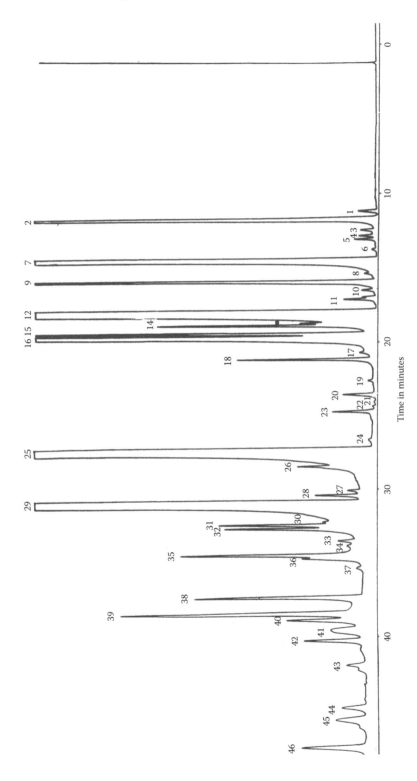

FIGURE 13.6 GC chromatogram on stainless steel capillary column coated with UCON LB 550X. 27, terpinene-4-ol; 28, unknown. (From Calvarano, M., *Essenz. Deriv. Agrum.* 38, 21–30, 1968.)

terpinene-4-ol (0.167–1.875); octyl acetate/α-terpineol (0.842–4.742); γ-terpinene/sabinene + β-pinene (0.661–1.279); and *trans*-sabinene hydrate acetate/α-terpineol (0.704–3.323). These authors observed that some reconstituted bergamot oils showed all the conventional parameters (volatile and nonvolatile composition; enantiomeric distribution of linalool and linalyl acetate) compatible with genuine oils. However, these samples presented values of the abovementioned ratios outside the limits reported for authentic oils, thus they can be considered adulterated. Verzera et al. (1988) observed that the percentage of octanol could be used to differentiate the cold-extracted oils, with a maximum of 0.010% of octanol, from recovered oils (ricicli, torchi, pulizia dischi) where octanol increased from 0.023 to 0.044%. In the same article, these authors observed that the oils recovered by distillation, either at atmospheric pressure or at decreased pressure, presented lower amounts of linalyl acetate and higher amounts of alcohols (linalool, terpinene-4-ol, α-terpineol) and of (*E*)- and (*Z*)-β-ocimene, than cold-pressed oils. They concluded that an increase of alcohols and of (*E*)- and (*Z*)-β-ocimene was indicative of the addition of distilled oils to the cold-pressed ones.

Mondello et al. (2000) developed a method by Fast GC that allowed the elution of some oxygen heterocyclic compounds in citrus oils in less than two minutes. This method allowed detection of the addition of lime oil residue to bergamot oil by the determination of the presence of herniarin and isopimpinellin (Figure 13.7).

Sweet orange oil, and therefore its terpenes, contain an average amount of 0.1% of δ-3-carene, with a maximum of 0.3% in some cases. This component is absent, or present at trace amounts (never exceeding 0.01%), in bergamot, lemon, mandarin, and bitter orange oils (Dugo et al. 2011b). The presence of sweet orange oil or its terpenes can be revealed in bergamot oil, as well as in lemon and mandarin (Dugo et al. 1992a) by the increased amount of δ-3-carene and of the ratios of

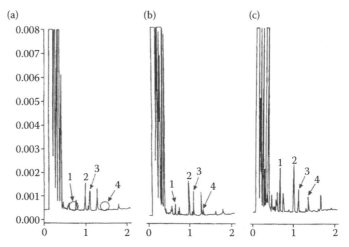

FIGURE 13.7 Fast GC chromatogram obtained for (a) genuine bergamot oil, (b) mixtures of bergamot and 20% of lime oil, (c) lime oil on a RTX-5 (10 m × 0.1 mm, 0.1 μm). (1) herniarin; (2) citropten; (3) bergapten; (4) isopimpinellin. (From Mondello, L., Zappia, G., Errante, G., and Dugo, G., *LC-GC Europe* 13, 495–502, 2000. Reproduced with permission.)

this component with α-terpinene and camphene. In Figure 13.8a through 13.8d are the chromatograms obtained by Dugo et al. (2011a) for a sample of bergamot oil, for mixtures of this oil with 5% and 10% of sweet orange oil, and of a sample of sweet orange oil. The adulteration is visible by a simple comparison of the chromatograms.

Lakszener and Szepsy (1988) used a selective detector for oxygenated compounds (O-FID) to differentiate natural essential oils from sophisticated ones. Figure 13.9 compares the chromatograms obtained by these authors for a genuine and a sophisticated oil. The two chromatograms are evidently different, but the identification of the components is not reported.

13.3.7 CHIRAL GAS CHROMATOGRAPHY

The study of the enantiomeric distribution of volatile components in citrus oils like bergamot by gas chromatography on columns with cyclodextrine as the

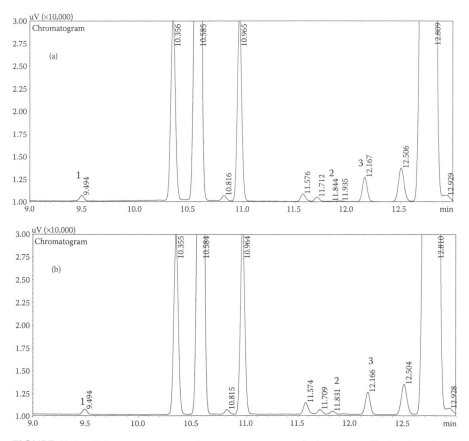

FIGURE 13.8 GC chromatogram of monoterpene zone of a bergamot oil (a), of a mixture of bergamot oil with 5% (b), or 10% (c) of sweet orange oil, and of a sweet orange oil (d). (1) camphene; (2) δ-3-carene; (3) α-terpinene. (From Dugo G., Bonaccorsi I., Russo M., and Dugo P., Unpublished results, 2011a.)

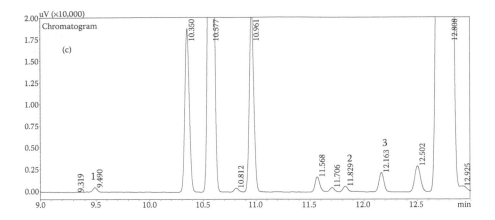

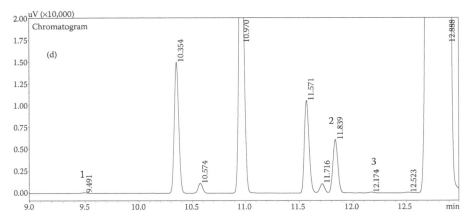

FIGURE 13.8 (CONTINUED) GC chromatogram of monoterpene zone of a bergamot oil (a), of a mixture of bergamot oil with 5% (b), or 10% (c) of sweet orange oil, and of a sweet orange oil (d). (1) camphene; (2) δ-3-carene; (3) α-terpinene. (From Dugo G., Bonaccorsi I., Russo M., and Dugo P., Unpublished results, 2011a.)

stationary phase started in the 1990s. It immediately showed its power to investigate the genuineness of the oils and to reveal adulterations until then undetectable by the commonly used analytical approaches. In fact, almost all the commercial samples analyzed during the first half of the 1990s showed for limonene (Mosandl et al. 1990; Hener et al. 1990; Weinreich and Nitz 1992; Mosandl 1995) and for linalool and linalyl acetate (Hener et al. 1990; Schubert and Mosandl 1991; Bernreuther and Schreier 1991; Weinreich and Nitz 1992; Neukom et al. 1993; Casabianca and Graff 1996) enantiomeric distributions incompatible with those later determined for genuine bergamot oils. Values of the enantiomeric distribution of linalool and linalyl acetate compatible with natural oils were reported only for few commercial samples (Bernreuther and Schreier 1991; Neukom et al. 1993; Ravid et al. 1994) and laboratory extracted oils (Schubert and Mosandl 1991; Weinreich and Nitz 1992). The enantiomeric distribution of linalool in tea

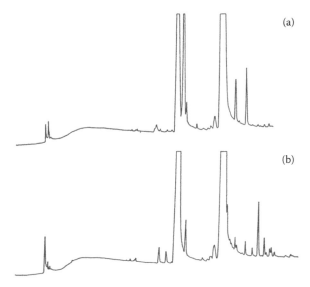

FIGURE 13.9 O-FID chromatogram of a synthetic (a) and a natural (b) bergamot oil. (From Lakszener, K., and Szepesy, L. *Chromatographia* 26, 91–96, 1988. Reproduced with permission.)

extracts was racemic or close to racemic in the samples analyzed by Neukom et al. (1993) and Casabianca et al. (1995). Similarly, the enantiomeric distribution of linalool and linalyl acetate determined by Becker (1995) in 37 out of 53 samples of perfumes was not compatible with the values obtained for these components in natural bergamot oils.

The analytical approaches used to determine the enantiomeric distribution of volatiles in bergamot were direct GC with chiral columns (Cotroneo et al. 1992; Ravid et al. 1994; Dellacassa et al. 1997; Costa et al. 2010); multidimensional GC (MDGC with a conventional column in the first dimension and in second dimension a column with stationary phase capable of discriminating chiral compounds, connected through an interface based on pressure balancing [line-T-piece]) (Mosandl et al. 1990; Hener et al. 1990; Casabianca et al. 1995; Casabianca and Graff 1996; Juchelka and Mosandl 1996; Mosandl and Juchelka 1997a, 1997b); MDGC with mechanical valve as interface (Mondello et al. 1997, 1998) or with an innovative system based on pressure balancing (Dugo et al. 2012) developed by Mondello et al. (2006); pre-separation by flash chromatography (Casabianca and Graff 1994) or by HPTLC (Mosandl and Juchelka 1997b) followed by multidimensional GC separation of the fractions. For the analysis of alcohols in essential oils of bitter and sweet orange, mandarin, and lemon, an HPLC-HRGC system was also used (Dugo et al. 1994a, 1994b; Mondello et al. 1996). For the analysis of limonene in lemon and mandarin oils (Dugo et al. 1992a, 1992b, 1993) two columns were used—one conventional, the second chiral, installed in series. The last method can also be applied to bergamot oils. It must be emphasized, however, that some of the earliest chiral separations of linalyl acetate were carried out on chiral columns of $Ni(HFC)_2$ (Hener et al. 1990) or by carrying out the analysis after hydrogenation of the compound separated by flash chromatography (Casabianca and Graff 1994).

Dugo et al. (1992a) and Cotroneo et al. (1992) analyzed 150 samples of genuine bergamot oils produced in Calabria, 8 samples of commercial linalool, and 20 samples of reconstituted oils obtained by addition to genuine oils of variable amounts of commercial linalool, and determined that in genuine samples of bergamot oil the S-(+)-enantiomer of linalool never exceeded 0.5% of the total amount of linalool. Values above this limit indicated adulteration by addition of synthetic or natural linalool of different origin than bergamot with high percentage of the S-(+)-enantiomer. These results were substantially confirmed by all successive studies, although a few authors found S-(+)-enantiomer values that were slightly higher (up to 1%) (Juchelka and Mosandl 1996). The same authors asserted, however, that the R-(−)-enantiomer of linalool in bergamot oil must be higher than 99% (Mosandl and Juchelka 1997a, 1997b).

In all the studies the enantiomeric distribution of linalyl acetate was determined in proven genuine bergamot oils (see Chapter 11, Table 11.1, this volume) with the S-(+)enantiomer never exceeding 0.4%. The only exception was in a sample from the Ivory Coast (Sciarrone 2009) where the S-(+)-linalyl acetate was 1%. Mosandl and Juchelka (1997a, 1997b) considered for the R-(−)-enantiomer of linalyl acetate the same limits established for linalool for bergamot oils, with this enantiomer always above 99% in genuine samples.

In oils recovered by distillation, the total amount of linalyl acetate decreases (König et al. 1997) while its enantiomeric distribution does not change unless in oils recovered in the past by inappropriate techniques (Mondello et al. 1998). On the contrary, linalool shows a tendency to racemization, more or less evidently, in recovered oils (Juchelka and Mosandl 1996; Mondello et al. 1996; König et al. 1997; Dugo et al. 2001, 2012; Sciarrone 2009); therefore its enantiomeric distribution results in a useful parameter to differentiate recovered oils from cold-extracted ones. In Figures 13.10a through 13.10c are the chromatograms relative to the enantiomeric separation of linalool and linalyl acetate in a genuine bergamot oil, in an adulterated sample, and of a mix of synthetic linalool and linalyl acetate (Dugo 1997, unpublished results).

The enantiomeric distribution of linalool and linalyl acetate represents a fundamental parameter of genuineness and quality of bergamot oil. However, this parameter alone is not sufficient to evaluate the adulteration of bergamot oils with linalool and linalyl acetate of natural origin, with enantiomeric distributions identical or similar to that of natural bergamot. Small dilutions with terpenes, without the addition of extraneous linalool and linalyl acetate, would not change their enantiomeric distribution.

The enantiomeric distribution of other volatiles in genuine bergamot oils was determined by numerous authors (see Chapter 11, Table 11.1, this volume). For some components the results are limited or the ranges of variations determined are too wide and therefore not characteristic and cannot be used for quality assessment; for other components the ranges of variation are narrow, so their enantiomeric distribution can be used for quality assessment. These are reported in the table below.

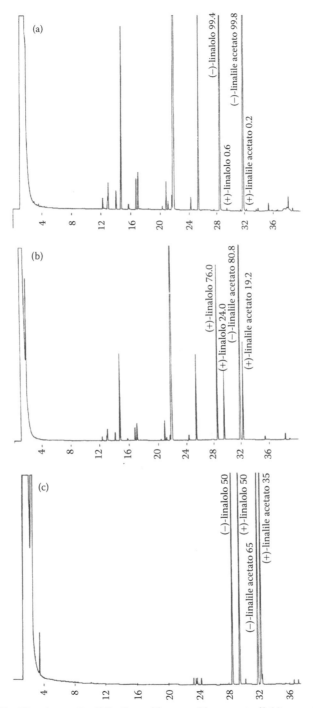

FIGURE 13.10 Direct enantio GC of a cold-pressed bergamot oil (a), an adulterated bergamot oil (b), and a mixture of synthetic linalool and linalyl acetate (c). (From Dugo, G., Unpublished results, 1997.)

Enantiomeric Excesses of Characteristic Volatiles in Bergamot Oil

R-(−)-α-Thujiene	>97.0
S-(−)-α-Pinene	>30.0
S-(−)-β-Pinene	>80.0
S-(−)-Sabinene	>60.0
R-(+)-Limonene	>94.0
R-(−)-Linalool	>98.0
R-(−)-Linalyl acetate	>99.0
S-(+)- (E)-Nerolidol[a]	>60.0

[a] Value reported uniquely by Mosandl and Juchelka (1997b)

The dilution of the most valuable citrus oils with cheap sweet orange terpenes leads to the increase in optical rotation. To correct this value, the addition of sweet orange terpenes is always followed by the addition of a commercial product, (−)-limonene. This addition brings the optical rotation of the adulterated oil back to the values expected for genuine bergamot, but the resulting enantiomeric distribution of limonene is not compatible with any of the natural citrus oils. Dugo et al. (1992a, 1992b, 1993) revealed this adulteration in lemon and mandarin oils. Similarly, this fraud can be revealed in bergamot. The enantiomeric excess of R-(+)-limonene in bergamot oil is never less than 94 and never more than 96. In sweet orange this excess is about 99. Value of enantiomeric excess of limonene in bergamot oils above 96 could be indicative of the addition of sweet orange terpenes without the addition of (−)-limonene.

Additional information on the enantiomeric distribution of volatile components in bergamot oil is reported in Chapter 11, this volume.

13.3.8   GAS CHROMATOGRAPHY–ISOTOPE RATIO MASS SPECTROMETRY (GC-IRMS)

Mass discrimination (kinetic isotope effect) and enantioselectivity are related to the biosynthetic pathway of the plant. Both phenomena can represent characteristic parameters of a single component to assess the origin and authenticity of natural flavors and fragrances.

13.3.8.1   State of the Art

One of the first systems for the determination of the isotope ratios $^{13}C/^{12}C$ and $^{15}N/^{14}N$ was developed by Matthews and Hayes (1978). This system was obtained by the online coupling of a gas chromatograph with a combustion chamber (CG-Combustion). Isotope ratio analysis of CO_2 and N_2 produced by combustion of GC-separated compounds was performed by a single collector mass spectrometer. Later, Barrie et al. (1984) improved the system by coupling to the GC-Combustion a double mass collector capable of registering simultaneously two successive masses and of continuous recording of their ratio (m + 1/m).

Isotope ratio mass spectrometry (IRMS) coupled with gas chromatography (GC) is a useful analytical tool to evaluate origin and authenticity of natural citrus oils as well of other natural flavors and fragrances as well as food. IRMS enables the measurements of deviation of isotope abundance ratios, from an agreed standard, by only a few parts per thousand for C, as well for other elements such as H, N, O, S. Prerequisite for GC-IRMS analysis is high GC resolution (baseline: $R_2 \geq 1.5$), enabling the quantitative transfer of high-purity compounds to the mass spectrometer, thus avoiding GC isotopic discrimination. Each compound must be converted in a gaseous species before the transfer into the ion source (i.e., determination of the $^{13}C/^{12}C$ ratio via IRMS is obtained by converting the C atoms of the analytes into CO_2 by using a combustion chamber and then by comparing the C isotope ratio of each analyte to that of a known reference). An adimensional quantity (δ) is used to express the isotope ratio value of a specific analyte in relation to the reference and is expressed in ‰ (Brenna et al. 1997). Standard deviation is generally between the fourth and sixth figures.

Most of the plants used for human consumption belong to the C3 group. In these plants the pathway of CO_2 fixation, known as Calvin cycle, results in the formation of a C3 body, 3-phosphoglycerate [3(PGA)]. When the IRMS analysis is performed on the metabolites of these plants the evaluation is critical. In fact, the values of $\delta^{13}C$ can overlap with those of synthetic compounds derived from fossil sources or CAM plants. To avoid this inconvenience the concept of an internal standard, i-STD, was introduced (Braunsdorf et al. 1993a; Mosandl 1995) based on the following criteria:

- The component selected as the internal standard should be characteristic of the matrix, but of low sensorial importance.
- The component must be present in a significant amount in the sample and must not undergo mass discrimination during sample preparation.
- It must be biogenetically related to the components investigated.
- It must be inert during storage and during the preparative processes applied to the matrix.
- The component selected as standard must not be a legally authorized additive.

This approach eliminates the influences, due to the geographic origin and climate conditions on $\delta^{13}C$ values, which occur during the CO_2 fixation step in photosynthesis. Thus it is possible to evaluate only the contribution of mass discrimination, which occurs in the enzymatic reaction of the secondary biogenetic pathway. Figure 13.11 provides an example of the reliability of this approach: application on a lemon essential oil (Mosandl 1995).

Enantio GC-IRMS also represents a unique method to reveal the addition of mixtures of natural chiral compounds with synthetic ones not detectable by enantio-GC or by IRMS measurements (Mosandl 1995).

Using the guidelines suggested by Braunsdorf et al. (1993a) and Mosandl (1995), the carbon isotope ratio has largely been applied to evaluate flavor and fragrance genuineness. The matrices studied, not including bergamot, were balm oils (Hener et al. 1995), coriander oils (Frank et al. 1995), orange oils (Braunsdorf et al. 1993b), mandarin oils (Faulhaber et al. 1997a, 1997b; Schipilliti et al. 2010), lemon (Schipilliti et al. 2011a), mentha piperita oils (Faber et al. 1995), volatile components from strawberry (Schumacher et al. 1995; Schipilliti et al. 2011b) and apple (Karl

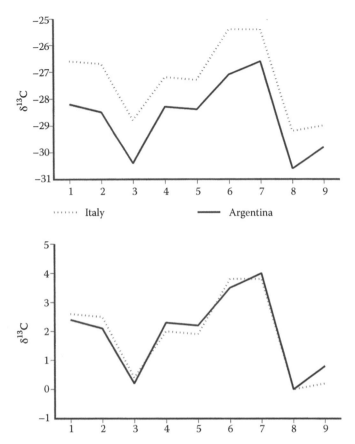

FIGURE 13.11 $\delta^{13}C$ fingerprint of biogenetically related compounds in lemon oils of different geographical origin (top graph). The graph at the bottom shows the $\delta^{13}C$ fingerprint of the samples obtained when using neryl acetate as internal standard. (1) β-pinene; (2) limonene; (3) γ-terpinene; (4) nerol; (5) geraniol; (6) neral; (7) geranial; (8) neryl acetate; (9) geranyl acetate. (From Mosandl, A., *Food Rev. Int.* 11, 597–664, 1995. Reproduced with permission.)

et al. 1994), and nerolì and lime oils (Bonaccorsi et al. 2011a, 2011b). Applications of IRMS to flavor and fragrances were revised by Meier-Augenstein (1999) and by Mosandl (2007).

13.3.8.2 Application on Bergamot Oils

Weinreich and Nitz (1992) studied the enantiomeric distribution and the carbon isotopic ratio by MDGC-IRMS of linalool and linalyl acetate in different essential oils to establish a method to determine their origin (to differentiate them). For linalool in bergamot oil, they reported values of $\delta^{13}C_{PDB}$ ‰ (−29.77 ± 0.1) lower than those obtained in lavender oil (−27.93 ± 0.1) and in coriander (−26.90 ± 0.1), and very similar to that obtained in geranium oil (−29.82 ± 0.1). Also for linalyl acetate, the value of $\delta^{13}C_{PDB}$ ‰ (−29.82 ± 0.1) obtained in bergamot oil was lower than that obtained in lavender oil (−28.41 ± 0.1).

Casabianca et al. (1995), in a study of the authentication of flavors including linalyl acetate from lavender and bergamot oils, asserted that the isotope ratio $^{13}C/^{12}C$ was not useful due to the great similarity with synthetic products, and found more reliable the ratio D/H determined by NMR.

Mosandl and Juchelka (1996, 1997a) used enantio-MDGC (see Section 13.3.7 of this chapter) and GC-C-IRMS to establish the genuineness of bergamot oils and unveil possible adulterations in commercial samples. They determined the ranges of authenticity of the isotope ratios determined in genuine samples for α-pinene, β-pinene+sabinene, γ-terpinene, limonene, myrcene, linalool, linalyl acetate, neryl acetate, and caryophyllene and the values of the same components in seven commercial oils. The adulteration of two commercial oils (samples 12 and 15) was highlighted based on the enantiomeric distribution, and/or on the values of isotopic ratios; the adulteration of sample 16 could not be unveiled based on enantiomeric distribution but was clearly evident based on isotopic ratios (Figure 13.12).

Recently, a new study on the stable isotope ratio of carbon determined on volatiles of bergamot oil was published by Schipilliti et al. (2011c). The genuineness assessment was established based on the values of $\delta^{13}C_{VPDB}$ determined on numerous Italian cold-pressed bergamot oils. These values were compared to those obtained from oils of different geographic origin, from commercial samples, from surely adulterated oils, and from distilled oils. This analytical approach was useful to differentiate bergamot oils of different geographic origin, and by the use of internal standard it was possible to confirm the authenticity of bergamot oils obtained from fruits cultivated in the Ivory Coast. In addition, this technique proved the results

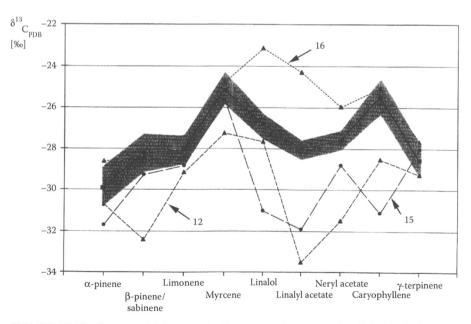

FIGURE 13.12 Commercial bergamot oils compared to the authenticity isotopic range obtained for genuine bergamot oils. (From Juchelka, D., and Mosandl, A., *Pharmazie* 51, 417–222, 1996. Reproduced with permission.)

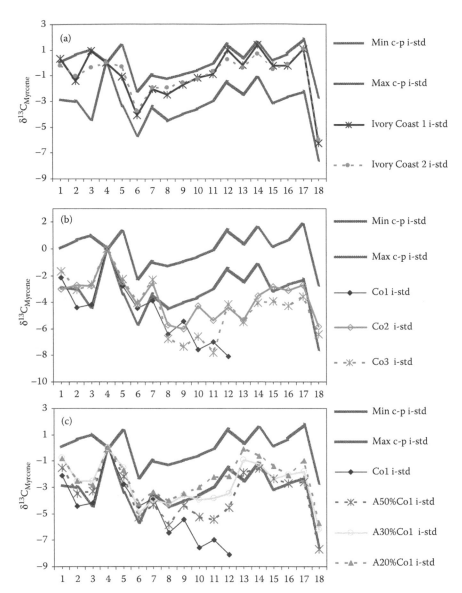

FIGURE 13.13 $\delta^{13}C_{Myrcene}$ values of Ivory Coast bergamot oils (a) of commercial berga-mot oils (b) and of self-adulterated bergamot oils (c) compared to the authenticity range determined for cold-pressed (c-p) bergamot oils. Compounds: (1) α-thujene; (2) α-pinene; (3) β-pinene; (4) myrcene; (5) limonene; (6) γ-terpinene; (7) linalool; (8) linalyl acetate; (9) α-terpinyl acetate; (10) neryl acetate; (11) geranyl acetate; (12) (*E*)-caryophyllene; (13) *trans*-α-bergamotene; (14) β-bisabolene; (15) 2,3-dimethyl-3-(4-methyl-3-pentenyl)-2-nor-bornanol; (16) campherenol; (17) α-bisabolol; (18) nootkatone. (From Schipilliti, L., Dugo, P., Mondello, L., Santi, L., and Dugo, G., *J. Essent. Oil Res.* 23(2), 60–70, 2011. Reproduced with permission.)

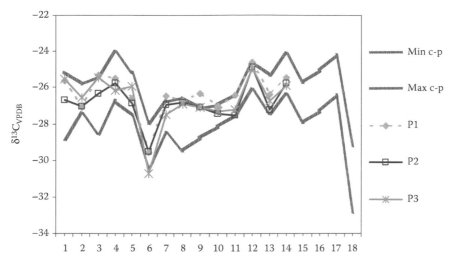

FIGURE 13.14 $\delta^{13}C_{VPDB}$ values of bergamot Peratoner oils compared to the authenticity range determined for cold-pressed bergamot oils. Compounds: (1) α-thujene; (2) α-pinene; (3) β-pinene; (4) myrcene; (5) limonene; (6) γ-terpinene; (7) linalool; (8) linalyl acetate; (9) α-terpinyl acetate; (10) neryl acetate; (11) geranyl acetate; (12) (*E*)-caryophyllene; (13) *trans*-α-bergamotene; (14) β-bisabolene; (15) 2,3-dimethyl-3-(4-methyl-3-pentenyl)-2-norbornanol; (16) campherenol; (17) α-bisabolol; (18) nootkatone. (From Schipilliti, L., Dugo, P., Mondello, L., Santi. L., and Dugo, G., *J. Essent. Oil Res.* 23(2), 60–70, 2011. Reproduced with permission.)

obtained by es-GC, unveiling the presence of adulteration in commercial samples and in samples prepared in laboratory. The values of $\delta^{13}C_{VPDB}$ determined in distilled samples matched, as predicted, the authenticity range determined for genuine cold-pressed ones. These results are illustrated in Figure 13.13a through 13.13c and in Figure 13.14.

In another article from the same research group (Dugo et al. 2012), cold-pressed Italian bergamot oils produced in the seasons 2008–2009, 2009–2010, and 2010–2011 were analyzed by GC-C-IRMS. Dugo et al. (2012) also analyzed by GC-C-IRMS different processed samples of bergamot oils (bergapten-free obtained by treatment with alkali and by distillation), recovered (Peratoner and "fecce" oils), and concentrated (crude and wax free) oils. The $\delta^{13}C_{VPDB}$ values determined for these samples matched the range of authenticity determined for cold-pressed oils. These results are illustrated in Figure 13.15a through 13.15c.

Hör et al. (2001) determined the hydrogen isotope ratio δ^2H_{SMOW} of linalool and linalyl acetate, synthetic and natural of different botanical origins, including bergamot, by gas chromatography-pyrolysis-isotope ratio mass spectrometry (GC-P-IRMS). The range of δ^2H_{SMOW} (isotope ratio $^2H/^1H$ expressed in ‰ deviation relative to the standard mean ocean water) determined for linalool isolated from bergamot (−273 to −294) was different from the values obtained for linalool isolated from coriander, ho-oil, laurel leaf, neroli, and orange oils and partially or totally overlapped with the ranges determined in lavender, lavandin, petitgrain, rose wood, and tea tree oils. The range determined for synthetic linalool (−207 to −301), however, overlapped, totally or partially, with all those of natural sources; thus based on this

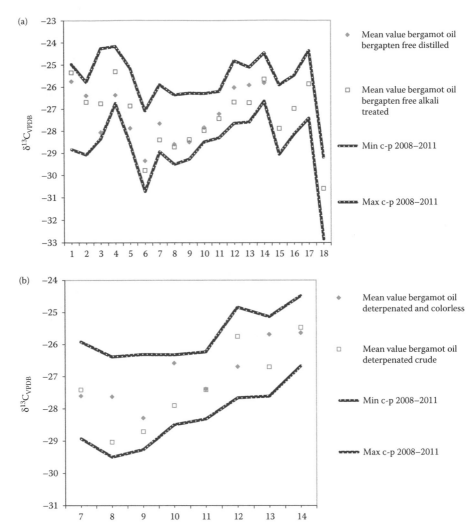

FIGURE 13.15 Carbon isotope ratio values relative to volatile components of bergapten-free distilled and alkali treated (a), deterpenated colorless and deterpenated crude (b), recovered (c) bergamot oils compared to the authenticity range from cold-pressed bergamot oils. Compounds: (1) α-thujene; (2) α-pinene; (3) β-pinene; (4) myrcene; (5) limonene; (6) γ-terpinene; (7) linalool; (8) linalyl acetate; (9) α-terpinyl acetate; (10) neryl acetate; (11) geranyl acetate; (12) (*E*)-caryophyllene; (13) *trans*-α-bergamotene; (14) β-bisabolene; (15) 2,3-dimethyl-3-(4-methyl-3-pentenyl)-2-norbornanol; (16) campherenol; (17) α-bisabolol; (18) nootkatone. (From Dugo, G., Bonaccorsi I., Sciarrone D., et al., *J. Essent. Oil Res.* 24: 93–117, 2012. Reproduced with permission.)

parameter it was not possible to differentiate the natural and synthetic samples. The range of δ^2H_{SMOW} linalyl acetate isolated from bergamot (−252 to 280) was different from those determined in synthetic linalyl acetate (−199 to −239) and from nerolì oil (−213 to −232) but totally or partially overlapped with those determined in lavender, lavandin, and petitgrain oils. Based on these results researchers could differentiate

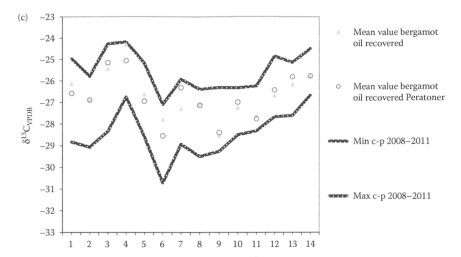

FIGURE 13.15 (CONTINUED) Carbon isotope ratio values relative to volatile components of bergapten-free distilled and alkali treated (a), deterpenated colorless and deterpenated crude (b), recovered (c) bergamot oils compared to the authenticity range from cold-pressed bergamot oils. Compounds: (1) α-thujene; (2) α-pinene; (3) β-pinene; (4) myrcene; (5) limonene; (6) γ-terpinene; (7) linalol; (8) linalyl acetate; (9) α-terpinyl acetate; (10) neryl acetate; (11) geranyl acetate; (12) (E)-caryophyllene, (13) trans-α-bergamotene; (14) β-bisabolene; (15) 2,3-dimethyl-3-(4-methyl-3-pentenyl)-2-norbornanol; (16) campherenol; (17) α-bisabolol; (18) nootkatone. (From Dugo, G., Bonaccorsi I., Sciarrone D., et al., *J. Essent. Oil Res.* 24, 93–117, 2012. Reproduced with permission.)

synthetic linalool from the natural one isolated from lavender and lavandin but not from those isolated from nerolì and petitgrain. Making a comparison between the $\delta^2 H_{SMOW}$ values of linalool and linalyl acetate of different natural origins, the authors observed a depletion of linalool between 7% and 30‰.

In addition to the studies carried out by GC-C-IRMS, Hanneguelle et al. (1992) used SNIF NMR and IRMS to characterize samples of linalool and linalyl acetate of natural and synthetic origin. They determined the isotope ratios $^{13}C/^{12}C$, expressed as δ, and the total ratio $\overline{D/H}$ determined by IRMS and the site-specific isotope ratio $(D/H)_i$ obtained by NMR and IRMS data. These authors did not observe significant differences for the values of $\delta^{13}C$ determined for synthetic and natural linalool and linalyl acetate. However, for the $\overline{D/H}$ values noticeable differences between synthetic and natural samples were seen. Values of site-specific isotopic ratios $(D/H)i$ and the statistical results based on principal component analysis (10 deuterium site-specific isotope ratios and $\delta^{13}C$) were useful to differentiate natural from synthetic samples but could not discriminate between bergamot and other natural sources investigated in this study.

13.4   FINAL REMARKS

The analytical tools available at this time, if applied properly, can successfully reveal almost all the economically significant adulteration practices used in bergamot oil.

However, it will always be necessary to improve the methodologies and develop new dedicated methods to reveal more subtle adulterations, which evolve in parallel to the depth of the analytical investigations. In these cases chiral and isotopic analyses performed by heart-cutting multidimensional chromatography appear to be the most promising techniques and, although not yet fully demonstrated, comprehensive chromatography techniques such as LC × LC could represent highly powerful tools still to be exploited for deep analyses of nonvolatiles in bergamot oil.

REFERENCES

Barrie, A., Bricout, J., and Koziet, J. 1984. Gas chromatography-stable isotope ratio analysis at natural abundance level. *Biomed. Mass Spectrom.* 11:583–588.

Becker et al. 1995. Ätherische öle-gepanschte seelen. *Öko-Test* 10:41–49.

Bernreuther, A., and Schreier, P. 1991. Multidimensional gas chromatography/mass spectrometry: A powerful tool for the direct chiral evaluation of aroma compounds in plant tissues. II. Linalol in essential oils and fruits. *Phytochem. Anal.* 2:167–170.

Bonaccorsi, I., McNair, H. M., Brunner, L. A., Dugo, P., and Dugo, G. 1999. Fast HPLC for the analysis of oxygen heterocyclic compounds of citrus essential oils. *J. Agric. Food Chem.* 47:4237–4239.

Bonaccorsi, I., Dugo, G., McNair, H. M., and Dugo, P. 2000. Rapid HPLC methods for the analysis of the oxygen heterocyclic fraction in citrus essential oils. *Ital. J. Food Sci.* 4:485–491.

Bonaccorsi, I., Sciarrone, D., Schipilliti, L., et al. 2011a. Composition of Egyptian nerolì oil. *Nat. Product. Commun.* 6:1009–1014.

Bonaccorsi, I., Sciarrone, D., Schipilliti, L., et al. 2011b. Multidimensional enantio gas chromatography/mass spectrometry and gas chromatography-combustion-isotopic ratio mass spectrometry for the authenticity assessment of lime essential oils (C. aurantifolia Swingle and C. latifolia Tanaka). *J. Chromatogr. A.* In press. http://dx.doi.org/10.1016/j.chroma.2011.10.038.

Bovalo, F., Cappello, C., Sardo, C., and Di Giacomo, A. 1985. A proposito dello spettro IR dell'essenza di bergamotto. *Essenz. Deriv. Agrum.* 55:36–47.

Braunsdorf, R., Hener, U., Stein, S., and Mosandl, A. 1993a. Comprehensive cGC-IRMS analysis in the authenticity control of flavours and essential oils. Part I: Lemon oil. *Z. Lebensm Unters. Forsch.* 197:137–141.

Braunsdorf, R., Hener, U., Przibilla, G., Piecha, S., and Mosandl, A. 1993b. Analytische und technologische Einflüsse auf das^{13}C/^{12}C-Isotopenverhältnis von Orangenöl-Komponenten. *Z. Lebensm Unters. Forsch.* 197:24–28.

Brenna J. T., Corso, T. N., Tobias, H. J., and Caimi, R. J. 1997. High precision continuous-flow isotope ratio mass spectrometry. *Mass Spectrometry Rev.* 16:227–258.

Calabrò, G., Currò, P., and Lo Coco, F. 1977. Studio spettrofluorimetrico delle essenze agrumarie. *Essenz. Deriv. Agrum.* 47:286–304.

Calvarano, M. 1961. Sulle cumarine dell'essenza di bergamotto. *Essenz. Deriv. Agrum.* 31:167.

Calvarano, M., and Calvarano, I. 1964. La composizione delle essenze di bergamotto. II. Contributo all'indagine analitica mediante spettrofotometria nell'UV e gascromatografia. *Essenz. Deriv. Agrum.* 34:71–92.

Calvarano, M. 1965. La composizione delle essenze di bergamotto. Nota III. *Essenz. Deriv. Agrum.* 47:473–484.

Calvarano, M. 1968. Variazioni nella composizione dell'essenza di bergamotto durante la maturazione del frutto. *Essenz. Deriv. Agrum.* 38:3–20.

Calvarano, M., and Calvarano, I. 1968. Applicazione della gascromatografia all'analisi delle essenze di bergamotto. *Essenz. Deriv. Agrum.* 38:21–30.

Calvarano, I., Calvarano, M., Gionfriddo, F., Bovalo, F., and Postorino, E. 1995. HPLC profile of citrus essential oils from different geographic origin. *Essenz. Deriv. Agrum.* 65:488–502.

Casabianca, H., and Graff, J.-B. 1994. Separation of linalyl acetate enantiomers: Application to the authentication of bergamot food products. *J. High Resolut. Chromatogr.* 17:184–186.

Casabianca, H., Graff, J.-B., Jame, P., Perrucchietti, C., and Chastrette, M. 1995. Application of hyphenated techniques to the chromatographic authentication of flavors in food products and perfumes. *J. High Resolut. Chromatogr.* 18:279–285.

Casabianca, H., and Graff, J.-B. 1996. Chiral analysis of linalol and linalyl acetate in various plants. *Riv. Ital. EPPOS* 7:227–243.

Chakraborty, D. P., and Bose, P. K. 1956. Paper chromatographic studies of some natural coumarins. *J. Ind. Chem. Soc.* 23:905–910.

Chambon, P., Huit, B., and Chambon-Mougenot R. 1969. Dosage spectrofluorimétrique du bergaptène et du citroptène dans les essence de bergamote. *Ann. Pharm. Franc.* 27:635–638.

Cieri, U. R. 1969. Characterization of the steam non-volatile residue of bergamot oil and some other essential oils. *J. Assoc. Off. Anal. Chem.* 52:719–728.

Costa, R., Dugo, P., Navarra, M., et al. 2010. Study on the chemical variability of some processed bergamot (*Citrus bergamia*) essential oils. *Flavour Fragr. J.* 25:4–12.

Cotroneo, A., Stagno d'Alcontres, I., and Trozzi, A. 1992. On the genuineness of citrus essential oils. Part XXXIV. Detection of added reconstituted bergamot oil in genuine bergamot essential oil by high resolution gas chromatography with chiral capillary columns. *Flavour Fragr. J.* 7:15–17.

Cultrera, R., Buffa, A., and Trifiro, E. 1952. Spectrophotometric analysis in the evaluation of lemon oils. *Conserve Deriv. Agrum. (Palermo)* 1(2): 18–20.

D'Amore, G., and Calapay, R. 1965. Le sostanze fluorescenti contenute nelle essenze di limone, begamotto, mandarino, arancio dolce, arancio amaro. *Rass. Chim.* 17: 264–269.

De Domenico, V. 1854. Sulla virtù medicamentosa dell'essenza di bergamotto. From a reprint held by Stazione Sperimentale per l'Industria delle Essenze, Reggio Calabria.

Dellacassa, E., Lorenzo, D., Moyna, P., Verzera, A., and Cavazza, A. 1997. Uruguayan essential oils. Part V. Composition of bergamot oil. *J. Essent. Oil Res.* 9:419–426.

Di Giacomo, A. 1972. I criteri seguiti dalla Stazione Sperimentale di Reggio Calabria per l'accertamnto della qualità degli oli essenziali agrumari estratti a freddo. *Essenz. Deriv. Agrum.* 42:232–242.

Di Giacomo, A., and Calvarano, I. 1974. Estudio del aceite de bergamotta por cromatografia en lamina delgada y espectrofluorimetria. *Essenz. Deriv. Agrum.* 44:329–339.

Di Giacomo, A. 1990. Valutazione della qualità delle essenze agrumarie "cold-pressed" in relazione al contenuto in composti cumarinici e psoralenici. *Essenz. Deriv. Agrum.* 60:313–334.

Donato, P., Russo, M., Bonaccorsi, I., and Dugo, P. 2013. Determination of bioflavonoids in bergamot (*C. bergamia*) peel oils by liquid chromatography coupled to tandem ion-trap time of flight mass spectrometry. *Flavour Fragr. J.* submitted.

Dugo, G., Lamonica, G., Cotroneo, A., et al. 1987. Sulla genuinità delle essenze agrumarie. Nota XVII. La composizione della frazione volatile dell'essenza di bergamotto Calabrese. *Essenz. Deriv. Agrum.* 57:456–534.

Dugo, G., Cotroneo, A., Verzera, A., et al. 1991. Genuineness characters of Calabrian bergamot essential oil. *Flavour Fragr. J.* 6:39–56.

Dugo, G., Lamonica, G., Cotroneo, A., et al. 1992a. High resolution gas chromatography for detection of adulterations of citrus cold-pressed essential oils. *Perfum. Flav.* 17(5): 57–74.

Dugo, G., Stagno d'Alcontres, I., Cotroneo, A., and Dugo, P. 1992b. On the genuineness of citrus essential oils. Part XXXV. Detection of added reconstituted mandarin oil in genuine cold-pressed mandarin essential oil by high resolution gas chromatography with chiral capillary columns. *J. Essent. Oil Res.* 4:589–594.

Dugo, G., Stagno d'Alcontres, I., Donato, M. G., and Dugo, P. 1993. On the genuineness of citrus essential oils. Part XXXVI. Detection of added reconstituted lemon oil in genuine cold-pressed lemon essential oil by high resolution gas chromatography with chiral capillary columns. *J. Essent. Oil Res.* 5:21–26.

Dugo, G., Verzera, A., Trozzi, A., et al. 1994a. Automated HPLC-HRGC: A powerful method for essential oils analysis. Part I. Investigation on enantiomeric distribution of monoterpene alcohols of lemon and mandarin essential oils. *Essenz. Deriv. Agrum.* 64:35–44.

Dugo, G., Verzera, A., Cotroneo, A., et al. 1994b. Automated HPLC-HRGC: A powerful method for essential oil analysis. Part II. Determination of the enantiomeric distribution of linalol in sweet orange, bitter orange and mandarin essential oils. *Flavour Fragr. J.* 9:99–104.

Dugo, G., 1997. Unpublished results.

Dugo, P., Mondello, L., Sebastiani E., et al. 1999. Identification of minor oxygen heterocyclic compounds of citrus essential oils by liquid chromatography-atmospheric pressure chemical ionization mass spectrometry. *J. Liq. Chrom. & Rel. Technol.* 22:2991–3005.

Dugo, G., Mondello, L., Cotroneo, A., Bonaccorsi, I., and Lamonica, G. 2001. Study on the enantiomeric distribution of volatile components of citrus essential oils by multidimensional gas chromatography (MDGC). *Perfum. Flav.* 27(1): 20–35.

Dugo, G., Bonaccorsi I., Russo M., and Dugo P. 2011a. Unpublished results.

Dugo, G., Cotroneo, A., Bonaccorsi, I., and Trozzi, A. 2011b. Composition of the volatile fraction of citrus peel oils. In: *Citrus Oils: Composition, Advanced Analytical Techniques, Contaminants, and Biological Activity*, eds. G. Dugo and L. Mondello, 1–162. London and New York: Taylor & Francis.

Dugo, G., Bonaccorsi, I., Sciarrone, D., et al. 2012. Characterization of cold-pressed and processed bergamot oils by using GC-FID, GC-MS, GC-C-IRMS, enantio-GC, MDGC, HPLC, and HPLC-MS-IT-TOF. *J. Essent. Oil Res.* 24:93–117.

Faber, B., Krause, B., and Mosandl, A. 1995. Gas chromatography-isotope ratio mass spectrometry in the analysis of peppermint oil and its importance in the authenticity control. *J. Essent. Oil Res.* 7:123–131.

Faulhaber, S., Hener, U., and Mosandl, A. 1997a. GC/IRMS analysis of mandarin essential oils. 1. $\delta^{13}C_{PDB}$ and $\delta^{15}N_{AIR}$ values of methyl N-methylanthranilate. *J. Agric. Food Chem.* 45:2579–2583.

Faulhaber, S., Hener, U., and Mosandl, A. 1997b. GC/IRMS analysis of mandarin essential oils. 2. $\delta^{13}C_{PDB}$ values of characteristic flavour components. *J. Agric. Food Chem.* 45:4719–4725.

Frank, C., Dietrich, A., Kremer, U., and Mosandl, A., 1995. GC-IRMS in the authenticity control of the essential of coriandrum-sativum L. *J. Agric. Food Chem.* 43:1634–1637.

Gildermeister, E., and Hoffmann, F. 1959. *Die Aetherischen Oele*, Vol. 5. Berlin: Akademie-Verlag.

Gionfriddo, F., Postorino, E., and Bovalo, F. 1997. On the authenticity of bergamot oil: HPLC profile of heterocyclic components. *Essenz. Deriv. Agrum.* 67:342–352.

Guenther, E. 1949. *The essential oils*, vol. 3. New York: D. Van Nostrand.

Günther, H. O., and Zigler, E. 1977. Formation of artefacts during thin layer chromatography of furanocoumarins of citrus oils. *Essenz. Deriv. Agrum.* 47:473–484.

Hanneguelle, S., Thibault, J. N., Naulet, N., and Martin G. J. 1992. Authentication of essential oils containing linalool and linalyl acetate by isotopic methods *J. Agric. Food Chem.* 40:81–87.

Hener, U., Hollnagel, A., Kreis, P., et al. 1990. Direct enantiomer separation of chiral volatiles from complex matrices by multidimensional gas chromatography. In *Flavour Science and Technology*, eds. Y. Bessiere and A. F. Thomas. Chichester, West Sussex, England: John Wiley & Sons.

Hener, U., Faullhaber, S., Kreis, P., and Mosandl, A., 1995. On the authenticity evaluation of balm oil. *(Melissa officinalis L.)*. *Pharmazie.* 50:60–62.

Hör, K., Ruff, C., Weckerle, B., König, T., and Schreier, P. 2001. Flavor authenticity studies by ^{2}H/^{1}H ratio determination using on-line gas chromatography pyrolysis isotope ratio mass spectrometry. *J. Agric. Food Chem.* 49:21–25.

Juchelka, D., and Mosandl, A. 1996. Authenticity profiles of bergamot oil. *Pharmazie* 51:417–422.

Karl, V., Dietrich, A., and Mosandl, A. 1994. Gas chromatography–isotope ratio mass spectrometry measurements of some carboxylic esters from different apple varieties. *Phytochem. Anal.* 5:32–37.

König, W. A., Fricke, C., Saritas, Y., Momeni, B., and Hohenfeld, G. 1997. Adulteration or natural variability? Enantioselective gas chromatography in purity control of essential oils. *J. High Resolut. Chromatogr.* 20:55–61.

La Face, D. 1959. Ricerche sull'essenza di bergamotto. I. Comportamento spettrofotometrico delle essenze pure nell'ultravioletto e nell'infrarosso. *Essenz. Deriv. Agrum.* 29:45–56.

Lakszener, K., and Szepesy L. 1988. Applications of the O-FID oxygenes analyzer in the cosmetic industry. *Chromatographia* 26:91–96.

Liberti, A., and Conte, G. 1956. Possibilità di applicazione della cromatografia in fase gassosa allo studio delle essenze. *Atti I Congresso Internazionale di Studi e Ricerche sulle Essenze*, Reggio Calabria, Italy, March.

Madsen, B. C., and Latz, H. L. 1970. Qualitative and quantitative in situ fluorimetry of citrus oil thin-layer chromatograms. *J. Chromatogr.* 50:288–303.

Matthews, D. E., and Hayes, J., M. 1978. Isotope-ratio-monitoring gas-chromatography-mass spectrometry. *Analytical Chem.* 50:1465–1473.

McHale, D., and Sheridan, J. B. 1989. The oxygen heterocyclic compounds of citrus peel oils. *J. Ess. Oil Res.* 1:139–149.

McHale, D. 2002. Adulteration of citrus oils. In: *Citrus. The Genus Citrus*, eds. G. Dugo and A. Di Giacomo, 496–517. London and New York: Taylor & Francis.

Meier-Augenstein, W. 1999. Applied gas chromatography coupled to isotope ratio mass spectrometry. *J. Chromatogr. A* 842:351–371.

Mondello, L., Stagno d'Alcontres, I., Del Duce, R., and Crispo, F., 1993. On the genuineness of citrus essential oils. Part 40. Composition of the coumarins and psoralens of Calabrian bergamot essential oil (*Citrus bergamia* Risso). *Flavour Fragr. J.* 8:17–24.

Mondello, L., Dugo, G., Dugo, P., and Bartle, K. D. 1996. On-line HPLC-HRGC in the analytical chemistry of citrus essential oils. *Perfum. Flav.* 21(4): 25–49.

Mondello, L., Catalfamo, M., Dugo, P., Proteggente, A. R., and Dugo, G. 1997. La gascromatografia multidimensionale per l'analisi di miscele complesse. Nota preliminare. Determinazione della distribuzione enantiomerica di componenti degli olii essenziali agrumari. *Essenz. Deriv. Agrum.* 67:62–85.

Mondello, L., Verzera, A., Previti, P., Crispo, F., and Dugo, G. 1998. Multidimensional capillary GC-GC for the analysis of complex samples. 5. Enantiomeric distribution of monoterpene hydrocarbons, monoterpene alcohols, and linalyl acetate of bergamot (*Citrus bergamia* Risso et Poiteau) oils. *J. Agric. Food Chemists* 46:4275–4282.

Mondello, L., Zappia, G., Errante, G., and Dugo, G. 2000. Fast GC and fast GC-MS for the analysis of natural complex matrices. *LC-GC Europe* 13:495–502.

Mondello, L., Casilli, A., Tranchida, P. Q., et al. 2006. Fast enantiomeric analysis of a complex essential oil with an innovative multidimensional chromatographic system. *J. Chromatog. A* 1105:11–16.

Morton, R. A. 1929. Radiation in connection with essential oils and perfumery chemicals. *Perfumery and Essential Oil Record.* 20:258–267.

Mosandl, A., Hener, U., Kreis P., and Schmarr, H-G. 1990. Enantiomeric distribution of α-pinene, β-pinene and limonene in essential oils and extracts. Part 1. Rutaceae and Gramineae. *Flavour Fragr. J.* 5:193–199.

382 *Citrus bergamia*

Mosandl, A. 1995. Enantioselective capillary gas chromatography and stable isotope ratio mass spectrometry in the authenticity control of flavors and essential oils. *Food Rev. Int.* 11:597–664.
Mosandl, A., and Juchelka, D. 1997a. Advances in the authenticity assessment of citrus oils. *J. Essent. Oil Res.* 9:5–12.
Mosandl, A., and Juchelka, D. 1997b. The bitter orange tree – a source of different essential oils. In *Flavour Perception*, eds. H-P. Kruse and M. Rothe, 321–331. Eigenverlag Deutsch Inst. F. Ernahrungsforsch.
Mosandl, A. 2007. Enantiomeric and isotope analysis. Key steps to flavour authentication. In: *Flavours and Fragrances: Chemistry, Bioprocessing and Sustainability*, ed. R. G. Berger. Berlin and Heidelberg: Springer-Verlag.
Neukom, H-P., Meier D. J., and Blum D. 1993. Nachweis von natürlichem oder rekonstituiertem bergamottöl in earl gray tees anhand der enantiomerentrennung von linalool und dihydrolinalool. *Mitt. Gebiete Lebensm. Hyg.* 84:537–544.
Presnel, A. K., 1953. Infrared spectroscopy of essential oils. *J. Soc. Cosmetic Chemists.* 4:101–109.
Ravid, U., Putievsky, E., and Katzir, I. 1994. Chiral GC analysis of enantiomerically pure (R) (−)-linalyl acetate in some Lamiaceae, myrtle and petitgrain essential oils. *Flav. Fragr. J.* 9:275–276.
Rodighero, G., and Caporale, G. 1954. The coumarins obtained from extracts of *Citrus bergamia*. *Atti Ist. Veneto Sci. Nat.* 112:97–102.
Sale, J. W., Winkler, W. O., Gnagy, M. J., et al. 1953. Analysis of lemon oils. *J. Assoc. Offic. Agr. Chem.* 36:112–119.
Schipilliti, L., Tranchida, P. Q., Sciarrone, D., et al. 2010. Genuineness assessment of mandarin essential oils employing gas chromatography-combustion-isotope ratio MS (GC-C-IRMS). *J. Sep. Sci.* 33:617–625.
Schipilliti, L., Dugo, P., Bonaccorsi, I., and Mondello, L. 2011a. Authenticity control on lemon essential oils employing gas chromatography-combustion-isotope ratio mass spectrometry (GC-C-IRMS). *Food Chem.* doi: 10.1016/j.foodchem.2011.09.119.
Schipilliti, L., Dugo, P., Bonaccorsi, I., and Mondello, L. 2011b. Headspace-solid phase microextraction coupled to gas chromatography-combustion-isotope ratio mass spectrometer and to enantioselective gas chromatography for strawberry flavoured food quality control. *J. Chomatogr. A* 1218:7481–7486.
Schipilliti, L., Dugo, P., Mondello, L., Santi. L., and Dugo, G. 2011c. Authentication of bergamot essential oil by gas-chromatography-combustion-isotope ratio mass spectrometer (GC-C-IRMS). *J. Essent. Oil Res.* 23(2): 60–70.
Schubert, V., and Mosandl, A. 1991. Chiral compounds of essential oils. VIII: Stereodifferentiation of linalol using multidimensional gas chromatography. *Phytochem. Anal.* 2:171–174.
Schumacher, K., Turgeon, H., and Mosandl, A. 1995. Sample preparation for gas-chromatography isotope ratio mass-spectrometry—an investigation with volatile components from strawberries. *Phytochem. Anal.* 6:258–261.
Sciarrone, D. 2009. Personal communication.
Semmler, F. W., and Tiemann, F. 1892. Ueber sauerstoffhaltige bestandtheile einiger aetherischer oele. *Ber.* 25:1180–1188.
Späth, E., and Socias, L., 1934. Über Bergaptol, einen neuen Inhaltsstoff des Calabrischen Bergamottöles (VIII. Mitteil. über natürliche Cumarine). *Ber.* 67B:59–61.
Stanley, W. L. 1959. Determination of menthyl salicylates in lemon oil. *J. Assoc. Offic. Agr. Chemists* 42:643–646.
Stanley, W. L. 1961. A test for chalcones in lemon oil. *J. Assoc. Offic. Agr. Chemists* 44:546–548.
Stanley, W. L., and Vannier, S. H. 1957a. Chemical composition of lemon oil. I. Isolation of a series of substituted coumarins. *J. Amer. Chem. Soc.* 79:3488–3491.

Stanley, W. L., and Vannier, S. H. 1957b. Analysis of coumarin compounds in citrus oils by liquid solid partition. *J. Assoc. Offic. Agr. Chemists* 40:582–588.

Stanley, W. L., and Jurd, L. 1971. Citrus coumarins. *J. Agric. Food Chem.* 19:1106–1110.

Theile, F. C., Dean, D. E., and Suffis, R. 1960. The evaluation of bergamot oil. *Drug and Cosmetic Ind.* 86:758–759, 837–840.

Van Os, D., and Dykstra, K. 1937. Examination of essential oils by measurement of absorption in the ultraviolet. *J. Parm. Chim.* 25:437–454, 485–501.

Vannier, S. H., and Stanley, W. L. 1958 Fluorometric determination of 7-geranoxycoumarin in lemon oil: Analysis of mixtures of grapefruit oil in lemon oil. *J. Assoc. Offic. Agric. Chem.* 41:432–435.

Verzera, A., Lamonica, G., Mondello, L., Trozzi, A., and Dugo, G. 1996. The composition of bergamot oil. *Perfum. Flav.* 21(6): 19–34.

Verzera, A., Trozzi, A., Stagno d'Alcontres, I., et al. 1998. The composition of the volatile fraction of Calabrian bergamot essential oil. *Riv. Ital. EPPOS* 25:17–38.

Weinreich, B., and Nitz, S. 1992. Influences of processing on the enantiomeric distribution of chiral flavour compounds. Part A: Linalyl acetate and terpene alcohols. *Chem. Microbiol. Technol. Lebensm.* 14:117–124.

14 The Composition of Bergamot Juice

Giovanni Dugo and Alessandra Trozzi

CONTENTS

14.1 INTRODUCTION

In the 1960s bergamot juice was still the unique source of "natural" citric acid produced in Italy. It was produced from juice extracted from December through March. The juice has a high yield and high acidity, expressed as monohydrated citric acid, of about 3.5%. The industrial process also included the recovery from the previously fermented juice of about 2% of ethyl alcohol. That type of production of citric acid was then dismissed, due to the progressive introduction of the "biological" production of citric acid from the molasses of beets or of sugar cane, which are very good substrates for the citric fermentation.

The juice dedicated to the production of citric acid was obtained by pressing the fruits, previously used for the extraction of the essential oil, locally named "bocce," by screw press. Strong pressure on the fruits led to the extraction of all the chemicals present in the albedo responsible for the bitter taste of the juice; in addition, the presence of scant amounts of residue of essential oil led to a rapid change in the taste, which appeared "girato" (inverted). The contact with the iron parts and the lack of deaeration also contributed to spoiling the final quality of the juice, causing a rapid darkening. More appropriately, this juice could be defined as an agro-industrial product, not suitable for drinking.

From the 1970s, in fact, a process was initiated to adjust the diagram of production to fit the needs of a more rational technology, diverting to the use of similar instrumentation that was applied to the production of other citrus juice extraction. It was thus confirmed that it was possible to obtain from bergamot a juice appreciated of its organoleptic characteristics, if the modern technology applied for other citrus was applied to it.

The juice can be used by itself, appropriately diluted and sweetened, or blended with other citrus or tropical juices. Even its bitter taste can be appreciated, integrating other peculiar characteristics for which citrus juices are usually well accepted by humans.

The aforementioned considerations led to increased interest in the study of this juice's composition. In addition to past investigations, which led to the acquisition of the knowledge of some principal components (sugars, organic acids, mineral compounds, vitamin C), there have recently been studies of the amino acids, carotenoids, flavonoids, and other properties.

14.2   SUGARS

Information on the content of this class of compounds and on its composition in Calabrian bergamot juice was reported by Calvarano (1958, 1961), Di Giacomo and Calvarano (1972, 1993), Di Giacomo (1974, 1989), Calvarano et al. (1995, 1996), Postorino et al. (2001), Montesano et al. (2003), and Cautela et al. (2008). In addition, Moufida and Marzouk (2003) analyzed juice of bergamot cultivated in Tunisia.

Calvarano (1958) determined total sugars and reducing ones in juices extracted industrially by Birillatrice machines and by screw presses during the productive season 1956–57. The results, expressed as g/L, for the two types of juice are reported below:

Extraction	Total Sugars	Reducing Sugars
Birillatrice	34.0–41.1	14.0–24.7
Screw press	32.5–34.5	21.8–27.4

In the same paper are the results relative to some samples extracted in laboratory from fruits harvested from August 1957 to February 1958. Sugars were not determined from August to October, but between October and February reducing sugars were less than 10 g/L to 16 g/L, while in the same period the total sugars ranged between 23.2 and 38.0 g/L.

A few years later the same author (Calvarano 1961) analyzed 19 samples of juice extracted in laboratory from fruits harvested from December 1960 to May 1961. The extraction was made by means of a technique similar to the industrial "birillatura." The reducing sugars in the 19 samples were 10.8–27.4 g/L, and the total sugars were 29.2–41.0 g/L. This study used paper chromatography to find glucose and fructose, and tentatively sucrose.

The variability interval of the total sugars in the samples analyzed by Calvarano (1958) is reported within the characteristic composition of bergamot juice by Di Giacomo and Calvarano (1972) along with the presence of glucose, fructose, and sucrose.

Di Giacomo (1974, 1989) and Calvarano et al. (1995) reported for sugars the same values determined by Calvarano (1961). Di Giacomo and Calvarano (1993), in a review on the nutritional properties of citrus juices, found a total sugar content between 25 and 40 g/L. The results obtained by Calvarano et al. (1996), Postorino et al. (2001), Cautela et al. (2008), and Montesano et al. (2003) related to the dosage of the single components of this class of compounds are summarized in Table 14.1.

Calvarano et al. (1996) studied the variation in sugar content between August and March during two consecutive seasons (1991–92 and 1992–93). This investigation was carried out on the three most common cultivars (cvs.) in Calabria (Fantastico, Femminello, and Castagnaro); the juice was obtained in laboratory, avoiding the presence of pulp by using a sieve with holes of 1.18 mm^2. The samples were collected monthly between August and October, and fortnightly between November and February, with the last sample collected in March. The single sugars were determined by enzymatic methods. The variability ranges of sugars determined in the juice produced from ripened fruits from November to March, which is the period when the bergamot fruits is industrially transformed, are reported in Table 14.1. At the end of their investigation, which was also focused on other components of the juice, Calvarano et al. (1996) asserted that it was not possible to highlight significant differences between the three cultivars analyzed and that the amount of sugars increased during the ripening of the fruits. In Figure 14.1, note that the amount of sugars in the juice of Castagnaro bergamot is lower than the others in almost all the periods of the seasons, and that although higher amounts were determined in the fruits harvested from November–December, the increase during ripening is not consistently determined.

Postorino et al. (2001) determined during the productive season 1998–99, using enzymatic methods, the glucose, fructose, and sucrose in industrial bergamot juice obtained from oil-free fruits by a rolling extractor and screw press, and from the whole fruits using a Citrostar machine. This extractor has a mechanism inspired by that of the in-line FMC machines. Samples obtained in the laboratory were analyzed simultaneously. The results obtained for the four types of juice were similar, although sucrose and fructose were slightly lower in the juice obtained by the screw press.

Also in Table 14.1 are results obtained by Cautela et al. (2008) relative to laboratory-extracted juice in a study aimed at unveiling adulteration of lemon juice by the addition of bergamot. The same table includes the results obtained by Montesano et al. (2003) on bergamot juice obtained with the Naviglio extractor (Naviglio 2003) from peeled fruits using methanol as a solvent. The methanol extracts after concentration were solved in aqueous methanol and fractionated by successive extraction with different organic solvents. The sugars were recovered from the residual water phase at the end of the procedure.

The results reported in Table 14.1 are in good agreement among each other and can be considered representative of the glucidic composition of Calabrian bergamot juice.

Total sugars were also determined in the juice extracted in laboratory from bergamot fruits cultivated in Tunisia (Moufida and Marzouk 2003). The value reported by these authors is about 76 mg/L, significantly higher than what was obtained from Calabrian bergamot juice.

If compared to the other citrus juices, the amount of sugars in bergamot is comparable with that of grapefruit (Postorino et al. 2001) and lemon (Di Giacomo and Calvarano 1972). These citrus fruits, in fact, present sugar amounts that are

TABLE 14.1
Sugars (g/L) in Calabrian Bergamot Juice

	1a	1b	1c	2	3a	3b	3c	3d	4
Total sugars	35.9–50.7	33.6–48.4	24.3–40.1	24–51					
Reducing sugars				10.8–27.4					
Sucrose	12.9–21.8	11.8–24.3	10.1–20.7	10.0–24.5	11.25–23.71	18.4–24.6	6.12–24.20	9.6–24.1	10–24
Glucose	11.7–14.0	7.9–12.5	6.4–10.3	6.0–14.0	10.03–14.80	9.11–13.5	5.05–13.83	9.1–15.1	9–14
Fructose	11.2–17.4	9.5–11.8	7.9–10.3	7.0–17.0	8.91–14.25	9.4–13.2	5.60–13.47	9.9–14.8	9–15
Glucose/fructose					1.01–1.12	0.94–1.22	0.90–1.19	0.92–1.06	0.94–1.06

Appendix to Table 14.1:

1. Calvarano et al. (1996). Laboratory extracted samples from November to March of the following years during two productive seasons (1991–92 and 1992–93); (a) 12 samples from the cultivar Femminello; (b) 12 samples from the cultivar Fantastico; (c) 12 samples from the cultivar Castagnaro.
2. Postorino et al. (1998). Industrial samples.
3. Postorino et al. (2001). Productive season 1998–99; (a) 6 industrial samples extracted by Citrostar machine; (b) 6 industrial samples extracted from the deoleated fruits ("bocce") by Speciale RS machine; (c) 5 industrial samples extracted from the deoleated fruits by Torchi; (d) 6 laboratory samples.
4. Cautela et al. (2008). Thirty samples laboratory extracted in January 2003, 2004, and 2005.

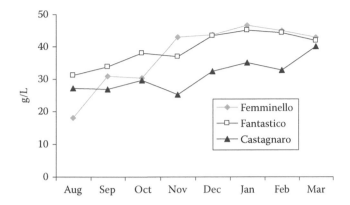

FIGURE 14.1   Total content of sugars (g/L) in bergamot juice, from the cvs. Femminello, Fantastico, and Castagnaro, during the ripening of the fruits. (From the results of Calvarano et al., *Essenz. Deriv. Agrum.* 66, 158–186, 1991.)

much lower than mandarin, clementine, and sweet orange juices (Di Giacomo and Calvarano 1972).

14.3   ORGANIC ACIDS

Calvarano (1958), within an investigation carried out by paper chromatography on samples of juice extracted industrially and in the laboratory, found the presence of three organic acids. These were identified based on their Rf values as citric, malic, and succinic acids. This author also assumed the presence of malonic acid and a second acid, which based on the Rf could have been oxalic or tartaric acid. Citric acid was present in all the samples analyzed and decreased with the ripening stage of the fruit. Succinic acid was not found in the samples extracted in the laboratory from fruits harvested in July and August, while malic acid was present in unripe fruits and gradually disappeared during ripening. The total acidity of the samples analyzed, expressed as g/L of monohydrated citric acid, varied in the samples industrially extracted in February and March 1957 from 25.5 to 49.8; in the samples extracted in laboratory from completely ripe fruits from November to February it decreased from 47.0 to 43.1; in the samples extracted in the laboratory from August to October it decreased from 59.0 to 47.5. Calvarano (1961) found in bergamot juice extracted in laboratory from December 1960 to March 1961 that the total acidity was between 31.5 and 68.8 g/L of monohydrated citric acid. Di Giacomo and Calvarano (1972) reported the range of variability of total acidity in bergamot juice at 29.5–61.0 g/L, which corresponds to the range of all the samples analyzed by Calvarano (1958).

   Di Giacomo et al. (1974), analyzing 26 samples of juice extracted in laboratory from fruits produced between November 1972 and February 1973, found that the acidity was between 42.8 and 60.5 g/L of monohydrated citric acid. Di Giacomo et al. (1993), in a review on the nutritional properties of citrus juices, reported for the same parameter the range of 30.0–67.0 g/L, probably taking into account the value found by Calvarano (1958) for juices extracted in laboratory for unripe fruits.

TABLE 14.2
Acidity Values (g/L of Citric Acid Monohydrate) in
Bergamot Juice Industrially Produced during the 1973–74
Season

Production Period	Juice Produced (Kg × 1000)	Acidity
12/17—12/29	113	50.4–59.5
1/2—1/10	90	51.8–59.5
1/12—1/18	130	52.0–56.7
1/21—1/31	165	47.9–53.9
2/1—2/10	80	45.5–55.5
2/11—2/25	103	45.0–49.7

Source: Di Giacomo, A., Le jus de la bergamotte. Atti del 12th Congresso Internazionale C.I.I.A.A. Atene, 687–694, 1974.

During the productive season 1973–74, Di Giacomo (1974) analyzed numerous lots of bergamot juice produced by the Birillatrice Speciale model RS200 machine. The results, representative of about 680 tons of juice and thus statistically significant, are summarized in Table 14.2. Evaluating the data reported in Table 14.2, the acidity values vary within a quite narrow range, from 45.0 to 59.5 g/L of monohydrated citric acid. In the period evaluated, malic acid resulted in variable amounts from 1.0 to 2.0 g/L. The same results are reported by Di Giacomo (1989) and by Calvarano et al. (1995).

Di Giacomo et al. (1975) determined the total acidity and the malic acid content in 23 samples industrially produced from December 1973 to February 1974. The values were between 41.9 and 57.0 g/L, and the amount of malic acid was between 1.05 and 1.99 g/L. These values were in good agreement with those previously reported by Di Giacomo (1974). Although some irregular values were detected, the amount of malic acid decreased during the productive season. The content of malic acid was judged as similar to those found in other drinkable citrus juices.

Additional studies on the organic acid occurrence in bergamot juice were carried out by Calvarano et al. (1996), Gionfriddo et al. (1996), Postorino et al. (1998, 2001), Montesano et al. (2003), and Cautela et al. (2008). These results are summarized in Table 14.3.

The results reported by Calvarano et al. (1996) (see Table 14.3) refer to the same samples described in Section 14.2 of this chapter, relative to their sugar content. Regarding the content of organic acids, samples of different cvs. did not exhibit significant differences. The authors also confirmed that, as in any other citrus juice, the amount of organic acids decreases during ripening of the fruits; in particular the decrease starts to be significant from January onwards. The cv. Fantastico showed an inversion of the content in March; the amount of isocitric acid decreased for the three cvs. during the entire period, but in Femminello it remains constant in March; malic acid increases from August to December, then decreases for the cvs. Femminello and Fantastico while the cv. Castagnaro reaches its maximum level in

TABLE 14.3

Organic Acid in Calabrian Bergamot Juices

	1a	1b	1c	2	3	4a	4b	4c	4d	5
Total acidity (g/L) of Citric acid monohydrate	39.0–48.3	38.4–56.3	39.7–53.8	35.7–45.4	35.7–57.0	39.4–45.8	39.0–47.7	38.1–44.2	40.0–50.5	39.9–51.5
Malic acid (g/L)	1.00–2.50	0.92–2.58	1.15–3.33	0.99–1.63	0.92–2.60	0.93–2.18	0.90–2.60	0.89–1.94	0.97–2.25	0.9–2.9
D-Lactic acid (mg/L)						44–53	38–44	43–49	48–57	
L-Lactic acid (mg/L)						43–49	38–48	38–44	43–49	
Isocitric acid (g/L)	0.34–0.61	0.33–0.70	0.31–0.59	0.279–0.338	0.300–0.700	0.360–0.510	0.330–0.470	0.303–0.430	0.360–0.550	0.364–0.531
Citric acid/Isocitric acid				88–143	80–145	86–111	101–141	98–129	92–140	80–137

Appendix to Table 14.3:

1. Calvarano et al. (1996). See point 1 of appendix to Table 14.1.
2. Gionfriddo et al. (1996). Thirteen industrial samples produced from December 1994 to April 1995.
3. Postorino et al. (1998). Industrial oils.
4. Postorino et al. (2001). See point 3 of appendix to Table 14.1.
5. Cautela et al. (2008). See point 4 of appendix to Table 14.1.

November. The behavior of malic and isocitric acids is reported in Figure 14.2. The industrial juice, produced between December and April and analyzed by Gionfriddo et al. (1996) (Table 14.3), presented an irregular variation of total acidity and of the content of isocitric acid, while malic acid behaved as in the samples obtained in the laboratory from December to March analyzed by Calvarano et al. (1996).

Postorino et al. (1998) reported the analytical profile of bergamot juice produced in Calabria, and later, Postorino et al. (2001) compared the composition of bergamot juices obtained by different industrial extraction technology and in the laboratory (Table 14.3). The Postorino et al. (2001) study did not observe significant differences in the composition of organic acids in different types of juice, and asserted that this fraction of bergamot juice is very similar to that of lemon juice. Actually, the juice of lemon generally presents a higher total acidity than bergamot juice (Di Giacomo and Calvarano 1972; Cautela et al. 2008). Cautela et al. (2008) reported for organic acids in bergamot juice a composition similar to what was previously determined, and Montesano et al. (2003) reported the same values found by Postorino et al. (1998).

In general, the values reported in Table 14.3 are in good agreement and provide a significant picture of the fraction of organic acids in bergamot juice.

14.4   AMINO ACIDS

The formol index (mL of NaOH 0.1 M per 100 mL of juice) provides a useful quantitative value of the total amount of amino acids in citrus juices. The formol index in Calabrian bergamot juice, industrial and extracted in laboratory, was determined by Calvarano (1961), Di Giacomo and Leuzzi (1973), Di Giacomo et al. (1973), Di Giacomo (1974), Gionfriddo et al. (1996), Calvarano et al. (1996), and Postorino et al. (1998, 2001). The results obtained by all these authors are summarized in Table 14.4. Di Giacomo (1974) observed lower values in the formol index in the late season juices; this phenomenon was not confirmed in successive studies. As can be seen in Table 14.4, Calvarano et al. (1996) and Postorino et al. (1998, 2001) found formol index values higher than what was previously determined.

The first studies on the composition of amino acids in bergamot juice were performed by Calvarano (1958) by paper chromatography after sample purification on ionic exchange resins. The analyses highlighted the presence of five amino acids (aspartic acid, glutamic acid, alanine, serine, and valine), among which the most abundant were glutamic acid and valine. Studies on the quantitative composition of amino acids in bergamot juice were performed by Di Giacomo and Leuzzi (1973), Di Giacomo et al. (1973), Leuzzi et al. (1991), Postorino et al. (2001), and Cautela et al. (2008). Their results are reported in Table 14.5.

The amounts of total amino acids and of the single components reported in Table 14.5 vary widely and are probably affected by the different analytical approaches. The total amounts of amino acids seem to be affected by the strength of the extraction procedure; lower amounts are detected in the juices extracted in laboratory while higher amounts are found in those industrially processed. The industrially processed juices obtained by screw press, the strongest extraction procedure, have the highest amount of amino acids (Postorino et al. 2001). However, it should be highlighted that the samples extracted in laboratory by Leuzzi et al. (1991) show

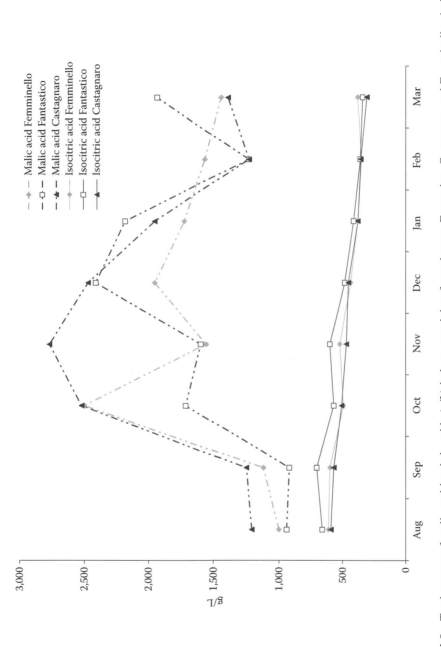

FIGURE 14.2 Total content of malic and isocitric acids (g/L) in bergamot juice, from the cvs. Fantastico, Castagnaro and Femminello, during the ripening of the fruits. (From the results of Calvarano et al., *Essenz. Deriv. Agrum.* 66, 158–186, 1991.)

TABLE 14.4
Formol Index (mL NaOH 1N/100 mL of Juice) in Bergamot Juice from Calabria

1	2a–2c	3	4a–4c	5	6	7a–7d
15.4–23.6	22.0 (a)	13.5–22.0	24.0–32.5 (a)	15.9–22.8	15–38	19.8–23.3 (a)
	15.0 (b)		28.0–37.5 (b)			17.0–27.7 (b)
	15.0 (c)		23.5–34.5 (c)			19.8–32.7 (c)
						20.0–30.0 (d)

Appendix to Table 14.4:

1. Calvarano (1961). Fifteen laboratory samples extracted from December 1960 to March 1961.
2. Di Giacomo et al. (1973). Three laboratory samples (see point 2 of appendix to Table 14.5).
3. Di Giacomo (1974). Industrial oils produced by Speciale machine RS200 during the 1973–74 productive season.
4. Calvarano et al. (1996). See point 1 of appendix to Table 14.1.
5. Gionfriddo et al. (1996). See point 2 of appendix to Table 14.3.
6. Postorino et al. (1998). See point 3 of appendix to Table 14.3.
7. Postorino et al. (2001). See point 3 of appendix to Table 14.1.

amounts of amino acids comparable to those extracted industrially and analyzed by Postorino et al. (2001). From the results published by Di Giacomo et al. (1973), it can be seen that the juice extracted in November presented higher amino acid content than those obtained in January and February. This behavior was not confirmed by Leuzzi et al. (1991) (see Figure 14.3).

As reported in Table 14.5, the most abundant amino acids in bergamot juice are aspartic acid, serine, proline, glutamic acid, alanine, asparagine, and γ-aminobutyric acid, which are always representative of 80% of the total amino acid amount. Among the values reported in Table 14.5 are the amount of γ-aminobutyric acid, usually one of the seven most abundant components of this class of compounds, reported at trace

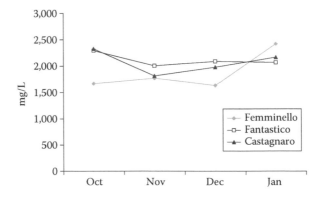

FIGURE 14.3 Total content of amino acids (mg/L) in bergamot juice during the ripening of the fruits. (From the results of Leuzzi et al., *Essenz. Deriv. Agrum.* 61, 310–322, 1991.)

TABLE 14.5

Amino Acids in Calabrian Bergamot Juice (mg/L)

	1	2a	2b	2c	3a	3b	3c	4a	4b	4c	4d	5
Arginine		28.3	20.2	46.3	28.2–74.9	22.0–89.9	28.8–141.5	22–90	72–79	81–90	25–32	21–35
Asparagine	+	186.0	51.7	96.7	111.6–184.5	132.3–209.3	84.0–163.1	132–209	139–161	254–288	79–90	77–91
Serine	+	354.1	153.6	175.2	289.2–351.1	220.6–313.7	261.2–407.1	221–314	346–377	375–433	171–191	169–194
Aspartic acid	+	416.0	162.4	232.9	359.7–522.3	312.7–412.3	385.8–486.4	322–412	355–393	260–291	207–237	201–241
Glutamic acid	+	109.6	130.0	102.5	223.5–300.4	181.9–287.9	241.1–345.8	182–288	261–292	222–247	152–176	144–182
Threonine		16.1	7.1	9.5	9.4–44.8	10.4–27.4	9.6–26.2	10–27	38–43	80–91	24–33	21–36
Glycine		8.3	5.1	8.1	10.9–47.4	18.6–40.1	17.7–30.7	19–40	10–15	12–16	6–8	7–8
γ-aminobutyric acid		80.9	99.1	tr	47.6–355.0	57.1–149.2	69.8–301.8	57–149	62–76	149–167	64–71	61–70
Alanine	+	134.2	70.4	72.1	111.0–147.9	85.8–155.6	101.2–143.1	86–156	174–202	172–201	84–93	74–97
Proline		211.5	145.0	163.4	150.0–424.1	122.2–502.9	182.3–441.2	122–503	260–450	374–416	217–238	219–243
Valine	+	11.6	12.9	12.4	19.9–29.5	17.0–27.4	19.2–32.9	17–27	14–17	14–17	11–14	9–12
Phenylalanine		11.8	13.3	34.9	17.9–23.3	19.8–23.2	18.1–37.9	20–23	15–21	12–19	16–21	13–29
Isoleucine		4.8	5.5	7.2	10.0–24.8	10.1–49.3	8.8–21.7	10–49	4–7	4–7	4–8	4–8
Leucine		6.7	5.8	7.3	5.5–14.0	11.2–48.0	10.1–28.6	11–48	7–13	7–12	6–10	5–12
Histidine		tr	tr	7.6	13.1–28.5	18.8–33.7	7.6–46.4	–	–	–	–	–
Lysine		9.4	8.2	11.1	97.7–133.8	85.7–239.6	85.5–126.1	4–17	7–12	11–16	7–12	7–12
Tyrosine		tr	5.7	6.5	12.6–31.5	5.3–32.4	5.1–21.4	5–32	8–11	8–11	5–7	5–7
Glutamine		–	–	–	–	–	–	–	23–32	51–63	21–27	22–25
Cysteine		tr	tr	tr	–	–	–	–	–	–	–	–
Ornithine		10.0	4.4	17.7	–	–	–	–	–	–	–	–
Total		1599.3	900.4	1011.4	2008.2–2293.7	1635.6–2422.6	1816.6–2328.9	1240–2384	1795–2201	2086–2385	1099–1268	

continued

TABLE 14.5 (CONTINUED)
Amino Acids in Calabrian Bergamot Juice (mg/L)

Appendix to Table 14.5:

1. Calvarano (1958). Five samples extracted by Birillatrice in February 1957, 4 samples extracted by Torchi in February and March 1957, 5 laboratory samples extracted from November 1957 to March 1958; paper chromatography. The same results are reported by Di Giacomo and Calvarano (1972).

2. Di Giacomo and Leuzzi (1973); Di Giacomo et al. (1973). Three laboratory samples extracted from (a) fruits picked in Ligonè in November 1972; (b) fruits picked in Arangea in January 1973; (c) fruits picked in Mortara in February 1973; isolation of amino acids from clarified juice by open column chromatography using a Dowex resin (HT form); analyses of amino acids by a Technicon Aminoacids Analyzer on a column (140 × 0.6 mm) packed with chromo-beads type A resin. The same results were reported by Di Giacomo and Leuzzi (1973), Di Giacomo (1989), Calvarano et al. (1995), and Postorino et al. (1998).

3. Leuzzi et al. (1991). Laboratory-extracted samples from November 1990 to January 1991 from fruits of the cultivars (a) Fantastico, (b) Femminello, (c) Castagnaro; isolation of amino acids from clarified juice by open column chromatography using a Dowex resin (HT form); derivatization of amino acids by 9-fluorene methyl orthochloroformate (FMOC reagent); analyses of amino acids by Varian Aminotag fitted with a Micro-Pak ODS column (150 × 4.6 mm) and a spectrofluorometric detector.

4. Postorino et al. (2001). From the 1998–99 productive season; (a) 6 samples extracted by Citrostar machine MU 2 (Bertuzzi); (b) 7 samples extracted from the deoleated fruits by Speciale RS machine; (c) 5 samples extracted from deoleated fruits by Torchi; (d) 6 samples laboratory extracted; (e) isolation of amino acids from clarified juice by open column chromatography using B10-RAD AC 50 W8-M resin; derivatization of amino acids to phenylthiocarbamyl derivatives; HPLC analyses of amino acids using the method described by Cohen et al. (1988).

5. Cautela et al. (2008). Laboratory samples from fruits picked in January 2003, 2004, and 2005; HPLC analyses of amino acids using the method described by Cohen et al. (1988).

TABLE 14.6
Distribution of Proline Derivatives in the Different Parts of the Bergamot Fruit (mg/Kg)

	Peel	Edible Part	Seed	Juice
Proline	185–798	155–479	1200–3616	200–440
4-Hydroxyproline	1.2–4.0	1.0–4.2	100–330	1.0–4.0
N-Methylproline	22.7–44.9	20.2–35.7	120–400	20.1–33.6
N,N-Dimethylproline	189–844	336–482	47–71	336–507
4-Hydroxyprolinebetaine	94.4–341.4	139–281	17–23	139–281

Source: Servillo, L., Giovane, A., Balestrieri, M. L., Cautela, D., and Castaldo, D., *J. Agric. Food Chem.* 59, 274–281, 2011.

levels by Di Giacomo et al. (1973) in a sample extracted in laboratory; the amount of lysine found by Leuzzi et al. (1991), in samples extracted in laboratory during the 1990–91 season, was about 10 times higher than what was previously determined, probably a mistake due to calculation or a misprint. The presence of cysteic acid and ornithine indicated by Di Giacomo et al. (1973), probably due to a wrong identification, was never reported in later studies.

In addition to the results reported in Table 14.5, the amino acid fraction of bergamot juice was the object of an investigation by Servillo et al. (2011) on the presence of proline and derivatives, probably linked to the plant stress and consequential adaptation periods, in different parts of the plant of bergamot. These results are reported in Table 14.6.

14.5 VITAMINS

14.5.1 VITAMIN C

In accordance with Gaudiano and Pruner (1950), vitamin C is present in the juice of bergamot in amounts of 0.35–0.50 g/L; in the epicarp there is from 1.0 to 1.5 g/L of ascorbic acid and an equal amount of dehydroascorbic acid, while in the mesocarp are present 0.50–0.60 and 0.30– 0.40 g/L of the two compounds, respectively.

The amount of vitamin C in bergamot juice was studied by Calvarano (1958, 1961), Calvarano et al. (1996), Gionfriddo et al. (1996), Postorino et al. (2001), and Cautela et al. (2008). Their results are reported in Table 14.7.

The values given in Table 14.7 for ripe fruits harvested from November to March are in good agreement. They indicate that the amount of vitamin C in bergamot juice extracted in the period when the fruits are normally sent to industrial processing is never lower than 300 mg/L. The only exceptions are the results presented by Gionfriddo et al. (1996) who found lower values in fruits harvested in January and February of the 1994–95 season. From the results obtained by Calvarano et al. (1996), the amount of vitamin C is higher in the juice of fruits harvested from August to October (unripe) and the amount determined from November to March is always

TABLE 14.7
Content (mg/L) of C_1 Vitamin in Calabrian Bergamot Juices

1a	1b	2	3a	3b	3c	4	5a	5b	5c	5d	6
416.9–524.8	482.0–556.9	380–500	310–364* 528–721**	337–431* 642–734**	312–396* 554–726**	220–389	385–570	325–433	360–475	380–440	378–465

* November–March; ** August–October.

Appendix to Table 14.7:

1. Calvarano (1958). (a) Nine industrial samples extracted by Brillatrice machine in February and March 1957; (b) seven laboratory samples extracted from October 1956 to February 1957.
2. Calvarano (1961). Nineteen laboratory samples extracted from December 1960 to March 1961.
3. Calvarano et al. (1996). Laboratory samples extracted from August to March of the following year during two productive seasons (1991–92 and 1992–93); (a) 12 samples from the cultivar Femminello; (b) 12 samples from the cultivar Fantastico; (c) 12 samples from the cultivar Castagnaro.
4. Gionfriddo et al. (1996). Thirteen industrial samples extracted from December 1994 to April 1995.
5. Postorino et al. (2001). Samples extracted during the productive season 1998/1999; (a) 6 industrial samples extracted by Citrostar machine; (b) 7 industrial samples extracted from the deoleated fruits by Speciale RS machine; (c) 5 industrial samples extracted by Citrostar machine; (b) 7 industrial samples extracted from the deoleated fruits by Torchi; (d) 6 laboratory samples.
6. Cautela et al. (2008). Thirty laboratory samples extracted in January 2003, 2004, and 2005.

around 50% of the initial content. Figure 14.4 shows the vitamin C variation in the three cvs. studied by Calvarano et al. (1996).

14.5.2 OTHER VITAMINS

The number of studies available in the literature on other vitamins determined in bergamot juice is low. According to Gaudiano and Pruner (1950) the juice of bergamot contains 2500, 400, and 0.1 μg/L of vitamin P, vitamin B_1, and vitamin B_2, respectively. Vitamin PP is present only at trace levels. Calvarano et al. (1995) and Di Giacomo (1989) reported for vitamin B_1 amounts ranging from 55 to 100 μg/L; Postorino et al. (1998) reported the following amounts of vitamins A and B_1:

Vitamin A (as β-carotene)	11–50 μg/L
Vitamin A (as retinol equivalent)	2–25 μg/L
Vitamin B_1	55–100 μg/L

14.6 PECTIC SUBSTANCES

Pectic substances can be chemically defined as polygalacturonides characterized by the nonuronide carbohydrate bound covalently to a linear chain of α-(1,4)-galacturonic acid units. The carboxyl groups of polygalacturonic acids may be partly esterified by methyl group and partly or completely neutralized by one or more bases. Important pectic substances include protopectin, pectinic acid, pectin, and pectic acid. These compounds have varying degrees of methyl ester content and neutralization.

In general, the amount of pectines is higher in the peel and in the membranes; the pectines present in the juice, although there in small amounts, play two important roles: they contribute to the formation of the "cloud," maintaining in suspension the fine fragments of the pulp, and contribute to the character that provides the "body" of the juice.

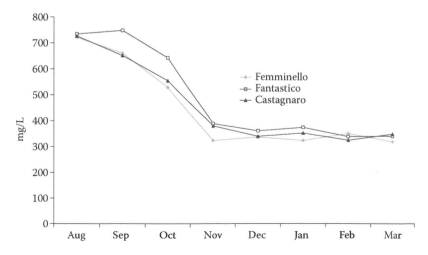

FIGURE 14.4 Vitamin C content (mg/L) in bergamot juice during a productive season. (From the results of Calvarano et al., *Essenz. Deriv. Agrum.* 66, 158–186, 1996.)

Pectic substances in bergamot were determined by Postorino et al. (2001) to evaluate how extraction technology affected the composition of the juice. In particular,
the total amount of pectines and the different fractions (water soluble, soluble in
hexamethaphosphate, soluble in alkaline solution) were found for samples extracted
in laboratory with a normal electric citrus juicer and compared with those extracted
industrially with different machines: Citrostar model MO 2, Speciale RS, and a continuous screw press. The results are reported in Table 14.8. The amount of total pectines and water-soluble pectines appears to be affected by the extraction technology,
with the highest values found in the juice extracted by the Speciale RS and the screw
press. The amounts of the other two classes of pectines did not appear to be affected.

Cautela et al. (2008) determined, in juice obtained in laboratory, an amount of
water-soluble pectines ranging between 247 and 261 mg/L.

14.7   INORGANIC COMPOUNDS

Information on the inorganic components of the juice of bergamot can be found in the
papers by Calvarano (1958, 1961), Di Giacomo and Calvarano (1972), Di Giacomo
(1974, 1989), Postorino et al. (1998, 2001), Leuzzi and Cimino (1991), Calvarano
et al. (1995, 1996), Gionfriddo et al. (1996), Montesano et al. (2003), and Cautela
et al. (2008). Calvarano (1958) and Calvarano et al. (1996) reported only the total
ashes. Calvarano (1958) found in juice obtained by screw press amounts ranging
from 2.9 to 3.3 g/L, in those obtained by the "birillatrice" values between 2.6 and
3.4 g/L, and in juice extracted in laboratory ashes amounts between 2.0 and 4.0 g/L.
Calvarano et al. (1996) found that in juices of the cvs. Femminello, Fantastico
and Castagnaro, the total ashes were within the ranges 2.79–4.80, 2.60–4.91, and
3.17–4.77 g/L, respectively.

Di Giacomo and Calvarano (1972), in a review on the components found in citrus,
reported for the total ashes the range 2.5–5.0 g/L. In the same article was reported the
presence in bergamot juice of sodium, potassium, and phosphates. Postorino et al.
(1998) reported in industrial bergamot juices a total ash amount ranging between 3.6
and 5.3 g/L, and for those prepared in laboratory values between 3.0 and 6.3 g/L.

TABLE 14.8
Pectic Substances (mg/L) in Juice of Bergamot Extracted by
Different Technologies

	Laboratory	Citrostar Mod. MU2	Extractor "Speciale RS"	Continuous Screw Press
Water soluble pectines	250–350	150–335	300–650	380–500
Hexamethaphosphate soluble pectines	183–234	130–220	180–200	110–150
Pectines soluble in NaOH	118–152	120–250	120–150	50–80
Total pectines	560–740	500–700	600–1000	580–830

Source: Postorino, E., Poiana, M., Pirrello, A., and Castaldo, D., *Essenz. Deriv. Agrum.* 71, 57–66, 2001.

TABLE 14.9
Inorganic Components of Calabrian Bergamot Juice

	1	2	3a	3b	3c	4	5	6a	6b	6c	6d	7
Total ash (g/L)	2.4–4.0	2.5–5.0	3.03–3.15	2.95–3.21	2.91–3.18	3.50–4.30	2.5–5.0	2.79–3.75	3.10–3.82	3.23–4.67	2.92–3.60	2.81–3.61
Sodium (mg/L)	7.8–17.0	7.8–21.0	15.9–23.8	33.6–41.3	26.9–35.8	17–33	10–30	10–35	15–30	10–20	11–25	9–26
Potassium (mg/L)	820–1300	820–3900	918–1233	898–997	700–809	1330–1500	1100–1500	1000–1660	800–1650	1150–1580	1150–1500	1169–1567
Calcium (mg/L)			60.5–64.9	76.5–80.6	79.5–85.4	140–330	140–330	65–95	95–130	175–380	65–85	57–98
Magnesium (mg/L)			83.5–105.0	59.8–82.8	41.6–82.5	92–122	90–125	67–113	80–120	67–140	67–116	67–123
Chromium (mg/L)			47.7–55.1	3.49–6.49	16.3–21.1							
Copper (mg/L)			0.89–1.04	0.59–0.85	0.52–0.74							
Iron (mg/L)			0.26–0.33	0.24–0.27	0.18–0.22	2.1–3.3	0.8–2.9					
Manganese (mg/L)			0.27–0.32	0.18–0.21	0.19–0.21							
Zinc (mg/L)			0.76–1.31	0.45–1.61	0.85–1.89							
Chloride as Cl (mg/L)			9.42–16.5	11.5–25.6	6.25–16.1							
Nitrate as N (mg/L)			0.39–0.64	0.38–1.81	0.37–0.88							
Phosphate as P (mg/L)	40.5–91.5	41.5–91.5	11.7–15.1	16.5–21.4	14.2–18.7							
Sulfate as S (mg/L)			3.67–6.62	2.77–5.82	2.12–4.55							

Appendix to Table 14.9:

1. Calvarano (1961). Nineteen laboratory samples extracted from December 1960 to March 1961. The results of calcium (1961) are also reported in a review by Galoppini et al. (1974).
2. Di Giacomo (1974, 1989), Calvarano (1995).
3. Leuzzi and Cimino (1991). Laboratory samples extracted from October 1990 to January 1991 from the cultivars (a) Castagnaro; (b) Fantastico; (c) Femminello.
4. Gionfriddo et al. (1996). Calabria, Italy. Thirteen industrial samples produced from December 1994 to April 1995.
5. Postorino et al. (1998). Industrial oils.
6. Postorino et al. (2001). Productive season 1998–99; (a) 6 industrial samples extracted by Citrostar machine (b) 7 industrial samples extracted from the deoleated fruits by Special RS machine; (c) 5 industrial samples extracted by Torchi; (d) 6 laboratory samples.
7. Cautela et al. (2008). Thirty laboratory samples extracted in January 2003, 2004, and 2005.

In this article were also determined the amount of chlorides expressed as chlorine with ranges, in the two types of juice, of 23.1–56.1 mg/L for the industrial juice and 6.6–58.0 mg/L for the laboratory juice. Table 14.9 shows the values of all the inorganic compounds determined in other articles.

Di Giacomo (1974) found in bergamot juice produced industrially values of total ashes and of sodium and potassium higher than those reported several years before by Calvarano (1961). The same ranges of variability proposed by Di Giacomo were confirmed successively in later studies by the same author (Di Giacomo 1989) and by Calvarano et al. (1995).

The results by Leuzzi and Cimino (1991) (Table 14.9), which represent the widest investigation on these components in bergamot juice, show that the cv. Castagnaro is characterized by the highest amount of potassium, magnesium, chromium, and manganese, while the cv. Fantastico showed the highest amount of chlorides, phosphates, and nitrates. The juice of the cv. Femminello is the one with the lowest content of inorganic compounds. The ratio between potassium and sodium is between 25 and 50, and the potassium amount in one liter of juice can provide 50% of the recommended daily amount. Leuzzi and Cimino (1991) asserted that the amount with inorganic cations associated with the inorganic anions such as nitrates, phosphates, sulphates, and chlorides represent approximately 16% of the total inorganics of the juice of Fantastico bergamot and 6% in that of Castagnaro and Femminello; the remaining part of the cations is combined with organic acids.

From a comparison of the results published by Gionfriddo et al. (1996), Postorino et al. (2001), and Cautela et al. (2008), it can be seen that the juice extracted industrially, mainly those obtained by the rolling extractor (Speciale machine) and by the continuous screw press, have the highest amount of calcium, which evidently originates from the albedo of the fruits. This phenomenon is not determined in the juices obtained by the Citrostar machine, which functions on a principle inspired by the in-line FMC extractors mechanism. In the juice obtained in laboratory by Montesano et al. (2003) with the Naviglio extractor (Naviglio 2003), the value of calcium is similar to that obtained by the screw press and Speciale machines; in the extractor developed by Naviglio (2003), the extraction is performed by pressing the whole fruits which are immersed in a solvent, leading to the presence of the components of the albedo in the juice.

REFERENCES

Calvarano, M. 1958. Ricerche sul succo di bergamotto (*Citrus bergamia* Risso). *Essenz. Deriv. Agrum.* 28:59–72.

Calvarano, M. 1961. Ricerche sul succo di bergamotto. II Nota. Flavonoidi, zuccheri e costituenti minerali. *Essenz. Deriv. Agrum.* 31:61–76.

Calvarano, M., Postorino, E., Gionfriddo, F., and Calavarano, I. 1995. Sulla deamarizzazione del succo di bergamotto. *Essenz. Deriv. Agrum.* 65:384–398.

Calvarano, M., Gioffrè, D., Calavarano, I., Lacaria, M., and Cannavò, S. 1996. Variazioni delle caratteristiche morfologiche e di composizione dei frutti di bergamotto nel corso della maturazione. *Essenz. Deriv. Agrum.* 66:158–186.

Cautela, D., Laratta, B., Santelli, F., et al. 2008. Estimating bergamot juice adulteration of lemon juice by high-performance liquid chromatography (HPLC) analysis of flavanone glycosides. *J. Agric. Food Chem.* 56:5407–5414.

Cohen, S. A., Meys, M., and Tarvin, T. L. 1988. *The PicoTag Method: A Manual of Advanced Techniques for Amino Acid Analysis.* Milford, MA: Waters Chromatography Division, Millipore Corp.

Di Giacomo, A. 1974. Le jus de la bergamotte. Atti del 12th Congresso Internazionale C.I.I.A.A. Atene, 687–694.

Di Giacomo, A. 1989. *Citrus bergamia* Risso. In *Il bergamotto di Reggio Calabria*, eds. A. Di Giacomo and C. Mangiola. Reggio Calabria: Laruffa.

Di Giacomo, A., and Calvarano, M. 1972. I componenti degli agrumi. Parte II—Succhi. *Essenz. Deriv. Agrum.* 42:353–363.

Di Giacomo, A., and Leuzzi, U. 1973. Gli aminoacidi dei succhi di agrumi con particolare riferimento al bergamotto. Simposio Internazionale Technicon "Automazione, conferme e progressi" Milano.

Di Giacomo, A., Leuzzi, U., and Micali, G. 1973. Sugli aminoacidi del succo di bergamotto. *Essenz. Deriv. Agrum.* 43:11–18.

Di Giacomo, A., Mammì De Leo M., and La Bruna, C. 1974. I carotenoidi del succo di bergamotto. *Essenz. Deriv. Agrum.* 44:254–258.

Di Giacomo, A., Mammì De Leo, L., and Siclari P. 1975. L'acido L-malico nel succo di bergamotto. *Essenz. Deriv. Agrum.* 45:144–148.

Di Giacomo, A., and Calvarano, M. 1993. I succhi di agrumi: componenti e proprietà alimentari. *Essenz. Deriv. Agrum.* 63:66–96.

Galoppini, C., Russo, C., and Pennisi, L. 1974. Sui costituenti minerali degli agrumi italiani. *Essenz. Deriv. Agrum.* 44:143–451.

Gaudiano, A., and Pruner, G. 1950. Sul contenuto vitaminico del bergamotto. *Rendiconti Istituto Superiore della Sanità* 13:142.

Gionfriddo, F., Postorino, E., and Bovalo, F. 1996. I flavoni glucosidici del succo di bergamotto. *Essenz. Deriv. Agrum.* 66:404–413.

Leuzzi, U., and Cimino, G. 1991. Mineral composition of juices derived from various bergamot cultivars. *Essenz. Deriv. Agrum.* 61:305–309.

Leuzzi, U., Lo Curto, R., and Di Giacomo, A. 1991. Amino acid composition of bergamot (*Citrus bergamia* Risso). Studies of the juices from three cultivars collected during ripening. *Essenz. Deriv. Agrum.* 61:310–322.

Montesano, D., Sgueglia, M. V., Naviglio, D., and Borbone, N. 2003. Valorizzazione del frutto di bergamotto (*Citrus bergamia*). *Ingredienti Alimentari* 2 (December):17–23.

Moufida, S., and Marzouk, B. 2003. Biochemical characterization of blood orange, sweet orange, lemon, bergamot and bitter orange. *Phytochemistry* 62:1283–1289.

Naviglio, D. 2003. Naviglio's principle and presentation of an innovative solid-liquid extraction technology: Extractor Naviglio. *Analytical Letters* 36(8):1647–1659.

Postorino, E., Gionfriddo, F., and Currò, P. 1998. Il succo di bergamotto. Atti XVIII Congresso Nazionale di Merceologia 1:533–542.

Postorino, E., Poiana, M., Pirrello, A., and Castaldo, D. 2001. Studio dell'influenza della tecnologia di estrazione sulla composizione del succo di bergamotto. *Essenz. Deriv. Agrum.* 71:57–66.

Servillo, L., Giovane, A., Balestrieri, M. L., Cautela, D., and Castaldo, D. 2011. Proline derivatives in fruits of bergamot (*Citrus bergamia* Risso et Poit.): presence of N-methyl-L-proline and 4-hydroxy-L-prolinebetaine. *J. Agric. Food Chem.* 59:274–281.

15 HMG-Flavonoids in Bergamot Fruits

Identification and Assay of Natural Statins by Recently Developed Mass Spectrometric Methods

Naim Malaj, Giselda Gallucci, Elvira
Romano, and Giovanni Sindona

CONTENTS

15.1 FLAVONOIDS: STRUCTURE, ACTIVITY, AND AVAILABILITY

Flavonoids are secondary metabolites of plants sharing characteristic structural phenolic moieties present as glycosylated derivatives or as aglycones (Stafford 1990; Andersen and Markham 2006). The relatively large number of flavonoids reported is a result of the many different combinations that are possible between polyhydroxylated aglycones with mono- and oligosaccharides. The most common sugar moieties include glucose, rhamnose, galactose, arabinose, xylose, and rutinose, or any combination of these. When the glycosylation site is on one of the hydroxyl groups of aglycone, members are called O-glycosides. The most diffused members of this group are those with the sugar moiety bound to the aglycone hydroxyl group at C-7, or in some cases, at C-3. C-glycosides (compounds with sugar moieties bound at one of the C-atoms of the aglycon moiety) have also been reported in various plants (Caristi et al. 2006). Diglycosylated members are further distinguished between rutinosides (rhamnosyl-α-1,6 glucose linkage) and neohesperidosides (rhamnosyl-α-1,2 glucose linkage), based on the site where rhamnose is linked to glucose (see Figure 15.1). Another group of compounds classified as polymethoxy flavonoids has also been reported in different plants (Bocco et al. 1998; Leuzzi et al. 2000).

Several health benefits (Prior 1997; Yochum et al. 1999; Pietta 2000; Drewnowski and Gomez-Carneros 2000; Le Marchand et al. 2000; Burda and Oleszek 2001; Rice-Evans 2001; Gross 2004; Kris-Etherton et al. 2004; Hooper and Cassidy 2006; Nichenametla et al. 2006) have been associated with this class of secondary

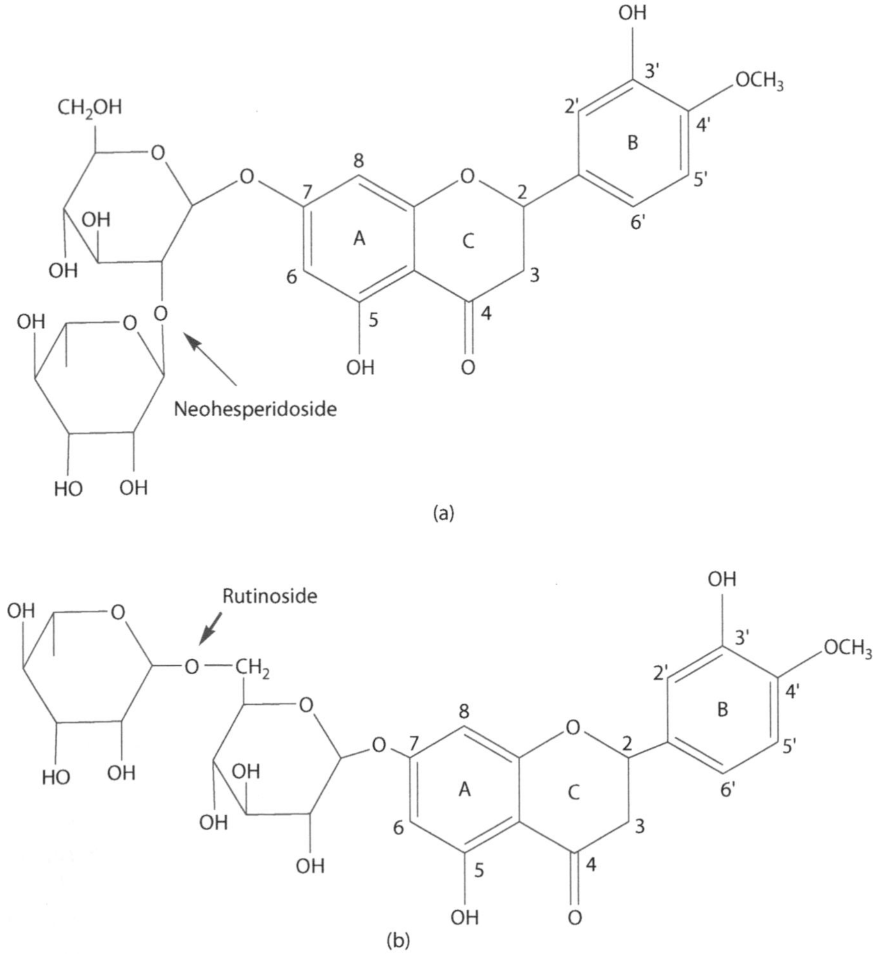

FIGURE 15.1 Neohesperidoside and rutinoside isomers of diglycosylated flavonoids. (a) Neohesperidin. (b) Hesperidin.

plant metabolites, deriving from the phenylpropanoid metabolism of the intermediates of shikimate pathway, such as antimicrobial (Cushnie and Lamb 2005) and anti-inflammatory activity (Kim et al. 2004) and beneficial effects on capillary fragility (Benavente-García et al. 1997). They may act as antiulcer remedies (Borrelli and Izzo 2000), reducing agents of prostate cancer (Maggiolini et al. 2002), and antihypertensive nutraceuticals (Chen et al. 2009). It should be mentioned, however, that most of the beneficial effects attributed to flavonoids are based on in vitro studies, and not much in vivo evidence of these positive proprieties is confirmed (Espín et al. 2007). The mechanism of action is largely unknown (Nijveldt et al. 2001), probably because most of the studies are focused on in vitro tests frequently involving much higher doses than those typical of human intake. Some epidemiological studies have, however, shown an inverse association between risk and intake level of particular flavonoids.

This class of compounds is present in fruits, vegetables, and cereals (Hollman and Arts 2000; Moufida and Marzouk 2003; Tripoli et al. 2007), and exploitation of their association to diet has rapidly increased over the past few years (Leuzzi et al. 2000; Ross and Kasum 2002; Franke et al. 2005), encouraged by media commercials about their healing effects.

Bergamot (*Citrus bergamia* Risso) is not as well known as some other members of *Citrus*. It was introduced in Calabria, a region of Italy with a peculiar pedoclimatic environment, in the middle of the eighteenth century, and its production is still mainly concentrated in that narrow area. The external part of the fruit's peel (flavedo) is yellow colored while the inner part (albedo), between flavedo and pulp, is lighter yellow to white. Bergamot juice was reported to contain high amounts of flavonoids. Table 15.1 lists all the flavonoids and related amounts described in literature. The concentration of flavonoids reported is sometimes discordant, which may be due to the different analytical techniques used for extraction and quantitation, or because of plant variety or fruit maturity.

Bergamot juice has been used in local folk medicine as a natural presidium to lower the blood cholesterol level. This effect has always been associated with the flavonoids, such as naringin, neoeriocitrin, and neohesperidin, present in the order of several hundred ppm, and to other minor species such as rhoifolin and neodiosmin present to a smaller extent. The association of the anticholesterolemic activity to the presence of classic *Citrus* flavonoids lacks scientific evidence, since it is not possible to explain why other fruits of the genus *Citrus* containing the same species do not exhibit the same healing effects.

15.2 THE HMG-FLAVONOIDS IN BERGAMOT TISSUE

The biochemical pathways controlling the multifaceted features of flavonoids are described in recently published accounts (Stafford 1990; Andersen and Markham 2006). The bergamot fruit flavedo (external part of the peel) is the only tissue of bergamot used as a source of essential oils in the cosmetic (for the production of perfumes, body lotions, soaps) and pharmaceutical (for the production of solutions with antiseptic and antibacterial proprieties) industries as well as the food industry (as an aroma source for production of sweets and liquors). Bergamot juice and albedo (the white tissue between flavedo and the pulp), on the contrary, have no important

TABLE 15.1

Flavonoids and Related Amounts of Hand-Squeezed Bergamot Juice (mg/L)

Compound	Amount (mg/L)[a]
Brutieridin	300–500[b]
Chrysoeriol 7-O-neohesperidiside	42.1–55.7
Chrysoeriol 7-O-neohesperidiside-4′-glucoside	11.6–13.2
Eriocitrin	13.4–15.6
Isovitexin	4.5–5.3
Lucenin-2	6.4–7.5
Lucenin-2 4′methyl ether	37.9–49.7
Melitidin	150–300[b]
Naringin	248.1–274.6
Neodiosmin	19.0–27.1
Neoeriocitrin	257.0–295.8
Neohesperidin	206.6–235.7
Orientin 4′-methyl ether	7.6–8.8
Rhoifolin	53.2–68.1
Rhoifolin 4′-glucoside	7.3–8.9
Scoparin	7.2–7.9
Stellarin-2	5.8–7.3
Vicenin-2	58.3–66.2

[a]Reported by Gattuso et al. 2006.
[b]Reported by Di Donna et al. 2009.

industrial applications and are actually waste products. The juice, likely for its bitter retro-taste, is not marketable in the way other *Citrus* juices are, although some have taken note of its remarkable content of recoverable flavonoids (Moufida and Marzouk 2003; Gattuso et al. 2006; Gattuso et al. 2007; Trovato et al. 2010). Two new active principles (Figure 15.2) containing the important 3-hydroxy-3-methylglutaryl (HMG) moiety were, in fact, isolated and characterized by LC-MS, MS/MS, and NMR methods (Di Donna et al. 2009). A real revolution in the bergamot productive chain is represented by the discovery of the HMG-flavonoids, which may act as natural statins.

Two new flavonoids, *brutieridin* (hesperetin 7-(2″-α-rhamnosyl-6″-(3⁗-hydroxy-3⁗-methylglutaryl)-β-glucoside)) and *melitidin* (naringenin 7-(2″-α-rhamnosyl-6″-(3⁗-hydroxy-3⁗-methylglutaryl)-β-glucoside)), are present in bergamot juice in concentrations of approximately 300 to 500 ppm and 150 to 300 ppm, respectively. The chemical structure of brutieridin (1) displays the 3-hydroxy-3-methylglutaric acid moiety (HMG, highlighted in Figure 15.2), besides the typical aglycon (hesperetin) and diglycoside (neohesperidoside). The same HMG functionality also characterizes the structure of melitidin (2).

One of the key steps of cholesterol biosynthesis is represented by the formation of MVA (4, Figure 15.2) from HMG-CoA (3, Figure 15.2) catalyzed by 3-hydroxy-3-methylglutaryl-CoA reductase (HMGR) (Figure 15.3).

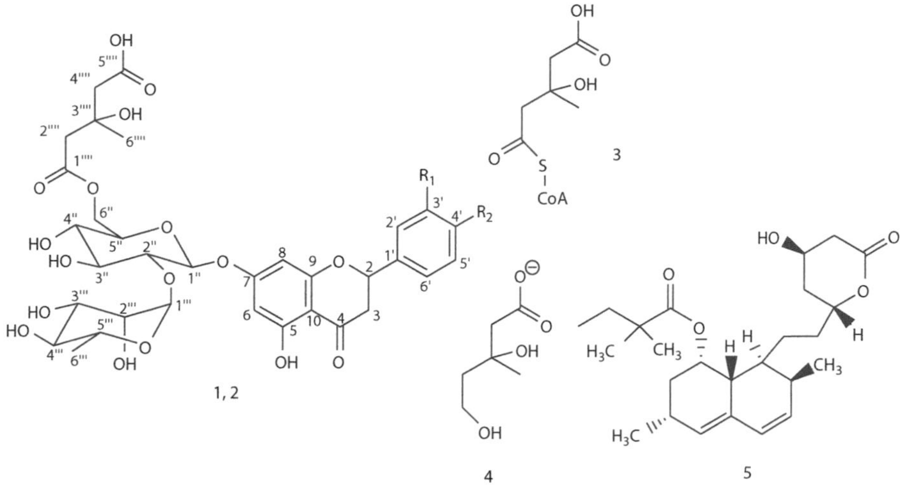

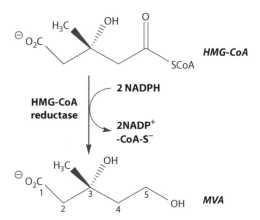

FIGURE 15.2 Brutieridin (1: R_1 = OH, R_2 = OCH$_3$), melitidin (2: R_1 = H, R_2 = OH), HMG-CoA (3), (3R)-3,5 dihydroxy-3-methylpentanoic acetate (4: MVA) and simvastatin (5).

The inhibition of the activity of this enzyme has been considered as the foremost remediation to high blood cholesterol level when it is not physiologically regulated. The commercially available statins are competitive antagonists of HMG-CoA in interaction with the active site of the HMGR enzyme (Brown and Goldstein 1986; Goldstein and Brown 1990; Farmer and Gotto 1994; Miller 1996; Endres et al. 1998; Ross et al. 1999; Istvan et al. 2000; Istvan and Deisenhofer 2001; Balbisi 2006).

The reduction of the thioester function of 3 (Figure 15.2) into the primary alcohol moiety of 4, likely through the correspondent aldehyde intermediate, is driven by the presence of a good (−)SCoA leaving group. Simvastatin (5, Figure 15.2), like other synthetic statins, likely behaves as a transition state analog, interfering with the

FIGURE 15.3 The reaction catalyzed by HMGR (the target of anticholesterolemic drugs [statins]).

achievement of the required configuration for the nucleophilic action of the NADPH reductive agent.

The structural feature of natural statins 1 and 2 found in bergamot juice is unique since the HMG moiety is directly esterified to the flavonoids and, in contrast with the natural substrate of cholesterol biosynthesis, shows a primary alcoholic moiety as an unsuitable candidate to behave as a leaving group in the reduction process. Therefore, the binding mode of the new inhibitor candidates to the active site of the HMGR enzyme was investigated. Density functional theory (DFT) was applied to verify the interactions of the two new compounds with the active site of the human HMGR enzyme and to acquire insights on their inhibitory capacity (Leopoldini et al. 2010). The model cluster used to simulate the active site of HMGR enzyme was built up starting from the 2.33 Å X-ray structure of human HMGR in complex with simvastatin inhibitor (Istvan and Deisenhofer 2001). A model was used to simulate the role of brutieridin/melitidin statin-like molecules, where a phenyl ring has replaced the flavanone moiety, and the presence of a monosaccharide only was considered, whereas simvastatin was evaluated as such.

The bonding network obtained through computations is in good agreement with that observed in the experimental crystal structure of simvastatin-HMGR complex (Istvan et al. 2000; Istvan and Deisenhofer 2001). The optimized structure of the complex between brutieridin/melitidin and the HMGR active site model cluster shows that most of the noncovalent bonds found for the binding of simvastatin to the HMGR active site are involved in the formation of the complex between brutieridin/ melitidin and the same enzyme, whereas the binding energy in the protein-like medium is slightly lower than exhibited by simvastatin. Therefore, it can be considered that the bergamot statins interact similarly with the HMGR active site as the synthetic analogs conventionally used in pharmacopoeia.

High levels of LDL cholesterol and triglycerides in serum, accompanied by a reduced HDL concentration, are associated with an elevated risk of coronary artery diseases (Miller 1996). In this context, the identification of new compounds from natural sources with anticholesterolemic proprieties but with no side effects would be of great interest in order to overcome the well-documented drawbacks associated with conventional statin therapy. Computational (Leopoldini et al. 2010) and in vitro kinetic results (Di Donna et al. 2009) have clearly indicated the capabilities of brutieridin and melitidin to act as inhibitors of HMGR, and consequently as modulators of serum cholesterol.

In vivo anticholesterolemic effects of brutieridin/melitidin-enriched bergamot extracts were compared with simvastatin in hypercholesterolemic rats (Dolce et al., unpublished results). Group of rats treated with simvastatin and with brutieridin/ melitidin-enriched bergamot extract exhibited, in serum, a significant decrease in total cholesterol compared to the untreated hypercholesterolemic group. HMGR inhibitors have also been shown to contribute to decreasing plasmatic triglyceride levels (Krause and Newton 1995; Scharnagl et al. 2001).

The ability of brutieridin/melitidin-enriched extract to decrease LDL levels in hypercholesterolemic rat serum was higher in respect to the simvastatin-treated group and, moreover, it comes together with another clinically very important effect, which is represented by a significant increase in HDL levels.

These findings confer to statin-like enriched bergamot dry extracts the peculiarity of a hypercholesterolemic controlling agent. It must be noted that the observed drop in LDL serum levels is accompanied by an increase of HDL content in the bergamot extract-treated rat group, whereas simvastatin did not show this positive effect.

15.3 RECENTLY DEVELOPED MASS SPECTROMETRIC METHODS IN ANALYSIS OF FLAVONOIDS

Liquid chromatography coupled with mass spectrometry (LC-MS) has so far been the analytic technique of choice for both qualitative and quantitative determinations of flavonoids in fruits and vegetables. Cuykens and Claeys have recently outlined the capability of this well-established methodology in the profiling of flavonoid content of natural matrices, from sampling to chromatographic separation and, finally, structure identification and assay (Cuykens and Claeys 2004). ESI, and all the traditional atmospheric pressure ionization (API) sources, such as APCI and APPI, still require extensive preparation steps before the sample can be dissolved in, or coated with, a specially selected suitable matrix that is amenable to and required by the analytical system. Typical procedures for sample preparation of vegetal matrices prior to LC-MS analysis of flavonoids include: (1) sample homogenization (chopping, milling); (2) extraction of the interested class of compounds from homogenized material by suitable solvents (methanol, ethanol, acetonitrile, acetone, chloroform, etc.) often under stirring or mild warming; (3) centrifugation/filtration; (4) removal of the nonpolar compounds from the previously filtrated fraction using petroleum ether, n-hexane, or other nonpolar solvents; (5) additional purification and preconcentration by SPE; (6) solvent removal or concentration (by rotary evaporator or nitrogen gas flow); and (7) dissolving the final dried extract.

The introduction of desorption electrospray ionization (DESI) (Takats et al. 2004; Takats et al. 2005) and direct analysis in real time (DART) (Cody et al. 2005) have allowed, for the first time, the direct analysis of samples in their native condition, bypassing most elements of the analytical system and transferring ions into the mass spectrometer without any sample-manipulation or sample-preparation steps. These ambient ionization techniques, in a single operational step, make it possible to generate and transfer ions from open ambient, where condensed phase samples are held, to the vacuum system where analysis takes place.

The development of DESI has created an awareness of the potentiality of open ambient environment analysis and sparked a new subfield in MS. The novelty of this concept is demonstrated by the introduction of 15 new methods from 2005 to 2008, summarized and described accurately by Cooks and coworkers (Venter et al. 2008).

Recently, two new ambient ionization methods, namely paper spray (PS) (Wang et al. 2010) and leaf spray (LS) (Liu et al. 2011), were introduced by the Cooks group. A brief description and their impact on direct analysis of flavonoids will be discussed in the following sections.

15.3.1 PAPER SPRAY MASS SPECTROMETRY (PS-MS)

Paper spray was developed as an ambient ionization method for direct mass spectrometric analysis of complex mixtures (Wang et al. 2010; Liu et al. 2010; Wang et al.

2011). In PS experiments the paper is cut to a sharp triangular tip, a small volume of solvent is added, and a high voltage is applied to the base of the triangle to generate charged gas phase ions for mass spectrometry. The experiments are performed in open air lab ambient simply by holding the tip of the paper in front of the mass spectrometer inlet (0.3–1 cm) and without any kind of pneumatic assistance for the transportation of analytes (Figure 15.4). Ions of a wide variety of compounds, including small organic compounds, peptides, and proteins, have already been detected operating in the appropriate conditions (Liu et al. 2010; Wang et al. 2010, 2011).

The use of paper to generate electrospray ions was reported previously by Fenn (2001), but no further investigation of its potential applications was done. The Cooks group investigated its potentialities for different analytical purposes. In fact, PS has now found applications for the analysis of therapeutic and illicit drug monitoring in whole blood and urine. It can also be used for the analysis of drugs and explosives on surfaces by wiping the surface of interest with a paper triangle, to which solvent is then added and high voltage applied (Wang et al. 2010). Sample collection by paper wiping followed by analysis using paper spray was also adapted for fast analysis of agrochemicals on fruit peel (Liu et al. 2010). Chromatography paper wetted with methanol was used to wipe a 10 cm² area on the peel of a lemon purchased from a grocery store. A triangle was cut from the center of the paper after methanol evaporation and used directly for paper spray by applying a 10 µL methanol/water solution. The spectra recorded show that the fungicide thiabendazole, originally on the lemon peel, (*m/z* 202 for protonated molecular ion and *m/z* 224 for sodium adduct ion), migrates to paper and could be identified easily by MS and confirmed using MS/MS analysis.

PS was used also for the analysis of food samples in liquid state. Liu et al. (2010), for instance, were able to detect and identify the presence of caffeine (in positive mode), and benzoate and acesulfame (in negative mode) simply by spraying the paper which was previously wetted with 10 µL of untreated cola drink.

15.3.1.1 PS-MS Determination of Flavonoids from Bergamot Juice

PS was recently applied for the detection of flavonoids from juices of different *Citrus* fruits without or with some simple and fast sample pretreatment (Malaj 2011). Bergamot juice (like all *Citrus* juices) is known to have a complex composition, including relatively high amounts of organic acids, sugars, flavonoids, amino acids, and soluble solids (Cautela et al. 2008). The high complexity of bergamot

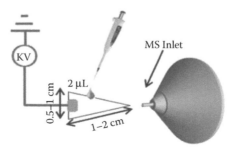

FIGURE 15.4   Schematic illustration of paper spray for MS analysis.

juice may pose problems of ion suppression even for the well-consolidated ionization techniques (such as ESI and APCI); it was, therefore, partially fractionated (by SPE) to get rid of water-soluble polar species before recovering the analytical sample. The target flavonoids of bergamot juice are listed in Table 15.2; their identification was performed by comparing the MS^n fragmentation pattern with literature data.

The role of the water-soluble removed fraction on the suppression of flavonoids' signal was found to be critical. Very low abundant peaks were, in fact, observed when bergamot juice was used without SPE treatment, and there were no signals at all for lemon, orange, and tangerine juices. The SPE-PS-MS^n spectra of the two HMG flavonoids present in juice are reported in Figure 15.5.

The PS method was also used to reveal the fraudulent addition of bergamot to lemon juice (Malaj 2011). Frequent frauds present in this product chain include false declaration of the geographic origin of the product, undeclared sugar addition, water addition in "not from concentrate" juices, and the addition of other not declared compounds (such as organic acids, coloring agents, etc.) (Cautela et al. 2008). Another common fraud is the undeclared addition of different juices from botanic-related fruits (Mouly and Gaydou 1993; Ooghe and Detavernier 1999). It has been reported that this is a very common practice with bergamot juice, considered a side-product of essential oil industries, which has had no remunerative market up to now. Thus, fraudulent operators use bergamot juice, with or without any treatment, to adulterate other *Citrus* juices that have a wider market; in particular lemon juice, which has organoleptic proprieties very similar to bergamot juice. The use of minor components of juices as markers, in particular, flavonoids, has been shown as a wise approach to assess the authenticity of fruit juices (Grandi et al. 1994; Mouly et al. 1994; Mouly et al. 1997). Cautela and collaborators (Cautela et al. 2008) developed an HPLC method based on the abundance of neohesperidin, naringin, and neoeriocitrin to detect and quantify the fraudulent addition of bergamot to lemon juice. That approach may overestimate or give false positives because the compounds used as

TABLE 15.2
Flavonoids and HMG-Flavonoids Identified by PS-MS in Bergamot Juice (and in Albedo by LS-MS)

Compound	[M-H]⁻
Brutieridin	753
Melitidin	723
Naringin	579
Neodiosmin	607
Neoeriocitrin	595
Neohesperidin	609
Rhoifolin	577
Stellarin-2/lucenin-2 4′-methyl ether	623
Vicenin-2	593

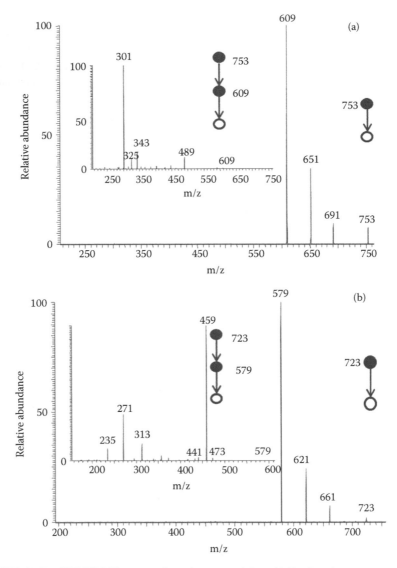

FIGURE 15.5 SPE-PS-MS spectra from bergamot juice. (a) Product ion mass spectrum (MS/MS) and MS³ (inset) of brutieridin. (b) MS/MS and MS³ (inset) of melitidin. (From Malaj, N., Modern mass spectrometric applications in the structure and function evaluation of active principles. Ph.D. diss. Università della Calabria, Bernardino Telesio School of Science and Technique, 2011.)

markers (naringin, neohesperidin, and neoeriocitrin) are present in small amounts in lemon juice as well. Brutieridin and melitidin, on the contrary, can be used as specific markers to detect and evaluate the fraudulent addition of bergamot to lemon juice by applying the fast PS-MS approach. The flavonoid fractions of different mixtures of lemon and bergamot juice were, therefore, checked by PS-MS to evaluate the presence of brutieridin and melitidin. Both markers were detected in product ion scan mode even at the low mixing level of 1% of bergamot added to lemon juice (Figure 15.6), but less sensitivity was observed in full-scan mode. Brutieridin and melitidin are, therefore, useful markers to assess fraudulent addition of bergamot to commercially available *Citrus* juices.

15.3.2 LEAF SPRAY MASS SPECTROMETRY (LS-MS)

Leaf spray can be considered an implementation of paper spray. The method, introduced by the same group that developed PS, considers the direct use of plant tissues to generate gas phase ions for mass spectrometric analysis (Liu et al. 2011). No other ionization device or support is needed since the vegetal tissue acts simultaneously as the ionization source and sample. Plant tissues can be used in leaf spray experiments just by cutting a sharp triangular tip, if needed, and spraying directly. The

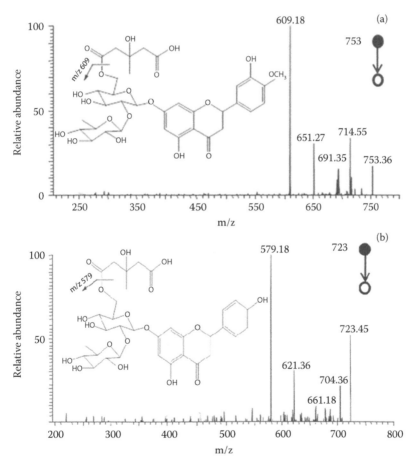

FIGURE 15.6 (a) Brutieridin and (b) melitidin detected in lemon adulterated with 1% bergamot juice. (From Malaj, N., Modern mass spectrometric applications in the structure and function evaluation of active principles. Ph.D. diss. Università della Calabria, Bernardino Telesio School of Science and Technique, 2011.)

mass spectrum is obtained by applying a high voltage to the base of the leaf after the addition of a small volume of solvent (Figure 15.7). Tissues which are naturally sufficiently juicy (high water content) do not require solvent addition, and the spray length is sufficient to perform full scan, tandem (MS/MS), and exact mass measurements.

The leaf spray approach is not limited to leaf material only, but any other parts of plants including the root, stem, flower, fruit, and seeds can be examined as well. Liu and colleagues (Liu et al. 2011) have, in fact, detected several compounds (amino acids, alkaloids, flavonols, carbohydrates, organic acids, fatty acids, and phospholipids) directly from different parts of potato, onion, cabbage, ginger, cranberry, gingko, and arabidopsis thaliana without any sample preparation but just adapting, when needed, the tip shape of the natural matrix.

LS mass spectrometric determinations of agrochemicals in fruit and vegetable tissue have been reported (Malaj 2011; Malaj et al. 2011, 2012), and this rapid approach has allowed the identification, and at least semi-quantitative determination, of a number of typical pesticides (such as diphenylamine, imazalil, thiabendazole, linuron, and acetamiprid) from a number of different fruit and vegetable samples (including apple, pear, orange, lemon, potato, carrot, cucumber, and eggplant). LS-MS showed excellent performance in rapid pesticide screening and in the discrimination between "organic" and "non-organic" products directly from fruit and vegetable tissue. The whole experimental procedure takes about 100 seconds to perform and gives limits of detection and quantitation for each pesticide well below the minimum residue levels imposed by the EU legislation, while displaying linear dynamic ranges of three orders of magnitude and precision better than 15%.

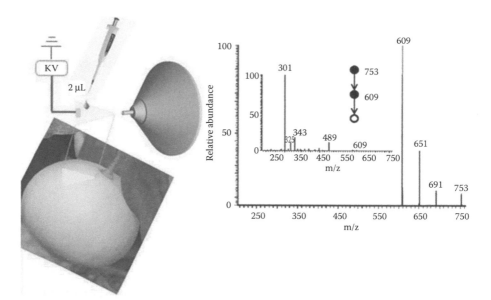

FIGURE 15.7 Leaf spray operation mode. Tissues that do not have a natural sharp tip are cut to a triangle and held by a high voltage clip in front of the inlet of a mass spectrometer. MS^n spectra of brutieridin directly from fruit flavedo are shown.

15.3.2.1 Analysis of Flavonoids from Bergamot Albedo by Leaf Spray Mass Spectrometry

The established LC-MS techniques available at any laboratory have the drawbacks of any analytical method that relies on sample preparation procedures. The LS approach was therefore checked as an alternative ambient ionization method in the analysis of flavonoids contained in plant tissues (Malaj 2011).

The albedo tissue of bergamot fruit was cut to a sharp triangular tip and sprayed without any further sample treatment. Flavonoids that were detected in bergamot juice by PS-MS (Table 15.2) were also the target of this LS approach, operating with the same instrument and keeping the same instrumental conditions as for PS-MS experiments.

It is well known that in *Citrus* fruits flavonoids are distributed mostly in the juice and albedo layer, and to a lesser extent in the flavedo. In fact, as the voltage is applied and the solvent is added to the albedo tissue triangle, typical MS/MS profiles of flavonoids under examination were observed. In Figure 15.8 are reported the MS^2 (and MS^3 as insets) mass spectra of (a) naringin and (b) neohesperidin, while in Table 15.2 are reported all the compounds detected and identified in bergamot albedo. On the other hand, when flavedo was used, characteristic MS/MS but very low intensity signals were observed, in agreement with previous literature reporting low flavonoid abundance in this tissue.

In conclusion, LS-MS is a fast, cheap, and very convenient approach to study flavonoids (and other secondary plant metabolites and chemicals) in fruit tissues (and in plants in general), without laborious time- and chemical-consuming sample preparations. In the future it is hoped that this new approach may be extended to a wide range of chemicals and applied in several other fields (e.g., to monitor seasonal variations of chemicals during plant/fruit development; to obtain biosynthetic information, etc.).

15.4 FUNCTIONAL FOOD FROM BERGAMOT WASTE TISSUE

The term "functional food," first used in Japan in the 1980s for food products fortified with particular constituents that possess advantageous physiological effects, is used in many different ways in different countries and by official bodies (Diplock et al. 1999; Directive 2002/46/EC 2002; Roberfroid 2002; Alzamora et al. 2005; Niva 2007; Niva and Makela 2007). The development of functional foods is currently one of the most intensive areas of food product development worldwide (Menrad 2003). Interest in functional foods has been prompted by a rapid expansion of scientific knowledge of the importance of a healthy diet as well as technical advances in the food industry (Biström and Nordström 2003). The possibility of extracting the HMG-flavonoids brutieridin and melitidin from bergamot tissue opens the door to applications of this *Citrus* genus for functional food preparations.

The albedo layer has never been investigated in detail to see whether it could provide value-added compounds, such as active ingredients. Both juice and albedo of different members of *Citrus* have very similar chemical composition, as shown by LC-UV/MS analysis (Malaj 2011). Figure 15.9 displays the chromatograms of

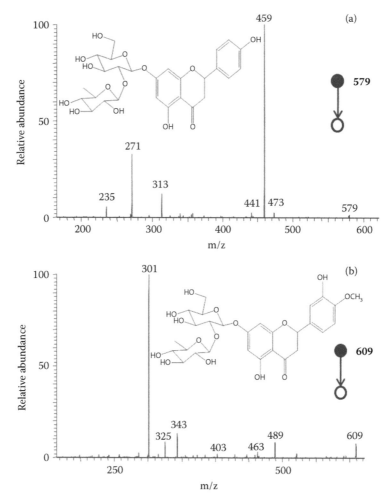

FIGURE 15.8 MS/MS of (a) naringin and (b) neohesperidin obtained from direct bergamot albedo spraying (LS-MS). (From Malaj, N., Modern mass spectrometric applications in the structure and function evaluation of active principles. Ph.D. diss. Università della Calabria, Bernardino Telesio School of Science and Technique, 2011.)

bergamot juice and albedo, showing the presence of the recently discovered HMG-flavonoids as well as other compounds.

Bergamot albedo has been examined in order to find a matrix allowing the recovery of brutieridin and melitidin without using toxic solvents (Di Donna et al. 2011), an important issue for the implementation of safety procedures matching food directives issued by international bodies (ICH 2009). The optimization of the extraction procedure was exploited using water as a solvent in different conditions, such as a microwave oven or water heated conventionally at different temperatures.

The results reported in Figure 15.10 allow us to conclude that there is no thermal instability in the conventional extraction of the two statin-like molecules with

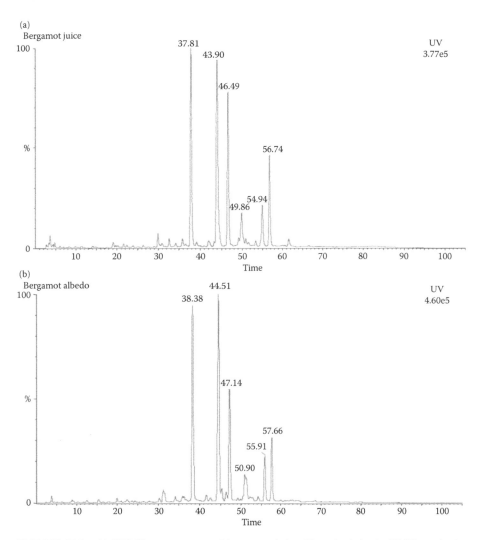

FIGURE 15.9 (a) UV Chromatogram of bergamot juice: Neoeriocitrin (rt 37.81); naringin (rt 43.90); neohesperidin (rt 46.49); neodiosmin (rt 49.86); melitidin (rt 54.94); brutieridin (rt 56.74). (b) UV chromatogram of methanol extract of bergamot albedo (peak identity as in bergamot juice).

warm to hot water. The thermal stability of brutieridin and melitidin indicates that both statin-like compounds may be easily available from whole, dried, or lyophilized albedo, or by simple extraction methods amenable to domestic cooking procedures. The possibility of preparing infusions by adding albedo to commercial tea bags and following domestic procedures for tea preparation has been examined. Infusions were analyzed by LC-MS/UV. Figure 15.11 confirms the relative amount of extracted active principles, and that the two statin-like molecules, brutieridin and melitidin, can be extracted easily by household procedures. For example, adding 0.5 g of dried albedo to a commercial tea bag (yellow label tea), which was left for four

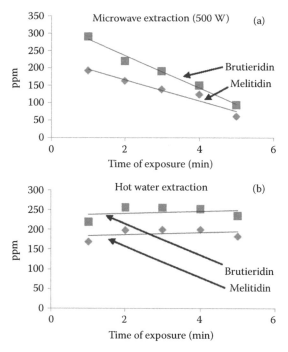

FIGURE 15.10 Brutieridin and melitidin content (ppm) after five independent experiments (1–5 min) by (a) microwave extraction; (b) warm-to-hot water extraction.

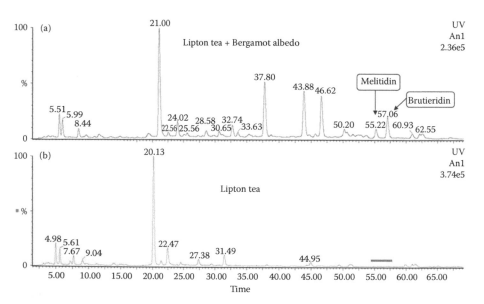

FIGURE 15.11 (a) UV profile of commercial black tea after addition of dry bergamot albedo. (b) UV profile of commercial black tea.

minutes in 200 mL of boiling water, concentrations of 10 and 7 ppm of brutieridin and melitidin, respectively, were found in so-prepared infusions (Di Donna et al. 2011). A similar procedure was considered for espresso coffee. Brutieridin and melitidin were present in extracts of commercial coffee capsules treated with lyophilized or oven-dried bergamot albedo and quantified by HPLC, thus showing the possibility of preparing anticholesterolemic coffee.

In other experiments, the use of bergamot albedo for preparation of milk and its derivatives was investigated. Seven grams of dried albedo were added to 500 mL of UHT milk, left under weak agitation for 10 minutes at 75°C, and then filtered (to remove the albedo); 0.5 mL of milk was then extracted according the method reported by Nagy et al. (2009) and analyzed by LC-MS/UV as for tea and coffee. After the confirmation of the presence of both active principles, the milk was cooled to 40°C, lactic ferments (*Lactobacillus bulgaricus* and *Streptococcus thermophilus*) were added, and then incubated for eight hours at 40°C. The yogurt obtained was analyzed every three days (for a total of three weeks) to follow the stability of brutieridin and melitidin during the fermentation process and its conservation at 4°C. The profiles shown in Figure 15.12 indicate that brutieridin and melitidin are unaffected by the fermentation process and were stable for the whole monitoring period. In fact, even on the 21st day of yogurt conservation at 4°C, the amounts of brutieridin and melitidin were invariable.

Bergamot juice is a very promising source of HMG-flavonoids, provided that simple US-FDA and EU-EMEA guidelines are followed for product manipulation. A lab scale procedure was developed whereby a fraction of whole bergamot juice enriched up to 60% in the HMG-flavonoids 1 and 2 can be easily isolated as a water solution (Figure 15.13) (Malaj 2011).

The enriched fraction of bergamot juice (Figure 15.13a) has been used to prepare the anticholesterolemic coffee where the relative amount of 1 and 2 was preserved.

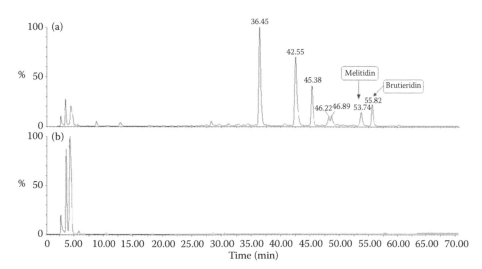

FIGURE 15.12 UV profiles obtained from LC/UV-MS of yogurt (after 21 days) from milk containing albedo (a) and yogurt made from milk without albedo addition (b).

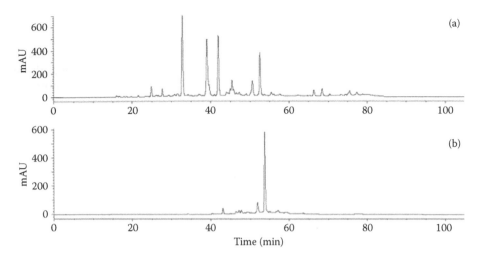

FIGURE 15.13   UV profile (HPLC) of a dry extract of bergamot juice before (a) and after (b) fractionation. The purified fraction was used to prepare the espresso coffee samples.

ACKNOWLEDGMENT

The authors are grateful to R. Graham Cooks, Purdue University, for his support on implementation of PS- and LS-MS techniques.

REFERENCES

Alzamora, S. M., Salvatori, D., Tapia, S. M., et al. 2005. Novel functional foods from vegetable matrices impregnated with biologically active compounds. *J. Food Eng.* 67:205–214.

Andersen, O. M., and Markham, K. R. 2006. *Flavonoids: Chemistry, biochemistry and applications*. Boca Raton, FL: Taylor & Francis.

Balbisi, E. A. 2006. Management of hyperlipidemia: New LDL-C targets for persons at high-risk for cardiovascular events. *Med. Sci. Monit.* 12:RA34–RA39.

Benavente-García, O., Castillo, J., Marín, F. R., Ortuño, A., and Del Río, J. A. 1997. Uses and properties of *citrus* flavonoids. *J. Agric. Food Chem.* 45:4505–4515.

Biström, M., and Nordström, K. 2003. Identification of key success factors of functional dairy foods product development. *Trends in Food Sci. Technol.* 13:372–379.

Bocco, A., Cuvelier, M., Richard, H., and Berset, C. 1998. Antioxidant activity and phenolic composition of *citrus* peel and seed extracts. *J. Agric. Food Chem.* 46:2123–2129.

Borrelli, F., and Izzo, A. A. 2000. The plant kingdom as a source of anti-ulcer remedies. *Phytother. Res.* 14:581–591.

Brown, M. S., and Goldstein, J. L. 1986. A receptor-mediated pathway for cholesterol homeostasis. *Science* 232:34–47.

Burda, S., and Oleszek, W. 2001. Antioxidant and antiradical activities of flavonoids. *J. Agric. Food Chem.* 49:2774–2779.

Caristi, C., Bellocco, E., Gargiulli, C., Toscano, G., and Leuzzi, U. 2006. Flavone-di-*C*-glycosides in *citrus* juices from southern Italy. *Food Chem.* 95:431–437.

Cautela, D., Laratta, B., Santelli, F., et al. 2008. Estimating bergamot juice adulteration of lemon juice by high-performance liquid chromatography (HPLC) analysis of flavanone glycosides. *J. Agric. Food Chem.* 56:5407–5414.

Chen, Z. Y, Peng, C., Jiao, R., et al. 2009. Anti-hypertensive nutraceuticals and functional foods. *J. Agric. Food Chem.* 57:4485–4499.

Cody, R. B., Laramée, J. A., and Durst, H. D. 2005. Versatile new ion source for the analysis of materials in open air under ambient conditions. *Anal. Chem.* 77:2297–2302.

Cushnie, T. P. T., and Lamb, A. J. 2005. Antimicrobial activity of flavonoids. *Int. J. Antimicrob. Agent.* 26:343–356.

Cuykens, F., and Claeys, M. 2004. Mass spectrometry in the structural analysis of flavonoids. *J. Mass Spectrom.* 39:1–15.

Di Donna, L., De Luca, G., Mazzotti, F., et al. 2009. Statin-like principles of bergamot fruit (*citrus bergamia*): Isolation of 3-hydroxymethylglutaryl flavonoid glycosides. *J. Nat. Prod.* 72:1352–1354.

Di Donna, L., Dolce, V., and Sindona, G. 2009. Patent No. CS2008A00019.

Di Donna, L., Gallucci, G., Malaj, N., et al. 2011. Recycling of industrial essential oil waste: Brutieridin and melitidin, two anticholesterolaemic active principles from bergamot albedo. *Food Chem.* 125:438–441.

Diplock, A. T., Aggett, P. J., Ashwell, M., et al. 1999. Scientific concepts of functional foods in Europe-consensus document. *Br. J. Nutr.* 81:S1–S27.

Directive 2002/46/EC of the European Parliament and of the Council, June 10, 2002.

Drewnowski, A., and Gomez-Carneros, C. 2000. Bitter taste, phytonutrients, and the consumer: A review. *Am. J. Clin. Nutr.* 72:1424–1435.

Endres, M., Laufs, U., Huang, Z., et al. 1998. Stroke protection by 3-hydroxy-3-methylglutaryl (HMG)-CoA reductase inhibitors mediated by endothelial nitric oxide synthase. *Proc. Natl. Acad. Sci. USA* 95:8880–8885.

Espín, J. C., García-Conesa, M. T, and Tomás-Barberán, F. A. 2007. Nutraceuticals: Facts and fiction. *Phytochem.* 68:2986–3008.

Farmer, J. A., and Gotto Jr., A. M. 1994. Antihyperlipidemic agents. Drug interactions of clinical significance. *Drug Saf.* 11:301–309.

Fenn, J. B. 2001. U.S. Patent 6,297,499.

Franke, A. A., Cooney, R. V., Henning, S. M., and Custer, L. J. 2005. Bioavailability and antioxidant effects of orange juice components in humans. *J. Agric. Food. Chem.* 53:5170–5178.

Gattuso, G., Barreca, D., Caristi, C., Gargiulli, C., and Leuzzi U. 2007. Distribution of flavonoids and furocoumarins in juices from cultivars of *Citrus bergamia* Risso. *J. Agric. Food Chem.* 55:9921–9927.

Gattuso, G., Caristi, C., Gargiulli, C., et al. 2006. Flavonoid glycosides in bergamot juice (*Citrus bergamia* Risso). *J. Agric. Food Chem.* 54:3929–3935.

Goldstein, J. L., and Brown, M. S. 1990. Regulation of the mevalonate pathway. *Nature* 343:425–430.

Grandi, R., Trifirò, A., Gherardi, S., Calza, M., and Saccani, G. 1994. Characterization of lemon juice on the basis of flavonoid content. *Fruit Process.* 11:335–359.

Gross, M. 2004. Flavonoids and cardiovascular disease. *Pharm. Biol.* 42:21–35.

Hollman, P. C. H., and Arts, I. C. W. 2000. Flavonols, flavones and flavanols-nature, occurrence and dietary burden. *J. Sci. Food Agr.* 80:1081–1093.

Hooper, L., and Cassidy, A. 2006. A review of the health care potential of bioactive compounds. *J. Sci. Food Agric.* 86:1805–1813.

ICH. 2009. Impurities: Guideline for residual solvents. ICH Topic Q3C (R4). International Conference on Harmonisation of Technical Requirements for the Registration of Pharmaceuticals for Human Use.

Istvan, E. and Deisenhofer, J. 2001. Structural mechanism for statin inhibition of HMG-CoA reductase. *Science* 292:1160–1164.

Istvan, E. S., Palnitkar, M., Buchanan, S. K., and Deisenhofer, J. 2000. Crystal structure of the catalytic portion of human HMG-CoA reductase: Insights into regulation of activity and catalysis. *The EMBO Journal* 19:819–830.

Kim, H. P., Son, K. H., Chang, H. W., and Kang S. S. 2004. Anti-inflammatory plant flavonoids and cellular action mechanisms. *J. Pharmacol. Sci.* 96:229–245.

Krause, B. R., and Newton, R. S. 1995. Lipid-lowering activity of atorvastatin and lovastatin in rodent species: Triglyceride-lowering in rats correlates with efficacy in LDL animal models. *Atherosclerosis* 117:237–244.

Kris-Etherton, P. M., Lefevre, M., Beecher, G. R., et al. 2004. Bioactive compounds in nutrition and health-research methodologies for establishing biological function: The antioxidant and anti-inflammatory effects of flavonoids on atherosclerosis. *Ann. Rev. Nutr.* 24:511–538.

Le Marchand, L., Murphy, S. P., Hankin, J. H., Wilkens, L. R., and Kolonel, L. N. 2000. Intake of flavonoids and lung cancer. *J. Natl. Cancer Inst.* 92:154–160.

Leopoldini, M., Malaj, N., Toscano, M., Sindona, G., and Russo, N. 2010. On the inhibitor effects of bergamot juice flavonoids binding to the 3-hydroxy-3-methylglutaryl-CoA reductase (HMGR) enzyme. *J. Agric. Food Chem.* 58:10768–10773.

Leuzzi, U., Caristi, C., Panzera, V., and Licandro G. 2000. Flavonoids in pigmented orange juice and second-pressure extracts. *J. Agric. Food Chem.* 48:5501–5506.

Liu, J., Wang, H., Cooks, R. G., and Ouyang, Z. 2011. Leaf spray: Direct chemical analysis of plant material and living plants by mass spectrometry. *Anal. Chem.* 83:7608–7613.

Liu, J., Wang, H., Manicke, E. N., et al. 2010. Development, characterization, and application of paper spray ionization. *Anal. Chem.* 82:2463–2471.

Maggiolini, M., Vivacqua, A., Carpino, A., et al. 2002. Mutant androgen receptor T877A mediates the proliferative but not the cytotoxic dose-dependent effects of genistein and quercetin on human LNCaP prostate cancer cells. *Mol. Pharmacol.* 62:1027–1035.

Malaj, N. 2011. Modern mass spectrometric applications in the structure and function evaluation of active principles. Ph.D. diss. Università della Calabria, Bernardino Telesio School of Science and Technique.

Malaj, N., Ouyang, Z., Sindona, G., and Cooks R. G. 2012. Analysis of pesticide residues by leaf spray mass spectrometry, submitted.

Malaj, N., Ouyang, Z., Sindona, G., and Cooks, R. G. 2011. Analysis of agrochemicals by direct vegetable tissue spray mass spectrometry. Results presented at the Turkey Run analytical chemistry conference, Marshall, Indiana.

Manicke, E. N., Yang, Q., Wang, H., Oradua, Sh., Ouyang, Z., and Cooks, R. G. 2011. Assessment of paper spray ionization for quantitation of pharmaceuticals in blood spots. *Int. J. Mass Spectrom.* 300:123–129.

Menrad, K. 2003. Market and marketing of functional food in Europe. *J. Food Eng.* 56:181–188.

Miller, J. P. 1996. Hyperlipidaemia and cardiovascular disease. *Curr. Opin. Lipidol.* 7:U18–U24.

Moufida, S., and Marzouk, B. 2003. Biochemical characterization of blood orange, sweet orange, lemon, bergamot and bitter orange. *Phytochem.* 62:1283–1289.

Mouly, P. P., and Gaydou, E. M. 1993. Column liquid chromatographic determination of flavanone glycosides in *citrus*. Application to grapefruit and sour orange juice adulterations. *J. Chromatogr. A* 634:129–134.

Mouly, P. P., Arzouyan, C. R., Gaydou, E. M., and Estienne, J. M. 1994. Differentiation of *citrus* juices by factorial discriminant analysis using liquid chromatography of flavanone glycosides. *J. Agric. Food Chem.* 42:70–79.

Mouly, P. P., Gaydou, E. M., Faure, R., and Estienne, J. M. 1997. Blood orange juice authentication using cinnamic acid derivatives. Variety differentiations associated with flavanone glycoside content. *J. Agric. Food Chem.* 45:373–377.

Nagy, K., Redeuil, K., Bertholet, R., Steiling, H., and Kussmann, M. 2009. Quantification of anthocyanins and flavonols in milk-based food products by ultra performance liquid chromatography-tandem mass spectrometry. *Anal. Chem.* 81:6347–6356.

Nichenametla, S. N., Taruscio, T. G., Barney, D. L., and Exon, J. H. 2006. A review of the effects and mechanism of polyphenolics in cancer. *Crit. Rev. Food Sci.* 46:161–183.

Nijveldt, R. J., van Nood, E., van Hoorn, D. E. C., et al. 2001. Flavonoids: A review of probable mechanisms of action and potential applications. *Am. J. Clin. Nutr.* 74:418–425.

Niva, M. 2007. All foods affect health: Understandings of functional foods and healthy eating among health-oriented Finns. *Appetite* 48:384–393.

Niva, M., and Makela, J. 2007. Finns and functional foods: Sociodemographics, health efforts, notions of technology and the acceptability of health-promoting foods. *Int. J. Consum. Stud.* 31:34–45.

Ooghe, W., and Detavernier, M. J. 1999. Flavonoids as authenticity markers of *Citrus sinensis* juice. *Fruit Process.* 8:308–313.

Pietta, P. G. 2000. Flavonoids as antioxidants. *J. Nat. Prod.* 63:1035–1042.

Prior, R. L. 1997. Antioxidant and prooxidant behavior of flavonoids: Structure activity relationships. *Free Radic. Biol. Med.* 22:749–760.

Rice-Evans, C. 2001. Flavonoid antioxidants. *Curr. Med. Chem.* 8:797–807.

Roberfroid, M. B. 2002. Global view on functional foods: European perspectives. *Br. J. Nutr.* 88:S133–S138.

Ross, J. A., and Kasum, C. M. 2002. Dietary flavonoids: Bioavailability, metabolic effects and safety. *Ann. Rev. Nutr.* 22:19–34.

Ross, S. D., Allen, I. E., Connelly, J. E., et al. 1999. Clinical outcomes in statin treatment trials: A meta-analysis. *Arch. Intern. Med.* 159:1793–1802.

Scharnagl, H., Schinker, R., Gierens, H., et al. 2001. Effect of atorvastatin, simvastatin, and lovastatin on the metabolism of cholesterol and triacylglycerides in HepG2 cells. *Biochem. Pharmacol.* 62:1545–1555.

Stafford, H. A. 1990. *Flavonoid metabolism.* Boca Raton, FL: CRC Press.

Takats, Z., Wiseman, J. M., and Cooks, R. G. 2005. Ambient MS by DESI: Instrumentation, mechanisms and applications in forensics, chemistry, and biology. *J. Mass Spectrom.* 40:1261–1275.

Takats, Z., Wiseman, J., Gologan, B., and Cooks, R. G. 2004. Mass spectrometry sampling under ambient conditions with desorption electrospray ionization. *Science* 306:471–473.

Tripoli, E., La Guardia, M., Giammanco, S., Di Majo, D., and Giammanco, M. 2007. *Citrus* flavonoids: Molecular structure, biological activity and nutritional properties: A review. *Food* Chem. 104:466–479.

Trovato, A., Taviano, M. F., Pergolizzi, S., et al. 2010. *Citrus bergamia* Risso & Poiteau juice protects against renal injury of diet-induced hypercholesterolemia in rats. *Phytother. Res.* 24:514–519.

Venter, A., Nefliu, M., and Cooks, R. G. 2008. Ambient desorption ionization mass spectrometry. *Anal. Chem.* 27:284–290.

Wang, H., Liu, J., Cooks, R. G., and Ouyang, Z. 2010. Paper spray for direct analysis of complex mixtures using mass spectrometry. *Angew. Chem. Int. Ed.* 49:877–880.

Wang, H., Manicke, N. E., Yang, Q., et al. 2011. Direct analysis of biological tissue by paper spray mass spectrometry. *Anal. Chem.* 83:1197–1201.

Yochum, L., Kushi, L. H., Meyer, K., and Folsom, A. R. 1999. Dietary flavonoid intake and risk of cardiovascular disease in postmenopausal women. *Am. J. Epidemiol.* 149:943–949.

16 Limonoids of Bergamot
Occurrence and Biological Properties

Roberto Romeo and Maria Assunta Chiacchio

CONTENTS

16.1 INTRODUCTION

Limonoids are highly oxygenated, modified triterpenoids dominant in the plant family Meliaceae and to a lesser extent in Rutaceae. The latter family has received considerable attention due to the bitter taste of most limonoids: Rutaceae includes the *Citrus* species of commerce. The term "limonoids" was derived from limonin, the first tetranortriterpenoid obtained from citrus bitter principles. Compounds belonging to this group provide a wide range of biological activities, such as insecticides, insect antifeedants, and growth-regulating activity on insects. Currently, limonoids are under investigation for their antiviral, antifungal, antibacterial, antineoplastic, and antimalarial therapeutic effects.

16.2 STRUCTURE OF LIMONOIDS

Limonoids, secondary metabolites in all citrus fruit tissues, are compounds of moderate polarity, being insoluble in water and hexane but soluble in DMSO, toluene, ethanol, and ketone. *Citrus* plants from the Rutaceae family are the main source of the natural limonoids, and to date 39 limonoids aglycones and 21 glycosides have been isolated. The major limonoid of bergamot is limonin, which is accountable for the bitterness of low-quality citrus juices (Higby 1938; Emerson 1949). Limonin (see Figure 16.1) was first identified as the bitter constituents of citrus seeds in 1841 by Bernay, but its structure remained unknown for more than 120 years. In the 1960s, the structure was finally determined by a combination of chemical methods and X-ray crystallography (Arigoni et al. 1960). Limonin chemical composition is $C_{26}H_{30}O_8$ with a molecular weight of 470.

Several hundred structurally defined limonoids have been described in literature since the characterization of the limonin (Yuan et al. 2009; Connolly and Hill 2010; Fang et al. 2010; Zhang et al. 2010). Although the limonoid family of modified triterpenoids verifies a wide spectrum of structural architectures originated from oxidation and rearrangement of the parent ring framework, the prototypical limonoid scaffold is characterized by a 4,4,8-trimethyl-17-furanylsteroid skeleton (Liao et al. 2009; Heasley 2011), wherein the four fused (androstane-type) rings are designated A through D, as shown in Figure 16.2A. All naturally occurring *Citrus* limonoids contain this substructure or are derived from such a precursor.

Limonoids that do not show any modification in the pentacyclic carbon skeleton may be regarded as intact limonoids. Intact limonoids show various degrees of oxygenation and/or unsaturation within the core structure, but all contain a 3′-substituted furan group linked at position C_{17} in the D ring. For this reason, limonoids are often regarded as tetranortriterpenoids. With respect to intact triterpenoids, which possess 30 carbon atoms ($C_{30}H_{48}$) derived from six isoprene (C_5H_8) units (e.g., squalene or lanosterol), in the limonoids, a class of C_{26} degraded triterpenes, four terminal carbon atoms of the intact lanosterol-type C_{17} side chain (C_{24}-C_{27}) have been oxidatively removed (hence "tetranor"), and the remaining carbon atoms are involved in the furanyl heteroaromatic ring system. Common oxygenation patterns found within

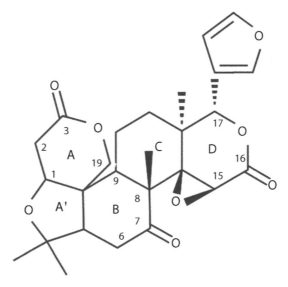

FIGURE 16.1   Structure of limonin.

intact limonoids include a C_{16} keto group; C_3, C_7, and/or C_{12} oxygen-containing functionalities; and a $\Delta^{14,15}$ double bond or epoxide (Figure 16.2).

Degraded limonoids (Figure 16.2b) show a more simplified structure characterized by the presence of fused bicyclic γ- or δ-lactones. *Seco*-limonoids are modified triterpenes that have undergone oxidative fission of one or more of the androstane rings. Representative examples of *seco*-limonoids are reported in Figure 16.2C.

Other members of the natural limonoid family show more complex structures, all amenable to wide and extended ring cleavage and/or C–C structural rearrangement processes. These compounds are simply categorized as highly oxidatively modified limonoids (see Figure 16.2D); their common feature is that they are derived from a furanylsteroid precursor. Their stereochemical features and skeletal connectivities are so diverse that it is often difficult to find a biogenetic relationship between these compounds and intact limonoids.

The structural variations of limonoids found in Rutaceae are less than in Meliaceae and are generally limited to the modification of A and B rings; the limonoids of Meliaceae are more complex, with a very high degree of oxidation and rearrangement exhibited in the parent limonoid structure.

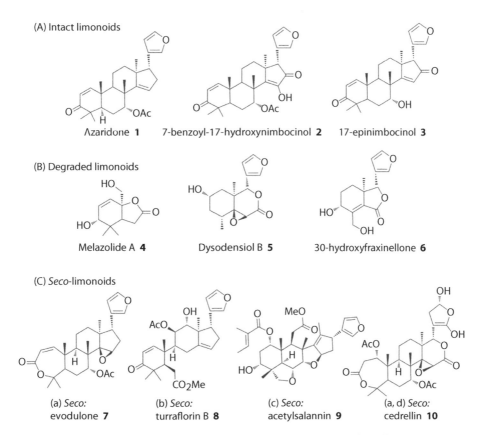

FIGURE 16.2 Examples of the four structural categories (A–D) of the limonoid aglycon family.

(D) Highly oxidative modified limonoids

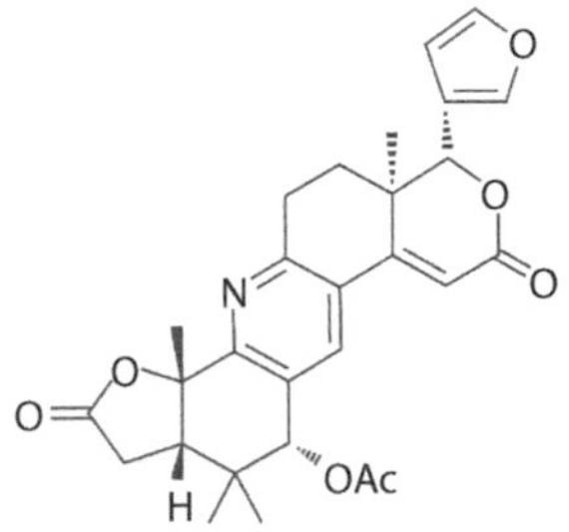

Xylogranatin G **11** 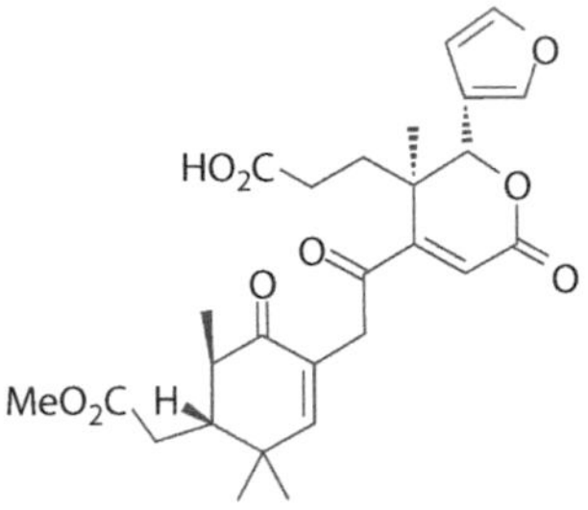   Xylogranatin R **12**

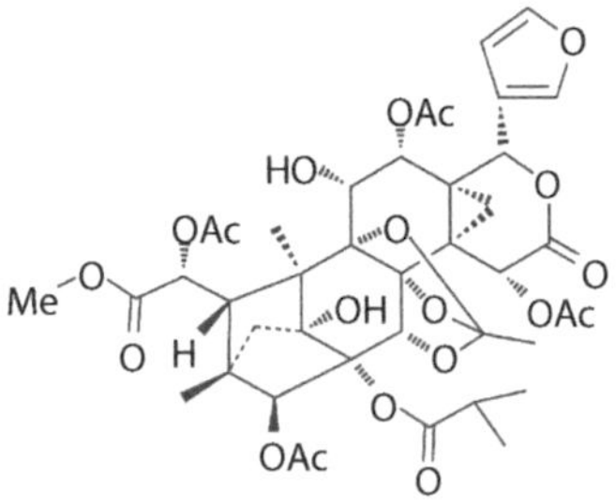

Tabularisin C **13** 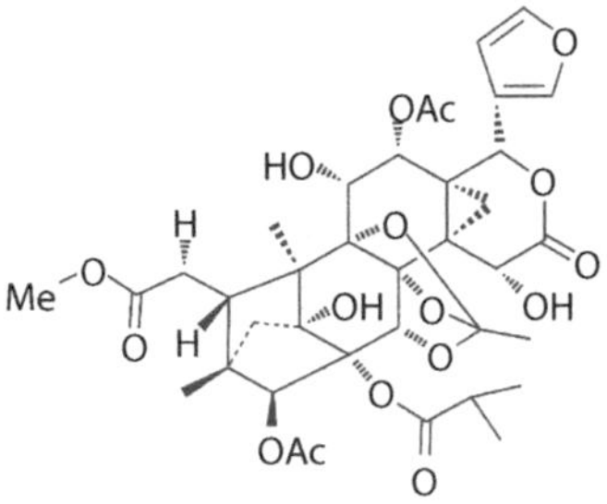   Tabularisin D **14**

FIGURE 16.2 (CONTINUED)   Examples of the four structural categories (A–D) of the limonoid aglycon family.

Finally, it should be noted that the widespread occurrence of glycoside derivatives (Manners 2007) of limonoids from all of these subcategories adds an additional degree of structural variation to this large family of natural products. Limonoids occur in fruits as water-insoluble "bitter" aglycones, and as soluble "tasteless" glucosides. Glucosides have an open D ring to accommodate an attached glucose moiety; in aglycones the D ring is closed (Figure 16.3). The interconversion of aglycones to glucosides occurs as the fruit ripens and is carried out by two enzymes, uridine diphosphoglucose-limonoid glycosyl transferase and limonoid D-ring lactone hydrolase (Hasegawa et al. 1991).

The most common limonoids present in citrus, and particularly in bergamot, are reported in Figure 16.4. The distribution of aglyconic and glycosidic forms of limonoids in bergamot (*Citrus bergamia*) seeds, albedo (inner part of fruit peel), flavedo (external part of fruit peel), pulp, and juice was evaluated (Esposito et al. 2006). The extracted limonoids were characterized by TLC and HPLC analyses. The presence of obacunone, limonine, nomilin, ichangin, and deacetylnomilin was confirmed.

Appreciable amounts of aglyconic limonoids were found in bergamot seeds. All the seeds contained 17-β-D-glucopyranosides of limonin, nomilin, obacunone, deacetylnomilin, nomilinic acid, and deacetylnomilinic acid. The total limonoid glucoside content ranged from 0.31% to 0.87% of the dry weight (Jayaprakasha et al. 2011). The concentration of nomilin glucoside was highest among the glucosides found in the seeds. All the seeds also contained the major neutral limonoid

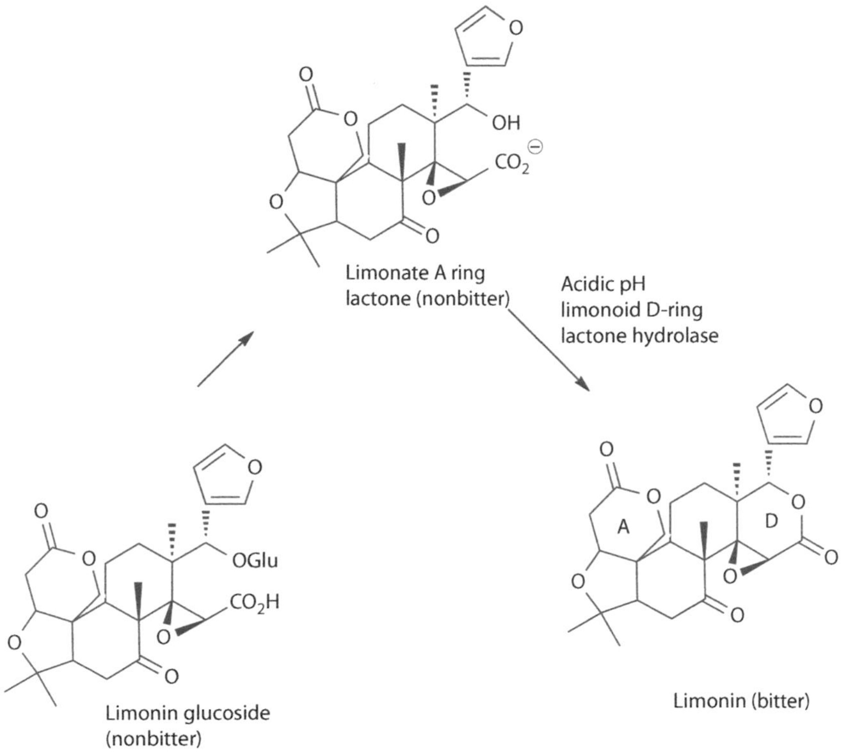

FIGURE 16.3   Interconversion of aglycones to glucosides.

aglycons commonly found in *Citrus*, namely limonin, nomilin, deacetylnomilin, obacunone, and ichangin. Limonin was the predominant aglycon, comprising 42%–89% of the total limonoids. The percentage of nomilin ranged from 8.9% to 48% of the total limonoids, and in a few cases was as great or greater than limonin. The relative nomilin concentration and the nomilin/limonin ratio were the most useful parameters in distinguishing the various citrus species (Rouseff and Nagy1982).

In the bergamot juice, the levels of glycosidic limonoids (about 4 g/kg) were 7–19 times higher than in juices from other *Citrus* species (Hasegawa et al. 1989; Herman et al. 1990; Fong et al. 1992). Since these substances may be useful in the prevention of human diseases and as natural repellents, by-products from bergamot processing could be a useful and cheap source for limonoid extraction.

16.3   BIOSYNTHESIS OF LIMONOIDS

Most of the biogenetic proposals are tentative because of the great variety of skeletal structures and the consequent lack of support of valid biosynthetic studies. The triterpenes containing a side chain at C_{17} are supposed to be biogenetic precursors (Endo et al. 2002; Suarez et al. 2002; Amit and Shilendra 2006), and hence are known as protolimonoids. The biosynthetic pathways start with cyclization of squalene and

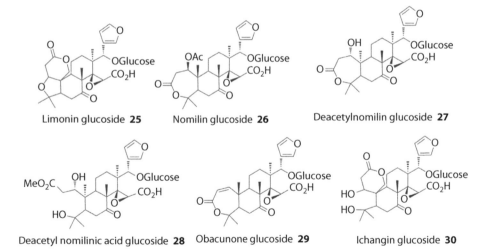

FIGURE 16.4 Detected limonoids in bergamot seed extract.

formation of a tetracyclic ion. Δ^7–Tirucallol (20 S) or Δ^7–euphol (20 R) (Figure 16.5), two epimers at C_{20}, are regarded as the ultimate biogenetic precursors.

The Δ^7–bond is epoxidized and then opened, inducing a Wagner-Merwein shift of Me_{14} to C_8, with formation of the OH_7 and the introduction of the double bond at $C_{14,15}$. This scheme accounts for both the ubiquitous presence of oxygen at C_7 and

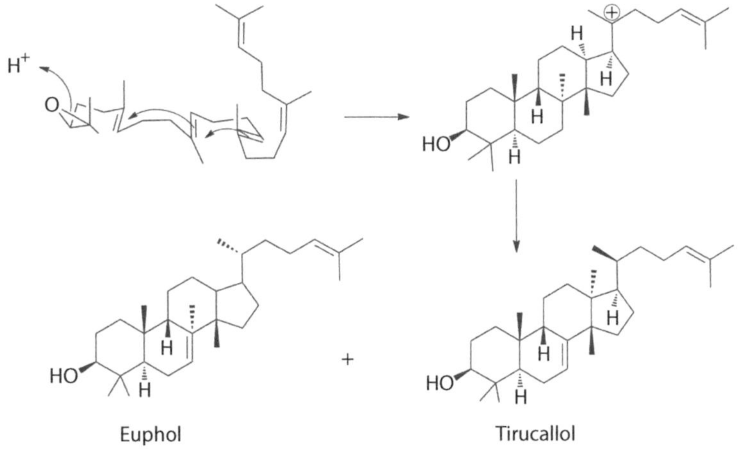

FIGURE 16.5 Proposed precursors of limonoids.

the correct stereochemistry of the C$_{30}$ methyl group (Tan and Luo 2011). Oxidative degradation in the C$_{17}$ side chain of either of these nuclei results in the loss of four carbon atoms, and formation of the 17 β-furan ring (Figure 16.6). That the latter step is accomplished after the formation of the 4,4,8-trimethyl-steroid skeleton is indicated by the occurrence of several protolimonoids. After the formation of the basic limonoid skeleton, a variety of oxidation and skeletal rearrangements in one or more of the four rings, A, B, C, and D, give rise to different groups of limonoids possessing a variety of triterpene skeletons.

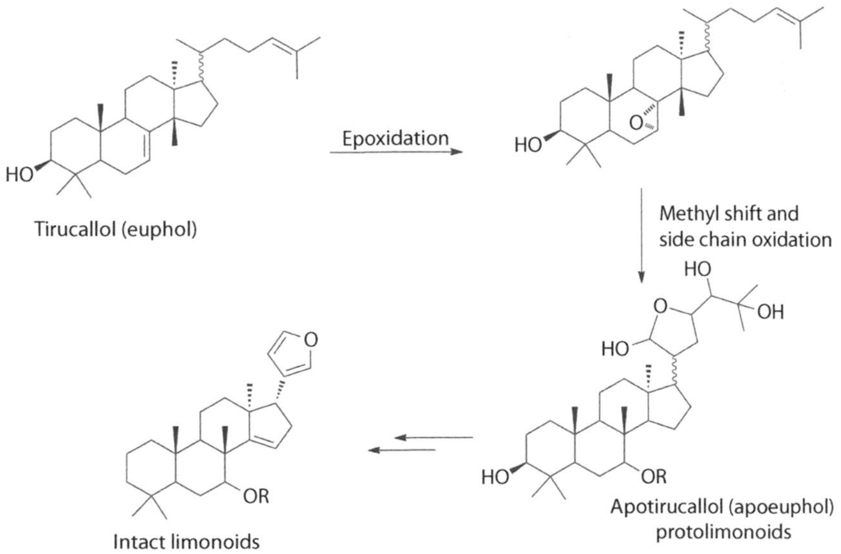

FIGURE 16.6 Proposed biosynthetic pathway to limonoids.

The limonoid biosynthetic pathways have been proposed based on radioactive tracer research (Hasegawa and Zerman 1992): *Citrus* seedlings have been used to prepare C^{14}-labeled nomilin from labeled acetate and mevalonate (Hasegawa et al. 1984). Deacetylnomilinic acid and nomilin are described as the most likely initial precursors of all the known Rutaceae limonoids (Roy and Saraf 2006). Radioactive tracers have shown that deacetylnomilinic acid converts into nomilin. Both deacetyl-nomilinic acid and nomilin may be biosynthesized from acetate, mevalonate, and/ or furanesyl pyrophosphate (Ou et al. 1988) in the phloem region of stem tissues and then are translocated to other plant tissues such as leaves, fruit tissues, and seeds.

Seed and fruit tissues are capable of biosynthesizing other limonoids starting from nomilin independently, by at least four different pathways (Endo et al. 2002; Moriguchi et al. 2003)—limonin, ichangensin, calamine, and 7-acetate limonoid. The first two routes are shown in Figure 16.7.

Five groups of enzymes are involved in the biosynthesis and biodegradation processes of limonoids in *Citrus* (Hasegawa and Miyake 1996). One group, present only in the phloem region of *Citrus* stem tissues, is dedicated to the production of nomilin. A second group, which converts nomilin to the other limonoid aglycones, occurs in all *Citrus* tissues including leaves, stems, fruit juice, fruit peels, and seeds regardless of maturity. Newly synthesized monolactones are converted to dilactones in the seeds during fruit growth by the limonoid D-ring lactone hydrolase, which catalyzes the lactonization of the D ring.

UDP-D-glucose-limonoid glucosyltransferase, which catalyzes the conversion of limonoid aglycones to their respective glucosides during maturation, occurs in fruit

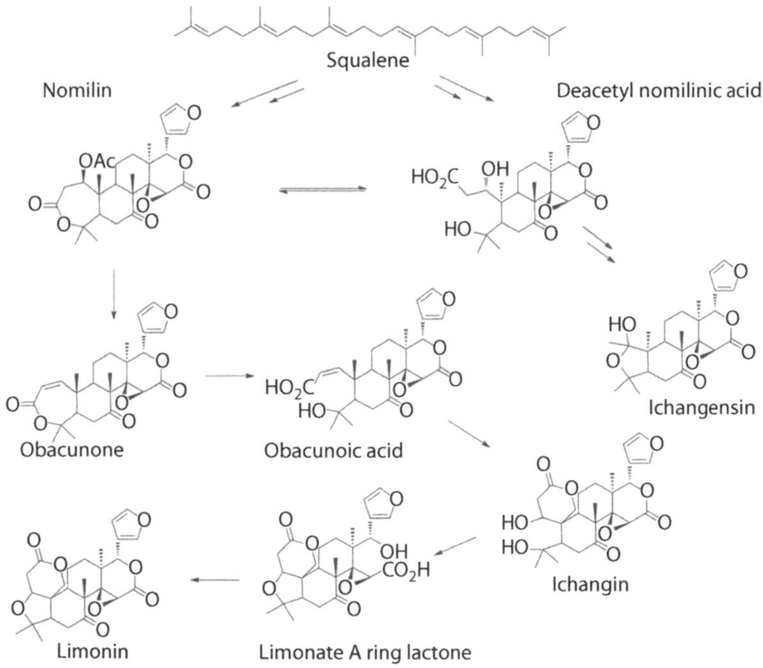

FIGURE 16.7   Biosynthetic pathway towards different structures of limonoids.

tissues and seed. The activity of limonoid glucoside β-glucosidase, which catalyzes the hydrolysis of limonoid glucosides to limonoid aglycones and glucose during seed germination, occurs only in seeds.

Major *Citrus* species accumulate limonin, nomilin, obacunone, and deacetyl-nomilin (Roy and Saraf 2006); *Citrus ichangensis* and relatives accumulate ichan-gensin (keto and ketal); *Fortunella* and related species accumulate calamine group limonoids, such as calamine and cyclocalamin.

Limonoid aglycones are endogenously converted into tasteless limonoid gluco-sides during fruit maturation (Endo et al. 2002). Recently a method combining solid-phase extraction and reversed-phase high-performance liquid chromatography has been described for the isolation of two key metabolites, limonoate and nomilino-ate A-ring lactones, in the limonoid biosynthetic pathway critical to citrus quality (Breksa et al. 2005).

16.4   ANALYSIS OF LIMONOIDS

The first contributions in the area of limonoid analysis were reported by Dreyer (1965a,b, 1967) with the exploitation of thin layer chromatography (TLC) methodol-ogy for the detection and the use of NMR for structural determination. Limonoid glucosides and limonoid aglycones can be monitored, in TLC analysis, with the Ehrlich reagent as a specific detection method (Dreyer 1965a,b) using a solvent sys-tem with a 5:3:1:1 mixture of ethyl acetate/methyl ethylketone/formic acid/water (Ohta et al. 1993; Manners 2007). A semiquantitative evaluation can be performed by comparison of the intensity of the response of individual glucosides to Ehrlich reagent, with the intensity of a limonoid glucoside standard.

For its accuracy and reproducibility, HPLC is undoubtedly the most widely analytical methodology used for limonoids. Normal phase and reverse phase tech-niques have been developed using both isocratic and gradient development protocols (Hasegawa et al. 1996). Thus, the main techniques for the detection and quantitative analysis of limonoids are represented by reverse-phase HPLC-UV, introduced by Fisher (Fisher 1975), and by reverse phase HPLC-MS, developed by Manners and Hasegawa (Manners and Hasegawa 1999). The first methodology uses C-18 bonded silica stationary media and acetonitrile, methanol, and aqueous acid as eluent (Fong et al. 1990; Herman et al. 1990; Ozaki et al. 1991; Ohta et al. 1993; Tian et al. 2006). The detection, performed by UV at 210–215 nm, allows detection limits near 1 μg.

In recent years, with the advent of mass spectrometric techniques coupled with LC, the resolution and precision of HPLC has improved. Several analytical methods couple HPLC with different versions of mass spectrometry (Tian et al. 2003; Manners et al. 2006). Thus, atmospheric pressure chemical ionization (APCI), electron spray ioniza-tion (ESI), LC-MS, and APCI tandem mass spectrometry (MS/MS) have allowed the detection and quantification of limonoid glucosydes and aglycones at nanogram levels. Figure 16.8 shows a normal-phase HPLC-EI/MS of standard citrus limonoid aglycones.

MS data can assist in the specific identification of limonoids by the examination of fragmentation patterns, which can be fruitfully used as an identification tool for unknown limonoid aglycones in *Citrus* extracts. ESI LC/MS, both in positive and

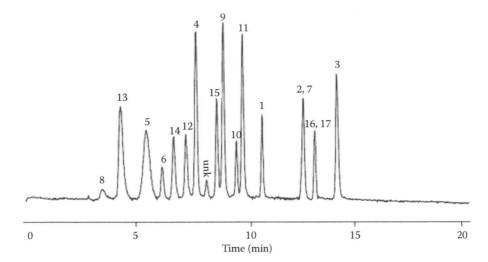

FIGURE 16.8 Normal-phase HPLC-EI/MS of standard citrus limonoid aglycones. (1) Limonin. (2) Nomilin. (3) Deacetylnomilin. (4) Obacunone. (5) Ichangensin. (6) Methyl isoobacunoate (7) Ichangin. (8) Methyl isoobacunoate diosphenol. (9) Methyl deacetylnomilinate. (10) 7α-limonol. (11) 7α-limonyl acetate. (12) 7α-obacunol. (13) Cyclocalamin. (14) Retrocalamin. (15) Calamin. (16) Deoxylimonin. (17) Deoxylimonol.

ion modes, leads to a qualitative detection of protonated and deprotonated molecular ions for limonoid glucosides (Tian and Ding 2000; Tian and Schwartz 2003). The same methodology has been usefully exploited, in negative ion mode, for a quantitative determination of limonoid glucosides up to 2 ng (Schoch et al. 2001).

Analogously, collision/activated dissociation (CAD) of protonated molecular ion of limonoid glucosides offers the possibility of detection of the molecular ion together with the acquisition of characteristic fragmentation patterns (Tian and Schwartz 2003; Raman et al. 2005; Jayaprakasha et al. 2011). Most of these methods require sample preparation protocols using organic solvent extraction, partitioning, and solid phase extraction.

Recently, capillary electrophoresis has been exploited for the analysis of limonin glucosides in seeds, peel, molasses, and pulp (Braddock et al. 2010). UV detection at 214 nm quantifies the amount of limonin glucoside with a limit of 2 mg/L. Structural assignments of limonoids have been supported by NMR spectroscopic methods (Ohochuku and Taylo 1969). In Figures 16.9 and 16.10, ^{1}H NMR spectra of limonin and nomilin are reported.

Extensive analyses ^{1}H and ^{13}C NMR have been applied for the elucidation of the structure of the new isolated compounds (Zhang et al. 2007; Natender and Khaliq 2008; Lin et al. 2009; Bacher et al. 2010). The 1D and 2D NMR techniques (Nuclear Overhauser experiments) provided a general approach to the determination of configuration of a large group of limonoids (Barrera et al. 2011; Kidambara et al. 2011; Liu et al. 2012).

Recently, separation and characterization of limonoids by the HPLC-NMR technique has been described. Analyses were carried out using reversed-phase gradient HPLC elution coupled to NMR (600 MHz) spectrometer in stopped-flow mode.

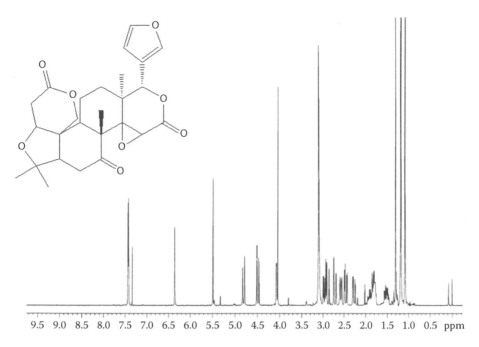

FIGURE 16.9 ¹H NMR spectra of limonin.

Structural attribution was attained by data obtained from ¹H and ¹³C NMR, DEPT, TOCSY, gHSQC, and gHMBC spectra, without the conventional isolation usually applied in the natural products studies (Schefer et al. 2006).

16.5 PHARMACOLOGICAL ACTIVITIES

In recent years, several plant-derived natural compounds have been screened in order to find lead compounds with novel structures or mechanisms of action. Several laboratories have recognized the biological activities of limonoids as antifeedants, inducers of gluthathione S-transferases, and anticancer, cholesterol-lowering, and antiviral compounds.

16.5.1 ANTICANCER ACTIVITY

Nutritional research on the health benefits of chemicals present in plant foods indicate that limonoids possess interesting anticancer activity and are free of any toxic effects. The antineoplastic properties of several limonoids, present in *Citrus* fruits as aglycones and glucosides, have been investigated in in vivo and in vitro studies (Miller et al. 1989, 1992, 1994a,b; Lam et al. 1994a,b, 2000). Obacunone, limonin, nomilin, and their glucosides, together with some other limonoid aglycones, inhibit chemically induced carcinogenesis in a series of human cancer cell lines, with remarkable cytotoxicity against lung, colon, oral, and skin cancer in animal test systems and human breast cancer cells (Miller et al. 1989; Tanaka et al. 2000a,b, 2001; Nakagawa et al. 2001; Silalahi 2002; Manners et al. 2003).

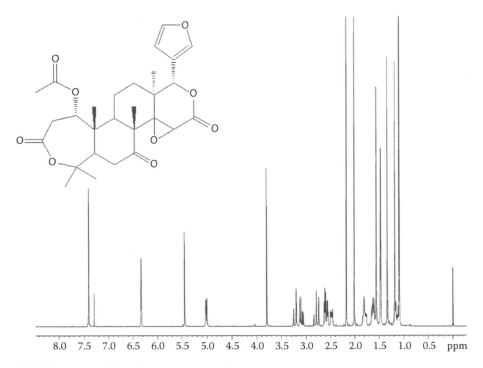

FIGURE 16.10   ¹H NMR spectra of nomilin.

Two Japanese studies reported that obacunone and limonin play an important role in the inhibition of azoxymethane-induced (AOM) colon tumorgenesis in a rat model system (Miyagi et al. 2000; Tanaka et al. 2000a,b). The aglycones were included as dietary supplements to rats exposed to AOM, and the effects were measured by changes in aberrant crypt foci (Tanaka et al. 2000a, 2000b). The obtained results show that the inclusion of limonoids in the diet at the AOM-induced initiation phase significantly reduces the incidence of colon adenocarcinoma (Manners 2007).

Strong evidence suggests that benzo[a]pyrene-induced forestomach neoplasia in mice was inhibited by limonin and nomilin (Lam et al. 1994a,b). The two limonoids were also shown to be effective in the inhibition of 7,12-dimethylbenz[a]anthracene (DMBA)-induced oral tumors (Miller et al. 1992; Lam et al. 1994a,b; Miller et al. 1994a,b, 2004).

There is much experimental evidence on the cancer chemopreventive properties of limonoids present in *Citrus* fruits as aglycones and glucosides. In particular, limonin 17-β-D-glucopyranoside and other limonoid glucosides are potential chemopreventive agents in orange juice that could account for the decreased colon tumorgenesis associated with feeding orange juice (Miyagi et al. 2000; Turner et al. 2006).

Several cell culture studies have been conducted to observe the effects of citrus limonoids on various cell lines. Limonoids have been shown to inhibit the growth of estrogen receptor–negative and receptor–positive human breast cancer cells in culture (Guthrie et al. 2000). Deacetylnomilin 16 was found to be the best inhibitor of estrogen receptor–negative cell growth (IC_{50} < 0.1 µg/mL), while its glucoside showed an activity about one-tenth that of deacetylnomilin 16. Deacetylnomilin and its glucoside were also found to be the most effective inhibitors of estrogen receptor–positive breast cancer cells. With respect to the cancer chemopreventive drug tamoxifen, deacetylnomilin was more than 1000 times more potent against estrogen receptor–negative cancer cells and more than 10 times more potent in inhibiting estrogen receptor–positive breast cancer cells.

Limonoids have also been evaluated as chemopreventives against neuroblastoma and Caco-colon carcinoma cells in vitro (Jacob et al. 2000; Poulose et al. 2005, 2006a,b). Limonin, nomilin, obacunone, deacetyllimonin, and the corresponding glucosides showed an interesting inhibitory activity. Glucosides were more effective than the aglycones at lower concentrations (5 µM), with clear evidence of induced apoptosis by an as-yet unknown mechanism of induction. Individual limonoid glucosides differ in efficacy as anticancer agents, and this difference may reside in structural variations in the A ring of the limonoid molecule.

Recent studies have investigated the chemopreventive effects of limonin glucoside on human pancreatic cancer cells using Panc-28 cells; IC50 values in the range of 18–42 µM in MTT assay have been reported (Patil et al. 2009; Patil et al. 2010). The involvement of apoptosis in induction of cytotoxicity was confirmed by expression of Bax, caspase-3, and p53.

Obacunone, limonin, and methyl nomilinate were evaluated for their potential biological effect on SW480 human colon adenocarcinoma cell proliferation, including mechanism of growth inhibition (Kim et al. 2010). The obtained results indicate that methyl nomilinate is the most potent inhibitor of cell metabolic activity in MTT assay. Although these limonoids did not affect apoptotic markers such as caspase-3 and PARP, methyl nomilinate demonstrated a significant induction of G0/G1 cell cycle arrest in SW480 cells (Chidambara et al. 2011).

The specific mechanism associated with the antineoplastic activity observed for limonoids in in vivo animal tests has not been fully assessed. The chemopreventive properties have in some cases been attributed to the induction of phase II enzyme activity, mainly the detoxifying glutathione S-transferase (GST). Several studies have evaluated the ability of limonoids to induce specific carcinogen-metabolizing enzymes, glutathione S-transferase and quinine reductase, in the liver and mucosa of the small intestine to detoxify chemical carcinogenesis (Lam and Hasegawa 1989; Miller et al. 1994a,b; Kelly et al. 2003; Ahmad et al. 2006).

In the liver of rats fed with diets containing limonin and nomilin, the activity of phase II enzyme glutathione S-transferase increased significantly in a dose-dependent manner, while nomilin and limonin were found to have no significant effect on the phase I enzyme Cytochrome P450. The data from these studies have suggested that the induction of GST activity may be correlated to the chemical structure. Certain rings in the limonoid nucleus may be critical to antineoplastic activity. In particular, the different anticancer agents may reside in structural variations in the A ring of the limonoid molecule (Poulose et al. 2005).

16.5.2 Antioxidant Activity

Radical scavenging and antioxidant activity have been reported for some limonoids, which are supposed to play a role in free radical reactions associated with tumor genesis in mammalian systems (Mandadi et al. 1999; Poulose et al. 2005; Sun et al. 2005; Yu et al. 2005). However, the argument is controversial, in fact a recent work on the antioxidant properties of limonin, nomilin, and limonin glucoside has determined that none of the three compounds tested possess any antioxidant capacity, consistent with their chemical structure lacking conjugated unsaturation and electron delocalization potential (Breksa and Manners 2007).

16.5.3 Anti-HIV Activity

Limonin and nomilin have been investigated for their capacity to act as retroviral agents (Battinelli et al. 2003; Sunthitikawinsakul et al. 2003). The effect of the two compounds on the growth of human immunodeficiency virus-1 (HIV-1) has been examined in a culture of human peripheral blood mononuclear cells (PMBC) and on monocytes/macrophages (M/M). Both compounds were found to inhibit viral replication in PMBC in the EC_{50} = 50–60 µM range. The limonoids were effective at inhibiting HIV-1 replication in infected M/M at concentrations of 20–60 µM. Finally, the authors reported that the mechanism of action could be attributed to inhibition of in vitro HIV-1 protease activity.

Citrus bergamia seed extracts have shown good ability in the inhibition of HIV and HTLV-1 reverse transcriptases; the limonoid components of bergamot seed extract are mainly responsible for the antiretroviral activity (Balestrieri et al. 2011). To better understand the mechanism of action, the two main components of the natural crude extract, limonin and nomilin, have been isolated and evaluated. Limonin inhibits the replication of HIV and HTLV-1 with a potency comparable to that of tenofovir and AZT, while nomilin is tenfold less potent.

The antiviral effect of limonin may be mainly attributed to the inhibition of the reverse trascriptase, especially HTLV-1-RT, while the anti-HIV activity only in part may be attributed to RT inhibition. The biological tests indicate that both limonin and nomilin are able to inhibit the two key enzymes, HIV-RT and HIV-PR, but with a different potency.

16.5.4 Antimicrobial Activity

Limonoids from several plants belonging to Meliaceae as well as the Rutaceae family were reported to have significant antifungal activity (Govindachari et al. 1995; Govindachari et al. 2000; Abdelgaleil et al. 2004; Abdelgaleil et al. 2005; Germano et al. 2005). Germano et al. (2005) reported that limonoids of *Trichilia emetica* can be considered responsible for activity against many bacterial strains.

Limonoids obtained from some *Khaya* species showed good antibacterial and antifungal activity (Abdelgaleil et al. 2004; Abdelgaleil et al. 2005). Limonin and nomilinic acid from *Citrus medica* were tested for their fungal activity against

Puccinia arachidis, a groundnut rust pathogen (Govindachari et al. 1995). These studies illustrate the importance of structural features on activity.

16.5.5 HYPOCHOLESTEROLEMIC ACTIVITY

Evidence of the potential hypocolesterolemic properties of *Citrus* limonoids has been reported. In vitro assay in a rabbit model (Kurowska et al. 2000a,b, 2004) have shown that these compounds, particularly limonin, contribute to the cholesterol-lowering action of *Citrus* juices by reduction of LDL levels. The same authors found that some limonoids decrease cholesterol release in cultured human cells.

Fourteen limonoid aglycones and glucosides have been examined for their ability to lower the structural protein of LDL cholesterol (apoB) in cultured HepG2 human liver cells (Kurowska et al. 2000a,b). Limonin reduces apoB production by more than 70%, while other limonoid aglycones show lower efficacy. In contrast, limonoid glucosides, including limonin glucoside, do not show any effect.

16.5.6 ANTIFEEDANT ACTIVITY

Antifeedant activity of limonoid aglycones, first reported in 1982 (Klocke and Kubo 1982), has been extensively investigated. Many reports describe the insecticidal activities of limonoids isolated from several plants from the Rutaceae and, especially, Meliaceae families (Alford et al. 1987; Hasegawa et al. 1988). The activity of limonin is about 10 times less than that of azadirachtin, a limonoid present in Meliaceae (Alford and Murray 2000). Limonoid glucosides have no antifeedant activity against insects.

More recently, Manners (2007) reviewed the activity of *Citrus* limonoids. Limonin and nomilin are particularly effective in disrupting larvae development of the corn earworm and fall army-worm. The results suggest that the insecticidal *Citrus* limonoids can function as an important component of an integrated pest-management program.

16.5.7 OTHER MISCELLANEOUS ACTIVITIES

Biswas et al. (2002) found that limonoids have anti-inflammatory, anti-arthritic, antipyretic, hypoglycemic, anti-gastric ulcer, spermicidal, diuretic, and other pharmacological properties. Raphael and Kuttan (2003) described the immunomodulatory activity of nomilin. In an in vitro study, limonoids isolated from *Swietenia humilis* showed concentration-dependent and nonreversible spasmogenic and uterotonic activity (Perusquia et al. 1998). Leishmanicidal activity has been reported for some seco-limonoids isolated from *Raputia heptaphylla* (Barrera et al. 2011). In vitro anti-sickling activity of a rearranged limonoid isolated from *Khaya senegalensis* has been reported (Fall et al. 1999). Ongoing research programs are examining the effects of limonin in human diseases. Limonin is supposed to be specifically directed towards protection of lungs for clearing congestive mucus (Rohr et al. 2002). Neuroprotective effects of limonin have also been described (Yoon et al. 2010). Furthermore, limonin has been tested as an anti-obesity agent in mice (Ono et al. 2011).

16.6 FUTURE PERSPECTIVES

The experimental results of *Citrus* limonoid biological activity, coupled with the high concentration of these compounds in citrus fruit and juices, suggest that citrus products can usefully contribute to human health (Manners 2007). Much experimental data show that citrus limonoids may provide interesting biological effects and, in particular, substantial anticancer actions. The compounds have been shown to be free of toxic effects; animals consuming a diet containing limonoid aglycones or glucosides suffered no toxic consequences. Moreover, in vivo studies support the evidence that citrus limonoids are not toxic to humans (Guthrie et al. 2000; Miller et al. 2000, 2006): Thus, potential exists for use of limonoids against human cancer in either natural fruit or in purified forms of specific limonoids.

The peculiar health benefits of citrus limonoids have led to a need for additional research to evaluate these compounds alone or in combination, with the aim to clarify their potential in commercial formulations. Due to their extreme bitterness, there is also the need to develop acceptable and versatile debittering methods to mitigate this effect in fruits and juices (Amit and Shilendra 2006).

With regard to the biological activities of limonoids, major efforts have to be targeted towards the characterization, quantification, and evaluation of the single derivatives as well as towards the design of suitable synthetic routes of the most interesting limonoids. The preparation of limonoids requires a multistep and laborious synthetic process. Although patents have been obtained for industrial scale methods for manufacturing limonoid glucosides contained in citrus fruits, the high cost and the complexity of these structures substantially preclude this type of approach. Biotechnology, tissue culture techniques, and optimization of extraction methods may be extensively investigated to enhance the production of limonoids to meet the increasing demand.

Finally, an important step is to assess the bioavailability of the compounds for humans—the absorption after ingestion, or the appearance in the blood and tissues, and for how long. In conclusion, the possible commercial exploitation of limonoids as nutraceuticals or components of functional foods involves an extensive knowledge of the chemistry and biochemistry of this group of compounds. More information is needed about the mode of action of these compounds and the specific correlation between the limonoid structure and the observed bioactivity.

REFERENCES

Abdelgaleil, S. A., Iwagawa, T., Doe, M., and Nakatani, M. 2004. Antifungal limonoids from the fruits of Khaya senegalensis. *Fitoterapia* 75:566–572.

Abdelgaleil, S. A., Hashinaga, F., and Nakatani, M. 2005. Antifungal activity of limonoids from Khaya ivorensis. *Pest Manag. Sci.* 61:186–190.

Ahmad, H., Li, J., Polson, M., et al. 2006. Citrus limonoids and flavonoids: Enhancement of phase II detoxification enzymes and their potential in chemoprevention. In *Potential health benefits of citrus*, eds. B. S. Patil, N. D. Turner, E. D. Miller, and J. S. Brodbelt, 130–143. Washington, DC: American Chemical Society.

Alford, A. R., and Bentley, M. D. 1986. Citrus limonoids as potential antifeedants for the spruce budworm (Lepidoptera: Tortricidae). *J. Econ. Entomol.* 79:35–38.

Alford, A. R., and Murray, K. D. 2000. Prospects for citrus limonoids in insect pest management. In *Citrus limonoids: Functional chemicals in agriculture and foods*, eds. M. A. Berhow, S. Hasegawa, and G. D. Manners, 201–211. Washington, DC: American Chemical Society.

Arigoni, D., Barton, D. H. R., Corey E. J., et al. 1960. The constitution of limonin. *Experientia* 16:41–49.

Bacher, M., Brader, G., Greger, H., and Hofer, O. 2010. Complete 1H and ^{13}C NMR data assignment of new constituents from *Severinia buxifolia*. *Magn. Reson. Chem.* 48:83–88.

Balestrieri, E., Pizzimenti, F., Ferlazzo, A., et al. 2011. Antiviral activity of seed extract from *Citrus bergamia* towards human retroviruses. *Bioorg. Med. Chem.* 19:2084–2089.

Barrera, C. A. C., Barrera, E. D. C., Falla, D. S. G., et al. 2011. Seco-limonoids and quinoline alkaloids from *Raputia heptaphylla* and their antileishmanial activity. *Chem. Pharm. Bull.* 59:855–859.

Barton, D. H. R., Pradhan, S. K., Sternhell, S., and Templeton, J. F. 1961. Triterpenois. XXV. Constitution of limonin and related bitter principles. *J. Chem. Soc.* 382:25.

Battinelli, L., Mengoni, F., Lichtner, M., et al. 2003. Effect of limonin and nomilin on HIV-1 replication on infected human mononuclear cells. *Planta Med.* 69:910–913.

Bernay, S., 1841. Limonin. *Annalen* 40:317–320.

Biswas, K., Chattopadhyay, I., Banerjee, R. K., and Bandpadhyay, U. 2002. Biological activities and medicinal properties of neem (*Azadirachta* indica). *Curr. Sci.* 11:1336–1347.

Braddock, R. J., and Bryan, C. R. 2001. Extraction parameters and capillary electrophoresis analysis of limonin glucoside and phlorin in citrus byproducts. *J. Agric. Food Chem.* 49:5982–5988.

Breksa, A. P., Zukas, A. A, and Manners, G. D. 2005. Determination of limonoate and nomilinoate A-ring lactones in citrus juices by liquid chromatography-electrospray ionization mass spectrometry. *J. Chromatogr. A* 1064:187–191.

Breksa, A. P., and Manners, G. D. 2007. Evaluation of antioxidant capacity of limonin, nomilin and limonin glucoside. *J. Agric. Food Chem.* 54:3827–3831.

Chidambara, K. N., Jayaprakasha, G. K., and Patil, B. S. 2011. Obacunone and obacunone glucoside inhibit human colon cancer (SW480) cells by the induction of apoptosis. *Food Chem. Toxicol.* 49:1616–1625.

Connolly, J. D., and Hill, R. A. 2010. Triterpenoids. *Nat. Prod. Rep.* 27:79–132.

Dreyer, D. L. 1965a. Citrus bitter principles. II: application of NMR to structural and stereochemical problems. *Tetrahedron* 21:75–87.

Dreyer D. L. 1965b. Limonoid bitter principles. In *Progress in the chemistry of organic natural products*, vol. 26, ed. L. Zechmeister, 191–244. New York: Springer.

Dreyer, D. L. 1967. Citrus bitter principles. VII. Rutaevin. *J. Org. Chem.* 32:3442–3445.

Emerson, O. H. 1949. The bitter principle of Navel oranges. *Food Tech.* 3:243–250.

Endo, T., Kita, M., Shimada, T., et al. 2002. Modification of limonoid metabolism in suspension cell culture of citrus. *Plant Biotechnol.* 19:397–403.

Esposito, C. Tridente, R. L., Balestrieri, M. L., et al. 2006. Distribuzione dei limonoidi nelle diverse parti del frutto di bergamotto. *Essenze Derivati Agr.* 36:11–18.

Fall, A. B., Vanhaelen-Faster, R., Vanhaelen, M. I., et al. 1999. In vitro antisickling activity of a rearranged limonoid isolated from *Khaya senegalensis*. *Planta Med.* 65:209–212.

Fang, X., Di, Z., Geng, Y., et al. 2010. Trichiliton A, a novel limonoid from *Trichilia connaroides*. *Eur. J. Org. Chem.* 1381–1387.

Fisher, J. F. 1975. Quantitative determination of limonin in grapefruit juice by high-pressure liquid chromatography. *J. Agric. Food Chem.* 23:1199–1201.

Fong, C. H., Hasegawa, S., Herman, Z., and Ou, P. 1990. Limonoid glucosides in commercial Citrus juices. *J. Food Sc.* 54:1505–1506.

Fong, C. H., Hasegawa, S., Coggins, C. W., et al. 1992. Contents of limonoids and limoni 17-B-D-glucopyranoside in fruit tissue of Valencia orange during fruit growth and maturation. *J. Agric. Food. Chem.* 40:1178–1181.

Germanò, M. P., D'Angelo, V., Sanogo, R., et al. 2005. Hepatoprotective and antibacterial effects of extracts from *Trichiliaemetica* Vahl (Meliaceae). *J. Ethnopharmacology* 96:227–232.

Govindachari, T. R., Suresh, G., Banumathi, B., et al. 1995. Antifungal activity of some B,D-seco limonoids from two meliaceous plants. *J. Chem. Ecol.* 25:923–926.

Govindachari, T. R., Suresh, G., Gopalakrishnan, G., et al. 2000. Antifungal activity of some tetranotriterpenoids. *Fitoterapia* 71:317–320.

Guthrie, N., Hasegawa, S., Manners, G. D., et al. Inhibition of human breast cancer cells by citrus limonoids. In *Citrus limonoids: Functional chemicals in agriculture and foods*, eds. M. A. Berhow, S. Hasegawa, and G. D. Manners, 164–174. Washington, DC: American Chemical Society.

Hasegawa, S., Bennet, R. D., and Maier V. P. 1984. Biosynthesis of limonoids in Citrus seedlings. *Phytochemistry* 23:1601–1603.

Hasegawa, S., Miyake, M., Robertson, G. H., and Berhow, M. 1988. Limonoids in miyamashi-kimi. *J. Japan Soc. Hort. Sci.* 67:835–838.

Hasegawa, S., Bennett, R. D., Herman, Z., et al. 1989. Limonoid glucosides in Citrus. *Phytochemistry* 28:1717–1720.

Hasegawa, S., Ou, P., Fong, C. H., et al. 1991. Changes in the limonoate A-ring lactone and limonin 17-b-D-glucopyranoside content of navel oranges during fruit growth and maturation. *J. Agric. Food Chem.* 39:262–265.

Hasegawa, S., and Herman, Z., 1992. Analysis of limonoids in Citrus seeds. In *Secondary metabolite biosynthesis and metabolism*, eds. R. J. Petrowski and S. P. McCormick, 305–318. New York: Plenum Press.

Hasegawa, S., and Miyake M. 1996. Biochemistry and biological functions of Citrus limonoids. *Food Rev. Int.* 12:413–415.

Hasegawa, S., Berhow, M. A., and Fong, C. H. 1996. Analysis of bitter principles in Citrus. In *Modern methods of fruit analysis*, eds. H. L. Linskens and J. F. Jackson, 59–80. Berlin: Springer-Verlag.

Heasley, B. 2011. Synthesis of limonoid natural products. *Eur. J. Org. Chem.* 19–46.

Herman, Z., Fong, C. H., Ou, P., and Hasegawa, S. 1990. Limonoid glycosides in orange juices by HPLC. *J. Agric. Food Chem.* 38:1860–1861.

Higby, R. H. 1938. The bitter constituents of Navel and Valencia oranges. *J. Am. Chem. Soc.* 60:3013–3018.

Jacob, R., Hasegawa, S., and Manners, G. D. 2000. The potential of citrus limonoids as anticancer agents. *Perish. Handl.* 102:6–8.

Jayaprakasha, G., Dandakar, D. V., Tichy, S. E., and Petil, B. S. 2011. Simultaneous separation and identification of limonoids from Citrus using liquid chromatography-collision-induced dissociation mass spectra. *J. Sep. Sci.* 34:2–10.

Kelly, C., Jewell, C., and O'Brien, N. M. 2003. The effect of dietary supplementation with the citrus limonoids, limonin and nomilin on xenobiotic metabolizing enzymes in the liver and small intestine of the rat. *Nutr. Res.* 23:681–690.

Kim, J., Jayaprakasha, J. K., Vikram, A., and Patil, B. S. 2010. Methyl nomilinate inhibits SW480 colon cancer cells growth through modulation of cell cycle regulators. Presented at ACS Meeting, Boston, August 22–26.

Klocke, J. A., and Kubo, I. 1982. Citrus limonoids by-products as insect control agents. *Entomol. Exp. Appl.* 32:299–301.

Kurowska, E. M., Borradaile, N. M., Spence, J. D., and Carroll, K. K. 2000a. Hypocholesterolemic effects of dietary citrus juices in rabbits. *Nutr. Res.* 20:121–129.

Kurowska, E. M., Banh, C., Hasegawa, S., and Manners, G. D. 2000b. Regulation of apo B production in Hepg2 cells by citrus limonoids. In *Citrus limonoids: Functional chemicals in agriculture and foods*, eds. M. A. Berhow, S. Hasegawa, and G. D. Manners, 175–184. Washington, DC: American Chemical Society.

Kurowska, E. M., Manthey, J. A., Borradaile, N. M., et al. 2004. Hydrolipidemic effects and absorption of Citrus polymethoxylated flavones in hamsters with diet-induced hypercholesterolemia. *J. Agric. Food Chem.* 52:2879–2886.

Lam, N. K. T., Li, Y., and Hasegawa, S. 1989. Effects of citrus limonoids on glutathione S-transferase in mice. *J. Agric. Food Chem.* 37:878–880.

Lam, L. K. T., Zhang, J., and Hasegawa, S. 1994a. Inhibition of chemically induced carcinogenesis by citrus limonoids. In *Phytochemicals for cancer prevention,* eds. H. A. J. C. Schut, M. T. Huang, C. T. H. Osawa, and R. T. Rosen, 209–219. Washington, DC: American Chemical Society.

Lam, L. K. T., Zhang, J., and Hasegawa, S. 1994b. Citrus limonoid reduction of chemically induced tumorigenesis. *Food Technol.* 48(11):101–108.

Lam, L. K. T., Hasegawa, S., Bergstrom, C., et al., 2000. Limonin and nomilin inhibitory effects on chemical induced tumorgenesis. In *Citrus limonoids: Functional chemicals in agriculture and foods*, eds. M. A. Berhow, S. Hasegawa, and G. D. Manners, 185–200. Washington, DC: American Chemical Society.

Liao, S.-G., Chen, H.-D., and Yue, J.-M. 2009. Plant orthoesters. *Chem Rev.* 109:1092–1140.

Lin, B.-D., Zhang, C.-R., Yiang, S.-P., et al. 2009. D-ring-opened phragmalin type limonoid orthoesters from the twigs of *Swietenia macrophylla. J. Nat. Prod.* 72:1305–1313.

Liu, J.-Q., Wang, C.-F., Li, Y., et al. 2012. Limonoids from the leaves of *Toona ciliate var. yunnanensis. Phytochem.* 76:141–149.

Mandadi, K. K., Jayaprakasha, G. K., Baht, N. G., and Patil, B. S. 1999. Red Mexican grapefruit: A novel source for bioactive limonoids and their antioxidant activity. *Z. Naturforsch[C]* 62:179–188.

Manners, G. D. 2007. Citrus limonoids: Analysis, bioactivity and biomedical prospects. *J. Agric. Food Chem.* 55:8285–8294.

Manners, G. D., and Breksa, I. A. 2004. Identifying citrus limonboid aglycones by HPLC-EI/MS and HPLC-APCI/MS techniques. *Phytochem. Anal.* 15:372–381.

Manners, G. D., and Hasegawa, S. 1999. A new normal phase liquid chromatographic method for the analysis of limonoids in citrus. *Phytochem. Anal.* 10:76–81.

Manners, G. D., Jacob, R. A., Breksa, I. A., et al. 2003. Bioavailability of citrus limonoids in humans. *J. Agr. Food Chem.* 51:4156–4161.

Manners, G. D., Hasegawa, S., Barnett, R. Y., and Wong, L. 2006. Long term clinic study on the potential toxicity of limonoids. In *Citrus limonoids: Functional chemicals in agriculture and foods.* eds. M. A. Berhow, S. Hasegawa, and G. D. Manners, ACS Symposium, Series 758, 82–94. Washington, DC: American Chemical Society.

Miller, E. G., Fanous, R., Rivera-Hidalgo, F., et al. 1989. The effect of citrus limonoids on hamster buccal pouch carcinogenesis. *Carcinogenesis* 10:1535–1537.

Miller, E. G., Gonzalez-Sanders, A. P., Couvillon, A. M., et al. 1992. Inhibition of hamster buccal pouch carcinogenesis by limonin 17 b-D-glucopyranoside. *Nutr. Cancer* 17:1–7.

Miller, E. G., Gonzalez-Sanders, A. P., Couvillon, A. M., et al. 1994a. Citrus limonoids as inhibitors of oral carcinogenesis. *Food Technol.* 48(11):110–114.

Miller, E. G., Gonzalez-Sanders, A. P., Couvillon, A. M. et al. 1994b. Inhibition of oral carcinogenesis by green coffee beans and limonoid glucosides. In *Food phytochemicals for cancer prevention*, eds. M. T. Huang, T. Osawa, C. T. Ho, and R. T. Rosen, 220–229. Washington, DC: American Chemical Society.

Miller, E. G., Record, R. T., Binnie, W. H., and Hasegawa, S. 1999. Limonoid glucosides: Systemic effect on oral carcinogenesis. In *Phytochemicals and phytopharmaceuticals*, eds. F. Shahidi and C. Ho, 95–105. Champaign, IL: AOAC Press.

Miller, E. G., Porter, J. L., Binnie, W. H., et al. 2004. Further studies on the anticancer activity of citrus limonoids. *J. Agric. Food Chem.* 52:4908–4912.

Miller, E. G., Gibbins, R. P., Taylor, S. E., McIntosh, J. E., and Patil B. S. 2006. Long-term screening study on the potential toxicity of limonoids. In *Potential health benefits of citrus*, eds. B. S. Patil, N. D. Turner, E. G. Miller, and J. S. Brodbelt, 82–94. Washington, DC: American Chemical Society.

Miyagi, Y., Om, A. S., Chee, K. M., and Bennink, M. R. 2000. Inhibition of azoxymethane-induced colon cancer by orange juice. *Nutr. Cancer* 36:224–229.

Moriguchi, T., Kita, M., Hasegawa, S. and Omura, M. 2003. Molecular approach to citrus flavonoid and limonoid biosynthesis. *Food Agric. Environment* 1:22–25.

Nakagawa, H., Duan, H., and Tahahishi, Y. 2001. Limonoids from *Citrus sudachi*. *Chem. Pharm. Bull.* 49:649–651.

Narender, T., Khaliq, T., and Shweta T. 2008. ^{13}C NMR spectroscopy of D and B, D-ring seco-limonoids of *Meliaceae* family. *Nat. Prod. Res.* 22:763–800.

Ohochuku, N. S., and Taylor, D. A. H. 1969. Chemical shift of the tertiary methyl groups in the NMR of some limonoids. *J. Chem. Soc. C* 6:864–870.

Ohta, H., Fong, C. H., Berhow, M. A., and Hasegawa, S. 1993. Thin-layer and high performance liquid chromatographic analyses of limonoids and limonoid glucosides in Citrus seeds. *J. Chromatogr.* 639:295–302.

Ono, E., Inoue, J., Hashidume, T., et al. 2011. Anti-obesity and anti-hyperglycemic effects of dietary citrus limonoid nomilin in mice fed a high-fat diet. *Biochem. Biophys. Res. Commun.* 410:677–681.

Orav, A., Vitak, A., and Vaher, M. 2010. Identification of bioactive compounds in the leaves and stems of *Aegopodium podograria* by various analytical techniques. *Procedia Chem.* 2:152–160.

Ou, P., Hasegawa, S., Herman, Z., and Fong, C. H. 1988. Limonoid biosynthesis in the stem of *Citrus limon*. *Phytochemistry*, 27:115–118.

Ozaki, Y., Fong, C. H., Herman, Z., et al. 1991. Limonoid glucosides in Citrus seeds. *Agric. Biol. Chem.* 55:137–141.

Patil, B. S., Yu, J., Dandekar, D. V. et al., 2006. Citrus bioactive limonoids and flavonoids extraction by supercritical fluids. In *Potential health benefits of citrus*, eds. B. S. Patil, N. D. Turner, E. D. Miller, and J. S. Brodbelt, 18–33. Washington, DC: American Chemical Society.

Patil, J. R., Jayaprakasha, G. K., Murthy, K. N. C., et al. 2009. Bioactive compounds from Mexican lime (*Citrus aurantifolia*) juice induce apoptosis in human pancreatic cells. *J. Agric. Food Chem.* 57:10933–10942.

Patil, J. R., Jayaprakasha, G. K., Murthy, K. N. C., et al. 2010. Characterization of *Citrus aurantifolia* bioactive compounds and their inhibition of human pancreatic cancer cella through apoptosis. *Microchem. J.* 94:108–117.

Perusquia, M., Hernandez, R., Jimenez, M. A., et al. 1997. Contractile response induced by a limonoid (humilinolide A) on spontaneous activity of isolated smooth muscle. *Phytother. Res.* 11:354–357.

Poulose, S. M., Harris, E. D., and Patil, B. S. 2005. Citrus limonoids induce apoptosis in human neuroblastoma cells and have radical scavenging activity, *J. Nutr.* 135:870–877.

Poulose, S. M, Harris, E. D., and Patil, B. S. 2006a. Cytotoxic and antineoplastic effects of citrus limonoids against human neuroblastoma and colonic adenocarcinoma cells. *FASEB* 20:A11–A12.

Poulose, S. M., Harris, E. D., Patil, B. S. 2006b. Antiproliferative effects of citrus limonoids against human neuroblastoma and colonic adenocarcinoma cells. *Nutr. Cancer* 56:103–112.

Raman, G., Cho, M., Brodbelt, J. S., and Patil, B. S. 2005. Isolation and purification of closely related citrus limonoid glucosides by flash chromatography. *Phytochem. Anal.* 16:155–160.

Raphael, T. J., and Kuttan, G. 2003. Effect of naturally occurring triterpenoids glycyrrhizic acid, ursolic acid, oleanolic acid and nomilin on the immune system. *Phytomedicine* 10:483–489.

Rohr, A. C., Wilkins, C. K., Clausen, P. A., et al. 2002. Upper airway and pulmonary effects of oxidation products of (+)-alpha-pinene, d-limonene, and isoprene in BALB/c mice. *Inhalation Toxicol.* 14:663–684.

Rouseff, R. L., and Nagy, S. 1982. Distribution of limonoids in Citrus seeds. *Phytochemistry* 21:85–90.

Roy, A., and Saraf, S. 2006. Limonoids: Overview of significant bioactive triterpenes distributed in plant kingdom. *Biol. Bull. Pharm.* 29:191–201.

Schefer, A. B., Braumann, U., Tseng, S. L.-H., et al. 2006. Application of high performance liquid chromatography-nuclear magnetic resonance coupling to the identification of limonoids from mahogany tree by stopped-flow 1D and 2D NMR spectroscopy. *J. Chromatogr. A* 1128:152–163.

Schoch, T. K., Manners, J. D., and Hasegawa, S. 2001. Analysis of limonoid glucosides from Citrus by electrospray ionization liquid chromatography-mass spectrometry. *J. Agric. Food Chem.* 49:1101–1108.

Silalahi, J. 2002. Anticancer and health protective properties of Citrus fruit components. *Asian Pac. J. Clin. Nutr.* 11:79–84.

Suarez, L. E., Menichini, C. F., and Monache, F. D. 2002. Tetranortriterpenoids and dihydrocinnamic acid derivatives from hortia colombiana. *J. Braz. Chem. Soc.* 13:339–344.

Sun, C., Chen, K., Chen, Y., and Chen, Q. 2005. Contents and antioxidant capacity of limonin and nomilin in different tissues of citrus fruit of four cultivars during fruit growth and maturation. *Food Chem.* 93:599–605.

Sunthitikawinsakul, A., Kongkathip, N., Kongkathip, B., et al. 2003. Anti HIV-1 limonoid: First isolation from *Clausena excavate. Phytother. Res.* 17:1101–1103.

Tanaka, T., Kohno, H., Honjo, S., et al. 2000a. Citrus limonoids obacunone and limonin inhibit azoxymethane-induced colon carcinogenesis in rats. *Biofactors* 13:213–218.

Tanaka, T., Kohno, H., Tsukio, Y., et al. 2000b. Citrus limonoids obacunone and limonin inhibit the development of a precursor lesion, aberrant crypto foci for colon cancer in rats. In *Citrus limonoids: Functional chemicals in agriculture and foods*, eds. M. A. Berhow, S. Hasegawa, and G. D. Manners, 145–163. Washington, DC: American Chemical Society.

Tanaka, T., Maeda, M., Kohno, H., et al. 2001. Inhibition of azoxymethane-induced colon carcinogenesis in male F344 rats by the citrus limonoids obacunone and limonin. *Carcinogenesis* 22:193–198.

Tian, Q., and Ding, X. 2000. Screening for limonoid glucosides in *Citrus tangerina* (Tanaka) Tseng by high-performance liquid chromatography electrospray ionization mass spectrometry. *J. Chromatogr. A* 874:13–19.

Tian, Q., Li, D. Barbacci, D., and Schwartz, S. J. 2003. Electron ionization mass spectrometry of citrus limonoids. *Rapid Commun. Mass Spectr.* 22:2517–2522.

Tian, Q., and Schwartz, S. J. 2003. Mass spectrometry and tandem mass spectrometry of citrus limonoids. *Anal. Chem.* 75:5451–5460.

Tian, Q., Miller, E. J., Jayaprakasha, G. K., and Patil, B. S. 2006. An improved HPLC method for the analysis of Citrus limonoids in culture media. *J. Chromatogr. B* 846:385–390.

Tian, Q., and Luo, X.-D. 2011. Meliaceous limonoids: Chemistry and biological activities. *Chem Rev.* 111:7437–7522.

Turner, N. D., Vanamala, J., Leonardi, T., et al. 2006. Comparison of chemoprotection conferred by grapefruit and isolated bioactive compounds against colon cancer. In *Potential health benefits of citrus*, eds. B. S. Patil, N. D. Turner, E. G. Miller, I. S. Brodbelt 936:121–129. Washington, DC: American Chemical Society.

Yoon, J. S., Yang, H., Kim, S. H., et al. 2010. Limonoids from *Dictamnus dasycarpus* protect against glutamate-induced toxicity in primary cultured rat cortical cells. *J. Mol. Neurosci.* 42:9–16.

Yu, J., Wang, L., Walzem, R. L., et al. 2006. Antioxidant activity of Citrus limonoids, flavonoids and coumarins. *J. Agric. Food Chem.* 53:2009–2014.

Yuan, D., Yang, S.-P., Zhang, C.-R., et al. 2009. Two limonoids, khayalenoids A and B with an unprecedented 8-oxa-tricyclo[4.3.2.02,7] undecane motif, from khayasenegalensis. *Org. Lett.* 11:617–620.

Zhang, C.-R., Fan, C.-Q., Zhang, L., et al. 2010. Chuktabrins A and B, two novel limonoids from the twigs and leaves of chukrasiatabularis. *Org. Lett.* 10:3183–3186.

Zhang, H., Chen, F., Wang, X., et al. 2007. Complete assignments of ^{1}H and ^{13}C NMR data for rings of A,B-seco limonoids from the seed of *Aphanamixis polystachya*. *Magn. Reson. Chem.* 45:189–192.

17 Chemical Contamination of Bergamot Essential Oils

Giacomo Dugo, Giuseppa Di Bella, and Marcello Saitta

CONTENTS

17.1 INTRODUCTION

Bergamot essential oil is widely used in the cosmetics industry for perfumes and sun-tanning lotions; in the pharmaceutical industry as flavoring, cicatrizing, and antiseptic; and in the confectionery industry as aromatizing for liqueurs, teas, sweets, and candied fruits. To ensure the quality and goodness of an essential oil, it is necessary that the product be free of any contaminants and residues within the limits set by importing countries. The term "contaminant" refers to all the classes of substances that do not belong to the natural composition of the essential oil. A contaminant may be voluntary (if it is deliberately added, as it could be in the case of a preservative or stabilizer) or involuntary (if it is inadvertently added, as in the case of a pesticide).

The first research on the contamination of bergamot oil dates back more than 30 years. Since that time, many classes of contaminants have been investigated. Bergamot trees grow mostly in the province of Reggio Calabria (Italy); attempts to implant bergamot in Florida, California, Louisiana, and French Guinea have not been successful. The only other place in the world where bergamot flourished is the Ivory Coast, but the oil quality is poorer than the Calabrian. In this review, the analytical methods and the data reported in literature regarding the contamination of Italian bergamot essential oils will be examined.

17.2 ORGANIC CONTAMINANTS

17.2.1 ORGANOPHOSPHORUS PESTICIDES

Historically, the first class of contaminants found in bergamot essential oils was organophosphorus pesticides. These compounds are essentially insecticides which interact with acetylcholinesterase and other cholinesterases. This results in disruption of insect nerve impulses, killing the insect or interfering with its ability to carry on normal functions. Organophosphate insecticides and chemical warfare nerve agents work in the same way. Organophosphates have an accumulative toxic effect to wildlife, so multiple exposures to the chemicals amplify the toxicity.

The first research into the presence of organophosphorus pesticide residues in bergamot essential oils was performed in 1978 (Leoni and D'Alessandro De Luca 1978). The method employed was laborious. Essential oils were dissolved in petroleum ether, then submitted to a partition procedure with acetonitrile (four extractions). Successively, a clean-up procedure on a silica gel column was necessary to purify the extract from the interferents. Finally, the purified samples were subjected to packed column gas chromatography with three different stationary phases and an alkaline flame ionization detector (AFID). The gas chromatographic conditions used are shown in Table 17.1. Three bergamot oils were analyzed under these conditions. All were contaminated by malathion; in two samples were recovered parathion-methyl and in one parathion-ethyl.

In 1992 a very simple method was used to analyze different essential citrus oils (see Table 17.1). Samples were directly introduced in a gas chromatograph equipped with a programmed temperature vaporizer (PTV), a 30-m capillary column, and a flame photometric detector (FPD) (Dugo et al. 1992). No extraction and purification was required; no recovery values were calculated; the FPD permitted discrimination between phosphorated and unphosphorated compounds because the latter gave negative peaks during the chromatographic runs. This method was applied to the analysis of nine samples of bergamot oils; all were contaminated by mevinphos and almost all by parathion-methyl, parathion-ethyl, and quinalphos (see Table 17.1); sulphotep, diazinon, malathion, and methidathion were recovered in a few samples. A further improvement in analysis was introduced with the use of a short (5 m) capillary column instead of a conventional-length one (Dugo et al. 1994). The gas chromatographic runs became very short, and 129 samples of bergamot essential oils were easily analyzed employing PTV and FPD. Figure 17.1 shows a typical FPD chromatogram obtained from the direct analysis of a bergamot oil. In 61 samples, organophosphorus pesticide residues were detected: methidathion was recovered in 33 samples, parathion-methyl in 18, and parathion-ethyl in 20 oils (see Table 17.1). In a few samples quinalphos was recovered, and in one sample fenitrothion and azinphos-ethyl were detected.

In a subsequent investigation, 26 bergamot essential oils produced in 1999 were analyzed using a GC equipped with PTV, FPD, and a 25-m capillary column (Di Bella et al. 2001a). In all these samples, the pesticide residues were below the detection limits (see Table 17.1). The same experimental conditions were adopted in the

TABLE 17.1

Organophosphorus Pesticide Residues in Bergamot Essential Oils

Number of Samples	Production Year	Pesticides (mg/L) Compound, Range and Mean[a]	Contaminated Oils (%)	Instrumental Conditions Adopted	Reference
3	Unknown	Parathion-Methyl n.d.–0.76 (0.31) Parathion-Ethyl n.d.–1.00 (0.33) Malathion 1.98–8.60 (4.27)	100	Col. 2 m × 4 mm, 3% OV-71; oven 198°C, AFID-ECD. Col. 1.8 × 4 mm, 3.8% SE-30; oven 175°C, AFID-ECD. Col. 1.8 × 4 mm, 2% DEGS; oven 195°C, AFID-ECD	Leoni and d'Alessandro De Luca 1978
9	1990–1991	Mevinphos 0.01–0.89 (0.46) Parathion-Methyl n.d.–19.57 (2.71) Parathion-Ethyl n.d.–6.83 (1.48) Quinalphos n.d.–4.58 (0.63)	100	Col. 30 m × 0.25 mm, SPB-5 0.25 μm; oven 75°C–170°C (30°C/min), hold 5', to 190°C (2°C/min), to 265°C (5°C/min). Inj. PTV 65°C–240°C, FPD 250°C	Dugo et al. 1992
129	1992	Parathion-Methyl n.d.–17.85 (0.21) Parathion-Ethyl n.d.–2.23 (0.05) Methidathion n.d.–19.44 (0.89)	47.3	Col. 5 m × 0.25 mm, SE-54 0.25 μm; oven 75°C–140°C (30°C/min), to 245°C (5°C/min). Inj. PTV 65°C–240°C, FPD 250°C	Dugo et al. 1994
26	1999	Residues below the detection limits	0	Col. 25 m × 0.32 mm, Mega 68 0.45 μm; oven 75°C (5') to 100°C (7.5°C/min) to 170°C (5') (2°C/min) to 250°C (10') (10°C/min). Inj. PTV 65°C–240°C, FPD 250°C	Di Bella et al. 2001
29	2000	Residues below the detection limits	0	Col. 25 m × 0.32 mm, Mega 68 0.45 μm; oven 75°C (5') to 100°C (7.5°C/min), to 170°C (5') (2°C/min) to 250°C (10') (10°C/min). Inj. PTV 65°C–240°C, FPD 250°C	Di Bella et al. 2004
30	2006–2007	Residues below the detection limits	0	Col. 30 m × 0.25 mm, RTX-5MS 0.25 μm; oven 100°C to 170°C (10')(25°C/min) to 195°C (10°C/min) to 280°C (5') (7°C/min). Inj. 260°C, FPD 260°C	Di Bella et al. 2009

[a] Calculated on all samples.

n.d.: not detected.

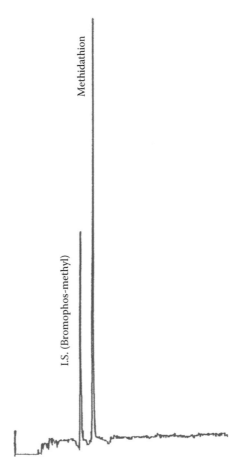

FIGURE 17.1 Chromatogram of organophosphorus pesticide residues in a bergamot sample carried out with a flame photometric detector. Negative peaks are due to unphosphorated compounds.

analytical determination of the organophosphorus pesticide residues in 29 bergamot essential oils produced in the year 2000 (Di Bella et al. 2004). Again, no pesticide residues were detected in these samples (see Table 17.1).

In the most recent research, performed on 30 bergamot oils produced in the years 2006–2007 (Di Bella et al. 2009), organophosphorus pesticide residues were always below the detection limits, confirming the trend observed in other earlier studies.

17.2.2 ORGANOCHLORINE PESTICIDES

Organophosphorus pesticides are not the only insecticides used on bergamot trees. Research about essential oil contamination led to the presence of organochlorine pesticides. These pesticides work by opening the sodium channels in the nerve cells of the insect. Their main problem is caused by the process of bioaccumulation wherein the chemicals, because of their stability and fat solubility, accumulate in organisms' fatty tissues. Also, organochlorine pesticides may biomagnify, which causes progressively higher concentrations in the body fat of animals farther up the food chain.

The analytical procedure proposed by Saitta et al. (2000) was adopted in all the organochlorine pesticide analyses carried out on bergamot essential oils. This procedure consists of oil clean-up on a silica gel column using dichloromethane as eluant and a subsequent GC analysis with an electron capture detector (ECD). The gas chromatographic conditions adopted are shown in Table 17.2. The study of organochlorine pesticide contamination in bergamot essential oils from 1991 to 1996 demonstrated the presence of the insecticides tetradifon, dicofol, and its decomposition product, 4,4'-dichlorobenzophenone (Saitta et al. 2000). The oil contamination percentage varied from 26.3% (1996) to 52.6% (1992), but the insecticide levels were not high, with maximum values of 1.18 mg/L for dicofol in 1995 and 0.57 mg/L for tetradifon in 1992 (see Table 17.2).

Similar results were found in two studies on bergamot oils produced in 1999 (Di Bella et al. 2001a) and 2000 (Di Bella et al. 2004); the oil contamination percentage was 58.0% in 1999 and 24.3% in 2000, and the maximum levels of dicofol and tetradifon were 0.58 mg/L and 0.10 mg/L in 1999, and 0.43 mg/L and 0.11 mg/L in 2000, respectively (see Table 17.2). Some years later, further research demonstrated the absence of organochlorine pesticide residues in 30 bergamot oils produced in 2006–2007 (Di Bella et al. 2009).

17.2.3 ORGANOPHOSPHORUS PLASTICIZERS

Contamination due to the presence of plasticizers is essentially caused by essential oil contact with plastic parts: the extraction of organic compounds like plasticizers is inevitable. It was demonstrated that a new plastic part (containers, pipes, etc.) releases large amounts of plasticizers, and these quantities gradually decrease to zero over time (Di Bella et al. 2001b). This is the basis of the extremely varied and inconsistent presence of plasticizers in essential oils.

The first class of plasticizers found in citrus essential oils was triarylphosphates (Saitta et al. 1997); these compounds were accidentally discovered during the organophosphorus pesticide analyses carried out with a flame photometric detector (Table 17.3). The high peak numbers (they include all the possible isomers of diphenyltolylphosphates, phenylditolylphosphates, tritolylphosphates, ditolylxylilphosphates, tolyldixylilphosphates, and trixylilphosphates), did not have the appearance of pesticide contamination. Figure 17.2 shows a chromatogram demonstrating the complex composition of a triarylphosphate industrial mixture. Triarylphosphates, with a maximum value of 2.01 mg/L (Saitta et al. 1997), were found in two samples of bergamot oils produced in 1995. This was the only time that triarylphosphates were found in bergamot oils; in 26 samples produced in 1999 (Di Bella et al. 2001a), 29 samples produced in 2000 (Di Bella et al. 2004), and 30 samples produced in 2006–2007 (Di Bella et al. 2009) these plasticizers were never found (Table 17.3). It is important to note that triarylphosphates are used in only a small number of plastic materials, such as PVC.

TABLE 17.2
Organochlorine Pesticide Residues in Bergamot Essential Oils

Number of Samples	Production Year	Pesticides (mg/L) Compound, Range and Mean[a]	Contaminated Oils (%)	Instrumental Conditions Adopted	Reference
20	1991	Dicofol n.d.–0.81 (0.20) Tetradifon n.d.–0.52 (0.08)	50.0	Col. 30 m × 0.32 mm, RTX-5 0.25 μm and 30 m × 0.32 mm, RTX-1701 0.25 μm; oven temp. 50°C (2') to 150°C (25°C/min) and to 270°C (4°C/min). Inj. 230°C, ECD 280°C	Saitta et al. 2000
19	1992	Dicofol n.d.–0.96 (0.24) Tetradifon n.d.–0.57 (0.09)	52.6	Col. 30 m × 0.32 mm, RTX-5 0.25 μm and 30 m × 0.32 mm, RTX-1701 0.25 μm; oven temp. 50°C (2') to 150°C (25°C/min) and to 270°C (4°C/min). Inj. 230°C, ECD 280°C	Saitta et al. 2000
19	1993	Dicofol n.d.–0.56 (0.10) Tetradifon n.d.–0.31 (0.04)	31.6	Col. 30 m × 0.32 mm, RTX-5 0.25 μm and 30 m × 0.32 mm, RTX-1701 0.25 μm; oven temp. 50°C (2') to 150°C (25°C/min) and to 270°C (4°C/min). Inj. 230°C, ECD 280°C	Saitta et al. 2000
20	1994	Dicofol n.d.–0.24 (0.04) Tetradifon n.d.–0.12 (0.01)	35.0	Col. 30 m × 0.32 mm, RTX-5 0.25 μm and 30 m × 0.32 mm, RTX-1701 0.25 μm; oven temp. 50°C (2') to 150°C (25°C/min) and to 270°C (4°C/min). Inj. 230°C, ECD 280°C	Saitta et al. 2000
20	1995	Dicofol n.d.–1.18 (0.11) Tetradifon n.d.–0.12 (0.01)	40.0	Col. 30 m × 0.32 mm, RTX-5 0.25 μm and 30 m × 0.32 mm, RTX-1701 0.25 μm; oven temp. 50°C (2') to 150°C (25°C/min) and to 270°C (4°C/min). Inj. 230°C, ECD 280°C	Saitta et al. 2000
19	1996	Dicofol n.d.–0.42 (0.05) Tetradifon n.d.–0.32 (0.03)	26.3	Col. 30 m × 0.32 mm, RTX-5 0.25 μm and 30 m × 0.32 mm, RTX-1701 0.25 μm; oven temp. 50°C (2') to 150°C (25°C/min) and to 270°C (4°C/min). Inj. 230°C, ECD 280°C	Saitta et al. 2000
26	1999	Dicofol n.d.–0.58 (0.05) Tetradifon n.d.–0.10 (0.02)	58.0	Col. 30 m × 0.32 mm, RTX-5 0.25 μm; oven temp. 150°C–230°C (2°C/min) and to 280°C (10°C/min). Inj. 250°C, ECD 280°C	Di Bella et al. 2001
29	2000	Dicofol n.d.–0.43 (0.07) Tetradifon n.d.–0.11 (0.01)	24.3	Col. 30 m × 0.32 mm, RTX-5 0.25 μm and 30 m × 0.32 mm, RTX-1701 0.25 μm; oven temp. 50°C (2') to 150°C (25°C/min) and to 270°C (4°C/min). Inj. 230°C, ECD 280°C	Di Bella et al. 2004
30	2006–2007	Residues below the detection limits	0	Col. 30 m × 0.25 mm, RTX-5MS 0.25 μm; oven temp. 150°C to 300°C (10') (4°C/min). Inj. 350°C, ECD 350°C	Di Bella et al. 2009

[a] Calculated on all samples.

n.d.: not detected.

TABLE 17.3
Organophosphorus Plasticizer Residues in Bergamot Essential Oils

Number of Samples	Production Year	Triarylphosphates (mg/L) Range and Mean[a]	Contaminated Oils (%)	Instrumental Conditions Adopted	Reference
4	1995	n.d.–2.01 (0.55)	50	Col. 25 m × 0.32 mm, Mega 68 0.45 µm; oven temp. 75°C (5') to 100°C (7.5°C/min) to 170°C (5') (2°C/min) to 250°C (10') (10°C/min). Inj. PTV 65°C–240°C, FPD 250°C	Saitta et al. 1997
26	1999	Residues below the detection limits	0	Col. 25 m × 0.32 mm, Mega 68 0.45 µm; oven temp. 75°C (5') to 100°C (7.5°C/min), to 170°C (5') (2°C/min), to 250°C (10') (10°C/min). Inj. PTV 65°C–240°C, FPD 250°C	Di Bella et al. 2001
29	2000	Residues below the detection limits	0	Col. 25 m × 0.32 mm, Mega 68 0.45 µm; oven temp. 75°C (5') to 100°C (7.5°C/min), to 170°C (5') (2°C/min), to 250°C (10') (10°C/min). Inj. PTV 65°C–240°C, FPD 250°C	Di Bella et al. 2004
30	2006–2007	Residues below the detection limits	0	Col. 30 m × 0.25 mm, RTX-5MS 0.25 µm; oven temp. 100°C–170°C (10')(25°C/min), to 195°C (10°C/min), to 280°C (5')(7°C/min). Inj. 260°C, FPD 260°C	Di Bella et al. 2009

[a] Calculated on all samples.
n.d.: not detected.

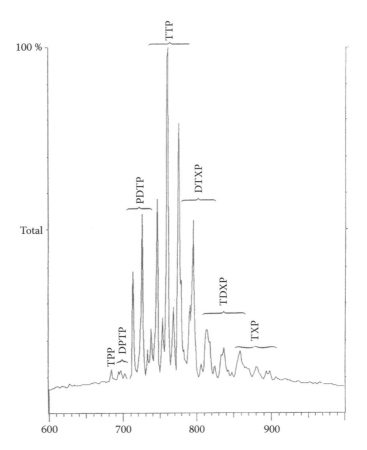

FIGURE 17.2 Chromatogram of a triarylphosphate mixture showing the different classes of the compounds. DPTP, diphenyltolylphosphates; DTXP, ditolylxylilphosphates; PDTP, phenylditolylphosphates; TDXP, tolyldixylilphosphates; TPP, triphenylphosphate; TTP, tritolylphosphates; TXP, trixylilphosphates.

17.2.4 ORGANOCHLORINE PLASTICIZERS

Having established the presence of phosphorated plasticizers in citrus essential oils, other types of plasticizers could undoubtedly contaminate bergamot oils. A class of plasticizers known as chloroparaffins consists of industrial mixtures of chloro-alkanes (C10–C20) with 40%–70% chlorine content. A typical chromatogram of standard chloroparaffins is shown in Figure 17.3.

Chloroparaffins could be analyzed employing the same method used for organochlorine pesticides (Di Bella et al. 2000); research conducted on 26 samples of bergamot oils (see Table 17.4) did not show chloroparaffin residues (Di Bella et al. 2001a). In a subsequent investigation regarding 29 bergamot oils produced in 2000 (see Table 17.4), the residues were below the detection limits (Di Bella et al. 2004). Chloroparaffin residues were detected in only 1 of the 30 samples produced in the years 2006–2007, confirming the randomness of this kind of contamination (Di Bella et al. 2009).

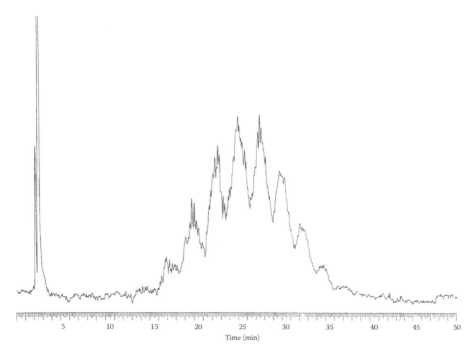

FIGURE 17.3 Chromatogram of standard chloroparaffins. The mixture is complex, with hundreds of compounds.

17.2.5 OTHER PLASTICIZERS: PHTHALATE ESTERS

Phthalate esters are the most commonly used class of plasticizers. The presence of phthalates in citrus essential oils has been known since 1999 (Di Bella et al. 1999); this analysis could not be carried out employing detectors like ECD or FPD because phthalate molecules contain only carbon, hydrogen, and oxygen. To avoid complex extraction and purification procedures, a simple method based on the use of a mass spectrometer (MS) as a GC detector, working with selected ion monitoring (SIM), permitted the injection of pure essential oil without any pretreatment. In the literature, there are only two phthalate residue analyses on bergamot oils. The first reviewed 26 samples produced in 1999 (Di Bella et al. 2001a); the second covered 29 samples produced in 2000 (Di Bella et al. 2004). In the first investigation, 76.9% of the oils were contaminated: diisobutyl phthalate reached 2.35 mg/L, di-*n*-butyl phthalate 4.45 mg/L, and bis-2-ethyl-hexyl phthalate 3.08 mg/L (see Table 17.5). In the second investigation, the results were similar: 66.2% of the oils were contaminated by diisobutyl phthalate with a maximum value of 2.40 mg/L, di-*n*-butyl phthalate with a maximum value of 3.20 mg/L, and bis-2-ethyl-hexyl phthalate with a maximum value of 2.90 mg/L (Table 17.5). In Figure 17.4 the chromatogram of a bergamot oil sample clearly shows the presence of diisobutyl phthalate and bis-2-ethyl-hexyl phthalate.

TABLE 17.4

Organochlorine Plasticizer Residues in Bergamot Essential Oils

Number of Samples	Production Year	Chloroparaffins (mg/L) Range and Mean[a]	Contaminated Oils (%)	Instrumental Conditions Adopted	Reference
26	1999	Residues below the detection limits	0	Col. 30 m × 0.32 mm, RTX-5 0.25 μm; oven temp. 150°C–230°C (2°C/min) and to 280°C (10°C/min). Inj. 250°C, ECD 280°C	Di Bella et al. 2001
29	2000	Residues below the detection limits	0	Col. 30 m × 0.32 mm, RTX-5 0.25 μm and 30 m × 0.32 mm, RTX-1701 0.25 μm; oven temp. 50°C (2') to 150°C (25°C/min) and to 270°C (4°C/min). Inj. 230°C, ECD 280°C	Di Bella et al. 2004
30	2006–2007	n.d.–5.0	3.3	Col. 30 m × 0.25 mm, RTX-5MS 0.25 μm; oven temp. 150°C–300°C (10') (4°C/min). Inj. 350°C, ECD 350°C	Di Bella et al. 2009

[a] Calculated on all samples.
n.d.: not detected.

17.2.6 POLYCHLORINATED DIBENZO-*P*-DIOXINS, POLYCHLORINATED DIBENZO-*P*-FURANS, AND POLYCHLORINATED BYPHENYLS

Polychlorinated byphenyls (PCBs) were widely used as dielectric and coolant fluids in transformers, capacitors, and electric motors. Due to PCBs' toxicity and classification as a persistent organic pollutant, many countries banned PCB production. Concerns about the toxicity of PCBs are largely based on compounds within this group that share a structural similarity and toxic mode of action with dioxin. Toxic effects such as endocrine disruption and neurotoxicity are also associated with other compounds within the group; polychlorinated dibenzo-*p*-dioxins (PCDDs) and polychlorinated dibenzo-*p*-furans (PCDFs) are well-known organic environmental contaminants, very stable and toxic. They are produced in some uncontrolled combustion processes,

TABLE 17.5
Phthalates Residues in Bergamot Essential Oils

Number of Samples	Production Year	Phthalates (mg/L) Compound, Range and Mean[a]	Contaminated Oils (%)	Instrumental Conditions Adopted	Reference
26	1999	DiBP n.d.–2.35 (0.56) DnBP n.d.–4.45 (0.41) Bis2EtHexP n.d.–3.08 (0.42)	76.9	Col. 30 m × 0.25 mm, RTX-5 0.25 µm; oven temp. 60°C–280°C (10′)(6°C/min). Inj. 250°C, X-line 280°C, MS EI 70 eV SIM	Di Bella et al. 2001
29	2000	DiBP n.d.–2.40 (0.51) DnBP n.d.–3.20 (0.33) Bis2EtHexP n.d.–2.90 (0.56)	66.2	Col. 30 m × 0.25 mm, DB-5MS 0.25 µm; oven temp. 60°C–275°C (14′) (15°C/min). Inj. 250°C, X-line 275°C, MS EI 70 eV SIM	Di Bella et al. 2004

[a] Calculated on all samples.
n.d.: not detected.

especially from municipal and industrial waste. PCDDs, PCDFs, and PCBs, dispersed as vapor and particulates into the atmosphere, fall down, contaminating soil and water. Analyses of these contaminants require the use of expensive, state-of-the-art instrumentation such as high resolution mass spectrometry (HRMS). Only one investigation was carried out on the presence of PCDDs, PCDFs, and PCBs in bergamot essential oils (Cautela et al. 2007); the analytical method used consists of three steps: extraction, clean-up, and quantification by GC-HRMS (Table 17.6). Fifteen bergamot oils showed high contamination percentages (86.7% for PCDDs and PCDFs, 100% for PCBs); Table 17.6 displays the levels of the single congeners detected. PCB contamination was quantitatively more important than PCDDs and PCDFs, but from a toxicologic point of view, the PCDD concentration was more relevant. The most abundant PCDD congener was the nontoxic Octachloro Dibenzo-*p*-Dioxin, thus highlighting a contamination due to a widespread pollution, typical of brush fires.

17.3 INORGANIC CONTAMINANTS

Many factors affect the presence of trace elements in citrus essential oils—nature of the soils, climatic conditions, genotype of the plant, agronomic techniques, storage, and extraction procedures (Bruno et al. 1978; Chiricosta et al. 1978; Di Giacomo 1988). Some important micronutrients (Cu, Mn, Se, Zn) are contained in citrus fruits, but knowing their quantities is critical because they are essential at low

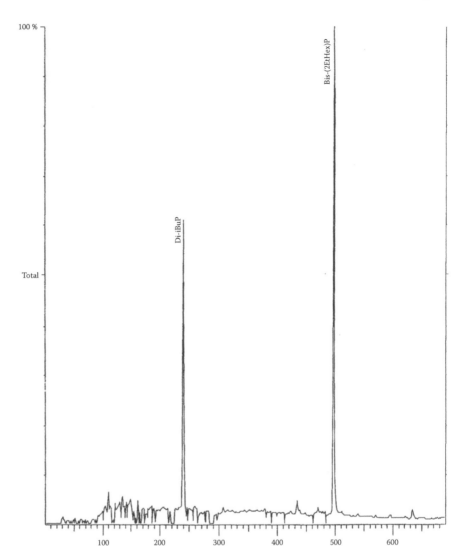

FIGURE 17.4 Chromatogram of phthalate residues in a bergamot sample (Di-iBuP diisobu-tyl phthalate, Bis-(2EtHex)P bis-(2-ethylhexyl) phthalate).

concentrations and become toxic at high levels (McLaughlin et al. 1999; Nielsen 1999). Other elements, such as Cd, Ni, and Pb, are potentially toxic, but their presence in essential oils is due only to contamination processes (Bruno et al. 1978). Since bergamot essential oils are used in very small quantities as flavorings for food, beverages, and cosmetics, it is not easy to estimate the metals' daily intake due to the ingestion through alimentary products or contact with the skin through the use of cosmetic products. This type of contamination is basically due to the contact between essential oils and metal surfaces, washing waters, or metal containers; contamination with metals like Fe, Cr, and Ni is inevitable (Kumar et al. 1994).

TABLE 17.6
PCDD, PCDF, and PCB residues in Bergamot Essential Oils

Number of Samples	Production Year	Concentrations (pg/g) Compound, Range and Mean[a]	Contaminated Oils (%)	Instrumental Conditions Adopted	Reference
15	2003–2005	2,3,7,8-TCDD n.d.–3.25 (0.61) 1,2,3,7,8-PeCDD n.d.–3.50 (0.42) 1,2,3,4,7,8-HxCDD n.d.–11.46 (1.50) 1,2,3,6,7,8-HxCDD n.d.–2.96 (0.79) 1,2,3,7,8,9-HxCDD n.d.–4.56 (0.76) 1,2,3,4,6,7,8-HpCDD n.d.–3.96 (1.10) OCDD n.d.–25.32 (5.61)	86.7	Col. 60 m × 0.25 mm, DB-5MS 0.25 μm; oven temp. 140°C (4') to 220°C (10°C/min), to 260°C (2.9') (28.5°C/min), to 310°C (12.5°C/min). Inj. 280°C, MS EI 35 eV SIM	Cautela et al. 2007
15	2003–2005	2,3,7,8-TCDF n.d.–2.35 (0.57) 1,2,3,7,8-PeCDF n.d.–2.66 (0.49) 2,3,4,7,8-PeCDF n.d.–2.10 (0.22) 1,2,3,4,7,8-HxCDF n.d.–2.90 (0.50) 1,2,3,6,7,8-HxCDF n.d.–2.60 (0.40) 2,3,4,6,7,8-HxCDF n.d.–3.96 (0.59) 1,2,3,7,8,9-HxCDF n.d.–2.86 (0.49) 1,2,3,4,6,7,8-HpCDF n.d.–15.20 (3.55) 1,2,3,4,7,8,9-HpCDF n.d.–2.33 (0.64) OCDF n.d.–5.20 (1.54)	86.7	Col. 60 m × 0.25 mm, DB-5MS 0.25 μm; oven temp. 140°C (4') to 220°C (10°C/min), to 260°C (2.9') (28.5°C/min), to 310°C (12.5°C/min). Inj. 280°C, MS EI 35 eV SIM	Cautela et al. 2007

continued

TABLE 17.6 (CONTINUED)
PCDD, PCDF, and PCB residues in Bergamot Essential Oils

Number of Samples	Production Year	Concentrations (pg/g) Compound, Range and Mean[a]	Contaminated Oils (%)	Instrumental Conditions Adopted	Reference
15	2003–2005	PCB-126 n.d.–182.25 (13.56)	100	Col. 60 m × 0.25 mm, DB-5MS 0.25 μm; oven temp. 140°C (1′) to 280°C (10′) (20°C/min), to 300°C (1.2′) (25°C/min). Inj. 280°C, MS EI 35 eV SIM	Cautela et al. 2007
		PCB-169 n.d.–0.56 (0.05)			
		PCB-114 n.d.–268.55 (44.06)			
		PCB-156 n.d.–10.26 (4.03)			
		PCB-157 n.d.–12.33 (1.66)			
		PCB-81 n.d.–3.20 (1.34)			
		PCB-77 n.d.–72.66 (22.05)			
		PCB-123 n.d.–61.29 (13.00)			
		PCB-118 n.d.–126.54 (40.70)			
		PCB-105 n.d.–265.52 (106.53)			
		PCB-189 n.d.			
		PCB-167 n.d.–16.58 (2.34)			

[a] Calculated on all samples.

n.d.: not detected.

The first determination of trace elements in a sample of bergamot oil was carried out with an acid extraction followed by a purification using the derivative potentiometric stripping analysis (dPSA) (La Pera et al. 2003); the analysis was performed to quantify at the same time Cd, Cu, Pb, and Zn (see Table 17.7). A more significant sampling relates to the years 1999–2000 (La Pera et al. 2005); the same analytical procedure was adopted, but in addition to the simultaneous determination of Cd, Cu, Pb, and Zn by dPSA, Mn (by dPSA), Se (by derivative cathodic stripping chronopotentiometry, dCSCP), and Ni (by derivative adsorptive stripping chronopotentiometry, dAdSCP) were also determined (see Table 17.7).

TABLE 17.7
Elements in Bergamot Essential Oils

Number of Samples	Production Year	Elements (ng/g) Range and Mean[a]	Instruments Used	Reference
1	1999–2000	Cd 22.5 Cu 375.8 Pb 75.6 Zn 785.1	dPSA	La Pera et al. 2003
26	1999	Cd n.d. Cu 75.7–440 (150) Mn 960–1550 (1178) Ni 225–380 (281) Pb 40.0–346.9 (147) Se n.d.–29.2 (12) Zn 110–807 (323)	dPSA for Cd, Cu, Pb, Zn dPSA for Mn dCSCP for Se dAdSCP for Ni	La Pera et al. 2005
29	2000	Cd n.d. Cu 90.0–310 (171) Mn 935–1440 (1187) Ni 216–401 (313) Pb 59.2282.3 (128) Se n.d.–23.7 (9.9) Zn 120–800 (307)	dPSA for Cd, Cu, Pb, Zn dPSA for Mn dCSCP for Se dAdSCP for Ni	La Pera et al. 2005
4	2003	Cd n.d. Cu 110–203 Mn 1021–1950 Ni 175–285 Pb 35.5–107 Se n.d. Zn 815–1050	dPSA for Cd, Cu, Pb, Zn dPSA for Mn dCSCP for Se dAdSCP for Ni	La Pera et al. 2005
4	2004	Cd n.d. Cu 77.2–198 Mn 255–512 Ni 60–100 Pb 26.9–102 Se n.d. Zn 650–1026	dPSA for Cd, Cu, Pb, Zn dPSA for Mn dCSCP for Se dAdSCP for Ni	La Pera et al. 2005

continued

TABLE 17.7 (CONTINUED)
Elements in Bergamot Essential Oils

Number of Samples	Production Year	Elements (ng/g) Range and Mean[a]	Instruments Used	Reference
30	2005–2006	Ag n.d.–8 (4)	ICP-OES;	Cautela et al.
		Al 499–1500 (955)	GFAAS for Cd and Pb;	2006
		Ba n.d.–362 (104)	FI-M/H-AAS for Hg	
		Be n.d.		
		Cd n.d.–8 (4)		
		Co n.d.–9		
		Cr n.d.–25 (14)		
		Cu 12–42 (29)		
		Fe 163–374 (269)		
		Hg n.d.		
		Mn n.d.–329 (88)		
		Ni n.d.–33 (17)		
		Pb n.d.–160 (46)		
		Sb n.d.–187 (63)		
		Sn n.d.–1520 (582)		
		Zn n.d.–1140 (418)		
15	2003–2005	Ag <0.1–8.2 (4.1)	GFAAS ; FI-M/H-AAS	Cautela et al.
		Al 499.4–1498.2 (955.7)	for Hg only	2007
		As <0.6–18.8 (4.3)		
		Ba <1–362 (104)		
		Be 0.04–2.21 (0.83)		
		Cd <0.02–8.37 (4.42)		
		Co <0.4–9.7 (5.3)		
		Cr <0.1–23.4 (14.1)		
		Cu 12.1–42.8 (29.3)		
		Fe 163–374 (269)		
		Hg <0.2		
		Mn <0.1–429.5 (88.2)		
		Ni <0.8–33 (17.2)		
		Pb <0.2–160 (46.1)		
		Sb <0.5–187 (63.3)		
		Sn <2–1520 (582)		
		Zn <0.3–1140 (418.4)		

[a] Calculated on all samples.
n.d.: not detected.

In the same paper, a few biological samples produced in the years 2003–2004 were analyzed. Results obtained showed the absence of cadmium from all samples, while manganese was found in up to a maximum value of 1950 ng/g. Selenium was found in low amounts in the 1999 and 2000 samples. Copper, nickel, lead, and zinc varied in amount; in the 2003 and 2004 samples, the zinc content was much higher than in the 1999 and 2000 samples (see Table 17.7). These results led the authors to conclude that this contamination mainly depended on environmental processes (soil contamination, air pollution) and food processing. Another paper on the mineral components of bergamot oils reports the use of inductively coupled plasma–optical emission spectrometry (ICP-OES) for 13 elements, graphite furnace atomic absorption spectrometry (GFAAS) for Cd and Pb, and flow injection analysis-mercury/hydride-atomic absorption spectrometry (FI-M/H-AAS) for mercury (Cautela et al. 2006). Samples were wet digested before the analyses (with the exception of samples for analysis of mercury); the results, obtained on 30 samples produced in the years 2005–2006, are reported in Table 17.7. The last research, carried out by GFAAS for 16 elements and FI-M/H-AAS for mercury, analyzed 15 samples of bergamot oil produced in the years 2003–2005 (Cautela et al. 2007). Although sampling and instrumentation were clearly different, for most of the elements analyzed the results are almost identical to those of previous research.

17.4 CONCLUSION

It is evident that the levels of contamination found to date are not a major concern. Organophosphorus and organochlorine pesticide residues are below the detection limits; organophosphorus and organochlorine plasticizer residues are rarely found; heavy metals, PCBs, PCDDs, and PCDFs depend on environmental factors and their levels are not dangerous; contamination by phthalates is widespread, but does not reach high levels. In the latter case, to eliminate this type of contamination, it would be sufficient to use phthalate-free plastics.

REFERENCES

Bruno, E., Chiricosta, S., and Licandro, G. 1978. Influenza del tipo di contenitore sulla concentrazione di metalli pesanti negli oli essenziali agrumari. *Essenz. Deriv. Agrum.* 48:265–281.

Cautela, D., Boscaino, F., Pirrello, A. G., and Castaldo, D. 2006. Determinazione dei principali componenti minerali presenti nell'olio essenziale di bergamotto "cold pressed." *Essenz. Deriv. Agrum.* 76:19–23.

Cautela, D., Castaldo, D., Santelli, F., et al. 2007. Survey of polychlorinated dibenzo-p-dioxins (PCDDs), polychlorinated bibenzo-p-furans (PCDFs), polychlorinated biphenyls (PCBs), and mineral components in Italian citrus cold-pressed essential oils. *J. Agric. Food Chem.* 55:1627–1637.

Chiricosta, S., Calabrò, G., and Bruno, E. 1978. Determinazione dei metalli presenti nell'olio essenziale di limone. *Essenz. Deriv. Agrum.* 48:107–120.

Di Bella, G., Saitta, M., Pellegrino, M. C., Salvo, F., and Dugo, G. 1999. Contamination of Italian citrus essential oils: Presence of phthalate esters. *J. Agric. Food Chem.* 47:1009–1012.

Di Bella, G., Saitta, M., Lo Curto, S., Visco, A., and Dugo, G. 2000. Contamination of citrus essential oils: The presence of chloroparaffin. *J. Agric. Food Chem.* 48:4460–4462.

Di Bella, G., Luppino, R., Faranda, et al. 2001a. Contaminazione da prodotti fitosanitari e plastificanti in oli essenziali di bergamotto. *Essenz. Deriv. Agrum.* 71:133–146.

Di Bella, G., Saitta, M., Lo Curto, S., et al. 2001b. Production process contamination of citrus essential oils by plastic materials. *J. Agric. Food Chem.* 49:3705–3708.

Di Bella, G., Saitta, M., La Pera, L., Alfa, M., and Dugo, G. 2004. Pesticide and plasticizer residues in bergamot essential oils from Calabria (Italy). *Chemosphere* 56:777–782.

Di Bella, G., Lo Turco, V., Rando, R., et al. 2009. Pesticide and plasticizer residues in citrus essential oils from different countries. *Nat. Prod. Comm.* 4:1–6.

Di Giacomo, A., 1988. Tecnologia dei prodotti agrumari, part 2, duplicated lecture notes; University of Reggio Calabria, 95–120.

Dugo, G., Famà, G., Saitta, M., and Stagno d'Alcontres, I. 1992. Sulla genuinità delle essenze agrumarie. Nota XLII. Determinazione di pesticidi organofosforici con rivelatore a fiamma fotometrica. *Essenz. Deriv. Agrum.* 62:127–146.

Dugo, G., Di Bella, G., Saitta, M., and Salvo, F., 1994. Sulla genuinità delle essenze agrumarie. Nota XLV. Dosaggio rapido di pesticidi organofosforici negli oli essenziali di bergamotto. *Essenz. Deriv. Agrum.* 64:234–247.

Kumar, R., Srivastava, P. K., and Srivastava, S. P. 1994. Leaching of heavy metals (Cr, Fe, and Ni) from stainless steel utensils in food simulants and food materials. *Bull. Environ. Contam. Toxicol.* 53:259–266.

La Pera, L., Saitta, M., Di Bella, G., and Dugo, G, 2003. Simultaneous determination of Cd(II), Cu(II), Pb(II), and Zn(II) in citrus essential oils by derivative potentiometric stripping analysis. *J. Agric. Food Chem.* 51:1125–1129.

La Pera, L., Lo Curto, R., Di Bella, G., and Dugo, G. 2005. Determination of some heavy metals and selenium in Sicilian and Calabrian citrus essential oils using derivative stripping chronopotentiometry. *J. Agric. Food Chem.* 53:5084–5088.

Leoni, V., and D'Alessandro De Luca, E., 1978. An important aspect of the health problem caused by pesticides: The presence of organophosphate insecticide residues in essential oils. *Essenz. Deriv. Agrum.* 48:39–50.

McLaughlin, M. J., Parker, D. R., and Clarke, J. M. 1999. Metals and micronutrients—food safety issues. *Field Crop Res.* 60:143–163.

Nielsen, F. H., 1999. Other trace elements. In *Modern Nutrition in Health and Disease*, eds. M. E. Shills, J. A. Olson, M. Shike, and C. A. Ross, 283–304. Baltimore, MD: Williams & Wilkins.

Saitta, M., Di Bella, G., Bonaccorsi, I., Dugo, G., and Dellacassa, E. 1997. Contamination of citrus essential oils: The presence of phosphorated plasticizers. *J. Essent. Oil Res.* 9:613–618.

Saitta, M., Di Bella, G., Dugo, G., Salvo, F., and Lo Curto, S. 2000. Organochlorine pesticide residues in Italian citrus essential oils, 1991–1996. *J. Agric. Food Chem.* 48:797–801.

18 Bergamot and Its Use in Perfumery

Norbert Bijaoui

CONTENTS

18.1 INTRODUCTION

Bergamot is the fruit of the bergamot tree, which is most common to the Calabrian region of the southern part of Italy. It has been cultivated since the Middle Ages. The tree is probably a hybrid between the lemon (*Citrus limon*) and bigaradier trees (*Citrus aurantium*).

Bergamot oil is very popular with perfumers because of its unique lemon and bitter orange signature. It is widely used in traditional as well as modern perfumery. Bergamot oil is timeless. Because of its high level of linalyl acetate and derivatives like linalool esters, bergamot is perceived as noble and suave as compared to citrus notes such as lemon, lime, and grapefruit, which are more citric in character. Moreover, its traces of methyl anthranilate and furocoumarin underline warm floral and sweet balsamic hay connotations, respectively. The oil blends remarkably well with any type of accord and, despite its characteristic freshness, bergamot oil lasts longer and is more mellow and richer when compared with other citrus notes.

Bergamot has been used for hundreds of years. Around 1740, King Stanislaus of Poland, who was a political refugee, settled in Nancy and introduced this precious oil in France. Around 1770, Paris and London were the centers of perfumery in Europe. Traditional perfumes based on musk became less fashionable and were replaced by fresh citrus scents, using bergamot for its sweet and intense smell, and amber for the dry down.

18.2 EAU DE COLOGNE

Looking back further in bergamot's history, one should first describe its earlier use in "Eau de Cologne." According to legend, it all started in Florence in 1221. In the convent adjacent to the church of Santa Maria Novella, Dominican monks were running a pharmacy. In the garden of this convent, plants and herbs were cultivated for internal consumption. Old recipes made of citrus and plants were prepared by the monks. Among them, the most famous was "Queen's water" or "Aqua da Regina," a concoction based on citrus, mainly bergamot, which was offered as a gift to Catherine de Medici. When Catherine de Medici moved to France to marry Henry of France (who became Henry II in 1547 after his father's death), she introduced this famous "Aqua da Regina" in the French court. Catherine had such a passion for fragrances that she asked her private perfumer, Renato Bianco, to follow her to Paris, where he opened a shop under the Pont Neuf and changed his name to "René le Florentin." René le Florentin became very successful with high society ladies, particularly in selling perfumed gloves.

In 1695, an Italian food merchant, Paul Féminis (1660–1736; see Figure 18.1), established a distillery business in Cologne and created the first alcoholic fragrant water. He named this water "Aqua Mirabilis," based on wine spirit, melissa, rosemary vinegar, bergamot, cedrat, lemon, and other secret key ingredients. This "Aqua" was famous for its digestive and hepatic values as well as for its analgesic and antiseptic properties.

Before his death, Féminis gave the secret formula to Jean-Antoine Farina, who continued manufacturing the famous "Aqua." In 1709, Jean-Marie Farina, a relative of Jean-Antoine Farina, decided to move to the city of Cologne to open a direct subsidiary selling products from Italy, such as soaps and fragrances, and continued to distribute the famous Aqua Mirabilis. In the same year, he established a new factory called "Johann Maria Farina gegenüber dem Jülichs-Platz."

In 1714, Jean-Marie Farina introduced his own water under the name of "Eau de Cologne" in acknowledgment of his new adoptive country (Figure 18.2). Following

FIGURE 18.1   Jean-Paul Féminis (a) and Jean-Marie Farina (b).

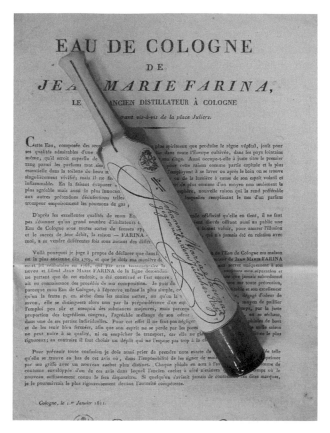

FIGURE 18.2 The bottle created by Johann Maria Farina (1685–1766).

the huge success of this fragrance, he decided to distribute it all over Europe, especially in the Royal Courts.

In 1792 Wilhelm Mühlens, the son of a Cologne banker, acquired the so-called "Aqua Mirabilis formula" through a Carthusian monk. He decided to build a factory to produce the Eau de Cologne 4711, named after the address of the shop in Cologne. In 1806, a member of the Farina family, another Jean-Marie Farina (1785–1864) (see Figure 18.1) opened a perfumery shop in Paris, 333 rue Saint Honoré. In order to validate the authenticity of his formula, he officially declared that Paul Féminis was the great-grandfather of his own grandfather.

Napoléon Bonaparte, who did not like heavy and powerful fragrances, commissioned Jean-Marie Farina to create a light and refreshing fragrance named "Eau Admirable dite de Cologne." Napoléon used it daily because he found this cologne stimulating and invigorating. A special bottle called "le Rouleau de l'Empereur" was designed for him so he could slip it into the side of his riding boots. He liked it so much that used up to 60 bottles per month. Napoleon loved hot baths using his favorite soap imported from England, "Brown Windsor," based on bergamot, petit grain, rosemary, lavender, thyme, and other oils.

Another example of the use of bergamot was the famous "Eau de Lubin," created at the end of the eighteenth century by Pierre François Lubin, a merchant who became a perfumer after being trained in Grasse in 1792 by Tombarelli. He created this cologne using a combination of bergamot, lemon, geranium, neroli, orange Portugal, citronellal, lavender, myrtle, and clove, and using benzoin and tolu balsam, musk, and civet as fixatives. This perfume played a great part in the success of the Lubin house. This type of cologne is still on the market today.

In 1853, Pierre-François-Pascal Guerlain created the "Eau de Cologne Impériale," a classic perfume packed in a cylinder bottle decorated with golden bees, from a formula based on bergamot oil (Figure 18.3).

Modern perfumery started in the late nineteenth century when famous perfumers like as François Coty, Aimé and Jacques Guerlain, Germaine Cellier, and Edmond Roudniska introduced the use of synthetic molecules in their creations. In 1889, continuing his father's work, Aimé Guerlain created an innovative accord for dandies, "Jicky," starting of a new trend, based on bergamot, lavender, orris, coumarin, balsams, and civet. Around 1906, he created "Après l'Ondée," followed by "L'Heure Bleue" in 1912, both containing bergamot as well as many floral and synthetic materials.

In 1925, Jacques Guerlain created Shalimar, one of the most famous Guerlain perfumes, in honor of the Hindu Princess Mumtaz Mahal, for whom was built the most beautiful temple in the world, the Taj Mahal. The Shalimar formula contains more than 30% bergamot. It initiated the so-called "Guerlinade," the new Guerlain fragrance signature. Of course, due to this Guerlinade Accord, bergamot was widely used in most of the fragrances created later by Guerlain.

However, Guerlain was not the only perfumer using bergamot at the beginning of the twentieth century. Other famous perfumers like François Coty, and later Edmond Roudnitska, also used large amounts of this oil. At the time, synthetic molecules such as vanillin, amyl salicylate, coumarin, heliotropin, ionones, and hydroxycitronellal participated in modern perfumery, becoming more popular. Before World War I, new brands like L.T. Piver, Roger & Gallet, Houbigant, and Paul Poiret joined ranks with the already-famous Guerlain and Coty brands. Bergamot oil was used in large quantities in most of the new creations.

18.3   MAIN ACCORDS CONTAINING BERGAMOT

In all the accords that dominated the "great classics," both feminine and masculine, bergamot always played a major role. Either natural or artificial, bergamot oil has been used extensively since the eighteenth century. In the citrus accords (colognes or "eaux fraîches"), or in heavier accords (such as fougere, chypre, or oriental), bergamot is one of the first materials selected by the perfumer to get freshness on the top of his composition, used alone or mixed with other citrus oils such as lemon, mandarin, orange, and grapefruit. Generally, a perfume is built according to a pyramid scheme: top notes, body or heart notes, and dry down notes. Contrary to building a house starting with a foundation, a perfume is generally created starting from the top notes, which are more volatile and play a major role in triggering the purchasing intent. In a second addition, the heart notes constitute the soul of the perfume, and

FIGURE 18.3 **(See color insert.)** The bottle of "Eau de Cologne Imperiale" created by Pierre-François-Pascal Guerlain (1853).

finally, the dry down notes or fixatives will remain after evaporation, representing the olfactive trail and enhancing its sensuality.

After the Eau de Cologne category, let's move now to the great accords created by perfumers, in the same way as musicians use musical accords (major, minor, tuned up, or tuned down) to create their melodies.

18.3.1 Citrus Accord

A citrus accord is obtained with citrus notes: lemon, orange, grapefruit, lime, mandarin, yuzu, and, of course bergamot. The Eau de Cologne is followed by the Eaux

Fraîches; the most popular of which is the Eau Sauvage by Christian Dior. During the 1960s, a new olfactory trend was born. Lancôme, Révillon, Rochas, Clarins, Hermès, Armani, and others served as successors to the Eau Sauvage, launching their own "Eaux Fraîches."

This trend continued in the 1990s with Acua di Bulgari (1992), considered a modern "Eau Fraîche," followed by Calvin Klein's "CK One" (1994), and "Chrome" by Azzaro (1996).

Finally, the 2000s were marked by "Thierry Mugler's Cologne," in which woody and musky ingredients have been added by the perfumer to strengthen the dry down.

In summary, bergamot is omnipresent in this citrus accord family.

18.3.2 Fougère Accord

It must be clear that the Fougère Accord has no relationship with the odor of the corresponding plant (fern in English). Such a name was given to a fragrance created in 1884 by Houbigant, called "Fougère Royale." This wonderful accord is compounded by different naturals and synthetics: bergamot, lavender, geranium, coumarin, oakmoss, and amylsalicylate. Five years later, in 1889, Guerlain was inspired by Houbigant to create its famous "Jicky." This olfactory family is the origin of many successful fragrances, such as "Canoé" by Dana (1935), "Moustache" by Rochas (1949), "Brut for Men" by Fabergé (1964), "Paco Rabanne pour Homme" (1973), "Azzaro pour Homme" (1978), and "Drakkar Noir" by Guy Laroche (1982).

18.3.3 Chypre Accord

As with its cousin Fougere, the origin of the "Chypre" name came from "poudre de chypre," created in the sixteenth century from oakmoss. In 1840, Guerlain was inspired by this material to create its "Eau de Chypre." Then François Coty launched his very successful "Chypre" in 1917, which stimulated Guerlain to launch in 1919 "Mitsouko," a typical fruity chypre thanks to the undecalactone discovery. This was followed by Millot's "Crêpe de Chine" (1925), "Zibeline" by Weil (1928), "Rumeur" by Lanvin (1932), "Femme de Rochas" and "Bandit" from Robert Piguet (1944), "Ma Griffe" by Carven (1946), and many other successful fragrances.

Later, this Chypre family was extended to other subfamilies; floral chypre: "Coriandre" by Jean Couturier (1973); leather chypre: "Cabochard" by Grés (1959); green chypre: "Miss Dior" (1947); and woody-aromatic chypre mainly used for Eaux de Toilette for men like "Halston Z14" (1976) and "Cerruti 1881" (1990).

Using the evernyl, a crystallized raw material reminding oakmoss odor and discovered by Roure Bertrand Dupont, the chypre accord was modernized with Paco Rabanne's creation, "Calandre" (1969).

18.3.4 Oriental Accord

This olfactive family was also initiated by François Coty, with the creation of "L'Ambre Antique" (1905), a new accord containing benzoin and tolu balsam mixed with coumarin and vanillin. In 1925, Guerlain brought freshness to this accord by adding an overdose of bergamot to create Shalimar, which was a tremendous success and continues to be to this day. In this important oriental family, we can also mention "Habanita" by Molinard (1924), "Tabu" by Dana (1931), "Must du Soir" by Cartier (1981), "Obsession" by Calvin Klein (1985), and "Dior Addict" (2002).

Combining this ambery accord with floral and spicy notes, Coty created "L'Origan" in 1905. The following year, Jacques Guerlain launched "Après l'Ondée," followed by "L'Heure Bleue" in 1912, both containing a fresh bergamot top note. This trend was further illustrated by "Bal à Versailles" from Jean Desprez (1962), "Oscar de la Renta" (1977), and "Insolence" by Guerlain (2006).

18.4 CONCLUSION

The list of all perfumes containing bergamot is so long that it would be much quicker to list the fragrances that do not contain this ingredient. Bergamot plays a crucial role in top notes, combining perfectly with spicy, floral, aldehydic, chypre, ambery, oriental, floral, fresh, woody, or fruity accords.

Bergamot for a perfumer is the golden yellow sun for a painter; the window's light for an architect; the key note for a musician; the brightness for a photographer. Either natural or artificial, bergamot oil is present in all perfumers' palettes. It is an essential ingredient.

19 Bergamot in Foods and Beverages

Terence Radford

CONTENTS

19.1 INTRODUCTION

The bergamot is classified botanically as *Citrus bergamia* Risso et Poiteau, and in some publications as a subspecies of orange, *Citrus aurantium* L. subsp. *bergamia*. It is a medium-sized evergreen tree that produces green to yellow pear-shaped fruit resembling oranges in size. The cold-pressed (CP) oil was traditionally extracted manually from the fruit peel using the sponge process, which has transitioned via the manual "bowl" process to the use of the "Calabrese" extractor, and finally to the "Pelatrice Speciale" (Di Giacomo and Mangiola 1989; Dugo et al. 1998; Verzera et al. 1998). Several fruit varieties have been used over the years but now it is mainly the "Fantastico." The cold-pressed oil finds extensive use in fragrances and it is an ingredient of many of the famous perfumes (Calkin and Jellinek 1994). It is used to a much lesser extent in flavor applications.

The history of the bergamot industry in Reggio Calabria, Italy, the principal area producing oil of the highest quality, is detailed in a comprehensive review (Di Giacomo and Mangiola 1989). Efforts have been made to cultivate bergamot in North and South America but the trials had limited success. The susceptibility of the tree to rapid changes in temperature and to inclement weather may have been issues. The only commercially successful introduction has been on the Ivory Coast but the oil is of lower quality than that produced in Calabria. Small quantities are produced intermittently in other countries including Argentina, Brazil, Turkey, and Uruguay.

19.2 BERGAMOT PRODUCTS

A number of potential flavoring materials derived from bergamot have been described and characterized (Arctander 1960; Di Giacomo and Mangiola 1989; Verzera et al. 1998). These are summarized in Table 19.1. Products readily available to the flavor industry are oils marketed as cold pressed, terpeneless, and bergapten-free. Utilization of the juice is limited by its taste, and by the economics of citric acid production, when compared with competing processes.

19.3 ECONOMIC FACTORS

The bergamot industry is of great importance to the Calabrian economy. In the last century the instability of the system led to large price fluctuations, which were accentuated by competition from adulterated oils, and synthetic products sold at low prices. Reliability was introduced by governmental intervention, which led to the formation of a consortium, "Consorzio del Bergamotto di Reggio Calabria," responsible for controlling the production and distribution of bergamot products. It was decreed

TABLE 19.1
Products Derived from Bergamot

Product	Origin	Process	Characteristics
CP oil	Mature fruit	Cold extraction	Best-quality oil
Bergamotella[a]	Fallen fruit	Cold extraction	Lower esters, high residue
Ricicli	Recycled water at end of day	Centrifugation	1% of total oil
Torchiati	Solid residues from second separator	Pressing	3%–4% of total oil
Pulizia dischi	Liquid residues ("feccie") from second separator	Decantation, centrifugation	0.5% of total oil
Distilled oil	Liquid waste from first separator	Reduced pressure distillation ("Peratoner")	3% of total oil
Bergapten-free oil	CP oil	High vacuum distillation or sodium hydroxide wash	Composition of volatile fraction similar to CP
Terpeneless oil	CP oil	High vacuum distillation	Low boilers and waxes absent
Juice	Fruit	Extraction	Bitter, acidic

[a] Bergamotela fruits, not suitable for the mechanical cold extraction, are subject to distillation.

that the quality of bergamot oils would be certified by "Stazione Sperimentale" in Reggio Calabria.

Production statistics have been summarized showing on a yearly basis the number of hectares cultivated, the amounts of fruit produced, and the kilograms of oil extracted for the period 1951–1997 (Crispo and Lamonica 1998). During the years 1994–1997 values for these parameters showed little variation. Oil production was at its peak during the period 1956–1970 with annual yields of approximately 175 metric tons (MT), and dropped to its lowest level of 35 MT in 1990–1991. Sources in the industry indicate that during the past decade the highest production was in 2004 (120 MT), and the lowest was in 2007 (65 MT). Production has been on an upward trend during the past five years and is now approaching the 2004 level.

Export statistics for the period 1976–1996 are available (Crispo and Lamonica 1998). These show that only a few countries import the oil, with France receiving more than 50% of the exported product in 1996, and the United States utilizing close to 3%. It can be inferred that just a few international companies account for most of the bergamot commerce.

19.4 COMPOSITION

Cold-pressed bergamot oil is composed of 93%–96% volatile material by weight and 4%–7% nonvolatiles (Mondello et al. 1994; Costa et al. 2010). The general approach to controlling the quality of the oil has been to analyze the volatile components by gas chromatography (GC) and the nonvolatiles by high performance liquid chromatography (HPLC). In recent years other techniques have been applied.

19.4.1 GAS CHROMATOGRAPHY

Excellent summaries of GC analyses of bergamot oil are available (Dugo et al. 1987; Verzera et al. 1996; Verzera et al. 1998). The most extensive work by far involved the analysis of more than 1500 authentic samples of CP oil (Dugo et al. 1987; Dugo et al. 1991; Verzera et al. 1996; Verzera et al. 1998). Oil samples derived from process residues were also included, along with examples of bergapten-free oil. Compositional data were determined from relative peak area percentages. Ranges observed for peaks present at upper levels of >1% are listed in Table 19.2, along with summations for selected compound classes. The CP oil has a lower limonene content and higher levels of oxygenates than other CP citrus oils. The variability implied by the wide ranges in Table 19.2 also contrasts with data summarized for other citrus oils (Dugo et al. 2002). This wide variation was ascribed to differences in cultivars, fruit maturity, growing areas, and weather conditions (Verzera et al. 1998). Oils listed in Table 19.1 obtained from residues or fallen fruit are sufficiently similar to the standard cold-pressed oils to be used as supplements (Arctander 1960; Verzera et al. 1998). However, the flavor attributes of these materials are questionable (Arctander 1960; Di Giacomo and Mangiola 1989). Data for the bergapten-free oils produced by distillation and sodium hydroxide treatment are surprisingly similar and mainly fit the ranges for the cold-pressed oils. Oils produced by distillation

TABLE 19.2
Percentage Content of Major Volatile Components of Bergamot Oils

Constituent	CP (Range)	Torchiati	Ricicli	Pulizia dischi	Distilled A.P.	Distilled R.P.	Bergapten-Free Distilled	Bergapten-Free NaOH
α-Pinene	0.72–1.84	1.12	1.08	1.27	1.38	0.45	1.16	1.16
Sabinene + β pinene	4.81–12.80	8.15	7.96	8.81	7.92	4.24	7.55	7.53
Limonene + β-phellandrene	25.38–53.19	32.07	30.19	31.39	35.04	23.14	34.24	34.54
γ-Terpinene	5.27–11.38	7.57	7.58	7.57	4.65	5.99	7.15	7.28
Linalool	1.74–22.68	9.46	10.50	8.22	25.40	36.85	13.57	11.44
Linalyl acetate	15.62–41.36	36.30	36.50	37.64	8.88	22.76	31.18	31.99
Monoterpene hydrocarbons	38.75–76.81	50.82	48.77	51.04	54.33	35.52	52.47	53.04
Sesquiterpene hydrocarbons	0.68–1.89	1.10	1.46	1.06	1.06	1.04	0.91	1.23
Alcohols	1.85–22.85	10.02	10.93	8.52	30.80	38.69	13.79	11.71
Esters	17.17–42.21	37.14	37.65	38.42	11.77	23.92	32.06	33.11

Source: Verzera, A., Trozzi, A., Stagno d'Alcontres, I., et al. *EPPOS* 25, 17–38, 1998.

A.P. = atmospheric pressure

R.P. = reduced pressure

of process by-products at either atmospheric or reduced pressure show higher levels of alcohols (linalool and terpinen-4-ol) and lower levels of esters (linalyl acetate).

The large compositional differences observed for different bergamot oil samples have organoleptic consequences. Both linalool and linalyl acetate are major flavor contributors (Arctander 1969). It can be expected that the two have a biosynthetic relationship, and depletion of linalool parallels increases in linalyl acetate as the season progresses (Verzera et al. 1998; Lawrence 1999). Organoleptic evaluation of flavors in model beverage systems is problematic when variance in test materials is great.

A further consequence of the inherent variability in the composition of bergamot oil is the difficulty it confers in quality control using GC. The traditional method for the quality control of essential oils and flavors has been to develop a GC database with ranges for each component. These ranges serve as limits for assessing the integrity of new material. This approach is only effective if the database is large enough and batch to batch homogeneity is sufficient. Both of these become critical when bergamot oil is being used as a flavor ingredient. The use of peak area ratios to detect adulteration has been recommended (Verzera et al. 1996). A drawback to the use of the large body of GC data published for bergamot over the years is that it is based on relative peak area percentages. Being semi-quantitative, these may vary significantly among laboratories due to differences in equipment and chromatographic conditions. Calibration of systems with standards to develop response factors provided quantitative data for lime and lemon oils, and this has been suggested as a routine for standardization of GC data (Chamblee et al. 1991; Chamblee and Clark 1993). In recent work on bergamot essential oils, quantitative approaches were adopted (Sawamura et al. 2006, Costa et al. 2010). Significant differences were noted between relative peak area values and weight percentage data (Costa et al. 2010).

The development of fast GC using narrow bore columns (David et al. 1999; Mondello et al. 2000) and its application to citrus oils (Mondello et al. 2002) have provided a significant advance in quality control evaluations of flavors. Since bonded stationary phases were introduced in the 1980s, it has been possible to use GC to analyze both the volatile components of citrus oils and some of the coumarins and psoralens traditionally regarded as nonvolatile. Many compounds in these classes give sharp GC peaks, with linear responses. Problems occur when large unsaturated substituents such as geranyl groups are present. This approach has been used routinely to check for adulteration of citrus oils. The drawback with conventional GC is that run times as long as 90 minutes are involved, which is mitigated to some extent by operating 24/7 as a routine. The application of fast GC improves run times by an order of magnitude. Its application to cold-pressed bergamot oil provides a rapid method for analyzing volatiles, as well as bergapten and citropten, which can be regarded as critical components of the nonvolatiles due to their reported biological activity. Merging databases developed for conventional GC with fast GC data is feasible if analyses are quantitative.

19.4.2 High Performance Liquid Chromatography

Since the pioneering work which established HPLC as an important method for detecting adulteration in citrus oils (McHale and Sheridan 1989), the technique has been widely applied in quality control analyses. The profile for cold-pressed bergamot oil is relatively simple with essentially four components, which are listed in Table 19.3 along with their ranges from quantitative analyses of 128 samples (Mondello et al. 1993). Important differences among the four compounds were noted recently for processed bergamot oils obtained from a commercial source (Costa et al. 2010), and the relevant data are also given in Table 19.3. HPLC profiles showed coumarins and psoralens to be absent in a colorless terpeneless sample, and present at enhanced levels in a colored terpeneless oil. The bergapten-free distilled oil showed low levels of nonvolatiles but the washed oil contained appreciable levels of bergamottin and 5-geranyloxy-7-methoxycoumarin. Terpeneless bergamot oils used by the flavor industry are usually produced by high vacuum distillation with removal of nonvolatile components to yield colorless material. Other procedures have been employed including chromatography, counter-current, and supercritical carbon dioxide extraction. The latter shows promise but due to the expense it has not been used commercially.

The nonvolatiles in bergamot have fixative properties which are important in fragrances and for the stability of flavors during storage. However, none of the four compounds listed have significant aromas.

19.4.3 Ancillary Methods

Various procedures have been used in the past decade to assess bergamot oil quality. The most successful has been gas chromatography with modified cyclodextrins as stationary phases to separate optical isomers (chiral-GC). This technique is applicable to the bergamot components listed in Table 19.2 and several less abundant compounds. Initially difficulty was experienced in separating the isomers of linalyl acetate. However, separation of linalool isomers was utilized in a study involving 150 samples of CP oil and 8 samples of commercial linalool (Cotroneo et al. 1992). The R(−) isomer represented >99% of the linalool present in the genuine oil samples, while all the samples of commercial linalool were racemic. Using mixtures of CP oil

---

TABLE 19.3

Content (W/W%) of Coumarins and Psoralens in Cold-Pressed (CP) and Processed Bergamot Oils

| | | Processed Oils[b] | | | |
| | | Terpeneless | | Bergapten-Free | |
Constituent	CP Oils[a]	Colored	Colorless	Distilled	Washed
Bergamottin	1.02–2.75	3.92	-	0.20	1.82
5-genaryloxy-7-methoxycoumarin	0.08–0.22	0.28	-	trace	0.13
Citropten	0.14–0.35	0.61	-	trace	trace
Bergapten	0.11–0.32	0.42	-	trace	trace

[a] Mondello et al. 1993
[b] Costa et al. 2010

and commercial linalool, it was shown that additions of the latter as low as 5% were easily detected.

Separation of linalyl acetate enantiomers was achieved using complexation chromatography with a specially prepared stationary phase (Mosandl and Schubert 1990), and a feasible but laborious approach was developed using flash column chromatography to isolate linalyl acetate in >99% purity, followed by hydrogenation to 3,7-dimethyl-3-acetoxyoctane, which was analyzed by chiral-GC with a standard stationary phase (Casabianca and Graff 1994). Some food and pharmaceutical products containing bergamot oil were analyzed.

A more effective method was provided by the development of new modified cyclodextrin phases, and by utilizing multidimensional gas chromatography (MDGC) with two columns either in coupled gas chromatographs or a commercially available dual oven instrument. Fractions separated on the first column by heart cutting are transferred to a chiral column for analysis. MDGC was used for chiral analysis of samples of CP bergamot oil (Juchelka and Mosandl 1996; Mondello et al. 1998) as well as samples representing most of the derived products listed in Table 19.1 (Mondello et al. 1998). Compounds monitored were β-pinene, sabinene, limonene, linalool, terpinen-4-ol, α-terpineol, and linalyl acetate. The results for monoterpene hydrocarbons confirmed previous observations that enantiomeric ratios are not the best indicators of adulteration, as values are comparable with other citrus oils. Enantiomeric ratios for terpinen-4-ol and α-terpineol showed large changes during the season, also negating their usefulness. Distinctive results were obtained for linalool and linalyl acetate, which both showed >99% of R(−) isomer in CP oils and also in "pulizia dischi," "torchiati," "ricicli," and bergapten-free oils. Distilled oils produced from waste materials, either at atmospheric or reduced pressure, showed lesser values for the R(−) isomer, suggesting some racemization occurred, either under distillation conditions or due to enzymatic activity.

A more direct and less expensive chiral-GC method became possible with the availability of diethyl-*tert*-butylsilyl-β-cyclodextrin as a stationary phase. Single-column analysis was reported for some processed bergamot oils, which included two types each of terpeneless and bergapten-free oils (Costa et al. 2010). The results were consistent with the previous unequivocal enantiomeric trends observed with linalool and linalyl acetate (Mondello et al. 1998).

A definitive approach to authenticating bergamot oils may be the use of gas chromatography coupled with isotope ratio mass spectrometry (GC/IRMS). This method is currently being used in the flavor industry with both single compounds and essential oils. It has been applied to bergamot oils (Juchelka and Mosandl 1996, Schipilliti et al. 2011). In a recent study, stable carbon isotope values ($\delta^{13}C$) were measured for samples of Italian CP bergamot oil, Italian peratoner distilled oil, Ivory Coast CP oil, commercial oils, and intentionally adulterated material (Schipilliti et al. 2011). Calibration for authenticity was established using $\delta^{13}C$ values for 18 components of the Italian CP oils. Carbon isotope ratios for the distilled oils were within the calibration limits. The Ivory Coast samples showed small deviations from the limits probably due to the different geography and cultivation practices. These differences were eliminated by using myrcene as an internal standard. Conversely, the commercial oils and the intentionally adulterated samples showed significant deviations from the limits even when corrections were made using the internal standard. Although a greater number of samples need to be analyzed by GC/IRMS, the technique has great potential for differentiating geographical origins of bergamot and various types of adulteration.

Deuterium NMR provides the basis of a different approach to determining the authenticity of linalool and linalyl acetate (Martin et al. 1993). This technique known as SNIF-NMR is used in the flavor industry to validate samples of natural benzaldehyde and natural cinnamaldehyde. The drawback to its use is that single compounds are needed in high purity (>90%). Isolation of linalyl acetate from bergamot oils by flash chromatography or distillation could provide the requisite material especially with terpeneless oils (Casabianca 1994). SNIF-NMR has been used with linalool from a number of essential oils, including bergamot (Martin et al. 1993).

A practical approach to controlling the quality of bergamot oils in the flavor industry is illustrated by a study in which a combination of GC/MS, HPLC, and chiral-GC was used to derive comprehensive data on cold-pressed and processed bergamot oils (Costa et al. 2010). HPLC data are summarized in Table 19.3. Quantitative values for selected volatile components are listed in Table 19.4. If allowances are made for differences in quantitation, data for CP and bergapten-free oils in Table 19.4 are consistent with that given in Table 19.2 and the trends are identical. The large increases in linalool and linalyl acetate for terpeneless oils, and the expected reductions in terpene hydrocarbons, have significant flavor consequences. The results from chiral-GC analyses were consistent among the samples and confirm their integrity.

19.5 BIOLOGICAL ACTIVITY

Various biological effects, both detrimental and beneficial, have been associated with CP bergamot oil. The preoccupation has been with its activity in fragrances, which has been related to the presence of the coumarins and psoralens, with bergapten

TABLE 19.4
Percentage Content of Major Volatile Components of CP and Processed Bergamot Oils

Constituent	CP	Terpeneless		Bergapten-Free	
		Colored	Colorless	Distilled	Washed
α-Pinene	0.95	0.01	<0.01	0.91	0.86
β-Pinene	5.08	0.02	0.01	4.85	4.62
Limonene	38.89	0.22	0.32	36.76	37.40
γ-Terpinene	5.62	0.11	0.46	4.78	5.25
Linalool	6.55	20.30	16.79	11.91	11.86
Linalyl acetate	37.00	72.13	77.09	34.57	34.02
Terpinen-4-ol	0.03	0.04	0.05	0.03	0.03
α-Terpineol	0.07	0.18	0.11	0.09	0.09

Source: Costa, R., Dugo, P., Navarra, M., et al., Flavour Fragr. J. 25, 4–12, 2010

pinpointed as the causative agent. A summary of the biological activity associated with bergapten has been published (Bisignano and Saija 2002). Negative effects probably derive from the ability of bergapten to form stable adducts with DNA and include phototoxic, mutagenic, and carcinogenic reactions. In contrast, bergapten positively affects psoriasis (PUVA therapy) and vitiligo, and alleviates stress and depression. Phototoxic reactions have also been reported for citropten (Ashford-Smith et al. 1983).

Little information is available on health issues resulting from ingested material containing CP bergamot oil. However, oral PUVA therapy is widely practiced, which suggests that foods and beverages containing the oil may elicit biological responses under certain conditions. A case was reported of a 44-year-old man who consumed up to four liters of black tea on a daily basis for 25 years, with no harmful effects except occasional gastric pain. On switching to Earl Grey tea (containing CP bergamot oil), he experienced muscle cramps, blurred vision, and other reactions (Finsterer 2002). All symptoms were resolved within a week after he reverted to black tea. He found the symptoms did not recur if he limited his intake of Earl Grey tea to one liter per day. The adverse effects were explained on the basis that bergapten is a selective potassium channel blocker (Wulff et al. 1998).

Another observation relevant to the ingestion of beverages containing CP bergamot oil is the so-called "grapefruit effect," where the juice inhibits some forms of the cytochrome P450 enzyme, causing enhanced bioavailability of drugs in the bloodstream (Bailey et al. 1998). Bergamottin, the principal nonvolatile of bergamot oil, has been implicated as a contributor (Girennavar et al. 2006).

19.6   REGULATORY ASPECTS

Based on the predominant use of bergamot oils in fragrances and the concerns with the phototoxicity of bergapten, it is understandable that the main restriction is a recommendation from IFRA that CP bergamot oil be present at a maximum level of 0.4% in leave-on products applied to skin exposed to the sun. It is suggested that bergapten be limited to 75 ppm in a fragrance compound used at a 20% level in a product, which translates to a maximum level of 15 ppm. Where additive effects are known with other ingredients, the maximum levels should be proportionally reduced.

In flavor applications both CP and terpeneless bergamot oils are approved for use as food additives by FDA and have GRAS status. FEMA issued a list of recommended dosages for bergamot oil concentrate in various foods and beverages in 1994. These range from maximum levels of 0.15 ppm in meat products to 200 ppm in gelatins and puddings.

As a preliminary to regulatory action, use levels of a commodity are generally compiled using information from the industry. It may be significant that the Senate Commission on Food Safety of the German Research Foundation, which is involved with the toxicological assessment of furanocoumarins in foodstuffs, elected to revise previous estimates of the average daily intake of furanocoumarins from flavored foods before taking any action (Guth et al. 2011). This decision was based on recent data that indicates distilled citrus oils containing no furanocoumarins predominate over CP oils in soft drinks.

19.7   ORGANOLEPTIC PROPERTIES

The extensive compositional data published for bergamot oils contrasts with the amount of organoleptic information available. This is often the case with essential oils and flavors. Invariably Arctander's work is the most definitive (Arctander 1960).

A notable exception is the work of Sawamura et al., in which gas chromatography-olfactometry (GC-O) was used to identify GC components that contribute to bergamot aroma (Sawamura et al. 2006). Aroma extraction dilution analysis was then employed to assign characteristic odor contributors, and to develop an aroma model. These and related sensory techniques have been summarized (Chamblee and Clark 1993; Acree 1997). A simpler approach with a capillary GC sniff port was used to pinpoint important contributors to lemon oil aroma (Chamblee and Clark 1993). The sensory work with bergamot oil led to the conclusion that four compounds, (Z)-limonene oxide, decanal, linalyl acetate, and geraniol, are characteristic aroma components. An aroma complex was formulated using quantitative GC data for these four compounds and eight others with propylene glycol as carrier. A panel of 15 members gave the formulated product a high rating. It was acknowledged, however, that the complex aroma of bergamot is due to many compounds.

A drawback to the GC-O approach is that GC peaks usually result from more than one compound, and only the major component is assigned. Minor components which may be major contributors to aroma may be overlooked. It is noteworthy that in an investigation in which 175 compounds were identified, oxygenated sesquiterpenes in trace levels were reported to be strong contributors to bergamot oil aroma (Ohloff 1994). Coumarin, methyl anthranilate, and indole, which all have strong odors, were also identified as trace components of the oil.

The normal approach to assessing the quality of essential oils and flavors is to use a combination of sensory and analytical tests to compare incoming material against a standard. Sensory testing is the most important quality control tool. In the flavor industry, trained sensory panels are used to evaluate the aroma by sniffing, and the taste in a model system. It is necessary to develop a common vocabulary, and descriptors are selected for both desirable attributes and off flavors. The wide application of bergamot oils in fragrances has resulted in emphasis being placed on odor evaluations, which are also relevant to flavors. A summary of organoleptic descriptors for bergamot oil aroma is given in Table 19.5.

The overall impression is that the CP oil exhibits a very rich, fresh, sweet, fruity topnote with an oily, herbaceous, citrus-like body, progressing to a balsamic dryout. It functions mainly as a fresh, sweet-fruity topnote in fragrances (Arctander 1960; Williams 1996) and flavors (Arctander 1960) but it also acts as a fixative and modifier.

The terpeneless oil used by the flavor industry generally contains only the volatiles. It has lost the freshness associated with the terpenes but it displays an herbaceous, woody, oily body with only a weak dryout. The bergapten-free oils display many of the characteristics of the CP oil but they have lost the fixative properties, and also they have a weak dryout.

The major components of bergamot oils, linalool and linalyl acetate, have important flavor characteristics (Arctander 1969). Both compounds contribute fresh, floral,

TABLE 19.5

Organoleptic Descriptors for Bergamot Oils

Product	Top Notes	Body	Dryout	Reference
CP	Extremely rich, fresh, sweet, fruity	Oily, herbaceous	Balsamic, tobacco-like	Arctander 1960
	Fresh, sharp, lemony, orangey	Herbaceous, pepper, orangey, slightly floral	Warm, like orange pith	Williams 1996
	Fresh, leafy, green, floral	Herbaceous, grapefruit	Vanilla-like, lime-like	Industry panel
Terpeneless	Weak freshness	Rich, intensified	Oily, slightly grassy	Arctander 1960
	Woody	Herbaceous	Weak lime	Industry panel
Bergapten-free	Sharp, fresh	Citrus, herbaceous, spicy, peppery, lavender-like	Little characteristic	Williams 1996
	Fresh, citrus	Herbaceous, citrus	Weak dryout	Industry panel

fruity, woody notes. Linalool, being a low boiler, is not lasting, but it has a creamy, floral taste. Linalyl acetate has a sweet, fruity, pear-like taste. Both exhibit lavender attributes. Olfactory descriptions have been recorded for the linalool enantiomers (Casabianca and Graff 1994). The R(−) enantiomer ia described as lavender and woody, while its S(+) counterpart is said to be sweet and petitgrain-like. Adulteration of the CP oil with racemic linalool clearly has flavor implications.

The sweetness and rich bodynote of bergamot oils is considered to result in part from the presence of high levels of linalyl acetate and linalool, combined with trace amounts of methyl anthranilate (Arctander 1960). The freshness may result from these terpenes and small amounts of citral and aliphatic aldehydes (Arctander 1960). The presence of trace amounts of high boilers such as some sesquiterpene alcohols may be critical contributors to bergamot aroma (Ohloff 1994).

19.8 USES IN FOODS

The CP oil is used in hard candy, baked goods, desserts, and jams. A famous hard candy in France is "Bergamotes de Nancy," which has been manufactured since 1850 in Nancy, a town in the Lorraine region. It consists mainly of sugar and bergamot oil.

Some recipes for cakes and cookies call for bergamot peel, which is hard to obtain, and bergamot oil is often used as a substitute. Small amounts of the oil are used to flavor desserts such as fruit dishes and halva, which is popular in many countries throughout the Middle East and the Far East. It may be a flour-based confection or it can be made out of nut butter, principally sesame, which gives it the consistency of a sweet firm paste.

Bergamot marmalade is gaining popularity as an alternative to Seville orange marmalade, which it resembles in bitterness but differs in its unusual taste. Published recipes call for using the fruit as the only bergamot flavor, but it is difficult to work with (Saunders 2000). This aspect combined with the scarcity of the fruit in Western markets may lead to the oil being added as a flavor enhancer, especially where combinations of citrus fruits are used to make marmalade.

The oil is also likely to be used in jams which incorporate mixed fruits, and where bergamots are not obtainable locally. Recipes are available online, for example for strawberry-rhubarb jam, where a few drops of oil are added to give a fresh citrus aroma to the final product. A bergamot jam prepared locally is available in Nancy and countrywide in Norway.

Other categories in which bergamot oil is used include meat products, frozen dairy, and chewing gum.

19.9   USES IN BEVERAGES

The most famous beverage containing bergamot oil is a tea blend known as Earl Grey. It is named after the second Earl Grey, British Prime Minister during the 1830s, who reputedly received the tea as a gift from a Chinese diplomat. Jacksons of Piccadilly claims to have originated the brand when a partner was given the recipe by Lord Grey in 1830. Since that time they have produced the Earl Grey blend on a continuous basis using China tea. However, Twinings is the company most often associated with Earl Grey tea, and they market several variants which combine bergamot oil with other ingredients such as lavender, lemon grass, jasmine, and Seville oranges, as well as green tea and various flowers. In 2011, Twinings reformulated their flagship brand, which led to vigorous protests on their website and extended debate in the media. A beverage known as "London Fog," which is popular in London in winter, is composed of Earl Grey tea, steamed milk, and vanilla syrup.

Earl Grey tea is used to flavor many types of cakes, confections and sauces. For the latter tea bags are used whereas for cakes and confectionaries the loose tea is added to melted butter or hot cream and filtered out after infusion. Many tea companies now market blends known as Earl Grey. An open question is whether they are flavored with natural bergamot oils or compounded flavors. Some data are available from analyses of commercial tea samples (Neukom et al. 1993; Casabianca and Graff 1994; Sponsler and Biedermann 1997). It is significant that from a total of 15 samples analyzed, all were found to contain racemic linalool. This contrasts with the results of a personal survey of Earl Grey tea in the United States, in which label claims for 8 out of 10 brands indicated CP bergamot oil as the only added flavor. The use of substitutes for authentic bergamot oils may be based on cost or it could be due to undetected economic adulteration by suppliers. An additional reason may be a concern for the negative biological effects associated with bergamot oils. However, the calming effects and relief of stress claimed for the use of bergamot oils in aromatherapy should be duplicated to a degree with hot tea. Any effects would be modified by the use of substitutes, as would aspects of the aroma. In the latter context, the difference in odor between the linalool enantiomers may be significant (Casabianca and Graff 1994).

A consequence of the high cost associated with bergamot oils, and the modest expense associated with carbonated beverages, is that these ingredients are rarely used in the soft drink industry. They find occasional use in citrus beverages, principally lemon, and in fruit juice drinks. The terpeneless oil with both terpenes and waxes removed is preferred for its better solubility. An important difference exists between tea, which has a pH near to neutral (~6.0–6.4), and soft drinks, which have acidic pHs (~2.5–4.0). Under the acidic conditions the major components of bergamot listed in Table 19.2 are solvolyzed to give α-terpineol or terpinen-4-ol as the principal products, which may then undergo further reactions (Clark and Chamblee 1992). In particular, linalyl acetate reacts within 24 hours at pH 3.0 to give α-terpineol and racemic linalool. It can be expected that trace components characteristic of bergamot flavor will also be affected by the acid conditions with the result that the initial flavor profile is transformed.

Commercial flavors that are used in foods and beverages may contain small amounts of bergamot oil. Examples are pear and candy flavors. An interesting use of the oil is to flavor tobacco.

19.10 FUTURE TRENDS

A growing concern in the flavor industry is that the authenticity and geographical integrity of essential oils and other natural flavors are slowly being compromised. This trend is accelerated where expensive ingredients such as bergamot oil are involved. A compounded material can never be the flavor equivalent of an authentic oil from Calabria.

A possible paradox has already been mentioned for Earl Grey tea. In analyses of more than 60 commercial samples of CP bergamot oil by chiral-GC (Casabianca and Graff 1994; Konig et al. 1997) and by GC-IRMS (Schipilliti et al. 2011), a large proportion were determined to be adulterated. The widespread application of available methods, however complex, is urgently needed throughout the flavor industry to combat adulteration.

The lack of organoleptic investigations of bergamot oil has been mentioned. An in-depth study of bergamot volatiles would be pertinent using the capillary sniff port technique (Chamblee and Clark 1992), or a variant, in combination with prefractionation by open column chromatography or HPLC using offline (Chamblee and Clark 1993) or coupled techniques (Mondello et al. 1994). As a preliminary step a survey of compounds reliably identified in the CP oil is needed, as the number of compounds in bergamot is often quoted as >300, but most GC analyses account for less than 80. The high boilers appear to be of particular interest (Arctander 1960; Ohloff 1994). The distillation process used to prepare terpeneless oils for flavor use may eliminate critical components, and this aspect needs to be studied.

Currently there is a multifaceted analytical approach to adulteration in bergamot oil which produces considerable data for interpretation. The application of computeraided pattern recognition techniques, such as determinant analysis, may offer a means of facilitating and quantifying the process.

ACKNOWLEDGMENTS

The author thanks Ms. Donna Hudson for help with literature searches and Dr. B. C. Clark for useful suggestions.

REFERENCES

Acree, T. E. 1997. GC/Olfactometry: GC with a sense of smell. *Anal. Chem. News Features*: 69(3): 170A–175A.

Arctander, S. 1960. *Perfume and flavor materials of natural origin*. Elizabeth, NJ: Author's edition.

Arctander, S. 1969. *Perfume and flavor chemicals (aroma chemicals). II.* Carol Stream, IL: Allured Publishing.

Ashford-Smith, M. J., Poulton, G. A., and Liu, M. 1983. Photobiological activity of 5,7-dimethoxycoumarin. *Experientia* 39(l): 262–264.

Bailey, D. G., Malcolm, J., Arnold, O., et al. 1998. Grapefruit juice-drug interactions. *Br. J. Clin. Pharmacol.* 46(2): 101–110.

Bisignano, G., and Saija, A. 2002. The biological activity of citrus oils. In *Citrus*, eds. G. Dugo and A. Di Giacomo, 602–630. London: Taylor & Francis.

Calkin, R. R. and J. S. Jellinek. 1994. *Perfumery practice and principles*. New York: John Wiley and Sons.

Casabianca, H., and Graff, J-B. 1994. Separation of linalyl acetate enantiomers: Application to the authentication of bergamot food products. *J. High Res. Chrom.* 17:184–186.

Chamblee, T. S., Clark, B. C., Brewster, G. B., et al. 1991. Quantitative analysis of the volatile constituents of lemon peel oil. Effects of silica gel chromatography on the composition of its hydrocarbon and oxygenated fractions. *J. Agri. Food Chem.* 39:162–169.

Chamblee, T. S., and Clark, B. C. 1993. Lemon and lime citrus essential oils. Analysis and organoleptic evaluation. In *Bioactive volatile compounds from plants, ACS Symp. Ser. 525*, eds. R. Teranishi, R. G. Buttery, and H. Sugisawa, 88–102. Washington DC: American Chemical Society.

Clark, B. C., and Chamblee, T. S. 1992. Acid-catalyzed reactions of citrus oils and other terpene-containing flavors. In *Off-flavors in foods and beverages*, ed. G. Charalambous, 229–285. Amsterdam: Elsevier Science Publishers B. V.

Costa, R., Dugo, P., Navarra, M., et al. 2010. Study on the chemical composition variability of some processed bergamot (*Citrus bergamia*) essential oils. *Flavour Fragr. J.* 25:4–12.

Cotroneo, A., Stagno d'Alcontres, I., and Trozzi, A. 1992. On the genuineness of citrus essential oils. Part XXXIV. Detection of added reconstituted bergamot oil in genuine bergamot essential oil by high resolution gas chromatography with chiral capillary columns. *Flavour Fragr. J.* 7:15–17.

Crispo, F., and Lamonica, G. 1998. Economical aspects of Italian bergamot. *EPPOS* (numero speciale):60–73.

David, F., Gere, D. R., Scanlon, F., et al. 1999. Instrumentation and applications of fast high-resolution capillary gas chromatography. *J. Chromatogr.* 842:309–319.

Di Giacomo A., and Mangiola, C. 1989. *Il bergamotto di Reggio Calabria,* Reggio Calabria: Laruffa.

Dugo, G., Lamonica, G., Cotroneo, A., et al. 1987. Sulla genuinita delle essenze agrumarie. Nota XVII. La composizione della frazione volatile dell'essenza di bergamotto calabrese. *Essenz Deriv. Agrum.* 57:456–544.

Dugo, G., Cotroneo, A., Verzera, A., et al. 1991. Genuineness characters of Calabrian bergamot essential oil. *Flavour Fragr. J.* 6:39–56.

Dugo, G., Crispo, F. and Gazea, F. 1998. *Citrus bergamia* Risso, from origins to nowadays: Cultivation and extraction technologies of the essential oil. *EPPOS* (numero speciale):17–30.

Dugo, G., Cotroneo, A., Verzera, A. et al. 2002. Composition of the volatile fraction of cold-pressed citrus peel oils. In *Citrus,* eds. G. Dugo and A. Di Giacomo, 201–317. London: Taylor & Francis.

Finsterer, J. 2002. Earl Grey tea intoxication. *Lancet* 359:1484.

Girennavar, B., Poulose, S. M., Jayaprakasha, G. K., et al. 2006. Furocoumarins from grapefruit juice and their effect on human CYP 3A4 and CYP 1B1 isoenzymes. *Bioorg. Med. Chem.* 14:2606–2612.

Guth, S., Habermeyer, M., Schrenk, D., et al. 2011. Update of the toxicological assessment of furanocoumarins in foodstuffs (update of the Senate Commission on Food Safety statement of 23/24 September 2004); opinion of the Senate Commission on Food Safety of the German Research Foundation. *Mol. Nutrition and Food Res.* 55:807–810.

Juchelka, D., and Mosandl, A. 1996. Authenticity profiles of bergamot oil. *Pharmazie* 51:417–422.

Konig, W. A., Fricke, C., Saritas, Y., et al. 1997. Adulteration or natural variability? Enantioselective gas chromatography in purity control of essential oils. *J. High Resolut. Chromatogr.* 20:55–61.

Lawrence, B. M., 1999. Bergamot oil. *Perfum. Flavor.* 24(5): 45–53.

Martin, G., Remaud, G., and Martin, G. J. 1993. Isotopic methods for control of natural flavors authenticity. *Flavour Fragr. J.* 8:97–107.

Mc Hale, D., and Sheridan, J. B. 1989. The oxygen heterocyclic compounds of citrus peel oils. *J. Essent. Oil Res.* 1:139–149.

Mondello, L., Stagno d'Alcontres, I., Del Duce, R., et al. 1993. On the genuineness of citrus essential oils. Part XL. The composition of the coumarins and psoralens of Calabrian bergamot essential oil (*Citrus bergamia* Risso). *Flavour Fragr. J.* 8:17–24.

Mondello, L., Bartle, K. D., Dugo, P., et al. 1994. Automated LC-GC: A powerful method for essential oils analysis. Part IV. Coupled LC-GC-MS (ITD) for bergamot oil analysis. *J. Microcol. Sep.* 6:237–244.

Mondello, L., Verzera, A., Previti, P., et al. 1998. Multidimensional capillary GC-GC for the analysis of complex samples. 5. Enantiomeric distribution of monoterpene hydrocarbons, monoterpene alcohols, and linalyl acetate of bergamot (*Citrus bergamia* Risso et Poiteau) oils. *J. Agric. Food Chem.* 46:4275–4282.

Mondello, L., Zappia, G., Errante, G., et al. 2000. Fast GC and fast GC-MS for the analysis of natural complex matrices. *LC-GC Europe* 13:495–502.

Mondello, L., Zappia, G., Dugo, P., et al. 2002. Advanced analytical techniques for the study of citrus oils. In *Citrus*, eds. G. Dugo and A. Di Giacomo, 179–200. London: Taylor & Francis.

Mosandl, A., and Schubert, V. 1990. Stereoisomere aromastoffe. XXXIX. Chirale inhaltsstoffe atherischer ole (1). Stereodifferenzierung des linalyl acetats ein neuer weg zur qualitatsbeurteilung des lavandelols. *Z. Lebensmitt. Unters. Forsch.* 190:506–510.

Neukom, H. P., Meier, D. J., and Blum, D. 1993. Detection of natural and reconstituted bergamot oil in Earl Grey teas by separation of the enantiomers of linalool and dihydrolinalool. *Mitt. Gebiete Lebensm. Hyg.* 84(5): 537–544.

Ohloff, G. 1994. *Scent and fragrances. The fascination of odors and their chemical perspectives.* Berlin: Springer-Verlag.

Saunders, R. 2010. *The blue chair jam cookbook.* Kansas City: Andrews McNeel.

Sawamura, M., Onishi, Y., Ikemoto, J., et al. 2006. Characteristic odour components of bergamot (*Citrus bergamia* Risso) essential oil. *Flavour Fragr. J.* 21:609–615.

Schipilliti, L., Dugo, G., Santi, L., et al. 2011. Authentication of bergamot essential oil by gas chromatography-combustion-isotope ratio mass spectrometer (GC-C-IRMS). *J. Essent. Oil Res.* 23:60–71.

Sponsler, S., and Biedermann, M. 1997. Natural vs artificial flavor identification. *Food Test. Analysis* 3(1): 8, 10, 15.

Verzera, A., Lamonica, G., Mondello, L., et al. 1996. The composition of bergamot oil. *Perfum. Flavor.* 21:19–34.

Verzera, A., Trozzi, A., Stagno d'Alcontres, I., et al. 1998. The composition of the volatile fraction of Calabrian bergamot essential oil. *EPPOS* 25:17–38.

Williams, D. G. 1996. *The chemistry of essential oils*. Weymouth, Dorset, England: Micelle Press.

Wulff, H., Rauer, H., Doring, T., et al. 1998. Alkoxypsoralens, novel nonpeptide blockers of Shaker-type K+ channels: Synthesis and photoreactivity. *J. Med. Chem.* 41:4542–4549.

20 Antimicrobial Activity of *Citrus bergamia* Risso & Poiteau

Francesco Pizzimenti, Antonia Nostro, and Andreana Marino

CONTENTS

20.1 INTRODUCTION

Current consumer trends and the increasing isolation of antibiotic resistant pathogens have led to a renewed scientific interest in plant-derived compounds. In this context, essential oils (EOs) have been recognized for their wide-spectrum antimicrobial activity and have been investigated by many researchers worldwide (Helander et al. 1998; Hammer et al. 1999; Dorman and Deans 2000; Elgayyar et al. 2001; Ceylan and Fung 2004; Bakkali et al. 2008). They could represent an interesting approach to control the emergence and the spread of pathogenic organisms as well as the development of resistance to the conventional antibiotics already in use. Although the antimicrobial effects of essential oils are extensively documented, their mode of action is poorly known. Considering the large number of different groups of active chemical compounds (phenols, terpenes, and aldehydes) present in EOs, it is most likely that their antibacterial activity is not attributable to one specific mechanism but that there are various targets in the microbial cell (Burt 2004). However, some studies have demonstrated that the cytoplasmic membrane is the target common to many oils. The hydrophobic nature and high volatility of the essential oils allows them to penetrate into the cell membrane to exert their biological effect.

In this regard, the mode of action of carvacrol, a monoterpene phenol present in several essential oils, has been the most studied. Carvacrol interacts with the cytoplasmatic membrane and aligns between the fatty acids chains, causing expansion and destabilization of the membrane structure increasing its fluidity and permeability. The oil acting on the permeability barrier of the cytoplasmic membrane could also cause leakage of various other substances, such as ions, ATP, nucleic acids, and amino acids (Sikkema et al. 1995; Ultee et al. 2000, 2002; Lambert et al. 2001; Burt and Reinders 2003; Di Pasqua et al. 2007).

The attention given to natural compounds has also increased studies on the biological activity of *Citrus* oils and their components. The genus *Citrus* has approximately 16 species in the family *Rutaceae*. *Citrus bergamia* Risso & Poiteau (Bergamot) is a typical fruit grown in Calabria, in southern Italy. Bergamot oil extracted from fruit peel is composed of several volatile and nonvolatile constituents recognized as safe (GRAS), and its scent and flavor lend to its use in food (Svoboda and Greenaway 2003). The major active volatile components are citral, limonene, and linalool, although limonene levels are the most abundant (32%–45%) (Moufida and Marzouk 2003; Svoboda and Greenaway 2003).

Citrus oils are known for their anti-infective properties. The first records of *Citrus* fruits having medicinal uses were by Theophrastus in fourth century B.C., when they were used as antidotes to poisons and inhaled to ease the throat (Fisher and Phillips 2008). In 1804, Francesco Calabrò, a physician born in Reggio Calabria, noticed that when workers engaged in the manufacture of bergamot oil had wounds on their hands, they healed spontaneously and with incredible speed.

Citrus oils now have a wide application in folk medicine and in the fragrance, food, and pharmaceutical industries. This chapter reviews literature data on the antimicrobial activity of *Citrus bergamia* oil and highlights future applicative perspectives.

20.2   ANTIMICROBIAL ACTIVITY

Citrus oils have been found to be inhibitory both by direct contact and vapor form against a wide range of microorganisms. In this context, *Citrus bergamia* oil and flavonoid-rich ethanolic fractions have shown good antibacterial, antimycotic, and antiviral activity (Statti et al. 2004; Romano et al. 2005; Fisher and Phillips 2006; Mandalari et al. 2007; Sanguinetti et al. 2007; Balestrieri et al. 2011). Studies *in vitro* have demonstrated its efficacy against microorganisms, common causes of food poisoning, and human infections.

20.2.1   BACTERIA

Citrus bergamia oil displayed a variable degree of antibacterial activity against different strains. Some authors have demonstrated the activity of this oil against *Campylobacter jejuni*, *Escherichia coli* O157, *Listeria monocytogenes*, *Bacillus cereus*, *B. subtilis*, *Staphylococcus aureus,* and *Arcobacter butzleri* (Fisher and Phillips 2006; Fisher et al. 2007; Lv et al. 2011). The oil and its major components— citral, limonene, and linalool—were assayed both when applied directly and in contact with vapor. These studies were carried out *in vitro* and in food systems such as chicken and cabbage leaf. The results demonstrated the higher susceptibility of gram-positive bacteria with minimal inhibitory concentration (MIC) values of 0.125–1%, v/v, than gram-negative bacteria, although *S. aureus* was the least susceptible. This trend was observed also for bergamot vapor treatment. Gram-negative bacteria were less sensitive to the bergamot oils, probably because of the lipopolysaccharide nature of the outer membrane, which limits diffusion of hydrophobic compounds (Vaara 1992; Oussalah et al. 2006). However, among gram-negative strains, *E. coli* O157 was inhibited with a MIC equal to 0.5%, v/v. In this regard, Somolinos et al. (2009) reported the *E. coli* cell envelope damage as an important event in citral inactivation. A strain-depending susceptibility was shown by three *Arcobacter* isolates (MIC values of 0.125%–2%, v/v). All bacteria tested were more susceptible *in vitro* than in cabbage leaf and chicken skin systems.

Bergamot oil was also active when combined with orange (*Citrus sinensis*) oil. Fisher and Phillips (2009a, 2009b) have demonstrated the efficacy of a bergamot:orange blend (1:1, v/v) and its vapors against vancomycin-resistant and vancomycin-sensitive strains of *Enterococcus faecium* and *E. faecalis*. The blend had MIC values of 0.25%–0.5% (v/v) and a minimum inhibitory dose (MID) of 50 mg/L. An increase in cell membrane permeability was also observed. The effect was more evident when the combination was applied in vapor form; in fact, the increase of cell permeability was 32–40 times higher than control. The authors suggested that one of the ways in which EO bring about their antimicrobial effect may be the uptake of the *Citrus* EO blend into Enterococcal cells, causing permeability of the cell membrane and reducing ATP synthesis. Although the vapors of this blend resulted in the same damage, they were probably acting via different mechanisms than the EO because of the increased permeability of the cell membrane, particularly the different morphological changes they induced (Fisher and Phillips 2009a).

Other bergamot combinations have been studied, including bergamot and oregano and bergamot and basil oils. According to the checkerboard test, the combinations were synergic on *S. aureus* with fractional inhibitory concentration indices (FICI) equal to 0.375 (Lv et al. 2011). There are some generally accepted mechanisms of antimicrobial interaction for synergy, such as sequential inhibition of a common biochemical pathway, inhibition of protective enzymes, combinations of cell wall active agents, and use of cell wall active agents to enhance the uptake of other antimicrobials (Santiesteban-Lopez et al. 2007). Therefore, it is necessary to clarify the specific modes of action of antimicrobial EOs and their constituents on the metabolic pathway of microorganisms (Goni et al. 2009). Scanning electron microscopy observation and the determination of cell constituents release demonstrated that EO combinations destroyed the integrity of the cell membrane, resulting in the death of microorganisms (Lv et al. 2011).

20.2.2 Mycetes

Bergamot oil has been tested for *in vivo* and *in vitro* antimycotic activity, and the results demonstrated its potential as an antifungal agent. Romano et al. (2005) evaluated the *in vitro* activity of three bergamot oils alone and with boric acid against 40 *Candida* clinical isolates. The distilled extract alone demonstrated higher activity (MIC from 0.312% to 2.5% v/v) than that of furocoumarine-free and natural oil for all isolates: *Candida albicans, C. glabrata, C. krusei, C. tropicalis*, and *C. parapsilosis*. The activity of each bergamot oil was potentiated when it was combined with sub-inhibitory concentrations of boric acid. Interestingly, the oils alone and in association were efficacious both on fluconazole-resistant and susceptible isolates.

Bergamot oil was also active against several common species of dermatophytes. Sanguinetti et al. (2007) assayed the same three oils versus 92 clinical isolates of *Tricophyton mentagrophytes, T. rubrum, T. interdigitale, T. tonsurans, Microsporum canis, M. gypseum*, and *Epidermophyton floccosum*. The distilled and furocoumarine-free extracts were more active than natural oil. In general, the oils had lower MICs (from 0.02% to 2.5% v/v) than those obtained on *Candida* species. Moreover, the antifungal activity was reported for *Kluyveromyces fragilis, Rhodotorula rubra, C. albicans, Hanseniaspora guilliermondii*, and *Debaryomyces hansenii* yeasts (Kirbaslar et al. 2009). The action of EOs on the fungal cell appears to be predominantly focused on cell membrane disruption but also on inhibition of the germination, fungal proliferation, and cellular respiration. Park et al. (2009) demonstrated that citral caused the growth inhibition and deformation of *T. mentagrophytes* hyphae. Moreover, it was able to form a charge transfer complex with an electron donor of fungal cells, resulting in fungal death.

20.2.3 VIRUS

The methanol extract obtained from the seeds of *Citrus bergamia* (BSext) and two compounds, purified limonin and nomilin, have been evaluated against HTLV-1 and HIV-1 by Balestrieri et al. (2011). The results showed the weak protease inhibition in infected peripheral blood mononuclear cells (PBMC). This activity was compared with reference compounds such as AZT and 3TC. BSext was about 20 times less effective than AZT in inhibiting HTLV-1 viral expression. At the same time, BSext was found to inhibit *gag* expression in HIV-1 infected cells, with an efficiency 4 times lower than that of AZT and 10 times lower than that of 3TC. As to the examined purified components of BSext, limonin was practically as efficient as the whole extract in protecting from HTLV-1 and HIV-1 infections. At the same time, it was more effective than nomilin in inhibiting both HTLV-1 tax/rex and HIV-1 *gag* expression in infected cells.

The exact mechanism by which these compounds exert their anti-retroviral action has not been fully defined. However, the authors investigated at least one of the potential mechanisms involved. To this purpose, they analyzed the effects of the examined products directly on HTLV-1 and, for comparison, HIV-1 RT enzymatic activities using cell-free and virus-free assays. The results demonstrated that all the natural compounds assayed are endowed with an RT inhibitory activity. Limonin was the most efficacious natural product in directly inhibiting both HTLV-1 and HIV-1 RT, in comparison with nomilin and BSext. It must be emphasized that the lower antiretroviral activity of BSext, limonin, and nomilin was compensated by their lower toxicity in comparison with the reference compounds. BSext was about 28 times less cytotoxic than AZT and about 20 times less cytotoxic than 3TC in

PBMC, in any case showing a more favorable selectivity index. Considering their low *in vitro* toxicity, these natural compounds deserve further study to compare their activity to that of antiretroviral drugs currently available.

20.3 FUTURE PERSPECTIVES

The essential oils and their constituents have interesting applicative prospects. This provides an incentive for researchers to study different antimicrobial formulations and materials for biomedical and food packaging applications. Their future applications could concern either natural preservation in the food industries or an alternative which supports the conventional antimicrobial protocols.

20.3.1 FOOD PRESERVATION

The survival of microorganisms in food can cause spoilage of products or infection and illness (Jacob et al. 2010). Increased consumer demand for healthy food products free of synthetic chemical residues has led to research on new antimicrobial agents from plants. In this regard, EOs have been studied in the preservation of a wide range of foods, including vegetables, fruit, dairy products, fish, meat, and poultry. However, their use is often limited as the effective antimicrobial concentrations may exceed the acceptable sensory levels (Hsieh et al. 2001).

In order to retain their activity and minimize their impact on the organoleptic properties of food, they can be incorporated within a polymer matrix and progressively released from the surface (McClements 1999). The essential oils can be incorporated into biodegradable material to increase shelf life and quality of food products. The incorporation of natural compounds into polymer matrices can also improve mechanical properties. Promising results with chitosan-essential oils composite films have been reported. The use of systems made up of biocompatible compounds, such as chitosan (Pedro et al. 2009) or Ca-alginate (Wang et al. 2009), enabled the controlled release of essential oils in the mouth (chitosan) or in the small intestine (Ca-alginate).

Bergamot essential oil has been investigated as a promising alternative to extend the shelf life of fruits (Sánchez-González et al. 2010). Recently, its incorporation into chitosan films has been highlighted as a food-preservative system. The bergamot essential oil incorporated into chitosan films resulted in a significant inhibitory effect on the growth of *Penicillium italicum*. The oil improved also the film's water vapor barrier properties. The highest activity was obtained in films with 3% w/w oil, which caused a reduction of two log units at the end of storage (12 days). Successively, the same authors have analyzed the effect of two polysaccharides matrices, hidroxypropylmethylcellulose and chitosan, containing 2% w/w bergamot oil on the physicochemical properties, respiration rate, and microbial counts of organic table grapes (storage for 22 days). The most highly recommended coating for table grapes was the chitosan-bergamot oil, as it had the highest antimicrobial activity against total mesophilic bacteria, yeasts, and molds. Moreover, the chitosan-bergamot oil was a good alternative for grape preservation as it reduced respiration rates and showed good control of water loss during storage (Sánchez-González et al. 2011).

20.3.2 ANTIFUNGAL EFFECTS

Dermatophytoses and candidiasis are common cutaneous fungal infections in humans with considerable incidence (Gupta et al. 1998; Kane and Summerbell 1999). Various conventional antifungal agents are used but the success of the therapeutic treatments is limited. The reason is due to the poor spectrum of the antifungal drugs and/or to increased fungal resistance as well as to their adverse effects and the expensive treatment. This problem has attracted the attention of the scientific community in the development of natural active compounds.

Pizzimenti et al. (2006) have evaluated the *in vivo* antifungal activity of "Peratoner," which is a waste product obtained from bergamot peel whose qualitative and quantitative profiles show an alcoholic fraction formed mainly of α-terpineol, terpinen-4-ol and linalool, and a hydrocarbon fraction. The study was carried out through Peratoner applications to Wistar rats' back skin infected with *C. albicans*. The Peratoner caused a gradual activity from 24 to 144 h, reducing the yeast load and infiltrating cells in the dermis. The authors hypothesize that α-terpineol, terpinen-4-ol, and linalool, which are present in rather high percentages compared to other lypophilic components, favor membrane permeability, which can repress or activate the aspartate proteinase of *C. albicans*. This type of inhibition could slow down the penetration process and the subsequent adherence, with formation of cavities on the corneocyte surface. In the study, a detachment process of the *C. albicans* cells from the corneocytes was observed. The topical treatment of Peratoner could counteract both the adhesion process and the formation of surface cavities, with consequent extrusion of blastoconidia from the outer strata of the epidermis. Previous studies demonstrated that α-terpineol, terpinen-4-ol, and linalool showed *in vitro* antifungal effects against *C. albicans* (Pizzimenti et al. 1998). Early recovery from the infection[*] could be related to the synergistic action of the oxidized essential oil components, formed mainly of α-terpineol, terpinen-4-ol, and linalool. They, when in contact with the corneocytes, could interact with junctional complexes by saturating the intercellular space, blocking a massive migration of the blastoconidia in the exfoliating layer. The presence of different chemical components in the Peratoner seems to produce a synergistic action, determining a valid antifungal effect without damaging the skin structure.

REFERENCES

Bakkali, F., Averbeck, S., Averbeck, D., and Waomar, M. 2008. Biological effects of essential oils—A review. *Food Chem. Toxicol.* 46:446–475.

Balestrieri, E., Pizzimenti, F., Ferlazzo, A., et al. 2011. Antiviral activity of seed extract from *Citrus bergamia* towards human retroviruses *Bioorgan. Med. Chem.* 19:2084–2089.

Burt, S. 2004. Essential oils: Their antibacterial properties and potential applications in foods—A review. *Int. J. Food Microbiol.* 94:223–253.

Burt, S., and Reinders, R. D. 2003. Antibacterial activity of selected plant essential oils against *Escherichia coli* O157:H7. *Appl. Microbiol.* 36:162–167.

---

[*] As hypothesized by Pizzimenti et al. (2006).

Ceylan, E., and Fung, D. Y. C. 2004. Antimicrobial activity of spices. *J. Rapid Meth. Aut. Mic.* 12:1–55.

Di Pasqua, R., Betts, G., Hoskins, N., et al. 2007. Membrane toxicity of antimicrobial compounds from essential oils. *J. Agric. Food Chem.* 55:4863–4870.

Dorman, H. J. D., and Deans, S. G. 2000. Antimicrobial agents from plants: Antibacterial activity of plant volatile oils. *J. Appl. Microbiol.* 88:308–316.

Elgayyar, M., Draughon, F. A., Golden, D. A., and Mount, J. R. 2001. Antimicrobial activity of essential oils from plants against selected pathogenic and saprophytic microorganisms. *J. Food Protect.* 64:1019–1024.

Fisher, K., and Phillips, C. A. 2006. The effect of lemon, orange and bergamot essential oils and their components on the survival of *Campylobacter jejuni, Escherichia coli* O157, *Listeria monocytogenes, Bacillus cereus* and *Staphylococcus aureus* in vitro and in food systems. *J. Appl. Microbiol.* 101:1232–1240.

Fisher, K., and Phillips, C. 2008. Potential antimicrobial uses of essential oils in food: Is citrus the answer? *Trends Food Sci. Tech.* 19:156–164.

Fisher, K., and Phillips, C. 2009a. The mechanism of action of a citrus oil blend against *Enterococcus faecium* and *Enterococcus faecalis*. *J. Appl. Microbiol.* 106:1343–1349.

Fisher K., and Phillips, C. 2009b. In vitro inhibition of vancomycin-susceptible and vancomycin-resistant *Enterococcus faecium* and *E. faecalis* in the presence of citrus essential oils. *Brit. J. Biomed. Sci.* 66:180–185.

Fisher K., Rowe, C., and Phillips C. A. 2007. The survival of three strains of *Arcobacter butzleri* in the presence of lemon, orange and bergamot essential oils and their components in vitro and on food. *Lett. Appl. Microbiol.* 44:495–499.

Goni, P., Lopez, P., Sanchez, C., et al. 2009. Antimicrobial activity in the vapour phase of a combination of cinnamon and clove essential oils. *Food Chem.* 116:982–989.

Gupta, A. K., Einarson, T. R., Summerbell, R. C., et al. 1998. An overview of topical antifungal therapy in dermatomycoses. A North American perspective. *Drugs* 55: 645–674.

Hammer, K. A., Carson, C. F., and Riley, T. V. 1999. Antimicrobial activity of essential oils and other plant extracts. *J. Appl. Microbiol.* 86:985–990.

Helander, I. M., Alakomi, H. L., Latva-Kala, K., et al. 1998. Characterization of the action of selected essential oil components on gram-negative bacteria. *J. Agr. Food Chem.* 46:3590–3595.

Hsieh, P. C., Mau, J. L., and Huang, S. H. 2001. Antimicrobial effect of various combinations of plant extracts. *Food Microbiol.* 18:35–43.

Jacob, C., Mathiasen, L., and Powell, D. 2010. Designing effective messages for microbial food safety hazards. *Food Control* 21:1–6.

Kane, J., and Summerbell, R. C. 1999. *Trichophyton, Microsporum, Epidermophyton,* and other agents of superficial mycoses. In *Manual of Clinical Microbiology,* eds. P. R. Murray, E. J. Baron, M. A. Pfaller, F. C. Tenover, and R. H. Yolkin, 1275–1294. Washington: ASM Press.

Kirbaslar, F. G., Tavman, A., Dulger, B., and Turker, G. 2009. Antimicrobial activity of Turkish *Citrus* peel oils. *Pakistan J. Bot.* 41:3207–3212.

Lambert, R. J. W., Skandamis, P. N., Coote, P., and Nychas, G. J. E. 2001. A study of the minimum inhibitory concentration and mode of action of oregano essential oil, thymol and carvacrol. *J. Appl. Microbiol.* 91:453–462.

Lv, F., Liang, H., Yuan, Q., and Li, C. 2011. In vitro antimicrobial effects and mechanism of action of selected plant essential oil combinations against four food-related microorganisms. *Food Res. Int.* 44:3057–3064.

Mandalari, G., Bennett, R. N., Bisignano, G., et al. 2007. Antimicrobial activity of flavonoids extracted from bergamot (*Citrus bergamia* Risso) peel, a byproduct of the essential oil industry. *J. Appl. Microbiol.* 103:2056–2064.

McClements, J. D. 1999. *Food emulsions: Principles, practices and techniques*. Boca Raton, FL: CRC Press.

Moufida, S., and Marzouk, B. 2003. Biochemical characterization of blood orange, sweet orange, lemon, bergamot and bitter orange. *Phytochemistry* 62:1283–1289.

Oussalah, M., Caillet, S., and Lacroix, M. 2006. Mechanism of action of Spanish oregano, Chinese cinnamon and savory essential oils against cell membranes and cell walls of *Escherichia coli* O157:H7 and *Listeria monocytogenes*. *J. Food Protect.* 69:1046–1055.

Park, M. J., Gwak, K. S., Yang, I., et al., 2009. Effect of citral, eugenol, nerolidol and a-ter-pineol on the ultrastructural changes of *Trichophyton mentagrophytes*. *Fitoterapia* 80:290–296.

Pedro, A. S., Cabral-Albuquerque, E., Ferreira, D., and Sarmento, B. 2009. Chitosan: An option for development of essential oil delivery systems for oral cavity care? *Carbohyd. Polym.* 76:501–508.

Pizzimenti, F. C., Mondello, M. R., Giampà, M., et al. 2006. In vivo morphological and anti-fungal study of the activity of a bergamot essential oil by-product. *Flavour Frag. J.* 21:585–591.

Pizzimenti, F., Tulino, G., and Marino, A. 1998. Antimicrobial and antifungal activity of ber-gamot oil. In *Program and Abstracts of the International Congress: "Bergamotto 98. Stato dell'arte,"* 38. Reggio Calabria, Italy: Laruffa.

Romano, L., Battaglia, F., Masucci, L., et al. 2005. In vitro activity of bergamot natural essence and furocoumarin-free and distilled extracts, and their associations with boric acid, against clinical yeast isolates. *J. Antimicrob. Chemother.* 55:110–114.

Sánchez-González, L., Chafer, M., Chiralt, A., and Gonzalez-Martinez, M. 2010. Physical properties of edible chitosan films containing bergamot essential oil and their inhibitory action on *Penicillium italicum*. *Carbohyd. Polym.* 82:277–283.

Sánchez-González, L., Chafer, M., Chiralt, A., and Gonzalez-Martinez, M. 2011. Study of the release of limonene present in chitosan films enriched with bergamot oil in food simu-lants. *J. Food Eng.* 105:138–143.

Sanguinetti, M., Posteraro, B., Romano, L., et al. 2007. In vitro activity of *Citrus bergamia* (bergamot) oil against clinical isolates of dermatophytes. *J. Antimicrob. Chemother.* 59:305–308.

Santiesteban-Lopez, A., Palou, E., and Lopez-Malo, A. 2007. Susceptibility of food-borne bacteria to binary combinations of antimicrobials at selected a (w) and pH. *J. Appl. Microbiol.* 102:486–497.

Sikkema, J., De Bont, J. A. M., and Poolman, B. 1995. Mechanisms of membrane toxicity of hydrocarbons. *Microbiol. Rev.* 59:201–222.

Somolinos M., García, D., Condón, S., Mackey, B., and Pagán, R. 2009. Inactivation of *Escherichia coli* by citral. *J. Appl. Microbiol.* 108:1928–1939.

Statti, G. A., Conforti, F., Sacchetti, G., et al. 2004. Chemical and biological diversity of Bergamot (*Citrus bergamia*) in relation to environmental factors *Fitoterapia* 75:212–216.

Svoboda, K., and Greenaway, R. I. 2003. Lemon scented plants. *Int. J. Aromather.* 13:23–32.

Ultee, A., Bennik, M. H. J., and Moezelaar, R. 2002. The phenolic hydroxyl group of carvacrol is essential for action against the food-borne pathogen *Bacillus cereus*. *Appl. Environ. Microbiol.* 68:1561–1568.

Ultee, A., Kets, E. P. W., Alberda, M., Hoekstra, F. A., and Smid, E. J. 2000. Adaptation of the food-borne pathogen *Bacillus cereus* to carvacrol. *Arch. Microbiol.* 174:233–238.

Vaara, M. 1992. Agents that increase the permeability of the outer membrane. *Microbiol. Rev.* 56:395–411.

Wang, Q., Gong, J., Huang, X., Yu, H., and Xue, F. 2009. In vitro evaluation of the activity of microencapsulated carvacrol against *Escherichia coli* with K88 pili. *J. Appl. Microbiol.* 107:1781–1788.

21 Pharmacological Activity of Essential Oils and Juice

Antonella Saija, Giuseppe Bisignano, and Francesco Cimino

CONTENTS

21.1 INTRODUCTION

Bergamot is a small and roughly pear-shaped citrus fruit that grows on small trees known as bergamots (*Citrus bergamia* Risso, Rutaceae). Bergamot plants, despite their uncertain species definition and phytogeographical origin, are mainly cultivated in Ionian coastal regions of southern Italy, Argentina, and Brazil. Bergamot essential oil is used in the cosmetic and perfumery industries, regardless of the presence on the market of synthetic surrogates. Bergamot juice has not reached the popularity of the other citrus juices in the daily diet for its organoleptic properties, but it is used to fortify fruit juice in place of synthetic additives. A peculiarity of bergamot is the considerable abundance and variety of bioactive compounds, present in all parts of the fruit and so recovered in large amounts in its essential oil, juice, and peels. Over the past few years, following the growing interest in antioxidant bioactive compounds and their dietary sources, bergamot juice has attracted attention as a result of its remarkable flavonoid content. Despite little clinical evidence proving their effects, bergamot derivatives (especially essential oil and juice) have been used for centuries, reputedly effectively, as a traditional medicine, in particular as digestive tonic and worming agents. Due to their strong antiseptic properties, they have also been employed to treat a variety of infections and inflammations of the respiratory, digestive, and urinary systems. However, in recent years much has been achieved with regards to knowledge of the biological properties of these phytocomplexes and, especially, of their active principles, some of which are therapeutic.

Herein we review the scientific evidence concerning the bioactivities of bergamot-derived products with potential utility for human health, as well as the therapeutic uses of some pure compounds isolated from them.

21.2 NEUROBIOLOGICAL PROPERTIES OF BERGAMOT ESSENTIAL OIL

21.2.1 CENTRAL EFFECTS

The data accumulated in the literature so far indicate that bergamot essential oil (BEO), obtained from the peel of *Citrus bergamia* Risso, is endowed with interesting neurobiological effects due, partially at least, to an interference with basic mechanisms involved in synaptic plasticity under physiological as well as pathological conditions (Bagetta et al. 2010).

A large number of biological effects induced by BEO may be related to its capability to modulate neurotransmitter functions. The BEO volatile fraction contains some monoterpene hydrocarbons able to stimulate glutamate release by transporter reversal and/or by exocytosis, depending on the dose administered (Morrone et al. 2007). High concentrations of BEO provoke glutamate release by a Ca^{2+}-independent carrier-mediated process, while lower concentrations stimulate exocytosis of the excitatory amino acid, possibly by activation of presynaptic receptors located on glutamate-releasing nerve endings. In fact, intraperitoneal administration of BEO in rats can significantly affect the extracellular concentration of aspartate, glycine, and taurine; furthermore, when perfused into the hippocampus, BEO produces a significant increase of extracellular aspartate, glycine, and taurine as well as of GABA and glutamate. These effects appear to be dependent on the glutamate transporters and on extracellular Ca^{2+}.

Rombolà et al. (2009) described the systemic effects of this phytocomplex on gross behavior and EEG activity recorded from the hippocampus and cerebral cortex of the rat. Systemic administration of BEO produces dose-dependent increases in rat locomotor and exploratory activity that correlate with significant changes in the EEG spectrum. These findings provide evidence that the behavioral and EEG spectrum power effects of BEO correlate well with its capability to affect exocytotic and carrier-mediated release of discrete amino acid neurotransmitters in the mammalian hippocampus, and thus lead to the hypothesis that BEO is able to interfere with normal and pathological synaptic plasticity.

Other findings concerning BEO neuroprotection against experimental brain ischemia are in agreement with this hypothesis. Corasaniti et al. (2007) have demonstrated that BEO can reduce neuronal damage caused in vitro by excitotoxic stimuli by preventing injury-induced engagement of critical cell death pathways. In fact, the protective effects of BEO on excitotoxic damage were demonstrated in human SH-SY5Y neuroblastoma cells exposed in vitro to N-methyl-D-aspartate (NMDA).

In addition to preventing accumulation of reactive oxygen species (ROS) and activation of the calcium-activated protease calpain, BEO counteracted the deactivation of the prosurvival kinase Akt and the consequent activation of the glycogen synthase kinase-3beta induced by NMDA. Results obtained using specific fractions of BEO have suggested that monoterpene hydrocarbons are responsible for neuroprotection afforded by BEO against NMDA-induced cell death.

Amantea et al. (2009) investigated the effects of systemic pretreatment with BEO on brain damage following permanent focal cerebral ischemia (induced by occlusion of the middle cerebral artery, MCAo) in rats. BEO is able to significantly reduce infarct size following permanent MCAo throughout the brain, especially in the medial striatum and the motor cortex. BEO does not affect basal amino acid levels, whereas it significantly reduces aspartate and glutamate efflux in the frontoparietal cortex following MCAo. These early neuroprotective effects are associated with a significant increase of p-Akt and in the phosphorylation of the deleterious downstream kinase GSK-3beta, whose activity is negatively regulated by p-Akt.

The molecules responsible for the neurobiological effects described here are very likely one of the compounds present in the BEO volatile fraction, such as linalool, while a contribution from bergapten or other furocoumarins present in BEO can be excluded (Bagetta et al. 2010).

21.2.2 ANXIOLYTIC PROPERTIES

The word "aromatherapy" combines two words: aroma (a fragrance or sweet smell) and therapy (a treatment) and is used to describe a wide range of practices involving odorous substances, although only aroma delivery through inhalation to induce psychological or physical effects can be correctly defined as aromatherapy. This complementary medicine is highly diffused in the industrialized countries to improve mood and mild symptoms of stress-induced disorders such as anxiety, depression, behavioral disturbances in dementia, and chronic pain (Edris 2007). Physiological and pharmacological studies, combined with phytochemical analyses, indicate that a variety of essential oils exert specific effects on the central nervous system (CNS), since their individual components of essential oils reach the blood, cross the blood-brain barrier, and enter the CNS following inhalation, dermal application, intraperitoneal (i.p.) or subcutaneous (s.c.) injection, and oral administration.

A number of essential oils are currently in use as aromatherapy agents to relieve anxiety, stress, and depression. Popular anxiolytic oils include those from several *Citrus spp* and also BEO, which is used widely in aromatherapy despite limited scientific evidence of its effectiveness.

Aromatherapy may be especially useful for populations that work under high stress, since workplace stress-related illness is a serious issue. Chang and Shen (2011) have tested the response to bergamot essential oil, when used for aromatherapy spray for 10 minutes, in 54 elementary school teachers (a high-stress working population) in Taiwan. Results showed that there were significant decreases in blood pressure and heart rate, also depending on several variables such as age, gender, and position.

Although blended essential oils are increasingly being used for improvement in the quality of life and for the relief of various symptoms in patients, the scientific evaluation of the aroma-therapeutic effects of blended essential oils in humans is rather scarce. Hongratanaworakit (2011) hypothesized that applying blended essential oil consisting of lavender and bergamot oils would provide a synergistic effect useful in treating depression or anxiety. In this study, blended essential oil was applied topically to the abdominal skin of 40 healthy volunteers. In the subjects treated with blended essential oils significant decreases in pulse rate and systolic and diastolic blood pressure were observed; furthermore, these subjects rated themselves as "more calm" and "more relaxed" than controls.

Concerning the mechanism of action, Morrone et al. (2007) showed that BEO significantly increases γ-aminobutyric acid levels in rat hippocampus, which explains its potential anxiolytic properties. Saiyudthong and Marsden (2011) investigated the effect of inhaled BEO administered to rats on both anxiety-related behaviors and stress-induced levels of plasma corticosterone in comparison with the effects of injected diazepam. Both BEO and diazepam exhibited anxiolytic-like behaviors and attenuated hypothalamic-pituitary-adrenal axis activity by reducing the corticosterone response to stress.

21.2.3 ANTINOCICEPTIVE PROPERTIES

Due to the increasing use of aromatherapy oils, the antinociceptive properties of *Citrus* essential oils have also been investigated by several authors. Sakurada et al. (2009) reported that the capsaicin-induced nociceptive response in mice is significantly reduced by intraplantar injection of BEO, while orange sweet essential oil has no effect. Among the monoterpene hydrocarbons found in BEO volatile fraction, linalool might be responsible for the antinociceptive effects of this phytocomplex. In fact, linalool possesses antinociceptive, antihyperalgesic, and anti-inflammatory properties in different experimental animal models. Linalool effects in neuropathic pain have been investigated by Berliocchi et al. (2009). Chronic administration of linalool is able to reduce mechanical allodynia (but not sensitivity to noxious radiant heat) following spinal nerve ligation as a model of neuropathic pain in mice. Mechanisms other than modulation of inflammatory processes may be mediating the protective effect of linalool in this model of neuropathic pain.

Sakurada et al. (2011) showed that i.p. and intraplantar pretreatment with naloxone hydrochloride, an opioid receptor antagonist, significantly reversed BEO- and linalool-induced antinociception. Pretreatment with naloxone methiodide, a peripherally acting μ-opioid receptor antagonist, resulted in a significant antagonizing effect on antinociception induced by BEO and linalool. Antinociception induced by i.p. or intrathecal morphine was enhanced by the combined injection of BEO or linalool. The enhanced effect of a combination of BEO or linalool with morphine was antagonized by pretreatment with naloxone hydrochloride. These results provide evidence for the involvement of peripheral opioids in the antinociception induced by BEO and linalool, suggesting that combined administration of BEO or linalool (acting at the peripheral site) and morphine may be a promising approach in the treatment of clinical pain.

21.3 PHOTOTOXICITY OF BERGAMOT OIL AND METHOXYPSORALEN

BEO has been used for centuries as a fundamental ingredient of many perfumes. Its widespread use depends on its particular fragrance, though that fragrance rapidly disappears when the oil is exposed to solar radiation.

Application of BEO-containing perfumes on skin areas exposed to solar radiation can cause undesired phototoxic effects (edema and long-lasting erythema) and, successively, a pronounced cutaneous pigmentation, which are the classic clinical symptoms of Berlock dermatitis. This pathological condition is due to 5-methoxypsoralen (5-MOP, or bergapten; see Figure 21.1), which is present in pure bergamot essence in concentrations up to 3600 ppm.

The phototoxic, photomutagenic, and photocarcinogenic properties of bergamot oil essentially lead back to 5-MOP and other components, such as bergamottine (5-geranoxy-psoralen), contained in it. These properties were exhaustively studied even in experimental conditions simulating solar exposure and described by various research groups (Averbeck et al. 1990; Dubertret et al. 1990; Morlière et al. 1990; Young et al. 1990), and can be attributed to the ability of 5-MOP to form stable adducts with DNA in the presence of UV radiation (Bisignano and Saija 2002; Bisignano et al. 2010) (Figure 21.2). On the other hand, bergamottine, which is the major UVB plus UVA radiation absorber in the bergamot oil, is rapidly degraded following UV radiation exposure. However, contrary to many furocoumarins, bergamottine does not strongly interact with DNA (Aubin et al. 1991; Morlière et al. 1991).

The International Fragrance Association (IFRA) recommends the use of bergamot oil in concentrations not higher than 0.4% in preparations to be applied on skin exposed to solar radiation (except for preparations that can be washed off the skin, such as soaps). In accordance with IFRA guidelines, it is recommended that the total levels of bergapten in consumer products should not exceed 0.0015% (15 ppm) for applications on areas of skin exposed to sunshine, excluding bath preparations, soaps, and other products which are washed off the skin. This concentration is equivalent to 0.0075% (75 ppm) in a fragrance compound used at 20% in the consumer product.

Since ancient times, many cultures worldwide found out independently that the topical administration of some photoactive natural products (mainly extracted from

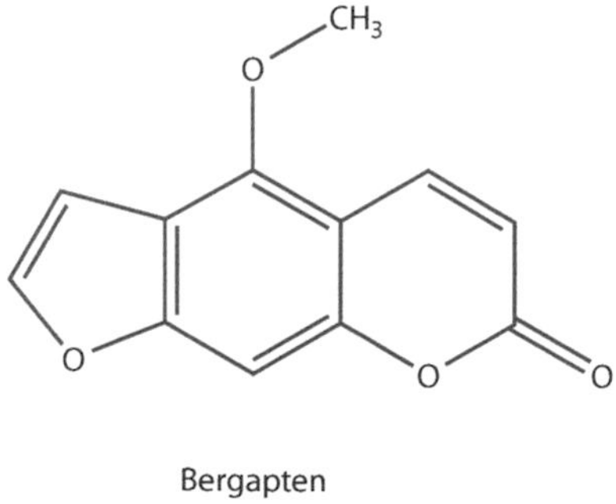

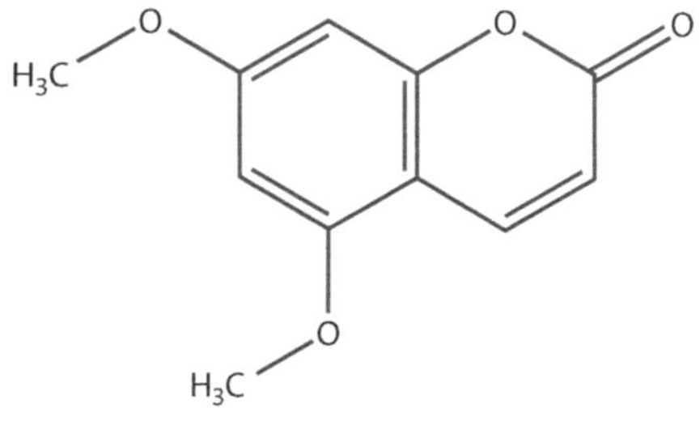

Bergapten   Citropten

FIGURE 21.1 Structural formula of bergapten (5-methoxypsoralen, 5-MOP) and citropten (5,7-dimethoxycoumarin).

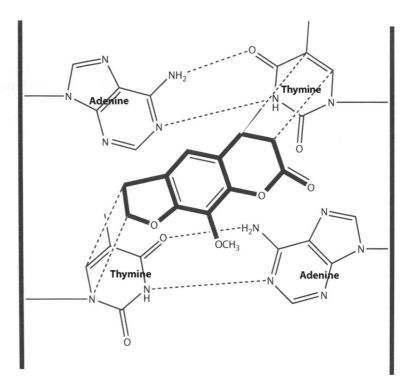

FIGURE 21.2 Stable adducts formed between 5-MPO and DNA in the presence of UV radiations.

plants), followed by exposure to sunlight, might be an effective treatment of some skin diseases, thus accidently giving birth to so-called photochemotherapy. Today it has become evident that phototherapy is the oldest "biological" therapeutic strategy, whose main target is the T-cell-mediated immunopathology of psoriasis (Schneider et al. 2008; Nagy et al. 2010). In the attempt to resemble nature by exploiting its teaching, during the last two centuries, scientists tried to rationalize this knowledge in order to develop more effective therapeutic strategies and to understand in depth the mechanisms of action involved, expanding the potential application of this therapy to a larger number of skin pathologies, including neoplasias.

The peculiar photosensitizing properties of 5-MOP also provide a phototherapeutic activity of this psoralen, which is being fully exploited in the so-called PUVA therapy (psoralen plus UVA). The administration (topic or systemic) of a psoralen (5-MOP, 8-MOP, or 4,5′,8-trimethoxypsoralen, TMP) together with UVA irradiation has been clearly demonstrated to be therapeutically efficacious in the treatment of two radically different pathological conditions: psoriasis and vitiligo.

Vitiligo is characterized by the spontaneous appearance, often progressive and persistent, of depigmented cutaneous patches; this pathology is due to an alteration in melanocyte functionality and thus in melanin production; T-lymphocytes might also be involved in melanocyte destruction characterizing this pathology by a sort of auto-immune mechanism. Psoriasis is a common, chronic, inflammatory, multisystem disease with predominantly skin and joint manifestations affecting approximately 2% of the population; it is characterized by raised, scaly, reddened cutaneous plaques, resulting from hyperproliferation of the epidermis and inflammation of both epidermal and dermal layers.

The rationale for the use of PUVA therapy in the treatment of vitiligo and psoriasis is rather complex. The photoactivated molecules of the psoralen form mono- or bi-functional adducts (crosslinks) between their 4',5' and/or 3,4 double bonds and the pyrimidine bases (thymine and cytosine) of DNA strands. These crosslinks are thought to prevent DNA replication and thus the hyperproliferative cutaneous state, a condition peculiar to psoriasis. Furthermore, the photoactivation of the psoralen causes generation of ROS, such as superoxide; this stimulates melanocyte proliferation and increases pigmentation, certainly fundamental factors in the beneficial effects of PUVA therapy in the treatment of vitiligo. Furthermore, 5-MOP may enhance the effects of UV exposure on cutaneous cytokine release. In particular, the efficacy of PUVA therapy is based on the capacity of psoralen to increase skin sensitivity to the beneficial effects of UV light (pigmentation and hypoproliferation, due to photoexcited psoralen molecules), and to provoke simultaneously a photoadaptation (pigmentation), thus protecting the skin from the toxic effects of UV radiation (erythema, edema, exfoliation).

PUVA therapy-induced melanogenesis seems to be consequent to an increased expression of tyrosinase (Mengeaud and Ortonne 1996) and to a melanocyte migration from hair follicles that is stimulated by cytokines released in the skin following UV irradiation. Moreover, PUVA therapy brings about a destruction of T-lymphocytes, resulting in a reduction of melanocyte destruction and thus in an inhibition of the further development of new vitiligo lesions.

Detailed papers on pharmacodynamic and pharmacokinetic properties, tolerability, and dosage schedules of the psoralens in PUVA therapy are widely present in international scientific literature (Morison 2004; Menter et al. 2010; Lapolla et al. 2011).

Recently, Amirnia et al. (2011) compared PUVA therapy with topical steroids in moderate plaque psoriasis. Although both PUVA therapy and topical steroids are equally efficient and cost-effective in moderate plaque psoriasis, the recurrence rate is higher in the latter group. Furthermore, retrospective studies revealed that the primary efficacy of PUVA may be superior to that of certain biologics, such as infliximab (Inzinger et al. 2011).

The process of choosing among potential treatment options for the treatment of moderate to severe psoriasis requires both the physician and the patient to weigh the benefits of individual modalities against their potential risks. Patel et al. (2009) reviewed the existing body of literature in order to define the known incidence of malignancy associated with PUVA and other therapies for moderate to severe psoriasis. Psoralen plus UVA, when given long term, is associated with increased risks of cutaneous squamous cell carcinoma and malignant melanoma. Stern (2009) reported the long-term risk of basal cancer cell with PUVA, particularly for patients exposed to PUVA when young. However, the majority of studies cited in these reviews lack the power and randomization of large clinical trials, as well as the long-term follow-up periods which would further substantiate the hypothetical link between these antipsoriatic treatment regimens and the potential for malignancy. Additionally, the increased risk of malignancy associated with psoriasis itself is a confounding factor.

PUVA therapy can induce chromosome damage in psoriatic patients. More exactly, psoriasis is accompanied by clastogenic factor-induced chromosomal breakage that increases, in a persistent way, during PUVA treatment (Emerit et al. 2011). Research demonstrating an increased incidence of skin cancer with PUVA therapy have been generally carried out on Caucasian patients, but the risk of nonmelanoma skin cancer with long-term PUVA therapy does not appear to increase in Asian and Arabian-African populations (Murase et al. 2005). Therefore, in phototherapy risk assessment, it is important to consider the patient's skin phototype and the potential protection that a more pigmented skin may confer.

PUVA treatment induced fibroblast senescence (as shown by increased expression cell senescence–associated β-galactosidase, shortened telomere length, and increased level of ROS), and irradiation with intense pulsed light after PUVA exposure partially rejuvenated the cells, demonstrating a protective effect (Wang et al. 2011). There is marked unpredictable interpatient variation in responses to PUVA photochemotherapy. Identification of molecular biomarkers of PUVA sensitivity may facilitate treatment predictability.

Exposure of human skin to UV radiation results in the generation of reactive intermediates and oxidative stress. Hepatic drug metabolizing and cytoprotective genes are induced as an adaptive response to xenobiotics and reactive intermediates; several of these genes are present in skin and their cutaneous expression and regulation is implicated in responses to UV. The findings by Smith et al. (2003) indicate that some genes may be associated with individuality in response to ultraviolet radiation. In fact, UV exposure elicits skin induction of cyclooxygenase-2, glutathione peroxidase (at a lower degree), glutathione-S-transferase (GST) P1, and the drug transporter multidrug resistance–associated protein-1. Several human GSTs, including GSTM1 and GSTT1, are polymorphic, and null polymorphisms have been associated with increased UVB erythemal sensitivity and skin cancer risk. Finally, GSTP1 and multidrug resistance–associated protein-1 appear to be significantly increased in psoriatic plaque, as well as CYP2E1 and heme oxygenase-1, implying a differential adaptive response to oxidant exposure in lesional psoriatic skin.

Since GSTs influence cutaneous defense against UVR-induced oxidative stress, they are candidate biomarkers of PUVA sensitivity. As reported by Ibbotson et al. (2011) in a study carried out on patients starting PUVA, the polymorphic human GSTs are associated with PUVA sensitivity; in fact, lower minimal phototoxic doses and higher serum 8-MOP concentrations were seen in GSTM1 null allele homozygotes compared with patients with one or two active alleles. According to these findings, exposure of mammalian cells to 8-MOP induced antioxidant and detoxifying gene expression, modulated by the same active motif in promoters of cytoprotective genes including GSTs, suggesting that these genes may be implicated in 8-MOP metabolism.

The melanocortin 1 receptor (MC1R) is a highly polymorphic G protein–coupled receptor. Inheritance of various MC1R alleles has been associated with a red hair/fair skin phenotype, increased incidence of skin cancer, and altered sensitivity to UV radiation. In particular, inheritance of the Val(60)Leu and Arg(163)Gln SNPs and of two or more MC1R SNPs were associated with increased PUVA erythemal sensitivity (Smith et al. 2007). Thus the MC1R genotype can influence PUVA erythemal sensitivity in patients with psoriasis and other common skin diseases. Besides psoriasis and vitiligo, other cutaneous pathologies could benefit from PUVA therapy and, more generally, from therapies based on the use of photosensitizing agents and UVA irradiation.

Cutaneous T-cell lymphoma (CTCL) is relatively benign in its early stages, but survival rates decrease dramatically as the disease progresses. As no curative treatment is currently available, the goal of therapy is preventing or delaying progression from early disease stages while minimizing long-term toxicity. PUVA, as well as extracorporeal photopheresis (ECP), are widely accepted types of photochemotherapy used for the treatment of CTCL (Geskin 2007; Stadler 2007; Schneider et al. 2008; Pothiawala et al. 2010a, 2010b; Raphael et al. 2011). Thus, both PUVA and ECP utilize a photosensitizing agent, which can be taken orally (PUVA) or added to the concentrated sample of white blood cells extracorporeally (ECP) prior to UVA exposure. More particularly, ECP is based on the biological effect of 8-MOP and UVA on mononuclear cells collected by apheresis, and reinfused into the patient (Oliven and Shechter 2001; Vagace Valero et al. 2003; Aubin and Mousson 2004). Both PUVA and ECP have been shown to be safe and effective for the treatment of CTCL. As a monotherapy, PUVA is preferentially used for treatment of patients at earlier stages with skin involvement alone, while ECP is usually used for patients with erythrodermic skin involvement in advanced stages with peripheral blood involvement.

However, no single agent, including PUVA, can control CTCL progression fully, so combination therapy is necessary to improve response rates. In addition, low-dose combination therapy may improve treatment safety and tolerability. A combination of PUVA and interferon alpha in early disease has been shown to be effective and well tolerated. Likewise, small studies of PUVA and bexarotene indicate good efficacy for this combination. Reduced doses of these combinations may also be effective as maintenance therapies following complete remission.

Concerning other cutaneous pathologies which could be successfully treated with PUVA therapy, Menichini et al. (2010) suggested that the phototoxicity of BEO might be considered as a treatment option in some cases of lentigo maligna or lentigo maligna melanoma. *C. bergamia* oil exhibits an antiproliferative activity against the malignant melanoma A375 cell line after exposure to UV irradiation. This seems related to the presence of bergapten (but not to that of citropten) in the bergamot oil.

Furthermore, Gleeson et al. (2011) recently reported a series of patients with cutaneous sarcoid successfully treated with topically applied psoralen and UVA. Cutaneous manifestations of sarcoidosis occur in 25%–30% of patients and are typically difficult to manage. Treatment of sarcoid is determined by the type and distribution of skin lesions and by the presence of systemic disease. Systemic prednisolone is the first-line treatment for patients with widespread, disfiguring cutaneous lesions or ulceration, but may be contraindicated or ineffective in up to one-third of cases, while topical treatments have fewer side effects but are often less effective because of low penetration. Sarcoidosis is characterized histologically by noncaseating granulomas, and it is believed that several immunopathological steps result in granuloma formation. The initiating step is thought to be exposure to an as-yet-unidentified antigen. Dendritic cells, including Langerhans cells, have been shown to be associated with granulomas in skin lesions. The mechanism of action of PUVA radiation in sarcoidosis can be postulated to be based on its immunosuppressive effect and capability to deplete epidermal Langerhans cells.

Finally, PUVA has become a common form of treatment for early-stage mycosis fungoides (MF) (Soung et al. 2005). PUVA for patients with early-stage MF is a safe and effective therapeutic modality with prolonged disease-free remissions; however, PUVA alone was not adequate for more advanced disease.

21.4 BERGAMOT-DERIVED COMPOUNDS AS POTENTIAL THERAPEUTIC AGENTS IN GENETIC DISEASES

The identification and characterization of potential therapeutic agents in hematological diseases such as β-thalassaemia and sickle cell anemia are based on several approaches, including pharmacologically mediated regulation of human γ-globin genes expression (El-Beshlawy et al. 2009; Testa 2009; Thein and Menzel 2009). Fetal hemoglobin (HbF) production is a generally accepted approach to ameliorate clinical hematological parameters of β-thalassemic patients; in fact, inducers of HbF (such as hydroxyurea) might be able to convert β-thalassemia patients (requiring a regular transfusion regimen) to transfusion-independent subjects. Novel HbF inducers are needed since some patients are refractory to the treatment and others develop resistance; not all of the patients are responders to single HbF inducer and the combined treatment with different HbF inducers is much more effective than the treatment with single compounds. HbF induction is clearly an important strategy in developing countries, where it is difficult to extend blood transfusion to all the population due to the unavailability of blood and the risk of infection.

Recently, two extracts obtained from epicarps of *Citrus bergamia* fruits and rich in coumarins and psoralens, as well as the main constituents detected in them (citropten and bergapten), were assayed for their capacity to increase erythroid differentiation of K562 cells and expression of γ-globin genes in human erythroid precursor cells. This was done using three experimental cell systems (human leukemic K562 cell line; K562 cell clones stably transfected with a pCCL construct carrying green-enhanced green fluorescence protein under the γ-globin gene promoter; and two-phase liquid culture of human erythroid progenitors isolated from healthy donors) (Guerrini et al. 2009). The results suggest that citropten and bergapten (Figure 21.1) are powerful inducers of differentiation and γ-globin gene expression in human erythroid cells and may be considered good candidates for potential therapeutic employment in hematological disorders, including β-thalassemia and sickle cell anemia.

The use of similar extracts from bergamot epicarps and of their main components, bergapten and citropten, has been investigated by Borgatti et al. (2011) as far as their capability to alter the expression of IL-8 associated with the cystic fibrosis (CF) airway pathology. CF airway pathology is a fatal, autosomal, recessive genetic disease characterized by extensive lung inflammation. After induction by tumor necrosis factor-α (TNF-α), elevated concentrations of several pro-inflammatory cytokines and chemokines (i.e., IL-6, IL-1β, IL-8) are released from airway epithelial cells. New therapeutic approaches, including medicinal plant extracts, have been studied and developed to reduce the excessive inflammatory response in the airways of CF patients. These bergamot extracts, as well as bergapten and citropten, appear able to affect the release of cytokines and chemokines released from cystic cultured fibrosis IB3-1 cells treated with TNF-α. These effects have been confirmed by evaluating mRNA levels and protein release in the CF cellular models IB3-1 and CuFi-1 challenged with TNF-α or exposed to heat-inactivated *Pseudomonas aeruginosa*. Thus bergapten and citropten are endowed with a significant capability to inhibit IL-8 expression, and further studies could verify their possible anti-inflammatory properties to reduce lung inflammation in CF patients.

21.5 HYPOLIPIDEMIC AND ANTIATHEROSCLEROTIC PROPERTIES OF BERGAMOT DERIVATIVES

Hyperlipidemia or hypercholesterolemia is an important risk factor for the development of atherosclerosis and coronary artery disease (Gielen et al. 2009). Increased concentrations of blood cholesterol bound to low-density lipoprotein (cLDL), total cholesterol (totChol), and triglycerides (TG), as well as insulin resistance, are strongly related to a high risk of cardiovascular diseases (CVD) (Jones 2008) (Figure 21.3).

The main therapeutic approach of hypercholesterolemia is based on statins that are inhibitors of 3-hydroxy-3-methylglutaryl-CoA (HMG-CoA) reductase, a key enzyme in cholesterol metabolism (Figures 21.4 and 21.5). However, many patients, in particular those with metabolic syndrome, do not achieve their recommended low-density and high-density lipoprotein (LDL, HDL) cholesterol target goals with statins. Moreover, statins induce side effects (including myalgia,

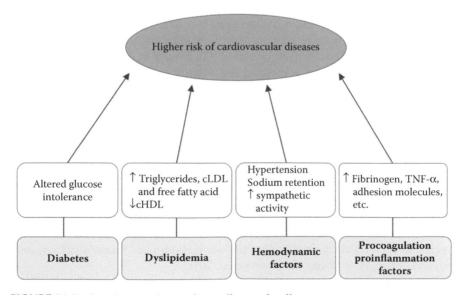

FIGURE 21.3 Leading risk factors for cardiovascular diseases.

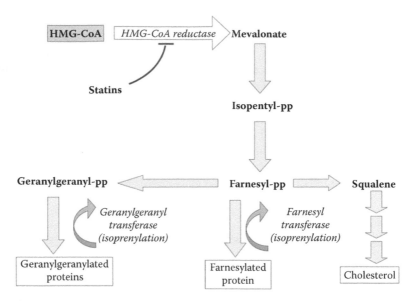

FIGURE 21.4 Hydroxymethylglutaryl-coenzyme A (HMG-CoA) is the precursor for cholesterol synthesis. Statins are inhibitors of HMG-CoA reductase, a key enzyme in cholesterol production. pp: pyrophosphate.

myopathy, or liver disease, and rhabdomyolysis in more severe cases) in more than 40% of patients (Alsheikh-Ali and Karas 2009; Joy and Hegele 2009). The traditional use of bergamot juice in the Calabrian region suggests that local residents have long known of its potential beneficial use in counteracting hypercholesterolemia and atherosclerosis.

Bergamot juice and albedo present a unique profile of flavonoid and flavonoid glycosides such as neoeriocitrin, neohesperidin, naringin, rutin, neodesmin, rhoifolin, and poncirin, differing from other citrus fruits also because of the particularly high content of these compounds (Dugo et al. 2005; Nogata et al. 2006). In particular, naringin (contained also in grapefruit) has already been reported to be active in animal models of atherosclerosis (Choe et al. 2001), while neoeriocitrin and rutin have been shown to inhibit LDL oxidation. Importantly, bergamot juice (BJ) is rich in neohesperidosides of hesperetin and naringenin, such as melitidin and brutieridin (identified to be structural analogues of statins) (Figure 21.5). These flavonoids possess a 3-hydroxy-3-methylglutaryl moiety with a structural similarity to the natural substrate of HMG-CoA reductase and are likely to exhibit statin-like proprieties (Di Donna et al. 2009; Leopoldini et al. 2010). In addition, naringin (the glycoside derivative of naringenin) has been shown to inhibit hepatic HMG-CoA reductase (Kim et al. 2004). Since bergamot albedo is a polluting waste in the production of BEO and is easily available by simple industrial processes, it might be of economic interest to set a chemical procedure to recover melitidin and brutieridin from water extracts of bergamot albedo (Di Donna et al. 2011).

Experimental evidence obtained in animal models of diet-induced hypercholesterolemia and renal damage (Miceli et al. 2007; Trovato et al. 2010) as well as in

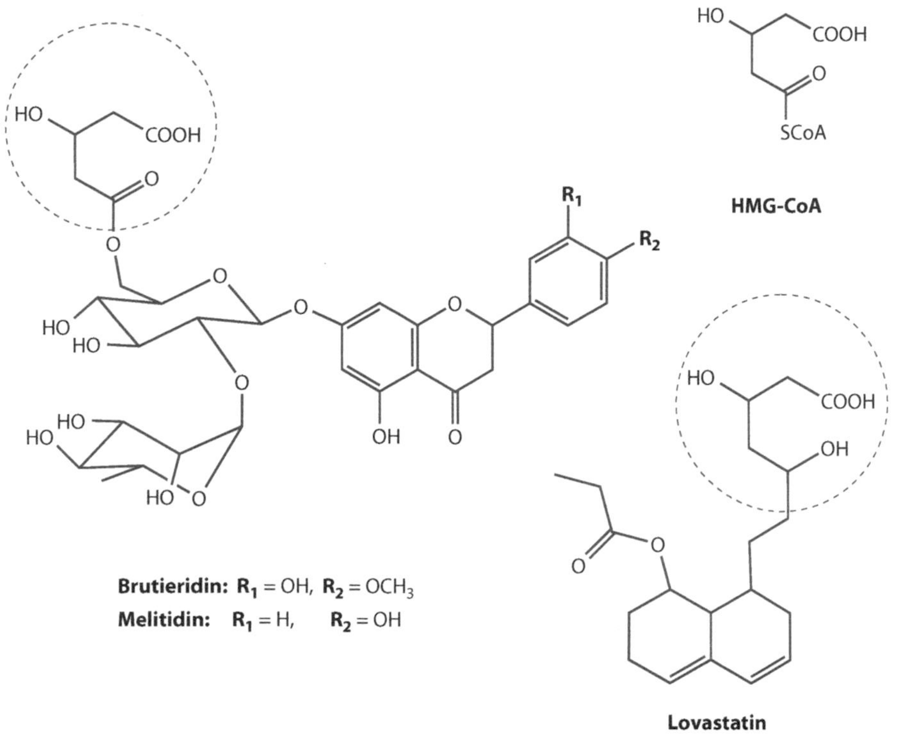

FIGURE 21.5 Structural formula of brutieridin and melitidin (found in bergamot albedo) and lovastatin (a statin). The circle evidences the hydroxymethylglutaryl (HMG)-moiety of these molecules, present also in HMG-CoA. Experimental studies have found that HMG-moiety of brutieridin and melitidin effectively binds with HMG-CoA reductase.

an experimental model of mechanical stress-induced vascular injury (Mollace et al. 2008) support the hypolipemic and vasoprotective effects of bergamot constituents.

The hypolipidemic effects of BJ and its protective effect on liver were investigated in rats fed a hypercholesterolemic diet by Miceli et al. (2007). Chronic administration of bergamot juice provoked a significant reduction in serum totChol, TG, and LDL levels and an increase in HDL levels in hyperlipidemic rats; moreover, histopathological observations showed a protection of hepatic parenchyma. In addition, fecal neutral sterols and fecal bile acid excretion increased after bergamot treatment. These results suggest that the hypocholesterolemic effect of BJ may be mediated by the increase in fecal neutral sterols and total bile acid excretion and also by its antioxidant properties. In the same experimental model, short-term diet-induced hypercholesterolemia causes renal damage (as demonstrated by accumulation of lipoperoxidation products and histological examination), although the renal glomerular function is not compromised; also, this damage appeared to be prevented by chronic administration of BJ (Trovato et al. 2010).

Mollace et al. (2011) have investigated, in a rat model of diet-induced hyperlipidemia, the effect of bergamot-derived polyphenolic fraction (BPF), standardized for its content in neoeriocitrin, naringin, neohesperidin, melitidin, and brutieridin, on totChol, cLDL, cHDL, TG, and blood glucose. Administration of BPF in diet-induced hypercholesterolemic rats produced a significant reduction in totChol, cLDL, and triglycerides, an effect accompanied by moderate elevation of cHDL. In agreement with these findings, BDF given orally for 30 consecutive days to patients suffering from isolated or mixed hyperlipidemia, either associated or not with hyperglycemia, produced a significant reduction in blood levels of totChol and cLDL without any side effects. Furthermore, metabolic syndrome patients presented with a highly significant reduction in blood glucose levels. Finally, BPF inhibited HMG-CoA reductase activity and enhanced reactive vasodilation. The authors suggested that melitidin and brutieridin in concert with naringin and other flavonone glycosides might be responsible for the effect of BPF in reducing cholesterol levels.

The antioxidant/radical scavenging properties of bergamot derivatives may be fundamental to explain their capability to protect against endothelial dysfunction. Since oxidative damage due to ROS overproduction has been proposed to play a pivotal role in the development of age-dependent diseases such as atherosclerosis, there is today increasing interest in the properties of some natural antioxidants, especially those found in edible plants, which might have a role in preventing these pathologies. In fact, recently Wei and Shibamoto (2010) have reported that bergamot oil exhibits a strong antioxidant activity in the aldehyde/carboxylic acid assay.

The BEO nonvolatile fraction has a protective effect on LOX-1 expression and free radical generation in common carotid artery injury induced by balloon angioplasty in rats (Mollace et al. 2008). Lectin-like oxyLDL receptor-1 (LOX-1) has recently been suggested to be involved in smooth muscle cell (SMC) proliferation and neointima formation in injured blood vessels. These results suggest that natural antioxidants from bergamot oil may be relevant in the treatment of vascular disorders in which proliferation of SMCs and oxyLDL-related endothelial cell dysfunction are involved.

Good amounts of bioactive compounds effective against endothelium dysfunction are contained not only in bergamot oil but also in its peel, which represents about 60% of the processed fruits and is regarded as primary waste. Trombetta et al. (2010) reported that two flavonoid-rich extracts from bergamot peel, endowed with strong antioxidant activity, are able to prevent alterations induced by the pleiotropic inflammatory cytokine TNF-α on human umbilical vein endothelial cells. This is demonstrated by monitoring intracellular levels of malondialdehyde/4-hydroxynonenal, reduced and oxidized glutathione, superoxide dismutase activity, and the activation status of nuclear factor-kappaB.

21.6   BERGAMOT DERIVATIVES AND CHEMOPREVENTION

Since several papers have demonstrated the capability of BEO or its components to modulate cellular signaling pathways, in particular those related to Akt, whose activity can regulate cell growth and proliferation, it is interesting to verify the effect on these phytocomplexes on proliferation of cancer cell.

Berliocchi et al. (2011) studied the potential effects of BEO on survival and proliferation of dividing cells using human SH-SY5Y neuroblastoma cell line. BEO triggered concentration-dependent mitochondrial dysfunction, cytoskeletal reorganization, cell shrinkage, DNA fragmentation, and both caspase-dependent and independent cell death. Analysis of cleavage products of poly-(ADP-ribose) polymerase revealed caspase-3 activation, but also activation of additional protease families. As a result of increased proteolytic activity, Akt protein levels decreased in BEO-treated cells.

Panno et al. (2009, 2010) found that bergapten by itself and PUVA are able to significantly affect proliferation in several lines of breast cancer cells (MCF-7, ZR-75, and SKBR-3). Bergapten induced a lowering of PI3K/Akt survival signal in hormone-dependent (estrogen-receptor positive) MCF-7 cells even in the presence of IGF-I/E2 mitogenic factors. Bergapten, and to a greater extent PUVA, upregulated mRNA and the protein content of p53. These findings suggest that bergapten and its photoactivated compound can exert antiproliferative effects and induce apoptotic responses in cancer cells.

Taken together, these data show that BEO can be lethal for dividing cells and may reduce the risk of unwanted cell proliferation after prolonged use (although caution is needed due to cytotoxicity of inappropriate BEO doses). This allows hypotheses on new uses for bergapten and its photoactivated derivative in cancer prevention and therapy.

Another interesting aspect in the study of plant chemopreventers is their capability to prevent carcinogen bioactivation. In fact, a large portion of human cancers are due to exposure to environmental and occupational carcinogens that need metabolic bioactivation in body tissues. The cytochrome P450 enzyme systems (CYPs) and N-acetyltransferase (NAT) can be used as biomarkers of human cancer susceptibility (Badawi et al. 1996).

Several recent reports have investigated the role of 5-MOP in relation to CYPs. 5-MOP significantly reduces the total CYP-specific content in the liver (Diawara et al. 2000). It is also an inhibitor and inactivator of CYP2A6 (Koenigs and Trager 1998a) and CYP2BI (Koenigs and Trager 1998b), and it was shown to produce mechanism-based inhibition of CYP3A4 (Ho and Saville 2001). The furan ring contained in 5-MOP has been suggested to be the moiety responsible for the inactivation of CYP1A (Cai et al. 1996).

The arylamine carcinogens represent one of the chemical classes known to induce tumors after metabolic activation by host enzymes. A major metabolic pathway for arylamines is N-acetylation, which is catalyzed by the host cytosolic arylamine NAT. In particular, 2-aminofluorene (AF) is then N-acetylated by NAT to 2-acetylaminofluorene (AAF), which is converted to N-hydroxy-AAF by isozymes of CYPs. Thus Lee and Wu (2005) have investigated the N-acetylation of AF to AAF by NAT in the stomach and colon of rats and in human stomach (SC-M1) and colon (COLO 205) tumor cell lines. Although 5-MOP increased the activity of NAT and also increased the further metabolism of AAF in the rat stomach and colon, it decreased the activity of NAT in SC-M1 and COLO 205 cells with a different time- and dose-dependent profile. Thus 5-MOP might play a role in human chemoprotection against xenobiotics in the human stomach and colon, reducing the risk of tumor formation if the host is exposed to the arylamine carcinogens or metabolically similar carcinogens.

REFERENCES

Alsheikh-Ali, A. A., and Karas, R. H. 2009. The relationship of statins to rhabdomyolysis, malignancy, and hepatic toxicity: Evidence from clinical trials. *Curr. Atheroscler. Rep.* 11:100–104.

Amantea, D., Fratto, V., Maida, S., et al. 2009. Prevention of glutamate accumulation and upregulation of phospho-Akt may account for neuroprotection afforded by bergamot essential oil against brain injury induced by focal cerebral ischemia in rat. *Int. Rev. Neurobiol.* 85:389–405.

Amirnia, M., Khodaeiani, E., Fouladi, R. F., and Hashemi, A. 2012. Topical steroids versus PUVA therapy in moderate plaque psoriasis: A clinical trial along with cost analysis. *J. Dermatolog. Treat.* 23:109–111.

Aubin, F., Humbert, P., and Agache, P. 1994. Effects of a new psoralen, 5-geranoxypsoralen, plus UVA radiation on murine ATPase positive Langerhans cells. *J. Dermatol. Sci.* 7:176–184.

Aubin, F., and Mousson, C. 2004. Ultraviolet light-induced regulatory (suppressor) T cells: An approach for promoting induction of operational allograft tolerance? *Transplantation* 77:S29–S31.

Averbeck, D., Dubertret, L., Young, A. R., and Morlière, P. 1990. Genotoxicity of bergapten and bergamot oil in *Saccharomyces cerevisiae. J. Photochem. Photobiol. B: Biol* 7:209–229.

Badawi, A. F., Stern, S. J., Lang, N. P., and Kadlubar, F. F. 1996. Cytochrome P-450 and acetyltransferase expression as biomarkers of carcinogen-DNA adduct levels and human cancer susceptibility. *Prog. Clin. Biol. Res.* 395:109–140.

Bagetta, G., Morrone, L. A., Rombolà, L., et al. 2010. Neuropharmacology of the essential oil of bergamot. *Fitoterapia* 81:453–461.

Berliocchi, L., Russo, R., Levato, A., et al. 2009. (–)-Linalool attenuates allodynia in neuropathic pain induced by spinal nerve ligation in c57/bl6 mice. *Int. Rev. Neurobiol.* 85:221–235.

Berliocchi, L., Ciociaro, A., Russo, R., et al. 2011. Toxic profile of bergamot essential oil on survival and proliferation of SH-SY5Y neuroblastoma cells. *Food. Chem. Toxicol.* 49:2780–2792.

Bisignano, G. and Saija, A. 2002. The biological activity of the citrus oils. In *Citrus. The Genus Citrus,* eds. G. Dugo and A. Di Giacomo, 602–630. London and New York: Taylor & Francis.

Bisignano, G., Cimino, F., and Saija, A. 2012. Biological activities of citrus essential oils. In *Citrus Oils: Composition, Advanced Analytical Techniques, Contaminants, and Biological Activity*, eds. G. Dugo and L. Mondello, 529–548. London and New York: Taylor & Francis.

Borgatti, M., Mancini, I., Bianchi, N., et al. 2011. Bergamot (*Citrus bergamia* Risso) fruit extracts and identified components alter expression of interleukin 8 gene in cystic fibrosis bronchial epithelial cell lines. *BMC Biochem.* 12:15.

Cai, Y., Bear-Dubowska, W., Ashwood-Smith, M. J., et al. 1996. Mechanism-based inactivation of hepatic ethoxyresorufin O-dealkylation activity by naturally occurring coumarins. *Chem. Res. Toxicol.* 9:729–736.

Chang, K. M., and Shen, C. W. 2011. Aromatherapy benefits autonomic nervous system regulation for elementary school faculty in Taiwan. *Evid. Based Complement. Alternat. Med.* 946537 (Article ID).

Choe, S. C., Kim, H. S., Jeong, T. S., et al. 2001. Naringin has an antiatherogenic effect with the inhibition of intercellular adhesion molecule-1 in hypercholesterolemic rabbits. *J. Cardiovasc. Pharmacol.* 38:947–955.

Corasaniti, M. T., Maiuolo, J., Maida, S., et al. 2007. Cell signaling pathways in the mechanisms of neuroprotection afforded by bergamot essential oil against NMDA-induced cell death in vitro. *Br. J. Pharmacol.* 151:518–529.

Di Donna, L., De Luca, G., Mazzotti, F., et al. 2009. Statin-like principles of bergamot fruit (*Citrus bergamia*): Isolation of 3-hydroxymethylglutaryl flavonoid glycosides. *J. Nat. Prod.* 72:1352–1354.

Di Donna, L., Gallucci, G., Malaj, N., et al. 2011. Recycling of industrial essential oil waste: Brutieridin and melitidin, two anticholesterolaemic active principles from bergamot albedo. *Food Chem.* 125(29):438–441.

Diawara, M. M., Williams, D. E., Oganesian, A., and Spitsbergen, J. 2000. Dietary psoralens induce hepatotoxicity in C57 mice. *J. Nat. Toxins* 9:179–195.

Dubertret, L., Morlière, P., Averbeck, D., and Young, A. R. 1990. The photochemistry and photobiology of bergamot oil as a perfume ingredient: An overview. *J. Photochem. Photobiol. B: Biol.* 7:362–365.

Dugo, P., Presti, M. L., Ohman, M., et al. 2005. Determination of flavonoids in citrus juices by micro-HPLC-ESI/MS. *J. Sep. Sci.* 28:1149–1156.

Edris, A. E. 2007. Pharmaceutical and therapeutic potentials of essential oils and their individual volatile constituents: A review. *Phytother. Res.* 21:308–323.

El-Beshlawy, A., Hamdy, M., and El-Ghamrawy, M. 2009. Fetal globin induction in beta-thalassemia. *Hemoglobin* 33:S197–S203.

Emerit, I., Antunes, J., Silva, J. M., et al. 2011. Clastogenic plasma factors in psoriasis-comparison of phototherapy and anti-TNF-α treatments. *Photochem. Photobiol.* 87:1427–1432.

Geskin, L. 2007. ECP versus PUVA for the treatment of cutaneous T-cell lymphoma. *Skin Therapy Lett.* 12:1–4.

Gielen, S., Sandri, M., Schuler, G., et al. 2009. Risk factor management: Antiatherogenic therapies. *Eur. J. Cardiovasc. Prev. Rehabil.* 16:S29–S36.

Gleeson, C. M., Morar, N., Staveley, I., and Bunker, C. B. 2011. Treatment of cutaneous sarcoid with topical gel psoralen and ultraviolet A. *Br. J. Dermatol.* 164:892–894.

Guerrini, A., Lampronti, I., Bianchi, N., et al. 2009. Bergamot (*Citrus bergamia* Risso) fruit extracts as γ-globin gene expression inducers: Phytochemical and functional perspectives. *J. Agric. Food Chem.* 57:4103–4111.

Ho, P. C., and Saville, D. J. 2001. Inhibition of CYP3A4 by grapefruit flavonoids furanocoumarins and related compounds. *J. Pharm. Pharmaceut. Sci.* 4:217–227.

Hongratanaworakit, T. 2011. Aroma-therapeutic effects of massage blended essential oils on humans. *Nat. Prod. Commun.* 6:1199–1204.

Ibbotson, S. H., Dawe, R. S., Dinkova-Kostova, A. T., et al. 2012. Glutathione S-transferase genotype is associated with sensitivity to psoralen-UVA (PUVA) photochemotherapy. *Br. J. Dermatol.* 166:380–388.

Inzinger, M., Heschl, B., Weger, W., et al. 2011. Efficacy of psoralen plus ultraviolet A therapy vs. biologics in moderate to severe chronic plaque psoriasis: Retrospective data analysis of a patient registry. *Br. J. Dermatol.* 165:640–645.

Jones, P. H. 2008. Expert perspective: Reducing cardiovascular risk in metabolic syndrome and type 2 diabetes mellitus beyond low-density lipoprotein cholesterol lowering. *Am. J. Cardiol.* 102:41L–47L.

Joy, T. R., and Hegele, R. A. 2009. Narrative review: Statin-related myopathy. *Ann. Intern. Med.* 150:858–868.

Kim, H. J., Oh, G. T., Park, Y. B., et al. 2004. Naringin alters the cholesterol biosynthesis and antioxidant enzyme activities in LDL receptor-knockout mice under cholesterol fed condition. *Life Sci.* 74:1621–1634.

Koenigs, L. L., and Trager, W. F. 1998a. Mechanism-based inactivation of P450 2A6 by furanocoumarins. *Biochemistry* 37:10047–10061.

Koenigs, L. L., and Trager, W. F. 1998b. Mechanism-based inactivation of cytochrome P450 2BI by 8-methoxypsoralen and several other furanocoumarins. *Biochemistry* 37:13184–13193.

Lapolla. W., Yentzer, B. A., Bagel, J., Halvorson, C. R., and Feldman, S. R. 2011. A review of phototherapy protocols for psoriasis treatment. *J. Am. Acad. Dermatol.* 64:936–949.

Lee, Y. M., and Wu, T. H. 2005. Effects of 5-methoxypsoralen (5-MOP) on arylamine N-acetyltransferase activity in the stomach and colon of rats and human stomach and colon tumor cell lines. *In Vivo* 19:1061–1069.

Leopoldini, M., Malaj, N., Toscano, M., Sindona, G., and Russo, N. 2010. On the inhibitor effects of bergamot juice flavonoids binding to the 3-hydroxy-3-methylglutaryl-CoA reductase (HMGR) enzyme. *J. Agric. Food Chem.* 58:10768–10773.

Mengeaud, V., and Ortonne, J. P. 1996. PUVA (5-methoxypsoralen plus UVA) enhances melanogenesis and modulates expression of melanogenic proteins in cultured melanocytes. *J. Invest. Dermatol.* 107:57–62.

Menichini, F., Tundis, R., Loizzo, M. R., et al. 2010. In vitro photo-induced cytotoxic activity of *Citrus bergamia* and *C. medica* L. cv. Diamante peel essential oils and identified active coumarins. *Pharm. Biol.* 48:1059–1065.

Menter, A., Korman, N. J., Elmets, C. A., et al. 2010. Guidelines of care for the management of psoriasis and psoriatic arthritis: Section 5. Guidelines of care for the treatment of psoriasis with phototherapy and photochemotherapy. *J. Am. Acad. Dermatol.* 62:114–135.

Miceli, N., Mondello, M. R., Monforte, M. T., et al. 2007. Hypolipidemic effects of *Citrus bergamia* Risso and Poiteau juice in rats fed a hypercholesterolemic diet. *J. Agric. Food Chem.* 55:10671–10677.

Mollace, V., Ragusa, S., Sacco, I., et al. 2008. The protective effect of bergamot oil extract on lecitine-like oxyLDL receptor-1 expression in balloon injury-related neointima formation. *J. Cardiovasc. Pharmacol. Ther.* 13:120–129.

Mollace, V., Sacco, I., Janda, E., et al. 2011. Hypolipemic and hypoglycaemic activity of bergamot polyphenols: From animal models to human studies. *Fitoterapia* 82:309–316.

Morison, W. L. 2004. Psoralen ultraviolet A therapy in 2004. *Photodermatol. Photoimmunol. Photomed.* 20:315–320.

Morlière, P., Huppe, G., Averbeck, D., et al. 1990. In vitro photostability and photosensitizing properties of bergamot oil. Effects of a cinnamate sunscreen. *J. Photochem. Photobiol. B: Biol.* 7:199–208.

Morlière, P., Bazin, M., Dubertret, L., et al. 1991. Photoreactivity of 5-geranoxypsoralen and lack of photoreaction with DNA. *Photochem. Photobiol.* 53:13–19.

Morrone, L. A., Rombolà, L., Pelle, C., et al. 2007. The essential oil of bergamot enhances the levels of amino acid neurotransmitters in the hippocampus of rat: Implication of monoterpene hydrocarbons. *Pharmacol. Res.* 55:255–262.

Murase, J. E., Lee, E. E., and Koo, J. 2005. Effect of ethnicity on the risk of developing nonmelanoma skin cancer following long-term PUVA therapy. *Int. J. Dermatol.* 44:1016–1021.

Nagy, E. M., Dalla Via, L., Ronconi, L., and Fregona, D. 2010. Recent advances in PUVA photochemotherapy and PDT for the treatment of cancer. *Curr. Pharm. Des.* 16:1863–1876.

Nogata, Y., Sakamoto, K., Shiratsuchi, H., et al. 2006. Flavonoid composition of fruit tissues of citrus species. *Biosci. Biotechnol. Biochem.* 70:178–192.

Oliven, A., and Shechter, Y. 2001. Extracorporeal photopheresis: A review. *Blood Rev.* 15:103–108.

Panno, M. L., Giordano, F., Palma, M. G., et al. 2009. Evidence that bergapten, independently of its photoactivation, enhances p53 gene expression and induces apoptosis in human breast cancer cells. *Curr. Cancer Drug Targets* 9:469–481.

Panno, M. L., Giordano, F., Mastroianni, F., et al. 2010. Breast cancer cell survival signal is affected by bergapten combined with an ultraviolet irradiation. *FEBS Lett.* 584:2321–2326.

Patel, R. V., Clark, L. N., Lebwohl, M., and Weinberg, J. M. 2009. Treatments for psoriasis and the risk of malignancy. *J. Am. Acad. Dermatol.* 60:1001–1017.

Pothiawala, S. Z., Baldwin, B. T., Cherpelis, B. S., Lien, M. H., and Fenske, N. A. 2010a. The role of phototherapy in cutaneous T-cell lymphoma. *J. Drugs Dermatol.* 9:764–772.

Pothiawala, S. Z., Baldwin, B. T., Cherpelis, B. S., Glass, L. F., and Fenske, N. A. 2010b. The role of maintenance phototherapy in cutaneous T-cell lymphoma. *J. Drugs Dermatol.* 9:800–803.

Raphael, B. A., Morrissey, K. A., Kim, E. J., Vittorio, C. C., and Rook, A. H. 2011. Psoralen plus ultraviolet A light may be associated with clearing of peripheral blood disease in advanced cutaneous T-cell lymphoma. *J. Am. Acad. Dermatol.* 65:212–214.

Rombolà, L., Corasaniti, M. T., Rotiroti, D., et al. 2009. Effects of systemic administration of the essential oil of bergamot (BEO) on gross behaviour and EEG power spectra recorded from the rat hippocampus and cerebral cortex. *Funct. Neurol.* 24:107–112.

Saiyudthong, S., and Marsden, C. A. 2011. Acute effects of bergamot oil on anxiety-related behaviour and corticosterone level in rats. *Phytother. Res.* 25:858–862.

Sakurada, T., Kuwahata, H., Katsuyama, S., et al. 2009. Intraplantar injection of bergamot essential oil into the mouse hindpaw: Effects on capsaicin-induced nociceptive behaviors. *Int. Rev. Neurobiol.* 85:237–248.

Sakurada, T., Mizoguchi, H., Kuwahata, H., et al. 2011. Intraplantar injection of bergamot essential oil induces peripheral antinociception mediated by opioid mechanism. *Pharmacol. Biochem. Behav.* 97:436–443.

Schneider, L. A., Hinrichs, R., and Scharffetter-Kochanek, K. 2008. Phototherapy and photochemotherapy. *Clin. Dermatol.* 26:464–476.

Smith, G., Dawe, R. S., Clark, C., et al. 2003. Quantitative real-time reverse transcription-polymerase chain reaction analysis of drug etabolizing and cytoprotective genes in psoriasis and regulation by ultraviolet radiation. *J. Invest. Dermatol.* 121:390–398.

Smith, G., Wilkie, M. J., Deeni, Y. Y., et al. 2007. Melanocortin 1 receptor (MC1R) genotype influences erythemal sensitivity to psoralen-ultraviolet A photochemotherapy. *Br. J. Dermatol.* 157:1230–1234.

Soung, J., Muigai, W., Amin, N., Stern, D. K., and Lebwohl, M. G. 2005. A chart review of patients with early stage mycosis fungoides treated with psoralen plus UVA (PUVA). *J. Drugs Dermatol.* 4:290–294.

Stadler, R. 2007. Optimal combination with PUVA: Rationale and clinical trial update. *Oncology* 21:29–32.

Stern, R. S. 2009. Putting iatrogenic risk in perspective: Basal cell cancer in PUVA patients and Australians. *J. Invest. Dermatol.* 129:2315–2316.

Testa, U. 2009. Fetal hemoglobin chemical inducers for treatment of hemoglobinopathies. *Ann. Hematol.* 88:505–528.

Thein, S. L., and Menzel, S. 2009. Discovering the genetics underlying foetal haemoglobin production in adults. *Br. J. Haematol.* 145:455–467.

Trombetta, D., Cimino, F., Cristani, M., et al. 2010. In vitro protective effects of two extracts from bergamot peels on human endothelial cells exposed to tumor necrosis factor-alpha (TNF-alpha). *J. Agric. Food Chem.* 58:8430–8436.

Trovato, A., Taviano, M. F., Pergolizzi, S., et al. 2010. *Citrus bergamia* Risso and Poiteau juice protects against renal injury of diet-induced hypercholesterolemia in rats. *Phytother. Res.* 24:514–519.

Vagace Valero, J. M., Alonso Escobar, N., De Argila Fernández-Durán, D., et al. 2003. Photopheresis: New immunomodulatory therapy for T-lymphocite mediated diseases. *An. Med. Interna* 20:421–426.

Wang, R., Liu, W., Gu, W., and Zhang, P. 2011. Intense pulsed light protects fibroblasts against the senescence induced by 8-methoxypsoralen plus ultraviolet-A irradiation. *Photomed. Laser Surg.* 29:685–690.

Wei, A., and Shibamoto, T. 2010. Antioxidant/lipoxygenase inhibitory activities and chemical compositions of selected essential oils. *J. Agric. Food Chem.* 58:7218–7225.

Young, A. R., Walker, S. L., Kinley, J. S., et al. 1990. Phototumorigenesis studies of 5-methoxy-psoralen in bergamot oil: Evaluation and modification of risk of human use in an albino mouse skin model. *J. Photochem. Photobiol. B: Biol.* 7:231–250.

22 Bergapten (5-MOP) from Plants to Drugs

Paul Forlot, Frederic Bourgaud, and Paul Pevet

CONTENTS

22.1 INTRODUCTION

The *Citrus bergamia* of southern Italy (Calabria) is well known for its fruits (bergamot) and the essential oil extracted from its pericarp. This oil is used in perfumery ("eau de cologne") as well as in popular medicine (Dugo and Di Giacomo 2002). The phototoxicity of bergamot oil (Berlock dermatitis) was first observed in 1916 by Freund, an Austro-Italian dermatologist, and attributed later to the presence of bergapten or 5-methoxypsoralen (5-MOP), a naturally occurring compound of the furocoumarin (psoralens) family that is also present in numerous plants. More recently, the mechanism involved in the phototoxicity was elucidated as well as its potential use in therapy for several skin diseases including psoriasis, vitiligo, and mycosis fungoides (photochemotherapy). Further studies on the pharmacology of bergapten allowed consideration of different biological effects of it in the absence of ultraviolet radiation. This review concerns the different aspects of these properties and their consequent use in cosmetics and medicine.

22.2   CHEMISTRY OF BERGAPTEN

Bergapten (5-MOP, 5-methoxypsoralen, 5-methoxy-2H-furo[3,2-g]chromen-2-one, majudin, herculin) belongs to the furocoumarin (psoralen) family, which occurs naturally in various plants like fig, celery, parsley, and citrus. It is structurally related to coumarin by a fused furan ring. The molecular formula of bergapten (CAS number 484-20-8) is $C_{12}H_8O_4$ with a molecular weight of 216.19. Furocoumarins can be angular (angelicin) or linear (psoralen) according to the position of the furan ring. This type of structure has an important influence on the biological properties of the psoralen.

22.3   PHARMACOLOGY OF BERGAPTEN PLUS UV LIGHT

22.3.1   PHOTOTOXICITY

Rosenthal (1925) described Berlock dermatitis as pendant-like streaks of pigmentation on the neck, face, arms, or trunk after application of "eau de cologne" and exposure to sunlight. The phototoxic ingredient proved to be bergapten, a natural furocoumarin present in the oil of bergamot (*Citrus bergamia*). Several cases were reported in the 1950s and 1960s following increased use of perfumes and the greater popularity of sunbathing. Since the introduction of debergaptenized bergamot oils, Berlock dermatitis has become rare. As bergapten is present in numerous plants, various phyto-dermatoses have been observed after contact or ingestion with these plants and exposure to sunlight (photo-phyto-dermatoses). Table 22.1 summarizes the plants known to be responsible for contact/oral phototoxicity induced by the presence of bergapten. Bergapten containing Rutaceae and Apiaceae are the most common plants able to induce phytophototoxicity.

22.3.2   PHOTOCHEMOPROTECTION

Topical or oral application of bergapten on human skin induces a phototoxic reaction in the presence of UVA light (320–400 nm) followed by an increase in skin pigmentation in the area exposed (Marzulli and Maibach 1983). The augmentation of the tanning responses in human was established by several controlled studies using the double-blind placebo method (Fitzpatrick et al. 1955). This concept of tanning as a defense against ultraviolet damage was called photochemoprotection and was the rationale for the development of topical formulations containing UVB sunscreens and oil of bergamot. The stimulation of natural melanin pigmentation acts on all aspects of the photo-induced pigmentary process: the increase in melanin production, melanin transfer toward the superficial layers of the skin, and of the active melanocytes (Levine et al. 1989; Kligman and Forlot 1992). 5-MOP plus UVA stimulates the cellular melanogenesis by increasing the activity and the synthesis of tyrosinase and modulates the expression of the independent melanogenic enzymes (Mengeaud and Puigserver 1994). At the same time, the beneficial protective effects induced against further UV damage has been largely studied and confirmed by different direct in vivo methods as well as direct ex vivo DNA analysis (Potten et al. 1993; Chadwick et al. 1994). 5-MOP sunscreen administered during the course of a tanning treatment results in changes in the skin that afford better protection as

TABLE 22.1

Bergapten in Plants with Risk of Topical/Oral Phytophototoxicity

Apiaceae	Rutaceae
Heracleum lacianatum (laciniste)	*Limonia acidissima* (elephant apple)
Heracleum sphondylium (cow parsnip)	*Casimiro aedulis* (white sapote)
Heracleum mantegazianum (giant hogweed)	*Ruta graveolens* (rue)
Ammimajus (bishops weed)	*Skimmia arborescens* (skimmia)
Anethum graveolens (dill)	*Dictammnus alba* (burning bush)
Coriandrum sativum (coriander)	*Fagarashinofolia* (wild lime)
Daucus carotta (wild carrot)	*Thamnosma montana* (turpentine broom)
Foeniculum vulgare (fennel)	*Citrus aurantifolia* (lime)
Petroselinum sativum (wild parsley)	*Citrus aurentium* (bitter orange)
Pastinaca sativa (parsnip)	*Citrus lemon* (lemon)
Apium graveolens (celery)	*Citrus bergamia* (bergamot)
Angelica archangelica (garden angelica)	*Citrus reticulata* (mandarin)
Glehnialittoralis (beishashen)	*Citrus acida* (lime)
Heracleum mantegazzianum (giant hogweed)	**Moraceae**
Leviscum officinale (levage)	*Ficus carica* (fig tree)
Petroselinum crispum (parsley)	*Ficus salicifolia* (wild fig tree)
Pimpinella anisum (anise)	
Fabaceae	
Glycyrizaglabra (licorice)	

measured by the UDS (unscheduled DNA synthesis) approach than the changes induced by the irradiation alone (Chadwick et al. 1994). Nevertheless, the use of bergamot oil containing sunscreen was controversial considering the phototoxic and photogenotoxic risks.

22.3.3 POTENTIAL RISKS ASSOCIATED WITH THE USE OF 5-MOP PLUS UVA

The exposure of human skin to solar light can lead to short-term responses (erythema and pigmentation) and long-term responses (carcinogenesis and aging) (Chadwick et al. 1994). These cutaneous effects are due to ultraviolet B (UVB, 290–320 nm) radiation, and to a lesser extent to the ultraviolet A (UVA, 320–400 nm) components of solar light (Gange et al. 1986; Parrish et al. 1986). UVB and UVA sunscreens introduced in sun-protecting preparations lower the amount of ultraviolet radiation and reduce the erythemal risk. However, it has been shown that sunscreens do not provide complete protection with respect to the photocarcinogenic risk if the exposure time is excessively prolonged when using these sunscreens (Parrish et al. 1978).

The mode of action of psoralens has been proposed as a covalent reaction with the pyrimidine bases of DNA upon irradiation with UVA. Photo-addition of bifunctional psoralens like 5-MOP begins with intercalation of the psoralen molecule between base pairs of complementary strands of DNA with photo-addition of either pyrone or furan rings of the psoralen to a pyrimidine base, resulting in a psoralen DNA monoadduct. The initial step is the formation of an intercalative complex between 5-MOP and the nucleic acids of the DNA in a dark reaction. On exposure to UVA (300–400 nm) radiation, the intercalated 5-MOP can react by a [2+2] cycloaddition at either the 3,4 double bond of the pyrone ring or the 4′,5′ double bond of the furan ring, with the 5,6 double bond of pyrimidine bases in DNA resulting in mono- and bi-adducts yielded in interstrand crosslinks (Kelfkens et al. 1972). The 4′,5′ (furan side) monoadducts can bind to the complementary DNA strand, forming an interstrand crosslink by further irradiation (Musajo and Rodighiero, 1972). The ability of cells to repair DNA crosslinks is a critical determinant of tumor sensitivity (Dall'Acqua et al. 1972). Monoadducts can easily be repaired in normal cells (excision repair), but the repair of crosslinks in DNA is more difficult and could induce long-term effects like mutagenicity and carcinogenicity (McHugh et al. 2001).

The photomutagenic and photocarcinogenic potentials of bergamot oil and bergapten have been largely studied in vitro and in vivo (Safaz et al. 1983). In the presence of solar simulating radiations, bergamot oil containing 5-MOP photoinduces lethal, mutagenic, and recombinogenic effects as effectively as 5-MOP alone. The genotoxic effects of bergamot oil can be strongly reduced by the addition of chemical filters. This was confirmed in vivo on volunteers after application of a bergamot-containing sunscreen and the study of suction blister fluid. The photogenotoxic potential assayed on the yeast *Saccharomyces cerevisiae* was found to be rather low (Averbeck et al. 1992). In addition, 5-MOP was shown to be photocarcinogenic in mouse skin (Moysan et al. 1993). This resulted in concern about the long-term safety of bergamot oil–containing sunscreens. However, UVB sunscreens and UVB + UVA sunscreens significantly inhibit 5-MOP–induced skin cancer in mice (Zajdela and Bisagni 1981; Young et al. 1981). Despite the encouraging results of all these studies, the European Commission on Cosmetology decided to not allow further use of bergapten (bergamot) in sunscreen preparation (18th Commission Directive 95/94/EC).

22.3.4 Photochemotherapy

In 1974, it was demonstrated that oral 8-methoxypsoralen in combination with newly developed artificial irradiators emitting high-intensity long-wave ultraviolet radiant energy (UVA) were highly effective in the control of psoriasis and mycosis fungoides. This treatment was called PUVA (psoralen + UVA) photochemotherapy (Young 1992). Numerous clinician investigators in the United States and Europe have carried out large clinical trials on the efficacy and safety of PUVA photochemotherapy (Parrish et al. 1974). 5-MOP (bergapten) has been also evaluated for photochemotherapy of psoriasis and vitiligo in different clinical studies (Hoenigsmann et al. 1987). Comparative clinical evaluation of 5-MOP and 8-MOP demonstrated that 5-MOP is as effective as 8-MOP in the treatment of psoriasis but 5-MOP showed a significant lower incidence of side effects (Hoenigsmann 1979). Nausea and/or vomiting, pruritus, and erythema were the most commonly reported adverse events in PUVA-treated patients; they occurred 2–11 times more with 8-MOP than with 5-MOP (Kalis et al. 1989; McNeely and Goa, 1998). 5-MOP was also shown to be effective and well tolerated in the photochemotherapy of mycosis fungoides (MF) in association with retinoic acid or alone (PUVA and RE-PUVA) (Thioly-Benssoussan et al. 1989). 5-MOP PUVA and 8-MOP PUVA MF treated patients did not differ significantly in terms of relapse rate or time to relapse and maintenance therapy (Wackernagel et al. 2006).

As far as the treatment of vitiligo is concerned, photochemotherapy using solar exposure on Indian patients with 5-MOP has been shown to be more efficient than with 8-MOP with much better long-term tolerance (Pathak et al. 1992).

Since the introduction of PUVA therapy in 1974, there have been questions concerning the cancerogenic potential of this therapy. Several studies have been performed to evaluate a potential risk of cutaneous cancer in patients receiving long-term PUVA treatment (Stern et al. 1979). After more than 30 years of use in many thousands of patients with psoriasis and vitiligo, the carcinogenic risk of PUVA is now regarded as low with the exception of patients who have been previously treated with ionizing radiation or inorganic arsenicals, or topical nitrogen mustard, and in skin phototypes I and II who are treated continuously for many years (Fitzpatrick 1989). More recent developments in terms of bioavailability of 5-MOP (new galenic forms such as liquid formulation or micronized bergapten) did improve the clinical efficacy and lower the UVA cumulative dose for clearing patient lesions (Tanew et al. 1988; Treffel et al. 1992).

It can be stated so far that PUVA 5-MOP is (1) a successful treatment for vitiligo, (2) an excellent treatment for the early stages of cutaneous T-cell lymphoma (MF), (3) a very effective alternative to methothrexate or UVB for the control of psoriasis, and (4) an effective alternative to the anti-malarials for the control of polymorphous light eruption.

22.4 PHARMACOLOGY OF 5-MOP IN THE ABSENCE OF UV LIGHT

In recent decades, different aspects of the biological effects of bergapten have been explored, and possible therapeutic advances examined for melatonin-dependent disorders as well as K+ channel deficiencies.

22.4.1 EFFECT OF BERGAPTEN ON MELATONIN SECRETION

N-acetyl-5-methoxytryptamine (melatonin) is considered as the best available index of pineal gland function. The secretion of melatonin undergoes a clear circadian rhythmicity with high plasma levels during the dark phase of the day. This rhythmicity is generated by the suprachiasmatic nuclei (NSC) of the hypothalamus, the circadian "clock." This clock is synchronized essentially by the day/night cycle. Although the role of melatonin remains unclear, it seems to be involved in the control of certain biological rhythms and their related physiologic functions (Arendt et al. 1981; Wetterberg et al. 1984). It has been shown that plasma levels of melatonin are significantly increased from the second hour after 5-MOP administration in human volunteers (Souetre et al. 1987). This increased secretion was more marked after evening than after morning administration. Since antidepressant drugs are directly affecting the brain pacemakers, it is logical to consider that to reset these "clocks" would be helpful to treat depressed patients. The data obtained after administration of 5-MOP in depressive patients showed a significant improvement of mood and an antidepressant activity equivalent to amitryptiline (Souetre 1989; Darcourt et al. 1995). In addition, 5-MOP induced sleep changes and helped correct most sleep disturbances seen with major depression (Schmittbiel et al. 1994; Feuillade et al. 1995). It was later shown that the increase in the plasma concentration of melatonin after administration of 5-MOP is not due to a stimulation of melatonin secretion by the pineal gland in vivo, but by inhibition of its degradation by the liver (Mauviard et al. 1992; Pevet and Mauviard 1992).

22.4.2 Bergapten as Potassium Channel Blocker

Multiple sclerosis (MS) is a chronic inflammatory autoimmune disease of the central nervous system characterized by demyelination and axon damage that result in disabling neurological deficits (Beeton and Chandy 2005). MS is considered as a primary inflammatory disease of central nervous system white matter with an inflammatory demyelination, perivascular/parenchymal infiltration of T lymphocytes and macrophages, and axonal damage that induce irreversible neurological deficits (Judge et al. 2006). The possible therapies for MS include treatment of symptomatic neurological deficits and immunomodulatory therapy to treat neuroinflammation and limit neurodegeneration. It has been considered rational to use inhibitors of the K_v potassium channels in immune cells as a therapy for MS, since the impairment of nerve conduction seen with MS has to do with uncovering of internodal potassium channels (Bethge et al. 1991). As symptomatic therapies, only two relatively nonspecific K_v channel blockers, 4-aminopyridine and 3,4-aminopyridine, have been tested clinically in patients with MS. Although clearly beneficial, the use of these two drugs is limited because they are potent convulsants.

When 5-MOP was tested on the excitation process at the nodes of Ranvier of intact myelinated nerve fiber, it was found to block the pathological activated internodal potassium channels K_v 1,3 and K_v 1,5, which may impair nerve function in MS (Bohulavizki et al. 1992; Bohulavizki et al. 1994). Therefore, suitable channel blockers like 5-MOP can alleviate deficits due to MS. This has been demonstrated by single tests under well-defined laboratory conditions (During et al. 2000; Koppenhöffer et al. 2001). With the dosage used, no unwanted side effects were detected. Preliminary clinical investigations have confirmed these results (Schober and Neundörfer, 2010). The 5-MOP and new 5-MOP derivatives without phototoxic potential may well prove to be particularly suitable K_v-channel blockers for symptomatic therapy of MS and other demyelinating diseases (Koppenhöffer et al. 2001).

22.5 OTHER BIOLOGICAL ACTIVITIES OF BERGAPTEN

22.5.1 Cytotoxicity

Bergapten and citropten have been shown to be effective inhibitors of interleukin-8 (IL-8) expression and could be proposed as potential anti-inflammatory molecules to reduce lung inflammation in patients with cystic fibrosis (Borgatti et al. 2011).

Apart from its proved clinical efficacy in the PUVA therapy of mycosis fungoides, a T-cell lymphoma of the skin, the activity of bergapten and ultraviolet irradiation on human cancer cells in vivo was first described by Lane-Brown and Forlot (1984) as photo-oncotherapy (POT). POT was clinically tested as an alternative modality for the treatment of large areas of chronic actinic dermatitis (AK), precancerous actinic keratosis, and some superficial skin cancers like basal cell carcinoma (BCC) and squamous cell carcinoma (SCC) in Australian patients. Topical 5-MOP + UVA led to remission of AKs, some BCCs, and SCCs with a total free time of 18 months. The chemopreventive role of 5-MOP on cancerous cell lines has been investigated by studying the regulation of proliferation and apoptosis human hepatocellular carcinoma cell HCC (J5) (Lee et al. 2003). Morphological analysis, cell viability assay, DNA analysis, and cell cycle analysis suggest that there are at least three ways the suppression effect of 5-MOP works: (1) killing J5 cells directly, (2) inducing apoptosis by arresting J5 cells at the G2/M phase of the cell cycle, and (3) inducing apoptosis through an independent pathway arresting the cell cycle. More recent studies showed that bergapten and its photo-activated compounds exert an anti-proliferative effect and induce apoptotic responses in MCF-7 breast cancer cells (Panno et al. 2009). Independent of UV irradiation, bergapten was also shown to generate membrane signaling pathways able to address apoptotic responses in these cells (Panno et al. 2010). Similar effects have been observed in blood lymphocytes of cattle infected with bovine leukemia virus (BLV) (Trombetta et al. 2010). In combination with ultraviolet light, bergapten was shown to be effective in preventing the proliferation of bladder carcinoma cells in vitro (Boguka-Kocka et al. 2004). Bergapten was also able to exhibit significant inhibition of the production of pro-inflammatory cytokines, namely tumor necrosis factor-alpha (TNF-alpha) and interleukin-6 (IL-6) (Keane et al. 1994). These progresses could permit new strategic approaches for breast cancer disease, hepatocellular carcinoma, and the prophylaxis of retroviral infections. More in vivo tests and clinical investigations are necessary to confirm these results.

22.5.2 BERGAPTEN AND OSTEOPOROSIS

The differentiation of osteoblast appears to be dependent on the bone morphogenetic protein (BMP), which plays a major role in the bone formation. The imbalance of bone formation and bone destruction plays a major role in osteoporosis and provokes a reduction in skeletal mass. It has been shown that bergapten-like imperatorin is able to enhance alkaline phosphatase activity, type I collagen synthesis, and bone nodule formation in cultured osteoblasts. Furthermore, local administration of bergapten and imperatorin in secondary spongiosa in animals (rats) increased bone volume, which was the consequence of the increase of BMP2 expression in local osteoblasts and of the bone formation (Tang et al. 2008). The two furocoumarins, bergapten and imperatorin, are involved in BMP activation and could represent good targets for the development of anti-osteoporosis treatment.

22.6 CONCLUSION

For more than two decades the pharmacology of derivates of *Citrus bergamia* (bergapten, essential oil, and flavonoids) has been extensively studied. Oil of bergamot was proposed as a photoprotective agent associated with sunscreens for protection of human skin against solar radiation. Although effective (DNA protection), this use was controversial due to the phototoxic and phototumorigenic potential of 5-MOP.

In medicine, bergapten demonstrated various properties associated or not associated with UV light (photochemotherapy of vitiligo, psoriasis, and T-cell skin lymphoma; melatoninergic effect and potassium channel blocking). Clinical studies are still progressing in these fields and some drugs are under development.

REFERENCES

Arendt, J., Symons A. M., and Land C. A. 1981. Pineal function in the sheep: Evidence for a possible mechanism mediating seasonal reproductive activity. *Experientia* 37:384–387.

Averbeck, D., Averbeck S., Dubertret L. et al. 1992. Genotoxic effects of bergamot oil in yeast (*Saccharomyces cerevisiae*). *Nouv. Dermatol.* 11:674–677.

Beeton, C., and Chandy, K. G. 2005. Potassium channels, memory T-cells, and multiple sclerosis. *Neuroscientist.* 11(6):550–562.

Bethge, E. W., Bohulavizki, K. H., Hansel, W. et al. 1991. Effects of some potassium channel blockers on the ionic currents in myelinated nerves. *Gen. Physiol. Biophys.* 10:225–244.

Boguka-Kocka, A., Rulka, J., Kocki, J. et al. 2004. Bergaptenapotosis induction in blood lymphocytes of cattle infected with bovine leukemia virus (BLV). *Bull. Vet. Inst. Pulawy.* 48:99–103.

Bohulavizki, K. H., Hansel, W., Kneip, A. et al. 1992. Potassium channel blockers from Ruta—A new approach for the treatment of multiple sclerosis. *Gen. Physiol. Biophys.* 11(5):507–512.

Bohulavizki, K. H., Hänsel, W., Kneip, A. et al. 1994. Mode of action of psoralens, benzofurans, acridinons, and coumarins on the ionic currents in intact myelinated nerve fibres and its significance in demyelinating diseases. *Gen. Physiol. Biophys.* 13(4):309–328.

Borgatti, M., Mancini, I., Bianchi, N. et al. 2011. Bergamot (*Citrus bergamia* Risso) fruit extracts and identified components alter expression of interleukin-8 gene in cystic fibrosis bronchial epithelial cell lines. *BMC Biochem.* 12(1):15.

Chadwick, C. A., Potten, C. S., Cohen, A. J., and Young, A. R. 1994. The time of onset and duration of 5-methoxypsoralen photochemoprotection from UVR induced DNA damage in human skin. *British J. Dermatol.* 13:483–494.

Dall'Acqua, F., Marciani, S., Ciavatta, L., et al. 1972. Formation of interstrand cross-links in the photoreaction between furocoumarins and DNA. *Naturforsch.* 26:561–569.

Darcourt, G., Feuillade, P., Bistagnin, V., et al. 1995. Antidepressant effect of 5-MOP: The melatonin synchronizer hypothesis. *Eur. Psychiatry* 10(3):142–154.

Dugo, G., and Di Giacomo, A. 2002. *Citrus: The genus citrus.* London and New York: Taylor & Francis.

During, T., Gerst, F., Hänsel, W., et al. 2000. Effects of three alkylpsoralens on voltage gated ion channels in Ranvier nodes. *Gen. Physiol. Biophys.* 19(4):345–364.

Feuillade, Ph., Bistagnin, Y., Belugou, J. L., et al. 1995. The effects of 5-methoxypsoralen on sleep in major depressive disorders: A comparative study versus amitriptyline. Personal communication.

Fitzpatrick, T. B., Hopkins, C. E., Blickenstaff, D. P., et al. 1955. Preliminary and short report: Augmented pigmentation and other responses of normal human skin to solar radiation following oral administration of 8-methoxypsoralen. *J. Invest. Dermatol.* 25(3):187–190.

Fitzpatrick, T. B. 1989. The psoralen story: Photochemotherapy and photochemoprotection. In *Psoralens: Past, present and future of photochemotherapy and other biological activities,* eds. T. B. Fitzpatrick, P. Forlot, and F. Urbach, 5–10. Montrouge, France: John Libbey Eurotext.

Freund E. 1916. Uber bisher noch nicht beschriebene Kunstike Hautverfarbung. *Dermatol. Wochenschr.* 63:931–933.

Gange, R. W., Park, Y. E., Auletta, M., et al. 1986. Action spectrum for cutaneous responses to ultraviolet radiation. In *The biological effects of UVA radiation,* eds. F. Urbach and R. W. Gange, 57–67. New York: Praeger.

Hoenigsmann, H., Jaschke, E., Gschnait, F., et al. 1979. 5-methoxypsoralen (bergapten) in photochemotherapy of psoriasis. *Brit. J. Derm.* 101:369–379.

Hoenigsmann, H., Wolf, K., Fitzpatrick T. B., et al. 1987. Oral photochemotherapy with psoralens and UVA (PUVA): Principles and practice. In *Dermatology in general medicine,* vol. 1, 3rd ed., eds. T. B. Fitzpatrick et al., 1513–1554. New York: McGraw-Hill Education.

Judge, S. I., Lee, J. M., Bever, C. T., et al. 2006. Voltage-gated potassium channels in multiple sclerosis: Overview and new implications for treatment of central nervous system inflammation and degeneration. *J. Rehab. Res. Devel.* 43(1):111–112.

Kalis, B., Sayag, S., and Forlot, P. 1989. Photochemotherapy (PUVA) of psoriasis: A double-blind comparative study of 5- and 8-methoxypsoralen. In *Psoralens: Past, present and future of photochemotherapy and other biological activities,* eds. T. B. Fitzpatrick, P. Forlot, and F. Urbach, 277–282. Montrouge, France: John Libbey Eurotext.

Keane, T. E., Velmirovich, B., Yue, K. T., et al. 1994. Methoxypsoralen phototherapy of transitional cell carcinoma. *Urology* 44(6):842–846.

Kelfkens, G., Van Weelden, H., De Gruijl, F. R., et al. 1972. The influence of dose rate on ultraviolet tumorigenesis. *J. Photochem. Photobiol. B* 10:41–50.

Kligman, A. M., and Forlot, P. 1992. Comparative photoprotection in humans by tans induced either by solar simulating radiation or after application of a psoralen-containing sunscreen. *Nouv. Dermatol.* 11:666–673.

Koppenhöffer, E., Fabre, G., and Forlot, P. 2001. New UV- and non-UV light-related pharmacological and potential therapeutic uses of bergapten. *Essenz. Deriv. Agrumar.* 71:231–236.

Lane-Brown, M. M., and Forlot, P. 1984. Photooncotherapy (POT): An alternative for the treatment of superficial skin cancers. *Ann. Dermatol. Venereol.* 111:851.

Lee, Y. M., Wu, T. H., Chen, S. F., and Chung, J. G. 2003. Effect of 5-MOP on cell apoptosis and cell cycle in human hepatocellular carcinoma cell line. *Toxicol. In Vitro* 17(3):279–287.

Levine, N., Scot, D., Owens, C., et al. 1989. The effect of bergapten and sunlight on cutaneous pigmentation. *Arch. Dermatol.* 125:1225–1230.

Marzulli, F. N., and Maibach, H. I. 1983. Perfume phototoxicity. *J Soc. Cosmetic Chem.* 21:695–715.

Mauviard, F., Pévet, P., and Forlot, P. 1992. The increase in plasma melatonin concentrations observed after administration of 5-methoxypsoralen (5-MOP) might be due to the 5-MOP-induced inhibition of melatonin clearance. In *Advances in pineal research,* vol. 6, eds. A. Foldes and R. J. Reiter, 99–102. London: Libbey.

McHugh, P. J., Spanswick, V. J., and Hartley, J. A. 2001. Repair of DNA interstrand crosslinks: Molecular mechanisms and clinical relevance. *Lancet Oncology* 2(8):483–490.

McNeely, W., and Goa, K. L. 1998. 5-Methoxypsoralen. A review of its effects in psoriasis and vitiligo. *Drugs* 56(4):667–669.

Mengeaud, V., and Puigserver, A. 1994. Mechanisms of melanogenic activity of 5-methoxypsoralen alone and with UVA (PUVA) on pigmented cells. Thesis, Travaux Universitaires, Université d'Aix en Provence.

Moysan, A., Morliere, P., Averbeck, D., et al. 1993. Evaluation of phototoxic and photogenotoxic risk associated with the use of photosensitizers in suntan preparations: Application to tanning preparations containing bergamot oil. *Skin Pharmacol.* 6:282–295.

Musajo, L., and Rodighiero, G. 1972. Mode of photosensitization of furocoumarins. *Photophysiology.* 7:115–147.

Panno, M. L., Giordino, F., Palma, K. G., et al. 2009. Evidence that bergapten, independently of its photoactivation, enhances gene expression and induces apoptosis in human breast cancer cells. *Current Canc. Drug Targ.* 9(4):469–481.

Panno, M. L., Giordino, F., Mastroianni, F., et al. 2010. Breast cancer cell survival is affected by bergapten combined with an UV irradiation. *FEBS Lett.* 584(11):2321–2326.

Parrish, J. A., Fitzpatrick, T. B., Tanenbaum, L., et al. 1974. Photochemotherapy of psoriasis with oral methoxalen and long wave ultraviolet light. *N. England J. Med.*, 291:1207–1212.

Parrish, J. A., Anderson, R. R., Urbach, F., et al. 1978. *Biological effects of ultraviolet radiation with emphasis on human responses to long-wave ultraviolet*, 157–175. New York: Plenum Press.

Pathak, M. A., Fitzpatrick, T. B., Mosher, D. B., et al. 1992. Photochemotherapy of vitiligo. Comparative effectiveness and tolerance of 5-methoxypsoralen and 8-methoxypsoralen in Indian patients. *Nouv. Dermatol.* 11:715–719.

Pevet, P., and Mauviard, F. 1992. Effect of 5-MOP on the circulating concentration of melatonin. *Nouv. Dermatol.* 11:729–735.

Potten, C. S., Chadwick, C. A., Young, A. R., et al. 1993. DNA damage in UV-irradiated human skin in vivo: Automated direct measurement by image analysis (thymin dimers) compared with indirect measurement (unscheduled DNA synthesis) and protection by 5-methoxypsoralen. *Int. J. Radiat. Biol.* 63(3):313–324.

Rosenthal, O. 1925. Berloque dermatitis: Berliner Dermatologische Gesellschaft. *Dermatol. Zeitschr.* 42:295.

Safaz, D., Zajdela, F., Barneque, C., et al. 1983. Evaluation of DNA crosslinks and monoadducts in mouse embryo fibroblasts after treatment with mono- and bi-functional furocoumarins and 365 nm (UVA) irradiation. Possible relationship to carcinogenicity. *Photochem. Photobiol.* 38(5):557–562.

Schmittbiel, A., Gross, M. S., Bujon-Pinard, P., et al. 1994. Chronobiology of depression: The seasonal depressions. Clinical aspects, physiology and specific treatments. *Ann. Med. Psychol.* 152(7):444–456.

Schober, S., and Neundörfer, B. 2010. Klinische studien zur symptomatischen Wirksamkeit and verträglichkeiteines Teeauszügesaus Rutagraveolensbei Multiple Sclerosis: Abschlussbericht. Personal communication.

Souetre, E., Salvati, E., Belugou, J. L., et al. 1987. 5-Methoxypsoralen increases plasma melatonin levels in human. *J. Invest. Dermatol.* 89:152–159.

Souetre, E., Salvati, E., and Darcourt, G. 1989. Melatonin and depression: A possible role of psoralen. In *Psoralens: Past, present and future of photochemotherapy and other biological activities*, eds. T. B. Fitzpatrick, P. Forlot, and F. Urbach, 301–326. John Libbey Eurotext.

Stern, R. S., Thibodeau, L. A., Kleinerman, R. A., et al. 1979. Risk of cutaneous carcinoma in patients treated with oral methoxalen photochemotherapy for psoriasis. *N. Engl. J. Med.* 300:809–813.

Tanew, A., Ortel, B., Rappensberger, K. et al. 1988. 5-Methoxypsoralen (Bergapten) for photochemotherapy. Bioavailability, phototoxicity, and clinical efficacy in psoriasis of a new drug preparation. *J. Amer. Acad. Dermatol.* 18(2):333–338.

Tang, C., Yang, S., Chien, C. C., and Fu W. W. 2008. Enhancement of bone morphogenetic protein-2 expression and bone formation by coumarin derivatives via p38 and ERK-dependent pathway in osteoporosis. *Eur. J. Pharmacol.* 579(1–3):40–49.

Thioly-Benssoussan, D., Berretti, B., and Grupper, C. 1989. PUVA and RE-PUVA in the treatment of mycosis fungoides: 58 cases of mycosis fungoides with 7 to 18 years follow up. In *Psoralens: Past, present and future of photochemotherapy and other biological activities*, eds. T. B. Fitzpatrick, P. Forlot, and F. Urbach, 293–300. John Libbey Eurotext.

Treffel, P., Makki, S., Humbert, P., et al. 1992. A new micronized 5-methoxypsoralen preparation. Higher bioavailability and lower UVA dose requirement. *Acta Dermato-venereologica.* 72(1):65–67.

Trombetta, D., Cimino, F., Cristani, M., et al. 2010. In vitro protective effect of two extracts from Bergamot peel on human endothelial cells exposed to tumor necrosis factor alpha (TNF-alpha). *J. Agric. Food Chem.* 58:8430–8436.

Wackernagel, A., Hofer, A., Legat, F., et al. 2006. Efficacy of 8-methoxypsoralen vs. 5-methoxypsoralen plus ultraviolet A therapy in patients with mycosis fungoides. *British J. Dermatol.* 154(3):519–523.

Wetterberg, L., Beckfriis, J., Kjellman B. F., et al. 1984. Circadian rhythms in melatonin and cortisol secretion in depression. *Adv. Biochem. Psychopharmacol.* 39:197–205.

Young, A. R., Gibbs, N. K., and Magnus, I. A. 1981. Modification of 5-methoxypsoralen phototumorigenesis by UVB sunscreens: A statistical and histologic study in the hairless albino mouse. *J. Invest. Dermatol.* 89:611–617.

Young, A. R. 1992. Experimental photocarcinogenesis of psoralens. *Nouv. Dermatol.* 11:683–687.

Zajdela, F., and Bisagni, E.1981. 5-Methoxypsoralen, the melanogenic additive in suntan preparations, is tumorigenic in mice exposed to 365 nm UV radiation. *Carcinogenesis* 2:121–127.

23 Uses of Juice and By-Products

Rosario Lo Curto

CONTENTS

23.1 INTRODUCTION

Bergamot is a *Citrus* grown primarily in Calabria, a region in southern of Italy; it is also grown in Turkey, Argentina, Brazil, Ivory Coast, and Uruguay, though in lesser amounts. Italian production is about 140,000 tons per year on a surface of about 1500 hectares, and fruit production cannot be considered important in a region that produces many other vegetables.

The main product of bergamot is the essential oil, known all over the world for its use in perfume formulations and, more recently, to flavor teas, chocolates, liquors, etc. The amount of oil extracted from the fruit is similar to that which can be obtained from all *Citrus* species, about 0.4%–0.6% w/w.

The whole fruit is about 35% juice. The juice has a low commercial value compared with other citrus juices. The main waste material remaining after oil and juice extraction accounts for 55%–60% of the whole fruit. It is made up of several heterogeneous materials commonly named *pastazzo*. Because of several beneficial characteristics for humans due to the presence of molecules with pharmaceutical as well as nutritional effects, the interest in this waste is increasing, which could lead to decreased disposal costs (Hardin et al. 2010). It must be pointed out that in *Citrus* fruit processing the amount of waste exceeds the amount of marketable products, essential oil, and juice, which is unique in industrial transformations.

The studies concerning the exploitation of waste materials coming from bergamot processing are rather few compared with those concerning other *Citrus* species more commonly used in industrial transformation, like lemon, sweet and blood orange, and mandarin (Arvanitoyannis 2008). This may be ascribed to the fact that bergamot, because it is mostly grown in a poor region of Italy that doesn't have an industrial background, traditionally was transformed in rather low-dimension plants. The few wastes resulting from these plants never created real disposal problems because wastes were simply discarded on the soil or used to feed ruminant herds. These low-dimension plants also could not support research addressed to find different and more profitable waste transformations. In fact, today some plants process bergamot fruits only partially: after oil extraction, the entire scraped fruit containing the juice is disposed of as ruminant feed. This behavior, linked mainly to bergamot juice marketing difficulties, implies a spontaneous resolution of waste disposal and juice marketing. However, for larger citrus transformation plants, the daily production of 100 tons of peel and juice is a source of major disposal problems that have engaged many researchers.

From the transformation point of view, bergamot fruit is processed in a very similar way to other fruits even though its oil is more accurately extracted, sometimes also from the wastes or the centrifuge sediments, thereby obtaining different oil qualities (Figoli et al. 2006).

In bergamot processing, the oil can be extracted before or together with the juice in relationship to the machinery employed, and the resulting pastazzo is discarded as previously stated. In the following paragraphs will be reported the systems used, also in the past, to exploit the wastes resulting from fruit processing.

23.2   THE BERGAMOT JUICE (BJ)

Unlike other *Citrus* juices, BJ has a sour taste which, together with its typical scent, prevents it from becoming widely popular (though in Calabria it was traditionally considered and consumed as a popular medicine) (Passalacqua et al. 2007). Its acidic taste comes from naringin, neoeriocitrin, and limonin, mainly in the albedo and membranes of the fruit. This flavor is not balanced by the total sugar contents, whose concentration (2.6–4.1 g/L), though higher than lemon juice (1.1–2.4 g/L), is lower than orange and mandarin juices (5.1–11.2 g/L and 7.6–10.6 g/L, respectively) (Di Giacomo and Calvarano 1972). As a consequence, bergamot juice tastes more acidic in the industrially extracted juices, if compared with a gently home-squeezed juice, because of the greater strength exerted by the squeezing machines on fruits, mainly when continuous presses are used (Postorino et al. 2001).

Comparing the mean content of some citrus juices in relationship to whole fruit weight, bergamot (about 33%—35%) is in the middle between high-content fruits (blood orange 50% and blond orange 46%) and low-content fruits (lemon 28% and bitter orange 26%). In BJ from the three more common cultivars, *Fantastico*, *Femminello*, and *Castagnaro*, can be found small amounts of free amino acids, mainly proline, increasing with the ripening degree of the fruit (Leuzzi et al. 1991).

Raw BJ never had great commercial value and often was mixed in small amounts with lemon juice, for the similar tastes of the two juices; today these fraudulent additions are easily found by high-performance liquid chromatography (HPLC), which allows quantification of some compounds typical of BJ, namely naringin, neohesperidin, and neoeriocitrin (Gionfriddo et al. 1996; Cautela et al. 2008). In the past, a method to make bergamot juice more palatable by removing the bitter compounds was carried out, but the method was not scaled up to industrial plants (Calvarano et al. 1995). Today small amounts of BJ are exported to Japan, where it is used to make up bitter drinks.

Attempts to exploit BJ started early last century, when the juice came to be used as a low-cost source of citric acid. In the Reggio Calabria country were allocated some plants which, by a rather simple technology, processed BJ in a similar way to that used in Sicily for lemon juice. From the juice was extracted raw calcium citrate by precipitation with lime, and from the precipitate the citric acid was further extracted. Subsequently this source of profit failed due to the cheaper direct synthesis of citric acid in surface fermentation by the mold *Aspergillus niger* and more recently in submerged fermentation by the yeast *Candida lipolytica* or other yeasts, in controlled conditions in a reactor.

Among the many proposals addressed to a real valorization of this product is its use as a natural antioxidant, protecting ascorbic acid during heat exposure of apricot and apple juices (Pernice et al. 2009). Other proposals concern the fermentation of the low sugar content with the aim to produce an alcoholic solution from which aromatic vinegars can be obtained (Caridi and Manganaro 1993, 1996; Caridi 2005).

Recently, more attention was paid to the research addressing the identification in BJ of several functional substances of pharmaceutical as well as nutraceutical interest for whose resolution HPLC-diode array detection (DAD) methods must be employed. The identification in BJ of several functional substances of pharmaceutical as well as nutraceutical interest has greatly increased research on this product.

It has been pointed out that the strong antioxidative activity of BJ is due to the uncommon level of its polyphenolic fraction. Flavonoids neohesperidin, naringin, and neoeritrocin are typically found in this fraction. While they are also present in other citrus juices, they are at their highest concentration in BJ. Moreover, two furocoumarins, bergapten and bergamottin, can be extracted that, in effect, are present in all the parts of the fruit (Giannetti et al. 2010). The cultivar Femminello, among the most commonly processed cultivars of bergamot, contains the highest levels both of flavonoids and furocoumarins, which can be extracted by a simple method (Gattuso et al. 2007a). The uncommonly high level of polyphenols led to a method for their extraction in solid phase by a membrane system (Conidi et al. 2011).

Of course the composition of the polyphenolic fraction, of great interest for its pharmacologic activity and effect on human welfare, has been deeply studied (Gionfriddo et al. 1996; Kawaii et al. 1999; Calabrò et al. 2004; Dugo et al. 2005; Nogata et al. 2006). For instance, naringin intake allows lowering of the haematic levels of total cholesterol and low-density lipoprotein (LDL) (Jung et al. 2003), and hesperidin lowers the haematic level of total cholesterol and the concentration of triglycerides (Kim et al. 2003). The uncommon composition of BJ also stimulated studies of ways to allow its assumption in a dry form in order to increase the concentration of the aforementioned components, and at the same time overcome taste problems. By spray-drying after a charge on malto-dextrin solution, polyphenol level and total antioxidant activity of the dry product are similar to the fresh juice (Picerno et al. 2011).

Besides furocoumarins, other flavonoids in bergamot juice include several flavanones and flavones. Flavanones are metabolized by gut microflora, which release aglycones that provide antioxidative, hepatolipidemic, and anti-inflammatory benefits to humans. Flavonoids in BJ can positively interfere with the metabolism of the platelet activation factor (PAF) (Balestrieri et al. 2004). The level of flavonoids in BJ could increase its use in food processing because, if compared with other citrus juices like lemon or orange, it is about 100 times more concentrated (Robards et al. 1997; Kawaii et al. 1999; Swantsitang et al. 2000; Nogata et al. 2006; Gattuso et al. 2007b).

It is estimated that there are many flavonoids in BJ. Due to the progress in separation methods (Dugo et al. 2005), 20 new flavonoids were observed. Gattuso et al. (2006) found eight new flavonoids; among these five are C-glycosidic (lucenin-2, stellarin-2, isovitexin, scoparin, and orientin-4'-methyl ether) and three are O-glycosides (rhoifolin 4'-O-glucoside, chrysoeriol 7-O-neoesperidoside-4'-O-glicoside, and chrysoeriol 7-O-neohesperidoside). Some neohesperidosides of hesperitin and naringenin can also be found, namely melitidin and brutieridin. These flavonoids show a 3-hydroxy-3-methylglutaryl acid moiety with a structure similar to the natural attack substrate of the HMG-CoA reductase, which interferes in the cholesterol metabolism. Moreover, their structure is similar to the structure of the statins used to control haematic cholesterol levels (Di Donna et al. 2009). The mechanism exerted by these neohesperidosides in controlling the high cholesterol levels was attributed to the inhibition of the HMGR enzyme commonly exerted by statins (Leopoldini et al. 2010).

By mass spectrometry (MS), besides the well-known flavonoids and furocoumarins, 20 new substances and 3 new acyl-flavanones were also identified; they are probably derivatives from neoeriotricin, naringin, and neohesperidin (Gardana et al. 2008). These authors believe that the main flavones that are present in BJ are vicenin-2, stellarin-2, rhoifolin, and neodiosmin and confirm bergapten and bergamottin as the main furocoumarins.

Among the health benefits of BJ is its hypolipidemic activity. A prolonged administration of BJ in rats resulted in a significant decrease of serum levels of cholesterol, triglycerides, and LDL, and an increase in high-density lipoprotein (HDL) level, besides a protective effect on hepatic parenchyma (Kris-Etherton et al. 2002; Miceli et al. 2007). The same positive effects were also evidenced in humans (Mollace et al. 2011) and attributed again to the high flavonoid content.

Finally, bergamot juice administration brings about a substantial reduction of plasmatic cholesterol, triglycerides, and LDH levels, and a contemporary increase of HDL levels in rats exposed to renal injury induced by hypercholesterolemic diets. These results are again attributed to the general antioxidative activities of BJ (Trovato et al. 2010).

23.3 THE BERGAMOT PEEL (BP)

It is interesting to remember a peculiar use of the bergamot fruit, that of course cannot be included among the exploitation systems of the peel. The external layer of the fruit, namely the flavedo and the external part of the albedo, obtained by accurate removing of the internal part through a little circular cut around the stem, was dried naturally. The bag was then filled with snuff or pipe tobacco, which was aromatized by the volatile compounds of the peel.

BP, or pastazzo, is the main waste produced during bergamot processing. As with all wastes coming from citrus processing, about 90% is made of a heterogeneous group of materials: mainly flavedo, the external green/yellow layer of the fruit, and albedo, the white part between the flavedo and the inner part of the fruit. However, other structures of the fruit can be found in the BP; among these are the sacs where the juice is contained, the walls of the parts containing the sacs, and the stones, whose number is variable in relationship to the cultivar. BP also contains fruit stems and the leaves not removed before oil extraction. Traces of oil from flavedo and sugars from juice are found in variable amounts in relationship to the efficiency of the devices used to process the fruits.

It is useful to differentiate between simple waste disposal, generally used in the past, and waste upgrading, or the actual methods or systems carried out to increase the value of BP, which in this way can now be considered a raw material.

Looking at transformation costs, BP is similar to any other citrus waste, though it has more cons than pros. Among the pros are the lack of collection cost because BP accumulates by the transformation plants; among the cons are the high water content, the great amounts produced in short times, the drying difficulties, and the low protein content. Among the cons must always be remembered that BP, being poor in simple sugars, cannot support yeast growth, microorganisms of high protein level but lacking enzymes capable of hydrolyzing the main components of BP, pectins and hemicelluloses.

As reported by Postorino et al. (2002) BP is more than 50% by weight of the whole fruit (about 55%). It is 50%–55% of de-oiled peel, 10%–15% of pulps, and 3%–5% of stones. The product shows a very high water content (about 82%–85%) that, also for the presence of glycidic residues, makes it easily fermentable and not storable. In fact, the presence of pectins in BP (17%–22% by weight) hampers the drying process until it reaches the the stability level (around 10%) unless high temperatures are used, like those obtained in rotating ovens. To accelerate the drying process, the BP can be treated with lime, transforming pectins into calcium pectate. The treated BP now can be pressed more easily by continuous press to lower water content, meanwhile avoiding the charring phenomena typical of the thermal drying of too-wet peels. Although this treatment favors the subsequent self-pasteurization of the BP, which can reach also 60°C (Di Giacomo et al. 1998), it gives rise to an alkaline liquor whose discharge is difficult and costly. Undergoing the high drying cost by heat can be justified only when the cost of oil used as the combustible increases so much that it makes it more convenient to use the dry peel as a fuel (van Heerden et al. 2002).

The first study of the composition, digestibility, and nutritional ability of BP was in 1949 (Brozzetti 1949). The material is still used today as a source of fiber for ruminants. To this aim the fresh peel generally is accumulated on a simple platform and left to ferment spontaneously for some months; during this period BP shows a weak increase in total protein concentration due to the growth of thermophilic bacteria inside and of mold on the surface. During the fermentation a very strong flavor is produced, and the mass changes to more and more resemble a paste, which ruminants appreciate very much.

Besides being a source of fiber, the whole BP was proposed as fertilizer and soil conditioner along with other agricultural wastes used for compost production. Nevertheless, composting implies high plant costs and needs a market for the final product, which is lacking in the zones where bergamot is cultivated. So, probably due to the presence of residual oil, which has a high bactericidal activity, compost has never been produced from BP, unlike other citrus peels (van Heerden et al. 2002).

To exploit BP by increasing its protein level and market value, some studies were carried out to produce ensilages by mixing BP with straw (Chies et al. 1979) and the not well-balanced nutritional characteristics of blends of BP and straw (Sinatra et al. 1988), but in these cases the processes were not further scaled up, whereas for other citrus peels it has been successfully done (Hernandez et al. 1975). In effect many microbial species, all belonging to the molds, are able to grow on BP. Two edible *Penicilli* (*P. camemberti* and *P. roqueforti*) were grown in solid-state fermentation to increase the protein level of the peel by the utilization of its C-sources (Scerra et al. 1999). Within a reactor, in controlled conditions, it is possible to grow other molds like *Geotrichum candidum* and *Tricoderma viride*, obtaining a peel/mold blend with a mean protein level of more than 25% as well as an enzymatic liquid rich in raw pectinases (Lo Curto 2005)

Further studies concern the utilization of BP as a source of several substances of industrial, pharmaceutical, and nutraceutical interest. One area of study is the extraction of food-grade pectins commonly used as jelling product in several food preparations. It has been observed how these pectins were similar to those extracted from grapefruit (Di Giacomo et al. 1986). Recently, pectins from BP showed anti-angiogenetic ability, and it has been speculated that they may provide an inhibitory effect of this substance on tumoral growth (Ferlazzo 2010).

Among the substances of pharmaceutical interest, BP flavonoids play an important role in human health (Mollace et al. 2011) because citrus peels generally are richer in comparison with other parts of the fruit (Manthey and Grohman 2001). A strict correlation was also demonstrated between phenolics' content and anti-oxidative ability of albedo extracts from several citrus fruit; in this list bergamot is placed in an intermediate position (Ramful et al. 2010). An extraction system of phytocomplexes from bergamot albedo was recently patented (Lombardo et al. 2011).

The amount of naringin in BP was determined by HPLC (Micali and Currò 1980; Calvarano et al. 1996b), demonstrating that contemporary extraction of naringin and pectins is possible. The extraction of food-grade pectins, with a gelling activity similar to the lemon pectin, the preferred industrial source of industrial pectin, was reported by Tripodo et al. (2007), who evidenced the contemporary extraction of rather high amounts of naringin.

It is possible also to prepare dihydrochalcones starting from the flavanones extracted from BP whose sweetening ability ranges 300–1100 times more than saccharose (Di Giacomo and Calvarano 1991; Wu et al. 1991). Subsequently, it has been possible to verify that the production of dihydrochalcones can also start utilizing flavonic precursors (Esposito et al. 2004).

Besides the well-known flavones, several minor flavones were found in BP, like apigenin, luteoilin, and diosmetin-derivates, and also minor flavanones like eridictyol, naringenin, and derivatives of hesperitin. The presence of di-C-glucosides of apigenin and diosmetin, whose presence was known in other citrus species, was also evidenced.

Among some substances of pharmacological interest are two new 3-hydroxymethylglutaryl flavonoid glycosides (Di Donna et. al 2009); the same AA subsequently succeeded in extraction from bergamot albedo of two new substances, brutieridin and melitidin, showing a statin-like activity; these substances also show anticholesterolemic activity but only in vitro (Di Donna et al. 2010).

From BP, as well as from BJ, it is possible to extract flavonoids showing interesting pharmacological activities, for example on the metabolism of the piastrinic activation factor (PAF) (Balestrieri et al. 2004). Alcoholic BP extracts show antimutagenic activities in vitro (Trombetta et al. 2007). These extracts show a strong protective activity in *Escherichia coli* against the mutagen 4-nitroquinoline-1-oxide; they are able also to modulate the cellular redox state in vitro induced by TNF-alpha human umbilical vein endothelial cells (HUVECs). The reported activities are attributed to the antioxidative and radical-scavenger abilities of the flavonoids in the extracts. Extracts from BP have protective effects on endothelial human cells exposed to the alpha tumoral necrotic factor (TNF-alpha) (Trombetta et al. 2010). In the alcoholic fraction from BP were high concentrations (2.1 mg/100 mL) of bergamottin, which was found 10 times more concentrated in the peel than in the whole fruit. This substance is known as a potential tumor-preventing agent able to hamper differentiation in HL-60 line cells coming from an acute myeloid leukemia (Kawaii et al. 1998).

Flavonoids extracted from BP show good antibiotic in vitro activities (Mandalari et al. 2007a). The antibiotic activity is specially exerted on gram-negative bacteria and increases after enzymatic deglicosilation of these substances. Previously (Mandalari et al. 2006b) the extraction from BP of pectic polysaccharides and low m.w. flavonoids was evaluated for its antibiotic activities. The same AA also showed an antiviral activity of BP extracts, both on rhinovirus HRV14, with a Protection Index more than 100 and, with a minor efficiency, on influenza virus (Inf A) and herpes virus (*H. simplex* 1 and 2) (Unpublished data).

An anti-inflammatory activity of bergamot fruit, as evidenced by inflammation reduction in vitro, was observed in patients with cystic fibrosis. This activity has been attributed mainly to the bergaptene and citropten (Borgatti et al. 2011).

It must be remembered among bergamot's various biological activities the prebiotic activity of its extracts—their ability to favor, at gut level, the settlement of beneficial bacteria. It has been shown, after treatment of BP with pectinolitic and cellulosolytic enzymes, not only the presence of oligosaccharides showing these activities, but also other natural antimicrobial substances which could be used to prolong food shelf-life (Faulds et al. 2006). Subsequently, the prebiotic activity of BP extracts was tested in vivo: the addition of oligosaccharides is able to increase, at gut level, the amount of bifidobacteria and lactobacilli and decrease clostridia (Mandalari et al. 2007b).

Recently Servillo et al. (2012) for the first time identified in all parts of the genus *Citrus* plants pipecolic acid and pipecolic acid betaine. They were particularly common in bergamot leaves, and present in lesser amounts in the albedo and juice.

23.4 THE BERGAMOT SEEDS (BS)

Even though, as previously reported, BS are found mixed with BP, there are some unique characteristics of these seeds when treated after separation from BP. Also BP, as with other parts of bergamot fruit, was revealed to be a source of several interesting substances. For instance, the presence of an extremely high content of proline and its derivatives (about 2,200 mg/kg) was observed, which is considered a probable anti-stress response of the plant due to climatic or osmotic environmental changes (Servillo et al. 2011).

From BS it is also possible to efficiently extract antioxidative factors and phenolic compounds; BS is one of the most interesting sources of glycosylated flavanones, naringin, and neohesperidin. Their extracts could be used to prevent oxidation in fruit juices and essential oils (Bocco et al. 1998).

The lipid content of citrus seeds is extremely variable, ranging from about 26% for blood orange, bergamot, and bitter orange, to 52% for sweet orange and 79% for lemon. Oleic acid displayed its highest content in bergamot oil seed (Saidani et al. 2004).

As previously pointed out, many vegetable products are sources of antiviral substances; some positively interfere with the multiplication processes of HIV-1 virus like inverse transcription, replication, and so on (Han et al. 2011). Raw methanolic extracts from BS have the main limonoids, limonin and nomilin, which inhibit in vitro, also at nanomolecular concentrations, the inverse transcriptase in HIV-1 and in HILV-1, with an activity similar to the pure substances (Ferlazzo 2010). Recently it was confirmed that limonoids isolated from BS, and particularly nomilin, inhibit the inverse transcriptase both of HTLV-1 and of HV1 (Balestrieri et al. 2011).

From all the studies previously reported about the by-products resulting from bergamot processing, it seems clear that now it is no longer possible to consider BJ, BP, and BS wastes to dispose of, but rather as raw materials for the production of many important substances, many of which probably are yet waiting to be discovered.

REFERENCES

Arvanitoyannis, I. A. 2008. *Waste management for the food industries.* Burlington, MA: Elsevier.

Balestrieri, E., Pizzimenti, F., Ferlazzo, A., et al. 2011. Antiviral activity of seed extract from *Citrus bergamia* towards human retroviruses. *Bioorg. Med. Chem.* 19:2084–2089.

Balestrieri, M. L., Balestrieri, C., Felice F., et al. 2004. Effetti di flavonoidi estratti da pastazzo di bergamotto sul metabolismo del Fattore di Attivazione Piastrinico (PAF). *Essenz. Deriv. Agrum.* 74:3–10.

Bocco, A., Cuvelier, M. E., Richard, H., and Berset C. 1998. Antioxidant activity and phenolic composition of citrus peel and seed extracts. *J. Agric. Food Chem.* 46:2123–2129.

Borgatti, M., Mancini, I., Bianchi, N., et al. 2011. Bergamot (*Citrus bergamia* Risso) fruit extracts and identified components alter expression of interleukin 8 gene in cystic fibrosis bronchial epithelial cell lines. *BMC Biochem.* 12:15–17.

Brozzetti, P. 1949. Ricerche sulla composizione chimica, la digeribilità ed il valore nutritivo del pastaccio di bergamotto. *Ann. Fac. Agraria,* Univ. Perugia, VI.

Calabrò, M. L., Galtieri, V., Cutroneo, P., et al. 2004. Study of the extraction procedure by experimental design and validation of a LC method for determination of flavonoids in *Citrus bergamia* juice. *J. Pharmacol. Biomed. Anal.* 35:349–363.

Calvarano, M., Postorino, E., Gionfriddo, F., and Calvarano I. 1995. Sulla deamarizzazione del succo di bergamotto. *Essenz. Deriv. Agrum.* 65:384–387.

Calvarano, M., Gioffrè, D., Calvarano, I., and Lacaria, M. 1996a. Variazioni delle caratteristiche morfologiche e di composizione dei frutti di bergamotto nel corso della maturazione. *Essenz. Deriv. Agrum.* 66:89–102.

Calvarano, M., Postorino, E., Gionfriddo, F., Calvarano, I., and Bovalo, F. 1996b. Naringin extraction from exhausted bergamot peels. *Perfumer & Flavorist* 21:1–4.

Caridi, A. 2005. Fruit vinegars based on citrus wines. International Symposium on Vinegars and Acetic Acid Bacteria, OL02-10, ISBN 88-901732-0-3. Reggio Emilia, Italy, May 8–12, p. 40.

Caridi, A., and Manganaro, R. 1993. Aceto da succo di bergamotto. 12th Conv. Sci. Soc. Ital. Microb. Gen. e Biotec. Microb., Camerino, Italy. September 26–29:107–108.

Caridi, A., and Manganaro, R. 1996. Prove di produzione di aceti aromatici da succo di bergamotto. *Essenz. Deriv. Agrum.* 66:376–388.

Cautela, D., Laratta, B., Santelli, F., et al. 2008. Estimating bergamot juice adulteration of lemon juice by high-performance liquid chromatography (HPLC) analysis of flavanone glycosides. *J. Agric. Food Chem.* 56:5407–5414.

Chies, L., Fasone, V., Arculeo, M., and D'Urso, G. 1989. Insilamento di miscele di pastazzo di bergamotto e paglia: Degradabilità potenziale ed effettiva. *Zootecnica e nutrizione animale* 5:409–411.

Conidi, C., Cassano, A., and Drioli, E. 2011. A membrane-based study for the recovery of polyphenols from bergamot juice. *J. Membrane Science* 375:182–190.

Di Donna, L., De Luca, G., Mazzotti, F., et al. 2009. Statin-like principles of bergamia fruit (*Citrus bergamia*): Isolation of 3-hydroxymethylglutaryl flavonoid glycosides. *J. Nat. Prod.* 72:1352–1354.

Di Donna, L., Gallucci, G., Malaj, N., et al. 2010. Recycling of industrial essential oil waste: *brutieridin* and *melitidin*, two anticholesterolaemic active principles from bergamot albedo. *Food Chem.* 125:438–441.

Di Giacomo, A., and Calvarano, M. 1972. I componenti degli agrumi. Parte II–I succhi. *Essenz. Deriv. Agrum.* 42:353–363.

Di Giacomo, A., Postorino, E., Castori, R., et al. 1986. Le pectine del bergamotto. *Essenz. Deriv. Agrum.* 56:212–219.

Di Giacomo, A., and Calvarano, M. 1991. Sulla preparazione dei flavonoidi e dei relativi diidrocalconi a partire dagli scarti dell'industria agrumaria. *Essenz. Deriv. Agrum.* 4:331–333.

Di Giacomo, A., Postorino, E., and Gionfriddo, F. 1998. Sulle scorze di agrumi essiccate in Italia per la produzione di mangimi. *Essenz. Deriv. Agrum.* 3:300–308.

Dugo, P., Lo Presti, M., Ohman, M., et al. 2005. Determination of flavonoids in citrus juices by micro-HPLC-ESI-MS. *J. Sep. Sci.* 28:1149–1156.

Esposito, C., Cautela, D., De Masi, L., et al. 2004. Sintesi di diidrocalconi ad elevata solubilità in acqua da precursori flavonici estratti da pastazzo di bergamotto. *Essenz. Deriv. Agrum.* 74:97–106.

Faulds, C., Mandalari, G., Bennett, R. N., et al. 2006. Production of prebiotics and antimicrobial agents from a by-product of essential oil extraction. In: *COST 928: Control and exploitation of enzymes for added-value products*, Reykjavik, Iceland, WG 1-3 meeting, O5-29.6–1.7.

Ferlazzo, A. 2010. Attività biologica dei limonoidi estratti dal *Citrus bergamia* e studio delle proprietà dei nanotubi di carbonio coniugati con cumarine ed ac. oleico. Ph.D. Thesis Ciclo XXIII, School of Pharmacy, University of Messina.

Figoli, A., Tagarelli, A., Mecchia, A., et al. 2006. Enzyme-assisted pervaporative recovery of concentrated bergamot peel oils. *Euromembrane* 199:111–112.

Gardana, C., Nalin, F., and Simonetti. P. 2008. Evaluation of flavonoids and furocoumarins from *Citrus bergamia* (bergamot) juice and identification of new compounds. *Molecules* 13:2220–2228.

Gattuso. G., Caristi, C., Gargiulli, C., et al. 2006. Flavonoid glycosides in bergamot juice (*Citrus bergamia* Risso). *J Agric. Food Chem.* 54:3929–3935.

Gattuso, G., Barreca, D., Caristi, C., Gargiulli, C., and Leuzzi, U. 2007a. Distribution of flavonoids and furocoumarins in juices from cultivars of *Citrus bergamia* Risso. *J. Agric. Food Chem.* 55:9921–9927.

Gattuso, G., Barreca, D., Gargiulli, C., Leuzzi, U., and Caristi C. 2007b. Flavonoid composition of Citrus juices. *Molecules* 12:1641–1673.

Giannetti, V., Boccacci Mariani, M., Testani, E., and D'Aiuto, V. 2010. Evaluation of flavonoids and furocoumarins in bergamot derivatives by HPLC-DAD. *J. Commodity Sci. Technol. Quality* 49:63–72.

Gionfriddo, F., Postorino, E., and Bovalo, F. 1996. I flavonoidi glicosidici del succo di bergamotto. *Essenz. Deriv. Agrum.* 66:404–416.

Han, H., He, W., Wang, W., and Gao, B. 2011. Inhibitory effect of aqueous dandelion extract on HIV-1 replication and reverse transcriptase activity. *BMC Complementary and Alternative Medicine* 11:112.

Hardin, A., Crandall, P. G., and Stankus, T. 2010. Essential oils and antioxidants derived from citrus by-products in food protection and medicine: An introduction and review of recent literature. *J. Agric. and Food Information* 11:99–122.

van Heerden, I., Cronjé, C., Swart, S. H., and Kotzé, J. M. 2002. Microbial, chemical and physical aspects of citrus waste composting. *Biores. Technol.* 81:71–76.

Hernandez, E., Belen Leborburo, M., Lequerica, J. L., Martin, F., and Lafuente, B. 1975. Aprovechamento de subproductos citricos. I. Enriquecimiento en proteinos del pienso de corteza de naranja mediante desarollo de levaduros. *Rev. Agroquim. Technol. Aliment.* 15:415–422.

Jung, U. J., Kim, H. J., Lee, J. S., et al. 2003. Naringin supplementation lowers plasma lipids and enhances erythrocyte antioxidant enzyme activities in hypercholesterolemic subjects. *Clin. Nutr.* 22:561–568.

Kawaii, S., Tomono, Y., Katase, E., Ogawa, K., and Yano, M. 1999. HL-60 differentiating activity and flavonoid content of the readily extractable fraction prepared from *Citrus* juices. *J. Agric. Food Chem.* 47:128–135.

Kim, H. K., Jeong, T. S., Lee, M. K., Park, Y. B., and Choi M. S. 2003. Lipid-lowering efficacy of hesperetin metabolites in high-cholesterol fed rats. *Clin. Chim. Acta* 327:129–137.

Kris-Etherton, P. M., Hecker, K. D., Bonanome, A., et al. 2002. Bioactive compounds in foods: Their role in the prevention of cardiovascular disease and cancer. *Am. J. Ed.* 113:71–88.

Leopoldini, M., Malay, N., Toscano, M., Sindona, G., and Russo, N. 2010. On the inhibitory effect of bergamot juice flavonoids binding to the 3-hydroxy-3-methylglutaryl-CoA reductase (HMGR) enzyme 2010. *J. Agric. Food Chem.* 58:10768–10773.

Leuzzi, U., Lo Curto, R. B., and Di Giacomo, G. 1991. Amino acid composition of bergamot (*Citrus bergamia* Risso) juice. Studies of the juices from three cultivars collected during ripening. *Essenz. Deriv. Agrum.* 61:310–322.

Lo Curto, R. B., Tripodo, M. M., Vaccarino, C., et al. 1992. Produzione in continuo di SCP da scorze di bergamotto. XI Conv. Scient. SIMGBM, Gubbio (I), October 4–7.

Lo Curto, A. Production of raw pectic enzymes from bergamot waste. 2005. Giornate di Chimica e Biotecnologie delle fermentazioni. Latina (I), May 5–7.

Lombardo, G., Malara, D., and Mollace, V. 2011. Phytocomplex from bergamot fruit, process of manufacture and use as dietary supplement and in the pharmaceutical field. U.S. Pat. Appl. 20110223271.

Mandalari, G., Bennett, R. N., Bisignano, G., et al. 2006a. Characterization of flavonoids and pectin from bergamot (*Citrus bergamia* Risso) peel, a major by-product of essential oil extraction. *J. Agric. Food. Chem.* 54:197–203.

Mandalari, G., Bennett, R. N., Kirby, A. R., et al. 2006b. Enzymatic hydrolysis of flavonoids and pectic oligosaccharides from bergamot (*Citrus bergamia* Risso) peel. *J. Agric. Food Chem.* 54:8307–8313.

Mandalari, G., Bennett, R. N., Bisignano, G., et al. 2007a. Antimicrobial activity of flavonoids extracted from bergamot (*Citrus bergamia* Risso) peel, a byproduct of the essential oil industry. *Appl. Microbiol.* 103:2056–2064.

Mandalari, G., Palop, C. N., Tuohy, K., et al. 2007b. In vitro evaluation of the prebiotic activity of a pectic oligosacharide-rich extract enzymatically derived from bergamot peel. *Appl. Microbiol. Biotechnol.* 73:1173–1179.

Manthey, J. A., and Grohman K. 2001. Phenols in citrus peel byproducts. Concentrations of hydroxycinnamates and polymethoxylated flavones in citrus peel molasses. *J. Agric. Food Chem.* 49:3268–3273.

Micali, G., and Currò, P. 1980. Determinazione mediante HPLC della naringina nel bergamotto. *Atti Soc. Pelorit. Sci. Fis. Mat. Nat.* 26:269–272.

Miceli, N., Mondello, M. R., Monforte, M. T., et al. 2007. Hypolipidemic effects of *Citrus bergamia* Risso et Poiteau juice in rats fed a hypercholesterolemic diet. *J. Agric. Food Chem.* 55:10671–10677.

Mollace, V., Sacco, I., Janda, E., et al. 2011. Hypolipemic and hypoglycemic activity of bergamot polyphenols: From animal models to human studies. *Fitoterapia* 82:309–316.

Nogata, Y., Sagamoto, K., Shiratsuchi, H., et al. 2006. Flavonoid composition of fruit tissues of citrus species. *Biosci. Biotechnol. Biochem.* 70:178–192.

Passalacqua, N. G., Guarrera, P. M., and De Fine, G. 2007. Contribution to the knowledge of the folk plant medicine in Calabria region (Southern Italy) 2007. *Fitoterapia* 78:52–68.

Pendino, G. M. 1998. Il bergamotto in terapia medica: attualità e prospettive. *Essenz. Deriv. Agrum.* 68:57–62.

Pernice, R., Borriello, G., Ferracane, R., et al. 2009. Bergamot: A source of natural antioxidants for functionalized juices. *Food Chem.* 112:545–550.

Picerno, P., Sansone, F., Mencherini, T., et al. 2011. *Citrus bergamia* juice: Phytochemical and technological studies. *Nat. Prod. Commun.* 6:951–955.

Postorino, E., Poiana, M., Pirrello, A., and Castaldo, D. 2001. Studio dell'influenza della tecnologia di estrazione nella composizione del succo di bergamotto. *Essenz. Deriv. Agrum.* 71:57–66.

Postorino, E., Finotti, E., Castaldo, D., and Pirrello A. 2002. La composizione chimica del "pastazzo" di bergamotto. *Essenz. Deriv. Agrum.* 72:15–19.

Ramful, D., Bahorun, T., Bourdon, E., Tarnus, E., and Arouma, O. 2010. Bioactive phenolics and antioxidant propensity of flavedo extracts of Mauritan citrus fruits: Potential prophylactic ingredients for functional food application. *Toxicology* 278:75–87.

Saidani, M., Dhifi, W., and Marzouk, B. 2004. Lipid evaluation of some Tunisian citrus seeds. *J. Food Lipids* 11:242–250.

Scerra, V., Caridi, A., Foti, F., and Sinatra M. C. 1999. Influence of dairy Penicillium spp. on nutrient content of citrus peel. *Animal Feed Sci. Technol.* 78:169–176.

Servillo, L., Giovane, A., Balestrieri, M. L., Cautela, D., and Castaldo, D. 2011. Proline derivatives in fruits of bergamot (*Citrus bergamia* Risso et Poit): Presence of N-methyl-L-proline and 4-hydroxy-L-prolinebetaine. *J. Agric. Food Chem.* 59:274–281.

Servillo, L., Giovane, A., Balestrieri, M. L., et al. 2012. Occurrence of pipecolic and pipecolic acid betaine (homostachydrine) in *Citrus* genus plants. *J. Agric. Food Chem.* 60:315–321.

Sinatra, M. C., Bittante, G., Celi, R., Fasone, V., and Chies L. 1988. Caratteristiche chimico-nutritive e digeribilità dell'insilato di bergamotto e delle miscele di pastazzo e di paglia 1988. XIII Conv. Naz. S.i.s. vet, September 29–30, Mantova 42:1219–1222.

Swantsitang, P., Tucker, G., Robards, K. and Jardine, D. 2000. Isolation and identification of phenolic compounds in citrus sinensis. *Anal. Chem.* 417:231–240.

Tripodo, M. M., Lanuzza, F., and Mondello, F. 2007. Utilization of a citrus industry waste: Bergamot peel. *Forum Ware International* 2:20–27.

Trombetta, D., Cimino, F., Cristani, M., et al. 2010. In vitro protective effects of two extracts from bergamot peels on human endothelial cells exposed to tumor necrosis factor-alpha (TNF-alpha). *J. Agric. Food Chem.* 58:8430–8436.

Trovato, A., Taviano, M. F., Pergolizzi, S., et al. 2010. *Citrus bergamia* Risso & Poiteau juice protects against renal injury of diet-induced hypercholesterolemia in rats. *Phytother. Res.* 24:514–519.

Trombetta, D., D'Arrigo, M., Ginestra, G., et al. 2007. Biological activity of bergamot peel extracts (*Citrus bergamia* Risso). 33 Congr. Naz. Soc. It. Farmacol., Cagliari (I), June 6–9.

Wu, H., Calvarano, M., and Di Giacomo, A. 1991. Sulla preparazione di naringina diidrocalcone e neoesperidina diidrocalcone in un impianto pilota. *Essenz. Deriv. Agrum.* 61:56–58.

24 Bergamot Oil Legislation

David A. Moyler

CONTENTS

24.1 INTRODUCTION

Citrus bergamia Risso et Poiteau fruit essential oil is cold pressed from the skin (epicarp) of the whole unripe green fruit by the Pelatrice (pronounced "pel-a-tree-chee") method (Di Giacomo and Di Giacomo 2002) of mechanically rasping the peel in a spray of water. The water is separated from the oil and reused for further spraying.

The chemical components of the oil are unique in the *Citrus* genus in that they contain high levels of linalool and linalyl acetate as well as the ubiquitous limonene and pinenes. It is these constituents together with the ultraviolet light–induced phototoxic furocoumarins that dictate the hazard assessment, toxicological properties, and legislation aspects relative to the maximum amounts tolerable in the final products that contain oil.

A detailed constituent list of the oil is well defined in Chapters 8 and 12, this volume and the excellent critical review articles of published data (Lawrence 1976–2012).

Toxicological testing carried out on the whole oil is reviewed and summarized in the online RIFM/FEMA scientific database and the equivalent end-points can also be calculated based upon the known toxicology of the constituents proportional to their typical quantity in the oil.

This chapter reviews the available data and global legislative status and indicates likely future developments, including the registration of this oil under the REACH regulations set out by ECHA in the EU.

24.2 COUMARINS AND PSORALENS

Coumarins and psoralens are biologically active constituents that occur naturally in various plants. Some exhibit a significant phototoxicity that could lead to UV light–induced skin effects. This includes the giant hogweed *Heracleum mantegaz-zianum*, which grows wild in many hedgerows. While it is popular with boys for making blowpipes, the unfortunate effect is an irritating rash on the lips and eyes in sunlight, caused by the FC in the hogweed sap.

Among the usual fragrance materials, there are many essential oils (NCS) produced from plants, notably citrus fruit peels, which represent a major consumption in fragrance manufacturing. While the content in steam distilled oils is tiny (<0.02%) due to the low volatility of FC, their concentration in cold-pressed oil is not negligible (e.g., bergamottin is 10–20 g/L) (Costa et al. 2010).

This family of bergamot constituents and their chemical structures are discussed in detail in Chapter 12 of this volume and not repeated here. However, whereas the chemical structure of bergaptene is not in doubt, in the past there has been some confusion about the numbering of the ring system. Under the IUPAC rules the numbering must embrace the whole of the ring system and must start from the furan oxygen. Thus bergaptene is named 4-methoxyfuro[3,2-g]chromen-7-one, although some references give it as either 5- or 8-methoxypsoralen. Neither psoralen nor bergaptene are acceptable names to IUPAC, but they are widely used throughout the essential oil and related industries.

Bergaptene occurs at a total level of about 3000 ppm in cold-pressed bergamot oil, and its removal is an important commercial process to enable the oil to be used safely in cosmetics and other consumer products with a potential of skin contact. Usually there is less concern with regard to rinse-off products simply due to different exposure scenarios.

The oil is vacuum distilled at low temperatures to separate the thermally labile monoterpene hydrocarbons (including limonene) and citral isomers. The temperature and vacuum are then increased to distill the linalool and linalyl acetate before the final stage of higher vacuum distillation of the sesquiterpene hydrocarbons, leaving the waxes and FC derivatives behind in the still body. It should be noted that molecular distillation (falling film evaporation) is to be avoided as it causes the FC derivatives to sublime in the still, a property which can be used to purify bergaptene standards for quantitative analysis.

The alternative process of freezing or breaking the ring structure of bergaptene in the distillation residues with sodium hydroxide before the distillation final stage is not currently used to make the so-called FCF (furocoumarin free) grade of oil.

The EC has not yet decided whether there is a need to further regulate these compounds beyond the existing directives. However, being aware of the SCCS opinion, several teams have undertaken the development of quantitative analytical methods to monitor FCs in NCS. One of the earlier methods was the collaborative HPLC study by the RSC essential oils analysis committee for the determination of bergaptene in oils of bergamot. Besides giving the methodology and results of an interlaboratory study, it also describes a method for the preparation and purification of a suitable reference standard (Analytical Methods Committee 1987).

A portion of the bergamot residue obtained after distillation of the oil under high vacuum was sublimed under a vacuum of 1 mm Hg to yield an off-white solid. 1 g of this was dissolved in 100 ml of 10% alcoholic KOH and refluxed for 2 hours. After cooling 400 mL was added and the mixture extracted with 2 × 100 mL of diethyl ether and then 2 × 100 mL chloroform, not using but recovering all the solvents. The phase was acidified by drop wise addition of 50 mL of concentrated sulphuric acid. A white flocculent precipitate was formed and the mixture heated on a boiling water bath for an hour to complete the reaction which reforms the original lactone. This was extracted and recrystallised to give chromatographically pure bergaptene with a melting point of 188°C.

The oxygen heterocyclic compounds present in all citrus oils including bergamot oil were identified (Dugo and McHale 2002).

Several publications have proposed modern instrumental methods for FC (Bonaccorsi et al. 1999; Dugo et al. 2000; Frerot et al. 2004; Hiserodt et al. 2012). The IFRA published a standardized quantitative method after a collaborative ring test validation trial. The paper considers the capabilities and limitations of current HPLC methodologies and serves as an excellent review for those analysts who are working in the field. By using UV detection at 310 nm, the group was able to separate 15 FC when present at >10 mg/L in NCS. They also reported that the noncompound specific nature of detection in the UV range is unable to overcome the effect of interferences arising from chromatographic coelutions, such as those encountered in complex commercial fragrance mixtures.

24.3   LEGISLATION

ISO standard 3520 relates to Bergamot oil.

The IFRA industry standard is 5 ppm of the bergaptenes in FCF-grade citrus oils and <1 ppm in finished consumer products. Where the level of bergaptene has not been determined by appropriate methods, the limits specified in the guidelines on individual oils should apply. Therefore, the fragrance industry via its voluntary system of IFRA Standards as well as several regulatory schemes, such as the EU Cosmetics Directive (1976) and the ASEAN Cosmetics Directive, have introduced restrictions on the FC content in cosmetics products. The current limitation of FCs in the EU and ASEAN Cosmetics Regulation (entry 358 in Annex II—list of substances which must not form part of the composition of cosmetics products) states that FCs (e.g., xanthotoxin and bergaptene) are prohibited except if they come from the normal sources of NCS used. In sun protection and bronzing products, those FCs resulting from natural oils shall be below 1 mg/kg. The SCCP, an advisory body of the European Commission, published in 2005 an outcome of another review of the data, an opinion that a concentration above 1 ppm of one of the FCs in any finished cosmetic product would be of concern.

The COE has a standard number 137 for flavor use of bergamot oil. FEMA has a listing number of 2153. JECFA carried out a bergamot oil safety review in 1976, the 23rd, and again in 1993, the 41st. RIFM have monographs 1091 for bergamot oil and 110 for FCF grade oil. The UN–GHS system of hazard communication and transportation does not have pictogram for photosensitization. However, it is recommended that a warning of potential for photosensitization is included in MSDS Section 11, if appropriate.

24.4 FLAVOR AND FRAGRANCE USES

In flavors the whole cold-expressed bergamot oil can be used in products like Earl Grey tea and sorbets, as there is no skin contact or exposure to UV light in the use of these. Table 24.1 illustrates the wide range of consumer product applications. The predicted average daily intake (PADI) is 25 ppm (RIFM/FEMA database).

In fragrance compounds used in consumer products with intended or foreseeable skin contact, a severe restriction on the total level of bergaptene is <15 ppm in the finished consumer product. This is applied by the IFRA and the FCF grade, rather than the expressed oil used. Categories of uses for cosmetics and fragrances are part of the IFRA standard.

24.5 SAFE USE IN CONSUMER PRODUCTS

The IFRA has a compliance testing regime where consumer products are taken randomly and in secret by a validated independent public analyst and tested to verify compliance with the IFRA standards. The combined IFRA/IOFI GHS taskforce publishes an annual "labeling manual" that sets out the regulatory labeling for all of the ingredients used in the industry. The current entry for bergamot oil is CAS registry 8007-75-8, EINECS 89957-91-5, EC 289-612-9. Please note that the last digit in these registrations is a check digit to confirm that there are no typographical errors in the rest of the numbers.

Percentage of total hydrocarbons is 55%, for aspiration hazard calculations of flavour and fragrance compounds.

UN transportation 1169 Extracts Aromatic Liquid, Class 3 (flammable), III packing group.

EU DSD: Xn, N; R 10-38-43-50/53-65, S 24-37-61-62.

(Xn—Harmful, N—Environmental hazard)

Risk phrases—R 10 Flammable, R 38 Irritating to skin, R 43 May cause sensitisation by skin contact, R 50/53 Very toxic to aquatic organisms—may cause long-term adverse effects in the aquatic environment, R 65 Harmful, may cause lung damage if swallowed.

TABLE 24.1
Expressed Bergamot Oil Average Food Product Use (in Parts per Million)

Product	Usual	Maximum	Daily Use (Grams)
Alcoholic drinks	89	91	33
Soft drinks	59	69	104
Baking	80	93	137
Chewing gum	3	6	0.2
Confectionery	89	95	6
Frozen dairy	45	49	26
Jelly puddings	174	200	20
Processed meats	0.2	0.2	78

Safety phrases—S 24 Avoid contact with skin, S 37 Wear suitable gloves, S 61 Avoid release to the environment, S 62 If swallowed, do not induce vomiting.

GHS global: FL 3, AH 1, SCI 2, EDI 2A, SS 1, EH A1 C1

CLP adaptation of GHS for the EU: FL 3, AH 1, SCI 2, EDI 2A, SS 1, EH A1 C1

(FL 3 Flammable liquid, AH 1 Aspiration hazard, SCI 2 Skin irritant, EDI 2A Eye irritant, SS 1 Potential skin sensitiser, EH A1 Environmental hazard acute 1, EH C1 Environmental hazard chronic 1)

Hazard statements—H 226 Flammable liquid and vapour, H 304 May be fatal if swallowed and enters airways, H 315 Causes skin irritation, H 317 May cause allergic skin reaction,

H 400 Very toxic to aquatic life, H 410 Very toxic to aquatic life with long lasting effects.

Precautionary statements—P 210 Keep away from sparks/open flames/hot surfaces, no smoking.

P233 Keep container tightly closed, P 241 Use spark proof electrical/ventilating/lighting equipment, P 242 Use only non-sparking tools, P 243 Take precautionary measures against static discharge, P 261 Avoid breathing dust/fume/gas/mist/vapour/spray, P 264 Wash (hands/face) thoroughly after handling, P 272 Contaminated work clothing should not be allowed out of the workplace, P 280 Wear protective gloves/protective clothing/eye protection/face protection,

P 301/310 If swallowed: immediately call a poison center or doctor/physician, P 302/352 If on skin: wash with plenty of soap and water, P 303/361/353 If on skin (or hair): remove/take off immediately all contaminated clothing. Rinse skin with water/shower, P 331 Do not induce vomiting, P 332/313 If skin irritation occurs get medical advice/attention, P 333/313 If skin irritation or rash occurs, get medical advice/attention, P 362 Take off contaminated clothing and wash before reuse, P 363 Wash contaminated clothing before reuse, P 391 Collect spillage,

P 403/235 Store in a well-ventilated place, keep cool, P 405 Store locked up, P 501 Dispose of contents/container in accordance with local/regional/national/international regulations.

GHS Signal word: DANGER

Bergamot oil is to be registered under the REACH regulations by EU manufactures and importers before the 2018 deadline. They cooperate for the preparation of the registration dossiers in a consortium, which has been formed as part of the EFEO REACH program. It is customary for these registrations that the company with the highest volume interest functions as lead registrant and submits the full dossier. In this case, a CSR has to be included because the volume exceeds 10 tons per year. Other manufacturers and importers with an annual volume of at least 1 ton are co-registrants and shall submit by the same deadline a limited reference. The registrations are submitted to ECHA using the IUCLID online system.

24.6 CONCLUSIONS

This chapter has given an overview of the global position on bergamot oil safety and shows that it is safe to use in consumer products under the use restrictions and purity criteria as applied by the flavor and fragrance industries.

24.7   ACRONYMS USED IN THIS CHAPTER

ACS	American Chemical Society (of the USA)
CAS	Chemical Abstracts System (of TSCA in the USA)
CLP	Classification Labeling and Packaging (EU's version of the GHS regulation)
COE	Council of Europe
CSR	Chemical Safety Report (REACH)
DSD	Dangerous Substances Directive (of the EU, replaced by CLP)
EC	European Commission
ECHA	European Chemicals Agency (based in Helsinki, Finland)
EFEO	European Federation of Essential Oils (trade associations)
EU	European Union
FAO	Food and Agricultural Organisation (of the UN)
FC	Furocoumarin
FCF	Furocoumarin Free
FEMA	Flavor Extract Manufacturers Association (of the USA)
HPLC	High Pressure Liquid Chromatography (analytical method)
IFRA	International Fragrance Association
IOFI	International Organization of the Flavor Industry
ISO	International Standards Organisation
IUCLID	International Uniform Chemical Information Database
IUPAC	International Union of Pure and Applied Chemistry
JECFA	Joint FAO/WHO Expert Committee on Food Additives
MSDS	Material Safety Data Sheet
NCS	Natural Complex Substance
PADI	Predicted Average Daily Intake (food consumption)
REACH	Registration Evaluation and Authorization of Chemicals (regulations of the EU)
RIFM	Research Institute for Fragrance Materials
RSC	Royal Society of Chemistry (of the UK)
SCCP	Scientific Committee for Consumer Products (now SCCS)
SCCS	Scientific Committee for Consumer Safety
SIEF	Substance Information Exchange Forum (REACH)
TSCA	Toxic Substances Control Act (USA)
UN-GHS	United Nations Global Harmonized System (of hazard communication)
UN WHO	United Nations World Health Organisation

REFERENCES

Analytical Methods Committee. 1987. Application of high-performance liquid chromatography to the analysis of essential oils. Part 1. Determination of bergaptene (4-methoxyfuro[3,2-*g*] chromen-7-one) (8-methoxypsoralen) in oils of bergamot. *The Analyst* 112:195–198.

ASEAN Cosmetics Association. 2012. http://aseancosmetics.org/default/asean-cosmetics-directive 2012.

Bonaccorsi, I. L., McNair, H. M., Brunner, L. A, Dugo, P., and Dugo, G. 1999. Fast HPLC for the analysis of oxygen heterocyclic compounds of citrus essential oils. *J. Agric. Food Chem.* 47:4237–4239.

Costa, R., Dugo, P., Navarra, M., et al. 2009. Study on the chemical composition variability of some processed bergamot (*Citrus bergamia*) essential oils. *Flavour Frag. J.* 25:4–12.

Di Giacomo, A., and Di Giacomo, G. 2002. Essential oil production. In *Citrus*, eds. G. Dugo and A. Di Giacomo, 114–147. London and New York: Taylor & Francis.

Dugo, P., Mondello, L., Dugo, L., Stancanelli, R., and Dugo G. 2000. LC-MS for the identification of oxygen heterocyclic compounds in citrus essential oils. *J.Pharm. Bio. Anal.* 2:147–154.

Dugo, P., and McHale, D. 2002. The oxygen heterocyclic compounds of citrus essential oils. In *Citrus*, eds. G. Dugo and A. Di Giacomo, 355–390. London and New York: Taylor & Francis.

European Commission. 1976. Official Journal of the EC, L262, 169.

Frerot, E., and Decorzant, E. E. 2004. Quantification of total furanocoumarins in citrus oils by HPLC coupled with UV, fluorescence, and mass detection. *J. Agric. Food Chem.* 52:6879–6886.

Hiserodt, R., and Chen, L. 2012. An LC/MS/MS method for the analysis of furocoumarins in citrus oil. In *Recent Advances in the Analysis of Food and Flavors*, Symposium Series Vol. 1098, eds. S. Toth and C. Mussinan, 71–88. Washington DC: American Chemical Society.

IFRA standards, http://www.ifraorg.org. Version 8, December 4, 2013.

Lawrence, B. M. 1979. Progress in essential oils. *Perfum. Flav.* 4(3):50–52.

Lawrence, B. M. 1982. Progress in essential oils. *Perfum. Flav.* 7(3):57–65.

Lawrence, B. M. 1983. Progress in essential oils. *Perfum. Flav.* 7(5):43–48.

Lawrence, B. M. 1987. Progress in essential oils. *Perfum. Flav.* 8(3):65–74.

Lawrence, B. M. 1988. Progress in essential oils. *Perfum. Flav.* 12(2):67–72.

Lawrence, B. M. 1991. Progress in essential oils. *Perfum. Flav.* 13(2):67–78.

Lawrence, B. M. 1994. Progress in essential oils. *Perfum. Flav.* 16(5):75–82.

Lawrence, B. M. 1999. Progress in essential oils. *Perfum. Flav.* 24(5):45–63.

Lawrence, B. M. 2002. Progress in essential oils. *Perfum. Flav.* 27(6):46–64.

Macmaster, A.P., Owen, N., Brussaux, S., et al. 2012. Quantification of selected furocoumarins by high-performance liquid chromatography and UV detection: Capabilities and limits. *J. Chromatography A* 1257:34–40.

RIFM/FEMA. 2013. Subscription database. www.rifm.org

Scientific Committee for Consumer Products (SCCP). 2005. Opinion 0942/05, adopted by the SCCP, December 2005.

Index